Mixed Cultures in Biotechnology

The McGraw-Hill Environmental Biotechnology Series

SERIES EDITORS

Published

LEVY & MILLER (EDS.) • *Gene Transfer in the Environment*

SAUNDERS & SAUNDERS (EDS.) • *Microbial Genetics Applied to Biotechnology*

NAKAS & HAGEDORN (EDS.) • *Biotechnology of Plant-Microbe Interactions*

EHRLICH & BRIERLEY (EDS.) • *Microbial Mineral Recovery*

LEVIN & STRAUSS (EDS.) • *Risk Assessment in Genetic Engineering*

Chapter 10. Mixed-Culture Interactions in Methanogenesis 261
Introduction 261
Definition of Microbial Interactions 262
Geochemical Parameters 263
Metabolic and Ecophysiological Context 268
Methanogenic Carbon Flow Coordination 277
Summary and Outlook 285
References 286

Chapter 11. Mixed Cultures in Detoxification of Hazardous Waste 293
Introduction 293
Ecological Role of Microbial Communities 295
Detoxification by Aerobic Mixed Cultures 300
Detoxification by Anaerobic Mixed Cultures 312
Biodetoxification by Novel Mixed-Culture Systems 325
Conclusion 328
References 329

Chapter 12. The Role of Consortia in Microbially Influenced Corrosion 341
Introduction 341
Economic Importance of MIC 343
General Microbiological Considerations 346
Interactions Among Consortial Members 351
Mechanisms 353
The Effects of Surface Perturbations on MIC Communities 356
Dissecting MIC Community Structure 362
Treatment Considerations 365
Summary 367
References 368

Chapter 13. Mixed Cultures in Biological Leaching Processes and Mineral Biotechnology 373
Introduction 373
Substrates 375
Organisms 385
Mixed Cultures 391
Environmental Parameters 394
Applications for Mineral Deposits and Ores 403
Concluding Remarks 418
References 420

Index 429

Contributors

Lakshmi Bhatnagar Michigan Biotechnology Institute, Lansing, Michigan; Department of Microbiology and Public Health, Michigan State University, East Lansing, Michigan. (CHAP. 11)

Linda F. Bisson Department of Viticulture and Enology, University of California, Davis, California. (CHAP. 3)

Nicholas J. E. Dowling Institute for Applied Microbiology, University of Tennessee, Knoxville, Tennessee. (CHAP. 12)

Babu Z. Fathepure Department of Civil & Environmental Engineering, Michigan State University, East Lansing, Michigan. (CHAP. 11)

Henry P. Fleming Food Fermentation Laboratory, U.S. Department of Agriculture, Agriculture Research Service, and Department of Food Science, North Carolina State University, Raleigh, North Carolina. (CHAP. 4)

Stoyan N. Groudev Higher Institute of Mining and Geology, Research and Training Center of Mineral Biotechnology, Sofia (1156), Bulgaria. (CHAP. 13)

C. W. Hesseltine Consultant, Peoria, Illinois. Retired from Department of Agriculture, Agricultural Research Service, Northern Regional Research Center, Peoria, Illinois. (CHAP. 1)

Eric A. Johnson Department of Food Microbiology and Toxicology, University of Wisconsin, Madison, Wisconsin. (CHAP. 6)

Bruce C. Kelley CRA, Roseville, New South Wales, Australia. (CHAP. 13)

Ralph E. Kunkee Department of Viticulture and Enology, University of California, Davis, California. (CHAP. 3)

Marc W. Mittelman Institute for Applied Microbiology, University of Tennessee, Knoxville, Tennessee. (CHAP. 12)

Richard E. Muck U.S. Department of Agriculture, Agricultural Research Service, U.S. Dairy Forage Research Center, Madison, Wisconsin. (CHAP. 7)

Henry J. Peppler Consultant, Whitefish Bay, Wisconsin. Retired from Universal Foods, Milwaukee, Wisconsin. (CHAP. 2)

Badal C. Saha Michigan Biotechnology Institute, Lansing, Michigan; Department of Food Science and Human Nutrition, Michigan State University, East Lansing, Michigan. (CHAP. 9)

Mary Ellen Sanders Consultant, Littleton, Colorado. (CHAP. 5)

Jürgen H. Thiele Gesellschaft für Biotechnologische Forschung mbH, D3300 Braunschweig, Federal Republic of Germany. (CHAP. 10)

Olli H. Tuovinen Department of Microbiology, Ohio State University, Columbus, Ohio. (CHAP. 13)

Paul J. Weimer U.S. Department of Agriculture, Agriculture Research Service, U.S. Dairy Forage Research Center and Department of Bacteriology, University of Wisconsin, Madison, Wisconsin. (CHAP. 8)

David C. White Institute for Applied Microbiology, University of Tennessee, Knoxville, Tennessee. (CHAP. 12)

Contents

Contributors viii
Preface ix

Chapter 1. Mixed-Culture Fermentations: An Introduction to Oriental Food Fermentation 1
Introduction 1
Advantages to Mixed-Culture Fermentations 3
Disadvantages to Mixed-Culture Fermentations 6
Mixed-Culture Fermentation Classification 7
Control of Mixed-Culture Fermentations 8
Future of Mixed-Culture Fermentations 10
References 15

Chapter 2. Breads from Mixed Cultures 17
Introduction 17
Microflora of Bread Doughs 20
References 33

Chapter 3. Microbial Interactions during Wine Production 37
Introduction 37
Natural Flora of Grapes 39
Population Dynamics during Fermentation 44
Factors Effecting Population Dynamics of Yeast Species 48
Factors Effecting Population Dynamics of Bacterial Species 52
Effect of Antimicrobial Agents on the Natural Wine Flora 55
Postfermentation Flora of Wines and Microbial Spoilage 57
Conclusion 61
References 61

Chapter 4. Mixed Cultures in Vegetable Fermentations 69
Introduction 69
Initiation of Fermentation 71
Primary Fermentation 77
Secondary Fermentation 93
Postfermentation 95
Summary 98
References 99

Chapter 5. Mixed Cultures in Dairy Fermentations 105
Historical Perspective 105
Mixed Dairy Cultures 111
Types of Interactions 114
Types of Mixed Starter Systems 126
Summary 128
References 128

Chapter 6. Microbiological Safety of Fermented Foods 135
Introduction 135
Importance of Foodborne Disease 136
Diseases Transmitted in Fermented Foods 137
Incidence of Foodborne Disease Transmitted in Fermented Food 141
Control of Pathogens and Toxins in Specific Fermented Foods 145
Fermented Dairy Products 151
Fermented Legume and Cereal Products 154
Mycotoxin Formation in Fermented Foods 158
Survival of Viruses in Fermented Foods 159
Mechanisms of Exclusion of Pathogens from Fermented Foods 160
Response of Starter Organisms and Pathogens to Osmotic and Acid Stresses 162
Future Research 163
References 164

Chapter 7. Silage Fermentation 171
Introduction 171
Important Microbial Species 172
Normal Ensiling Dynamics 179
Factors Effecting Ensiling Dynamics 184
Models of Silage Fermentation 192
Research Needs 196
References 198

Chapter 8. Use of Mixed Cultures for the Production of Commercial Chemicals 205
Introduction 205
Production of Commercial Chemicals by Mixed Cultures 207
Summary and Conclusion 227
References 228

Chapter 9. Mixed Cultures in Enzymatic Degradation of Polysaccharides 233
Introduction 233
Enzymatic Degradation Mechanisms and Application in Biomass Conversion 234
Conclusion 255
References 255

Mixed Cultures in Biotechnology

J. Gregory Zeikus
Michigan Biotechnology Institute
Lansing, Michigan

Department of Biochemistry
Department of Microbiology and Public Health
Michigan State University
East Lansing, Michigan

Eric A. Johnson
Department of Food Microbiology and Toxicology
University of Wisconsin
Madison, Wisconsin

McGraw-Hill, Inc.
New York St. Louis San Francisco Auckland Bogotá
Caracas Hamburg Lisbon London Madrid Mexico
Milan Montreal New Delhi Paris
San Juan São Paulo Singapore
Sydney Tokyo Toronto

Library of Congress Cataloging-in-Publication Data

Mixed cultures in biotechnology / [edited by] J. Gregory Zeikus, Eric A. Johnson.
p. cm. — (The McGraw-Hill environmental biotechnology series)
Includes bibliographical references and index.
ISBN 0-07-072844-5
1. Mixed cultures (Microbiology) 2. Microbial biotechnology.
I. Zeikus, J. Gregory, date. II. Johnson, Eric A., date.
III. Series: Environmental biotechnology.
TP248.25.M58M59 1991
660'.62–dc20 90-26554
CIP

ISBN 0-07-072844-5

1 2 3 4 5 6 7 8 9 0 DOC/DOC 9 7 6 5 4 3 2 1

The sponsoring editor for this book was Trev Léger, the editing supervisor was Joseph Bertuna, and the production supervisor was Pamela A. Pelton. It was set in Century Schoolbook by McGraw-Hill's Professional Publishing composition unit.

Printed and bound by R. R. Donnelley & Sons Company.

Preface

Mixed-culture biotechnology is integral to many important industrial processes used throughout the world. Also, mixed-culture fermentations, including those used in the production of food and beverages, have been in use since antiquity. Moreover, new fermentation processes, such as those used in the manufacture of fine chemicals and for the degradation of xenobiotics and pesticides, are being developed in response to new opportunities and demands. Mixed cultures, in fact, offer the advantages of combining in a single process many genetic traits from a variety of organisms—and without employing recombinant DNA. This permits the development and application of complex processes which would be difficult or impossible with pure cultures.

The biological interactions involved in most mixed-culture processes are complex and not fully understood. In this book experts on diverse types of mixed cultures identify common principles shared by various types of fermentations. Also, they emphasize the fundamental roles and interactions of the microorganisms involved. The book is intended to convey principles for improving present-day fermentation processes and for facilitating the development of new biotechnology.

We wish to express our appreciation to the contributors for their care and expertise in preparing their chapters, to Arnold L. Demain for inspiring this project, and to the McGraw-Hill editorial staff.

Eric A. Johnson
J. Gregory Zeikus

ABOUT THE EDITORS

J. Gregory Zeikus is president of the Michigan Biotechnology Institute in Lansing, Michigan and distinguished professor in the Department of Biochemistry and in the Department of Microbiology and Public Health at Michigan State University.

Eric A. Johnson is associate professor in the Department of Food Microbiology and Toxicology at the University of Wisconsin, Madison.

Chapter

1

Mixed-Culture Fermentations: An Introduction to Oriental Food Fermentations

C. W. Hesseltine

Agricultural Research Service
U. S. Department of Agriculture
Northern Regional Research Center
Peoria, Illinois

Introduction

Mixed-culture fermentations are those in which the inoculum always consists of two or more organisms. As opposed to this, pure cultures consist of only one strain of microorganism growing by itself and excludes inocula in which two or more strains of the same species are used. Mixed cultures can consist of known species to the exclusion of all others, or they may be composed of mixtures of unknown species. The mixed cultures may be all of one microbial group—all bacteria—or they may consist of a mixture of organisms of fungi and bacteria or fungi and yeasts or other combinations in which the components are quite unrelated. All of these combinations are encountered in oriental food fermentations.

A few special situations occur in nature where a single microorganism may be growing alone. The interior of stems, roots, leaves, and seeds are normally sterile. When these parts become infected with a pathogen, this organism may grow in pure culture in the host. Likewise, in extreme environmental conditions such as hot springs or very alkaline environments, a single microorganism may be responsible for the fermentation of a substrate.

The earliest studies of microorganisms were those made on mixed cultures by van Leeuwenhoek (1684). Micheli, working with fungi in 1718, reported his observations of the germination of mold spores on cut surfaces of melons and quinces. Brefield in 1875 obtained pure cultures of fungi, and Koch in 1878 obtained pure cultures of pathogenic bacteria. The objective of both Brefield's and Koch's studies was to identify pathogenic microorganisms. They wanted to prove what organism was actually responsible for a particular disease. Thus, part of Koch's fame rests on his discovery of the cause of tuberculosis. A second interest in pure cultures was the use of a particular microorganism to produce an organic compound by fermentation. For a more detailed discussion of the history of pure cultures, see Chap. 1 in Quayle and Bull (1982) and Buller (1915). The history of the discovery of methods for pure-culture techniques is important because without these methods, the classification of the components of mixed cultures could not have occurred.

An early paper on mixed-culture food fermentation was an address by Macfadyen at the Institute of Brewing, London, in 1903 entitled, "The Symbiotic Fermentations," in which he referred to mixed-culture fermentations as "mixed infections." Probably this expression reflected his being a member of the Jenner Institute of Preventive Medicine. About half of his lecture was devoted to mixed-culture fermentations of the Orient. Among those described were Chinese yeast, *koji*, Tonkin yeast, and *ragi*. In these preparations fungi and yeast were present, and in 1902 they were used commercially to produce alcohol in France.

Mixed cultures are the rule in nature; therefore, one would expect this condition to be the rule in fermented foods of relatively ancient origin. Soil is a mixed-organism environment with protozoa, bacteria, fungi, and algae growing in various numbers and kinds, depending upon the nutrients available, the temperature, and the pH of the soil. Soil microorganisms relate to each other, some as parasites on others, some forming substances essential to others for growth, and some which have no effect on each other. Four special mixed-culture situations exist in nature. The first is the lichens in which fungi and algae grow in such close relationship that groups of cells of the two organisms serve as propagules for reproduction. They are important because they can grow under extreme climatic conditions to form soil from the breakdown of rock.

A second example occurs in a mixed-plant *Rhizobium* interaction in which the bacterium fixes nitrogen in nodules formed on the plant's roots. Thirdly, if we did not have mixed cultures, cellulose might never be completely broken down to water and CO_2 in nature. If the animal and plant debris could not be degraded for the lack of mixed cultures, life would become impossible. Winogradsky (1949) demonstrated that mixed cultures better ferment cellulose than a single microorganism. Besides cellulose, there are many other complex compounds formed by plants and animals. To prevent these from accumulating, series of microorganisms degrade the compounds in various steps to simpler materials which can be used by other organisms. A fourth important mixed-culture fermentation is the activities of bacteria, protozoa, and fungi in the rumens of ruminant animals. Many of the interactions and growth of microorganisms in these fermentations are also found in fermented foods.

Advantages to Mixed-Culture Fermentations

Mixed-culture fermentations offer a number of advantages over conventional fermentations using a single pure culture.

1. Product yield may be higher. Yogurt is made by the fermentation of milk with *Streptococcus thermophilus* and *Lactobacillus bulgaricus*. Driessen (1981) demonstrated that when these species were grown separately, 24 and 20 mmol of acid was produced; together with the same amount of inoculum, a yield of 74 mmol was obtained. The number of *S. thermophilus* cells increased from 500×10^6 per milliliter to 880×10^6 per milliliter with *L. bulgaricus*.

2. The growth rate may be higher. In mixed culture, one microorganism may produce needed growth factors or produce essential growth compounds such as carbon or nitrogen sources beneficial to a second microorganism. It may alter the pH of the medium, thereby improving the activity of one or more enzymes. Even the temperature may be elevated and promote growth of a second microbe.

3. Mixed cultures offer more protection against contamination. In nature pure cultures rarely exist and consequently are less likely to become contaminated. Typically one organism will attack the primary organic material and the by-products are in turn attacked by secondary organisms. Koji mold (several strains used together) typically produces amylases and proteases which break down starch in rice and protein in soybeans. In the *miso* and *shoyu* fermentations, these compounds are in turn acted on by lactic acid bacteria and yeast. In such processes, especially with a heavy inoculum of selected strains, con-

tamination does not occur even when the fermentations are carried out in open pans or tanks allowing air contamination to occur.

4. In some mixed cultures a remarkably stable association of microorganisms may occur. Even when a mixture of cultures is prepared by untrained individuals working under atrociously dirty conditions, such as in *ragi*, mixtures of the same fungi, yeasts, and bacteria remain together even after years of subculture. Probably the steps in making the starter were established by trial and error and the process was handed down from one generation to the next, with the conditions being such that this mixture could compete against all contaminants.

5. Compounds made by a mixture of microorganisms often complement each other and work to the exclusion of unwanted microorganisms. In many of these associations various chemicals are produced that complement each other to the exclusion of undesirable microorganisms. For example, in some food fermentations yeast will produce alcohol and lactic acid bacteria will produce lactic acid and other organic acids and change the environment from aerobic to anaerobic. Inhibiting compounds are thus formed, the pH lowered, and anaerobic conditions are developed excluding most undesirable molds and bacteria.

6. Mixed cultures permit better utilization of the substrate. The substrate for fermented food is always a complex mixture of carbohydrates, proteins, and fats. Mixed cultures possess a wider range of enzymes and are able to attack a greater variety of compounds. Likewise, with proper strain selection they are better able to change or destroy toxic or noxious compounds which may be in the fermentation substrate.

7. Mixed cultures are able to bring about multistep transformations that would be impossible for a single microorganism. An example is the *miso* fermentation where *Aspergillus oryzae* strains are used to make *koji* to convert rice starch to sugar; yeast in turn uses the sugar to make small amounts of alcohol, and bacteria use the sugars to make acid and flavoring substances.

8. Mixed cultures can be maintained indefinitely by unskilled people with a minimum of training. If the environmental conditions can be maintained, i.e., temperature, mass of fermenting substrate, length of fermentation, and kind of substrate, it is easy to maintain a mixed-culture inoculum indefinitely and to carry out repeated successful fermentations.

9. In mixed-culture fermentations phage infections are reduced. In

pure-culture commercial fermentations involving bacteria and actinomycetes, invariably an epidemic of phage infections occurs, and the infection can completely shut down production. Since mixed cultures have a wider genetic base of resistance to phage, failures do not often occur because one strain is wiped out; a second or third phage-resistant strain in the inoculum will take over and continue the fermentation.

10. Simultaneous changes in the substrates may occur. Many different changes in the substrate may occur at the same time, which would not be possible if only one microorganism was used. Harrison (1978) cites an example of a stable enrichment culture in which four types of bacteria were present, one of which used methane; a second used methanol, but not methane; and two bacterial species would not grow on either methane or methanol. It was suggested that the role of the two heterotrophic bacteria was to remove complex products of growth and lysis. Methanol was removed by one autotrophic organism down to the level where it was not inhibiting to the methane-utilizing organism.

11. Mixed-culture fermentations enable the utilization of cheap and impure substrates. In any practical fermentation the cheapest substrate is always used, and this will not be of the highest purity and often will be a mixture of several materials in large quantities. For example, in the production of biomass, a mixed culture is desirable that not only attacks the cellulose but also starch and sugar. Cellulolytic fungi along with starch and sugar-utilizing yeasts would give a more efficient process, producing more biomass in a shorter time.

12. Mixed cultures can provide necessary nutrients for optimal performance. Many microorganisms, such as the cheese bacteria, which might be suitable for production of a fermentation product, require growth factors to achieve optimum growth rates. To add the proper vitamins to production adds complications and expense to the process. Thus, the addition of a symbiotic species which supplies the growth factors is a definite advantage in many types of fermentations.

Harrison (1978) mentioned a specialized advantage in single-cell protein fermentation. When methanol is the substrate, foaming is a large problem. However, when three cultures were used together, foaming did not occur, although it was a problem when each culture was grown separately, and an antifoaming compound had to be added during fermentation.

Disadvantages to Mixed-Culture Fermentations

Mixed-culture fermentations also have some disadvantages.

1. The products formed by mixed-culture fermentation may be more variable than in pure-culture fermentations. In mixed-culture fermentations, products may vary in amount and composition. The importance of the amounts of product produced may depend on what is wanted as an end product. If, on one hand, removal of waste is a problem, variation in amount of product may not be very important. On the other hand, if the objective is to obtain a specific product in maximum yield, then variation in yield becomes very important.
2. Scientific study of mixed cultures is difficult. Obviously it is more difficult to study the fermentation if more than one microorganism is involved. That is why most biochemical studies are conducted as single-culture fermentations because one variable is eliminated.
3. Defining the product and the microorganisms employed becomes more involved in patent and regulatory procedures.
4. Contamination of the fermentation is more difficult to detect and control.
5. When two or three pure cultures are mixed together, it requires more time and space to produce several sets of inocula rather than just one.
6. One of the worst problems in mixed-culture fermentation is the control of the optimum balance among the microorganisms involved. This can, however, be overcome if the behavior of the microorganisms is understood and this information is applied to their control.

The balance of organisms brings up the problem of the storage and maintenance of the cultures. Lyophilization presents difficulties because, in the freeze-drying process, the killing of different strains' cells will be unequal. It is also difficult, if not impossible, to grow a mixed culture from liquid medium in contrast to typical fermentations on solid medium, without the culture undergoing radical shifts in population numbers. According to Harrison (1978), the best way to preserve mixed cultures is to store the whole liquid culture in liquid nitrogen below 80°C. The culture, when removed from the frozen state, should be started in a small amount of the production medium and checked for the desired fermentation product and the normal fermentation time. Subcultures of this initial fermentation, if it is satisfactory, may then be used to start production fermentations.

It is desirable to determine the species in the mixed culture and to isolate each component. When this is done, the component pure cultures need to be put back together at several different population levels and then determine what the required strains are and the numbers needed to give the desired product. Some years ago we obtained a culture of the tea fungus, a mixed culture used to start a refreshing drink made by the fermentation of tea leaves and sucrose (Hesseltine, 1965). When the starter was examined, it was found that two yeasts and a bacterium were the major components (*Acetobacter* sp. NRRL B-2357, *Zygosaccharomyces bisporus* NRRL YB-4810, and *Candida* sp. NRRL YB-4882).

Some other bacteria were contaminates because they were in such low numbers in relation to the three major populations. These were discarded. When each of the three species was grown separately, no product was formed. Likewise, when each combination of two cultures was grown in the tea extract, no suitable beverage was formed. However, when the three species were grown together, the polymer film formed on the surface of the tea liquid and the desired beverage was formed. In this instance it was better to lyophilize the cultures separately and then start with mixtures to give a pure mixed-culture inoculum.

Recently three books have appeared that deal wholly or in part with mixed-culture fermentations. These are Bushell and Slater, *Mixed Culture Fermentations* (1981), Bull and Slater, *Microbial Interactions and Communities* (1982), and Quayle and Bull, *New Dimensions in Microbiology: Mixed Substrates, Mixed Cultures, and Microbial Communities* (1982).

Mixed-Culture Fermentation Classification

Mixed cultures can be classified as to their components according to Hesseltine (1983).

1. *Monoculture.* These are fermentations in which a single microorganism strain is employed to give a product. The Indonesian fermentation of cooked soybeans by the mold *Rhizopus oliosporus* to give the food *tempeh* is one example. Another fermentation of rice by *Monascus purpureus* to produce the natural coloring (purplish-red) pigment employs but a single strain as far as is known. Other monoculture fermentations in the Orient include the peanut food, *ontjom*, prepared by the fermentation of peanut press cake by

Neurospora intermedia and the production of Chinese cheese (*sufu*) by *Actinomucor elegans* using tofu as the fermentation substrate.

2. *Multiculture.* In multiculture fermentations more than one microorganism belonging to two different species is used. *Ragi* is a starter culture used for a variety of fermentations based on starch either from cassava or rice. This starter material is produced under various names in Indonesia, Indochina, China, India, Korea, and the Philippines. Two fungi are always present, including species of *Rhizopus* and *Amylomyces.* These are accompanied by certain yeast and bacterial species (Hesseltine et al., 1988). The main bacterium present, *Pediococcus*, may not be necessary.

3. *Unimulticulture.* In these fermentations two or more strains of the same species are used in the fermentation. In the production of *koji*, often two or more strains of *Aspergillus oryzae* are used in the inoculum in order to produce the desired enzymes for the fermentation or rice and soybeans or wheat and soybeans in making *miso* or *shoyu.* In another example, two strains of *Lactobacillus acidophilus* are employed to make a soybean yogurt. One strain produces the desired flavor; the second produces quantities of lactic acid.

4. *Polyculture.* In these food fermentations there are many different microorganisms, and the species specifically required in the fermentation is completely or partially unknown. Examples of these food fermentations are the fish fermentations of Indochina and Thailand in which fish are heavily salted and allowed to ferment (actually decay) until the fish is liquefied and the scales and bones removed by pressing to yield a soy sauce–like condiment widely used in this area. Because the environment is controlled by the high salt content, pathogenic and food-poisoning bacteria do not develop.

With respect to time of inoculation, one type of inoculation occurs with all organisms introduced simultaneously. In other cases they are added sequentially. Thus in some fermentations *koji* is first made, and bacteria and yeast cultures are added later in such food fermentations as in the preparation of *shoyu* and *miso.*

Control of Mixed-Culture Fermentations

Commercially mixed cultures are used in making such foods as cheeses, fermented milks, *miso, shoyu, sake, ragi*, single-cell protein, and fermented meats. The question arises as to how one controls mixed-culture fermentations.

1. Microorganisms may be used in a series in which just one organism is introduced at a time in a fermentation. After growth and production of a metabolic product, the whole culture is pumped into a second tank in which a second organism produces a new product from that made by the first. The Symba fermentation process (Jarl 1969) is used to make a single-cell protein from starchy wastes from potato processing factories. The initial fermentation is carried out by *Saccharomycopsis fibuligera*, which produces enzymes for the saccharification of the starch into glucose and maltose. After a proper holding time, the mash without treatment is pumped into a second tank which is inoculated with *Candida utilis*. This yeast uses the glucose and maltose to grow and produce biomass, though it cannot use starch.

2. In some cases it is desired to produce an organic acid after an enzyme active at an alkaline or neutral pH has had a chance to bring about a desired change in the substrate. This may be accomplished by using a heavy inoculum of a lactic acid bacterium to lower the pH.

3. The temperature may be adjusted so that one organism will grow slowly while a second may grow rapidly. According to Fukushima (1986), in the *shoyu* fermentation, glutaminases are required to liberate glutamic acid, one of the most important flavor components in *shoyu*. Glutaminases are inhibited greatly by an acidic pH. In this case the temperature of the *moromi* mash is kept low for the first 1 to 2 months, and the lactic acid bacterium *Pediococcus halophilus* grows slowly and the pH goes down gradually, allowing the glutaminases to act on the wheat and soybeans.

4. The growth of one organism may be controlled by the oxygen available for growth. *Koji* is produced by the molding of a starch substrate under aerobic conditions. This process involves a relatively thin layer of substrate and requires turning the molding substrate to bring the bottom to the top. However, *koji* is harvested at an early stage of mold growth before sporulation and is then put into the fermentation tanks where oxygen is low, with the result that mold growth is stopped and the yeast and bacteria, which can tolerate low oxygen levels, then take over.

In the starter for various starch-based fermentations, the microorganisms involved are able to grow in the absence of oxygen, provided CO_2 is supplied in small amounts. The inoculum (*ragi*) can therefore be used in fermentations involving rice or cassava without aeration to produce food products or alcohol (Hesseltine et al. 1988).

5. The mixed-culture fermentation may have whole groups of organisms excluded with the use of inhibitors. Thus tetracyline may be

incorporated into a fermentation and will exclude the growth of bacteria but permit growth of fungi, including yeasts. By contrast, there are compounds which permit the growth of bacteria but will exclude the development of fungi. For example, the antibiotic actidione may be used effectively to select only bacteria. These agents are not useful where only two or more known organisms are used. Obviously, such selective agents are useful if one wants a mixed, unclassified group of fungi or bacteria to be used.

6. The adjustment of the pH, often combined with chemical and physical control agents, may be very useful. Some microorganisms at the extreme range of pH may almost grow in a pure-culture state in nature. For example, soybean soapstock with a pH of as high as 11.5 supports growth of a few bacteria, fungi, and yeast, including strains of *Beauvaria, Candida*, and *Actinomucer*. At the other end of the scale, there are fungi such as *Aspergillus niger* which will grow at very low pH. However, more important is the use of less extreme pHs with fungi which generally tolerate low pH, whereas bacteria generally grow best near neutral pH.

7. Another control mechanism is to remove the product as it is being produced. Often, the accumulation of the desired product will not allow the fermentation to continue at a maximum rate. In the alcohol fermentation, when ethanol reaches a certain level, the yeast will stop producing even though plenty of substrate is present.

8. The reduction of a single nutrient from its maximum level for growth of one microorganism will, in turn, permit a second to grow which would otherwise be swamped by growth of the first.

Future of Mixed-Culture Fermentations

Mixed-culture fermentations will continue to be used in traditional processes such as the soybean and dairy fermentations. As long as mixed-culture fermentations will give no greater yields than monoculture fermentations and will not lower the cost, mixed cultures will not be used. Harrison (1978), discussing the production of single-cell protein from hydrocarbons, pointed out seven advantages in the use of mixed cultures.

1. Mixed cultures give a substantial increase in yield over monocultures.
2. Mixed-culture growth rates are much higher and the methane-utilizing microorganisms individually grow poorly or not at all.

3. Mixed cultures are more stable and recover more rapidly when growth conditions are disturbed.
4. Foaming does not occur with mixed cultures, but can be a major problem when a single microorganism is used.
5. Mixed cultures are more resistant to contamination.
6. The cheapest available substrates are not very pure. Mixed cultures do far better on these cheap substrates than monocultures.
7. Many microorganisms that might be suitable for single-cell protein production require a vitamin supply.

Turning to the future of mixed-culture fermentations in a more general way, a number of fields besides fermented foods (where mixed cultures will continue to be used) are described below.

1. Cellulose is one of the most common carbon compounds in nature. It has been studied extensively as a fermentation substrate to produce single-cell protein and fermentable sugars. Much of the work has been done with monocultures, especially fungi. Since cellulose will also probably have starch and lignin present in the plant material, it seems that mixed cultures made up of microorganisms known to attack these materials offer the best approach to cellulose utilization. In nature cellulose is rapidly broken down by a mixture of bacteria and fungi, especially *Basidiomycetes*. For example, I have seen corn fields receiving minimum tillage in which the soil surface was covered with leaves, husks, and stalks (from the previous year) at planting time, but by harvest all the litter had disappeared. Cellulose destroyers undoubtedly were helped by the large amount of nitrogen added at planting time; this supplied the nitrogen that enhanced the growth of the cellulose-degrading microorganisms.

2. A number of papers have appeared describing the use of mixed cultures to bring about the transformation of steroids to make desired drugs. The earliest work used sequential steroid transformations in which each step was done by a separate organism. The Squibb group (Ryu et al. 1969, Lee et al. 1969) proposed a fermentation in which two microorganisms were used together. Mixed-culture fermentation allowed a reduction in the number of steps and a consequent reduction in cost and time. The Squibb group used *Arthrobacter simplex* and *Streptomyces roseochromogenes* to bring about a double transformation of such compounds as 9_{α}-fluorohydrocortisone. They also found that 20-ketoreductase, responsible for the production of unwanted byproducts, was repressed in the mixed-culture fermentation.

Mixed culture may be useful in other transformations in which two or more steps are needed to give the desired compound. They have their limitations, depending on the completeness of the conversion, the value of the product, the volume of the fermentation, and the availability of suitable cultures.

3. Mixed cultures are finding use in the production of industrial ethanol. Distillery yeasts typically produce alcohol up to 10 to 12 percent before alcohol inhibits further growth and ethanol production. Yeasts are available which, although slower in growth, will continue to produce ethanol to about 22 percent at low temperatures. Recently there has been a great deal of work on a process to produce ethanol using mixed cultures to ferment starchy materials without the use of malt and without substrate sterilization. In this instance a yeast is combined with a suitable *Rhizopus* strain that will degrade raw starch to sugar, which then can be converted to alcohol. This process allows for a great saving in heat to solubilize the starch. Another use of mixed culture would be to use compounds such as cellodextrins as the fermentation substrate. Freer and Wing (1985) found that coinoculation with *Candida wickerhamii* and *Saccharomyces cerevisiae* in the ratio of 57:1, produced 12 to 45 percent more ethanol than either culture produced alone from crude cellodextrin.

A commercial ethanol fermentation used in the Orient makes use of a number of cultures (Hesseltine, 1981). The inoculum is prepared by grinding wheat into flour, maintaining the flour at a specific moisture level, pressing the flour into solid cakes, and incubating the cakes at below 40°C. The indigenous flora are selected and serve as the inoculum for the production of ethanol from grain sorghum. The moistened grain is fermented with the wheat cake inoculum, and the alcohol is steam-stripped from the solid substrate. As many as three ethanol fermentations are made from the same lot of sorghum. This process has many advantages, but could be greatly improved by the use of a more defined culture mixture.

4. Mixed cultures seem ideal for the degradation of toxic substances such as substituted aromatic compounds. Where toxic compounds occur in waste effluents from processing plants, the use of mixed cultures, which together will degrade these chemicals, offers excellent possibilities for commercial use. The destruction of toxic compounds by mixed cultures would be cheaper than other means, and a number of related compounds could be destroyed at the same time. An advantage would be that the toxic chemicals would reduce competition from other microorganisms.

The technology visualized would be to develop a mixed-culture inoculum either by finding pure cultures that attack the toxins and combining these or by enrichment to develop a mixed culture from nature. The mixed-culture inoculum could be freeze-dried, or better, stored under liquid nitrogen. The inoculum could be built up from this stock on liquid medium containing the chemical to be destroyed. Then this seed culture would be introduced into the effluent holding ponds just before processing of the toxic compound begins. If the process is to go on continuously, no further attention would be needed once the mixed culture is established. By introducing a flora initially, the clean-up should begin immediately and would eliminate the long period of time required to develop a degrading flora.

Goulding et al. (1988) recently studied the biodegradation of substituted benzene with a mixture of several microorganisms. Their complex culture consisted of five strains of pseudomonads, one klebsiella, four rhodococci and two fungi. Degradation of many of the halogen-substituted benzenes, phenols, and benzoic acids were evaluated, and 19 out of 24 compounds tested were completely removed in periods of 48 to 168 hours. For example, pentachlorophenol was completely removed when levels of 200 mg/L were tested.

5. Xanthan gum is a commercial extracellular polysaccharide produced by *Xanthomonas campestris*. It is used as a thickening agent in foods and as a crude product in oil-well-drilling muds to enhance oil recovery by controlling the mobility of water floods to sweep crude oil out of rock formations toward a production well where it may be pumped to the surface. Xanthan gum is a highly stable polysaccharide not easily degraded by most microorganisms.

There is a need for xanthanase that will degrade xanthan. Such an enzyme can alter the xanthan structure and, therefore, alter viscosity. Two objectives of studies on xanthanase involve, first, an understanding of the nature of aerobic and anaerobic degradation of xanthan gum in oil fields, and secondly, development of a viscosity breaker of xanthan in hydraulic fracture fluids used to stimulate the flow of gas and oil from geological formations of low porosity.

Recently Cadmus et al. (1988) have, by culture enrichment, discovered a microbial consortium that produces a high-temperature, salt-tolerant xanthanase. This apparent mixture of two enzymes is suitable as a viscosity breaker for xanthan-based hydraulic fracture fluids. The enzyme complex is active to a temperature of 65°C in the presence of up to 10% sodium chloride, and is not affected by Ca or Mg ions. The degradation products are pyruvic acetal of mannose, and branched oligosaccharides that result from a lipase type of activity formed by a cellulase-like depolymerase. Interestingly, no single-colony isolate or pairs of colonies degrade xanthan—only the mixed culture containing at least a *Bacillus* species was required to form the enzyme. The mixed culture is maintained on agar slants, in liquid broths, or is lyophilized and grown on a xanthan salts medium.

6. As noted above the extensive uses of mixed-culture fermentations for dairy and meat products are well known as to the type of cultures used and the fermentation process. However, there are a large number of food fermentations based on plant substrates such as rice, wheat, corn, soybeans, and peanuts where mixed cultures of microorganisms are used and will continue to use combinations of microorganisms within the near future.

One example of the complex sequential interaction of two fermentations, and which employs fungi, yeast, and bacteria, is the manufacture of *miso*. This oriental food fermentation product is based upon the fermentation of soybeans, rice, and salt to make a pastelike fermented food. *Miso* is used as a flavoring agent and as a base for *miso* soup. There are many types of *miso*, ranging from a yellow, sweet *miso* (prepared by a quick fermentation) to a dark, highly flavored *miso*. The type depends upon the amount of salt, the ratio of cereals to soybeans, and the duration of the fermentation.

The *miso* fermentation begins with the molding of sterile, moist, cooked rice that is inoculated with dry spores of *Aspergillus oryzae* and *A. soyae*. The inoculum consists of several mold strains combined with each strain producing a desired enzyme(s). The molded rice is called *koji* and is made to produce enzymes to act upon the soybean proteins, fats, and carbohydrates in the subsequent fermentation.

After the rice is thoroughly molded, accomplished by breaking the *koji* and mixing, the *koji* is harvested before mold sporulation starts, usually in one or two days. The *koji* is mixed with salt and soaked and steamed soybeans. This mixture is next inoculated with a new set of microorganisms, and the four ingredients are now mashed and mixed. After the production of *koji* with molds, the paste is then placed in large concrete or wooden tanks for the second fermentation. The inoculum consists of osmophilic yeasts (*Saccharomyces rouxii* and *Candida versatilis* and one or more strains of lactic acid bacteria, typically *Pediococcus pentosaceus* and *P. halophilus* (Hesseltine 1983). Conditions in the fermentation tanks are anaerobic or nearly so, with the temperature maintained at 30°C. The fermentation is allowed to proceed for varying lengths of time depending on the type of *miso* desired, but is typically one to three months. The fermenting mash is usually mixed several times, and liquid forms on the top of the fermenting mash.

The initial inoculum is about 10^5 microorganisms per gram. Typically 3300 kg of *miso* with a moisture level of 48 percent is obtained when 1000 kg of soybeans, 600 kg of rice, and 430 kg of salt is used. When the second fermentation is completed, aging is allowed to take place. A number of other mixed-culture fermentations are similar to the *miso* process, including *shoyu* (soy sauce) and *sake* (rice wine).

A legitimate question can be asked as to the future prospects for the use of mixed cultures in food fermentations. What will be the effect of genetic engineering on the use of mixed cultures? Would engineered organisms be able to compete in mixed culture? Many laboratories are busy introducing new desirable genetic material into a second organism. The characteristics being transferred may come from such diverse organisms as mammals and bacteria and may be transferred from animals to bacteria. In general the objective of this work involves the introduction of one desirable character, not a number. For instance, strains of *Escherichia coli* have been engineered to produce insulin. However, I suspect that it may be a long time, if ever, before a single organism can produce the multitude of flavors found in foods such as cheeses, soy sauce, *miso*, and other fermented foods used primarily as condiments. The reason for this is the fact that a flavoring agent such as *shoyu* contains literally hundreds of compounds produced by the microorganisms, products from the action of enzymes on the substrate, and compounds formed by the nonenzymatic interactions of the products with the original substrate compounds.

To put such a combination of genes for all these flavors into one microorganism would be almost impossible at the present. Secondly, the cost of producing the food, which is relatively inexpensive as now produced, would be economically prohibitive even if it were done. At present the use of mixed cultures in making fermented foods from milk, meat, cereals, and legumes will continue to be the direction in the future.

Harrison (1978) in his summary of the future prospects of mixed culture fermentations very succinctly concluded as follows:

> No claim for novelty can be made for mixed cultures: They form the basis of the most ancient fermentation processes. With the exploitation of monocultures having been pushed to its limits it is perhaps time to reappraise the potential of mixed culture systems. They provide a means of combining the genetic properties of species without the expense and dangers inherent in genetic engineering which, in general terms, aims at the same effect.

References

Bull, A. T., and J. H. Slater (eds.): *Microbial Interactions and Communities*, Academic Press, 1982.

Buller, A. H. R.: Micheli and the discovery of reproduction in fungi, *Trans. Roy. Soc. Canada*, Ser. III, 9:1–25, 1915.
Bushell, M. E., and M. E. Slater (eds.): *Mixed culture fermentations*, published for the Society for General Microbiology, No. 5, Academic Press, 1981.
Driessen, F. M.: Protocooperation of yogurt bacteria in continuous culture, in *Mixed Culture Fermentations* (M. E. Bushell and J. H. Slater), Academic Press, London, 1981, pp. 99–120.
Cadmus, M. C., M. E. Slodki, and J. J. Nicholson: High temperature, salt-tolerant Xanthanase. *J. Ind. Microbiol.* 4:127–133, 1989.
Freer, S. N., and R. E. Wing: Fermentation of cellodextrins to ethanol using mixed-culture fermentations. *Biotechnol. Bioeng.* 27:1085–1088, 1985.
Fukuskima, D.: Soy sauce and other fermented foods of Japan, in *Indigenous Fermented Food of Non-Western Origin* (C. W. Hesseltine and H. L. Wang), J. Cramer, Berlin, 1986, pp. 121–149.
Goulding, C., C. J. Gillen, and E. Bolton: Biodegradation of substituted benzenes. *J. Appl. Bact.* 65:1–5, 1988.
Harrison, D. E. F.: Mixed cultures in industrial fermentation processes. *Adv. Appl. Microb.* 24:129–164, 1978.
Hesseltine, C. W.: A millennium of fungi, food, and fermentation. *Mycologia* 57:149–197, 1965.
Hesseltine, C. W.: A microbe's view of fermentation. *Devel. Indust. Microb.* 22:1–18, 1981.
Hesseltine, C. W.: Microbiology of oriental fermented foods. *Ann. Rev. Microbiol.* 37: 575–601, 1983.
Hesseltine, C. W.: Fungi, people and soybeans. *Mycologia* 77:505–525, 1985.
Hesseltine, C. W., R. Rogers, and F. G. Winarno: Microbiological studies on amylolytic oriental fermentation starters. *Mycopathologia* 101:141–155, 1988.
Jarl, K.: Symba yeast process. *Food Technol.* 23:1009–1012, 1969.
Lee, B. K., D. Y. Ryu, R. W. Thoma, and W. E. Brown: Induction and repression of steroid hydroxylases and dehydrogenases in mixed culture fermentations, *J. Gen. Microbiol.* 55:145–153, 1969.
Macfadyen, A.: The symbiotic fermentations. *J. Fed. Institutes of Brewing* 9:2–15, 1903.
Quayle, J. R., and A. T. Bull (eds.): New dimensions in microbiology. Mixed substrates, mixed cultures, and microbial communities. *Phil. Trans. R. Soc. London*, B-297:447–639, 1982.
Ryu, D. Y., B. K. Lee, R. W. Thoma, and W. E. Brown: Transformation of steroids by mixed cultures. *Biotechnol. Bioeng.* 11:1255–1270, 1969.
Winogradsky, S.: *Microbiologie du Sol: Problèmes et Methodes.* Masson et Cie., Paris, 1949.

Chapter

2

Breads from Mixed Cultures

Henry J. Peppler

5157 N. Shoreland Avenue
Whitefish Bay, Wisconsin

Introduction

Bread and many breadlike products, which are made from fermented doughs in much of the world today, originated thousands of years ago. Along with fermented milks, cheese, wine, beer, and soybean-based foods, these foods are among the oldest microbially modified products. Indigenous breadmaking was commonplace in Biblical times, and in ancient Egypt public bakeries provided wheat and barley breads made from leavened doughs. Among the civilizations that followed, notably Babylon, Greece, and Rome, bread was a major source of energy and nutrition. These civilizations also advanced the art of breadmaking by milling finer flour, devising better baking ovens, and managing the perpetuation of crude leavening preparations. Their subsistence, as well as that of cultures rooted in northern and western Europe and North America, was based on the cultivation of wheat, rye, barley, and oats. Elsewhere, rice became the staple food of the Orient, corn (maize) in Central and South America, and grain sorghum in Africa and southern Asia. Today wheat, rice, and sorghum are grown in greatest abundance.

Wheat, the premier bread grain with its superior nutritive and textural values, is produced in selected regions of the world. Its importation by nongrowing countries is substantial. Over the past 20 years, for example, the annual wheat imports of Africa have increased from 4 million to more than 16 million metric tons, mainly for the purpose of upgrading the protein-deficient diets of many Africans, where indigenous starch crops are the staple foods that are consumed largely as flat breads made from spontaneously fermented doughs (Boeh-Ocansey, 1989). Of the numerous detailed publications of the historical development of breadmaking, the reader is referred to Jacob (1944), Tannahill (1973), Darby (1977), Campbell-Platt (1987), and Pyler (1988).

Among the wide assortment of breads common today, there are those made from flour or meal ground from grains (wheat, rye, oats, corn, millet, rice), and the cooked or steamed batters prepared from ground or finely macerated starch crops (cassava, yams, taro, plantain), or legumes, such as black gram, soybean, and cowpea (see Table 2.1). Whatever the substrate and however it is converted to a dough and processed for cooking or baking, mixed populations of microorganisms will be present, usually as natural inhabitants of the raw ingredients used (Rogers and Hesseltine 1978, Kent-Jones and Amos 1967, Carlin 1958). Their fermentative activities will affect, to varying de-

TABLE 2.1 Cereal Flours, Starch Crops, and Legume Meals Used in Breadmaking

Cereals	
Wheat	*Triticum aestivum*, *T. vulgare*
Rye	*Secale cereale*
Rice	*Oryza sativa*
Sorghum	*Sorghum vulgare*
Barley	*Hordeum vulgare*
Oats	*Avena sativa*
Corn	*Zea mays*
Millet	*Panicum miliaceum*
Buckwheat	*Fagopyrum esculentum*
Triticale	Wheat-rye hybrid
Teff	*Eragrostis tef*
Starch crops	
Cassava	*Manihot utilissima*
Yams	*Ipomoea* sp.; *Dioscorea* sp.
Plantain	*Musa paradisiaca*
Taro	*Colocasia esculenta*
Breadfruit	*Artocarpus altilis*
Legumes	
Black gram (urd)	*Phaseolus mungo*
Gram (chick pea)	*Cicer arietinum*
Cowpea	*Vigna unguiculata*
Soybean	*Glycine max*
Pigeon pea	*Cajanus cajan*
Groundnut (peanut)	*Arachis hypogeae*
Winged bean	*Phaseolus tetragonolobus*

gree, the texture, flavor, color, and nutritional characteristics of the end product. The microbial species and their concentration in the mixed fermentation may change with the different flours used, the nature of the liquid and other additions to the dough, as well as the manner of handling the dough before it is steamed, baked, or fried.

Fermentations during breadmaking are most often due to the metabolic activities of a relatively few species of yeast and certain lactic acid–forming bacteria. When these fermenters produce carbon dioxide from sugars available in or added to the dough, they are known as leavens, and the accumulation and distribution of the gas honeycombs and lightens the dough mass. Several types of microbial leavens may be used in making bread:

1. Mixed yeast and bacteria allowed to grow spontaneously in the batters and doughs
2. Sour starters for home and bakery use that are perpetuated by frequent transfer to fresh substrate in-house after suitable development of spontaneous fermentations in flour slurries, or by reseeding with proprietary cultures of known species and purity
3. Commercial sourdough starter cultures obtained either as monocultures of proven strains of heterolactic lactobacilli, *Candida milleri* (the unusual yeast dominant in San Francisco sourdough and Italian *pandoro* cake), or dehydrated mature sourdough
4. Bulk and packaged baker's yeast, *Saccharomyces cerevisiae*, marketed worldwide as relatively pure strains in the form of compressed yeast (about 30 percent solids) and active dry yeast (about 92 percent solids) (Schuldt 1984, Reed and Peppler 1973).

The intentional addition of known-culture concentrates of yeast and/or bacteria to bread doughs helps to gain control of the microbial environment, especially the regulation of fermentation rate and the physical changes affecting texture, flavor, and appearance of the baked loaf. Their primary function in breadmaking is twofold: production of carbon dioxide and acid production, mainly lactic acid and acetic acid.

In retrospect, it is noteworthy that fermentation as a microbiological process gained recognition in the mid-1800s with Pasteur's experimental evidence that living lactic acid bacteria and yeast are the agents causing fermentation. By late 1800 the pure-culture concept had gained momentum. The broad front of activity surrounding the isolation of pure cultures and their characterization and classification, led to the founding of two new sciences: biochemistry and microbiology (Brock 1975). E. C. Hansen (1896) pioneered pure-culture studies of brewer's and baker's yeast. Lister, Beijerinck, Leichmann, and others at the turn of the century isolated and classified the lactic acid bacteria involved in traditional mixed fermentations of milk, milk products, vegetables, and bread doughs.

Baker's yeast became the first microbe available to the baking industry as a marketed commodity. As far back as the 1850s and into the early 1900s, yeast for bakeries was skimmed from grain-malt mashes fermented primarily for their yield of ethanol. In this yeast concentrate, also known as *barm*, lactic acid–producing bacteria in substantial numbers invariably accompanied the harvested crop of yeast. Modern baker's yeast technology, which dates from the early 1920s, is now well established on every continent (Reed and Peppler 1973). Estimates of world production (Oura et al. 1982) indicate totals in excess of 1,400,000 metric tons of fresh yeast (about 30% dry matter).

Bacterial pure cultures suitable for breadmaking are propagated and marketed chiefly in Europe and North America. They are usually mixed with yeast in the leavening of specialty white and rye breads and crackers (Spicher 1983, Sugihara 1985). Although the advantageous use of available biological control agents is widespread in the food industry, the empiric processes which rely on natural, spontaneous microbial activities are still practiced in many countries. Recent publications have catalogued and updated the enormous fermented food field: a dictionary and guide by Campbell-Platt (1987), and other reviews, including those by Rehm and Reed (1983), Steinkraus (1983), Rose (1982), and Wood (1981).

This chapter is concerned with the occurrence and function of microbial populations involved in doughs made with different cereal flours, legume meals, and starch crop macerates. A few examples of the dough systems commonly prepared in different geographical areas are shown in Table 2.2.

Microflora of Bread Doughs

White and variety breads

Satisfactory white bread is readily made from a dough consisting of wheat flour, water, baker's compressed yeast, and sodium chloride in the proportion of 100:65:2:2, respectively. Other ingredients are usually added to the basic recipe for the purpose of modifying flavor and nutrition. Among these are sugar (glucose or sucrose), nonfat milk solids, and fat added at the rate of 4:4:3 parts per 100 parts of flour, respectively. After developing the dough with proper mixing of the in-

TABLE 2.2 **Sourdough Breads and Indigenous Breadlike Foods**[a]

Product	Base substrate	Region	Microflora[b]
San Francisco sourdough shepherder bread	Wheat	W. N. America	L, Y
Rye bread	Rye/wheat	Europe, N. America	L, Y
Nan, lavash, sangak	Wheat	Middle East, S. Asia, N. Africa	L, Y
Idli	Rice/wheat/black gram	S. India	L, Y
Dosa	Black gram/rice	India	L, Y
Puto	Rice/wheat/corn	E. & S.E. Asia, S. India	L, Y
Kenkey	Corn	W. & S. Africa, Ghana	L, Y
Corn bread	Corn	Brazil, N. America	L, Y
Lafun	Cassava/yam/ plantain	N. Africa, Nigeria	L
Kisra	Sorghum	N. Africa, Sudan	L, Y
Injera	Teff/corn/sorghum	Ethiopia, Somalia	L, Y

[a]Compiled from Campbell-Platt (1987), Wood (1981), Steinkraus (1983).
[b]L = lactic acid bacteria; Y = yeasts.

gredients, it is kneaded, shaped, allowed to ferment (up to 3 h at 32°C for straight doughs), scaled to pan, proofed, and baked.

Because of its content of gluten protein complex, a properly made wheat flour dough forms an extensible elastic and cohesive mass which retains leavening gases throughout the required handling steps of making the loaf or rolls. Yeast action results in the production and migration of carbon dioxide into a network of cellular compartments which approach a volume of about 120 in^3 per pound of loaf. These changes increase dough volume, lighten the loaf, and improve its ultimate palatability. Other metabolic by-products of fermentation, and the baking process, contribute to the aroma and flavor of the baked loaf.

Wheat flour contains around 0.5% fermentable sugars, mainly glucose, fructose, sucrose, maltose, and utilizable glucofructosans in amounts that are adequate for rapid and sustained fermentation (Saunders et al. 1972). In the absence of added sugar, yeast readily metabolizes glucose, then fructose and maltose, as shown in Fig. 2.1. More maltose becomes available in the later dough stages through the hydrolytic action of flour amylases on the starch contained in fractured starch granules. This sustained carbon dioxide generation accounts, in part, for the increase in volume as the loaf attains baking temperature. The volume change from the unbaked to the baked loaf is known as oven-spring.

Biochemically the yeast fermentation process proceeds by way of the Emden-Mayerhof glycolytic pathway, yielding approximately

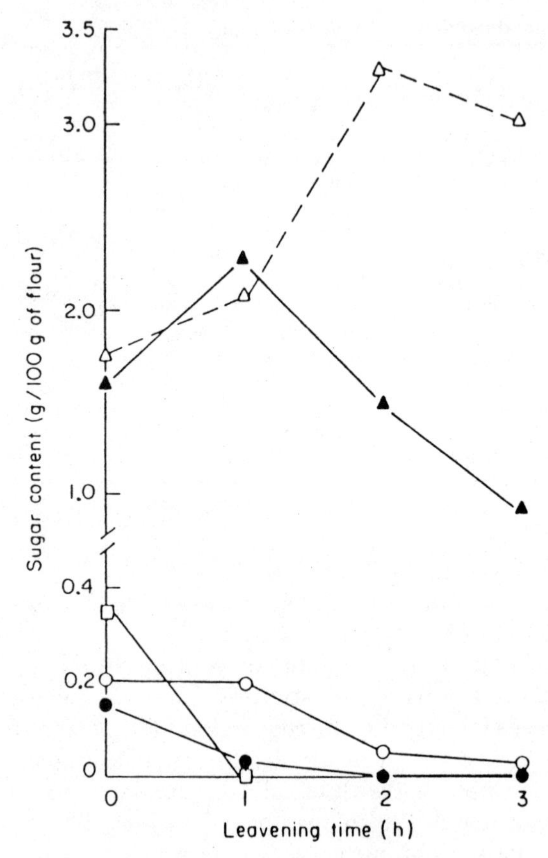

Figure 2.1 Changes in sugar concentrations of fermentation of a straight dough: sucrose (□), glucose (●), fructose (○), and maltose in a dough during test leavening and of maltose (△ in a nonyeasted dough and ▲ in yeasted dough). (*From Suomalainen et al., 1972.*)

0.47 g CO_2 (270 mL) and 0.48 g ethanol per gram of glucose fermented. Other by-products include glycerol, succinic acid, and small amounts of aldehydes, ketones, ethyl esters, and other compounds (Oura et al. 1982, Linko et al. 1962). While some of the more volatile compounds evaporate during baking, new carbonyls are formed via the Maillard reaction (Magoffin and Hoseney 1974, Linko et al. 1962).

End products of bacterial fermentation also accumulate during dough leavening and may enhance the flavor of white pan bread (Johnson and Sanchez 1973, Kent-Jones and Amos 1967). Lactic acid bacteria always accompany commercial baker's yeast. The higher counts, over 100 million cells per gram of yeast, are found in compressed yeast (Reed and Peppler 1973). Flour and nonfat milk also add their bacterial mixtures to the dough fermentation (Pyler 1988, Rogers and Hesseltine 1978, Kent-Jones and Amos 1967). Lactic and acetic acids are present in white bread doughs and are the main residual acids in the finished bread. Lactic acid is a colorless, nearly odorless nonvolatile liquid with a strong acid taste. Acetic acid is characterized as a weak acid with a fruity odor, and dilutions have a pleasant taste. Jackel (1969) reported levels of 50 to 100 mg lactic acid and 10 to 50 mg acetic acid per pound of bread. The ethyl esters of these acids may also augment overall bread flavor (Johnson and Sanchez 1973).

Breads baked from doughs fermented with pure cultures of baker's yeast and lactic acid bacteria, which were isolated from compressed yeast, scored lower in flavor than the loaves made from doughs fermented by the same commercial yeast sample with its associated bacterial contaminants (Carlin 1958). When specific mixtures of pure cultures were used to prepare straight doughs, the flavor score of the bread approached the score of those loaves made with the initial control lot of baker's yeast (see Table 2.3). At best, large numbers of the right bacterial species can modify bread flavor significantly. In general, because of the high variability of types and cell populations occurring in the microflora of dough ingredients, their influence on the flavor and aroma of conventional white bread is at best inconsistent and unclear. However, in the production of sourdough breads, specific heterofermentative lactobacilli are known to be key flavor-promoting agents.

Whole-wheat bread and mixed-grain breads, which are blends of wheat flour and various other cereals (see Table 2.1), rank second in popularity to white pan bread (Anon. 1989). Ranked third, based on

TABLE 2.3 Flavor Score of Bread Made with Straight Doughs Fermented with Pure Cultures of Baker's Yeast and *Lactobacillus* sp.

Dough inoculum	Yeast[a]	Bacteria[a]	Bread flavor score[b]
A. Compressed yeast, commercial	330	30	7.0
B. Yeast, pure culture	320	nil	4.0
C. *Lactobacillus* sp. plus B	370	5	5.0
D. *Lactobacillus* sp., double inoculum, plus B	250	10	6.4

[a]Viable counts: $\times 10^6$ per gram dough.
[b]Rated on the sixth day, hedonic scale: higher score indicates better flavor.
SOURCE: From Carlin, 1958.

value of products shipped, are hearth-baked breads of the French, Italian, and Vienna types, followed by lesser volumes of rye, pumpernickel, and specialty breads (raisin, potato, low-calorie, oat bran, etc.).

Generally, production of variety breads employs much the same technology and leavening practices used for conventional white bread except for those hearth-baked varieties seeded with natural or commercial sourdough starters.

Sourdough breads

The formulation, development, and perpetuation of sourdough leavening and flavoring systems, with desirable and reproducible dough and bread characteristics, have been the concern of bakers since ancient times. During the recent quarter-century or so, however, the interrelationships of yeast and bacteria mixtures in sourdoughs have been investigated extensively. Two major types of bread doughs have received the most attention: (1) some sourdough rye breads favored in Europe (Spicher 1983, 1986), and (2) white hearth breads produced in North America (Sugihara 1985).

Rye breads. Among traditional bread products, rye breads exhibit the broadest range in color, taste, and texture. The different formulations with distinctive characteristics have resulted in ethnic and regional varieties, including American, Bohemian, German, Jewish, Polish, and Swedish rye breads. Crumb color varies from tan to dark brown, taste ranges from low acid to sour, which may be modified by such flavorants as caraway seeds, molasses, dairy products, and nondiastatic malt. Rye breads are most popular in areas of greatest rye cereal production: USSR, Poland, Germany, and Scandinavia (Oura et al. 1982). In these nations, and in the United States, biological sourdoughs are preferred for the production of full-flavored quality rye breads. In Germany alone, 36 different types of rye bread are marketed (Lorenz 1981).

It is relatively easy to prepare rye sourdough starters: rye flour and water are mixed in equal proportions and fermentation is allowed to develop spontaneously at 28 to 30°C for several days. Such starter sponges become acidic and gassy due to the growth of indigenous bacteria and yeasts in the flour slurry. Gram-positive lactic acid bacteria and yeasts eventually dominate the fermentation. A typical pattern is shown in Fig. 2.2. Such single-stage sourdough starters are generally weak in leavening power, and therefore, they require the addition of baker's yeast, about 1 percent, to the final dough make-up. Many satisfactory modifications of the basic flour-water mixture have been devised, varying from one-stage sponges of short or long maturing or rip-

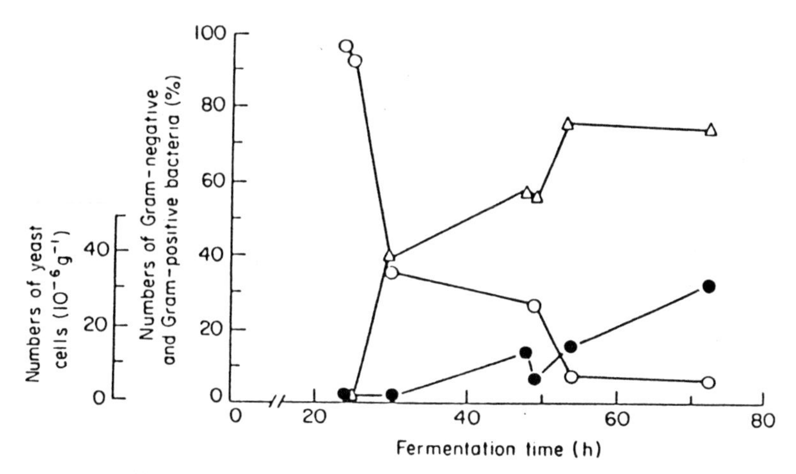

Figure 2.2 Changes in the contents of yeasts (●), gram-positive bacteria (△), and gram-negative bacteria (○) during spontaneous fermentation of sourdough. (*From Kosmina, 1977.*)

ening periods, to complex, multistage systems (Spicher 1983, 1986). Extensive investigations of numerous rye sourdoughs collected from German bakeries have established that the important and dominant souring bacteria are *Lactobacillus brevis, L. fermentum, L. plantarum*, and *L. casei* (Spicher 1983, 1986). The first 2 organisms are heterofermentative species, fermenting sugars, mainly to lactic acid and acetic acid, and some ethanol and carbon dioxide, and traces of other components. Their metabolic products are responsible for the flavor of sour rye bread (Spicher and Stephan 1982, Spicher et al. 1979). *L. plantarum* and *L. casei* are homofermentative species, which ferment sugars to lactic acid exclusively and contribute to crumb elasticity. Many other species of lactic acid bacteria have been isolated from bakery-prepared sourdough sponges; up to 9 species were identified in 1 starter (Spicher and Schröder 1978). Table 2.4 lists the microorganisms considered to be the principal fermenters in rye and wheat sourdoughs. According to Kosmina (1977) high-quality rye bread, in terms of aroma and crumb structure, can be produced with mixtures of pure cultures of *L. brevis* and *L. plantarum*. Although somewhat higher levels of salt are used in rye doughs, the fermentation of these 2 species is not retarded since both are less sensitive to salt than other heterolactics (Stamer et al. 1971).

The rate of acidification during sourdough development and the indicated ratio of lactic acid to acetic acid is shown in Fig. 2.3. A ratio of 4.0:1.0 of lactic to acetic acid is considered to have the optimum effect on rye bread flavor. Figure 2.4 shows the increase in bacteria and yeast populations in a spontaneous rye sourdough and compares the

TABLE 2.4 Sugars Fermented by the Principal Bacteria and Yeasts in Sourdoughs

Organism	Sugars fermented[a]						
	GLU	GAL	FRU	MAL	LAC	SUC	RAF
Saccharomyces cerevisiae	+	+	+	+	−	+	+
Candida krusei	+	−	−	−	−	−	−
C. milleri	+	+	+	−	−	+	+
Lactobacillus brevis[b]	+	+	+	+	+	+	+
L. fermentum[b]	+	+	+	+	+	+	+
L. sanfrancisco[b]	+	+	−	+	−	−	−
L. plantarum[c]	+	+	+	+	+	+	+
L. casei[c]	+	+	+	+	−	+	−
L. delbrueckii[c]	+	−	+	+	−	+	−
Leuconostoc mesenteroides[b]	+	+	+	+	+	+	+

[a]GLU = glucose, GAL = galactose, FRU = fructose, MAL = maltose, LAC = lactose, SUC = sucrose, RAF = raffinose.
[b]Heterofermentative species.
[c]Homofermentative species.
SOURCE: According to Kreger-van Rij (1984) and Sneath (1986).

changes in pH and titratable acidity. Growth accelerates in early hours until the dough pH is near 4.0, then both populations become stationary while acidity continues to increase. Although some sourdough heterolactics generate significant leavening power, it is generally inadequate for rye doughs. Baker's yeast is always added at the dough stage. Yeasts in rye sourdoughs vary with the kind of ingredients and culture conditions used. In the 44 starters examined by Spicher et al. (1979), *Candida krusei* (and related species), *Candida*

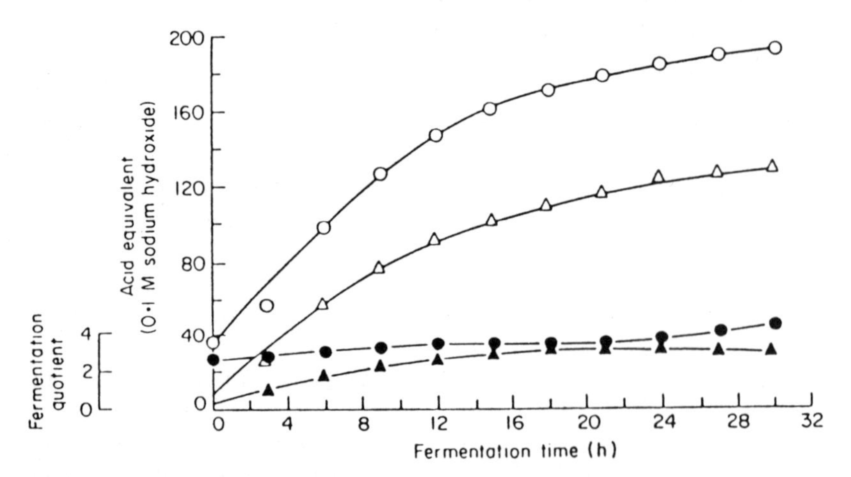

Figure 2.3 Formation of total acid (○), lactic acid (△), and acetic acid (▲), and fermentation quotient (●) during the fermentation of a rye sourdough. (*From Kosmina, 1977.*)

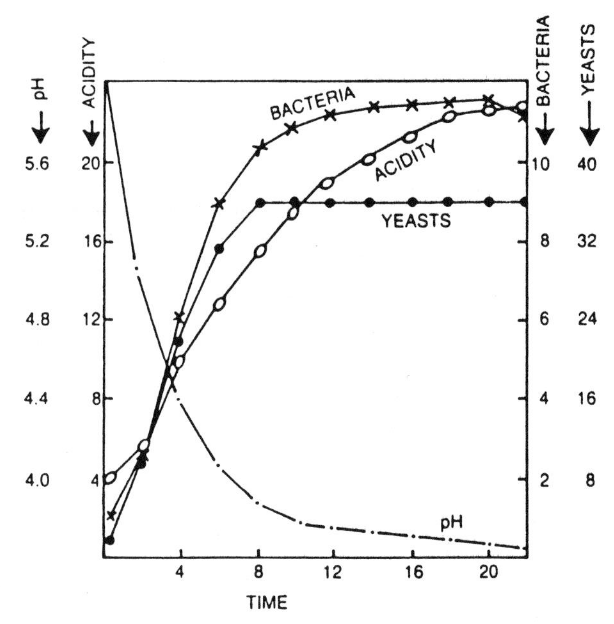

Figure 2.4 Multiplication of yeast and bacteria and changes in pH values and acidity during fermentation of rye sourdough. (*From Rohrlich and Stegeman, 1958.*)

milleri, and *Saccharomyces cerevisiae* were isolated most frequently (see Table 2.4). Quantitatively, the principal population of spontaneous rye sourdoughs consists of 3.7 to 7.5 × 10^9 bacteria, and 1.0 × 10^6 to 3.0 × 10^8 yeast per gram of dough (Spicher 1983).

The seeding of sourdough sponges may now be done with pure cultures of several starter bacteria; however, the practice is not in general use. Currently available cultures include *L. brevis, L. plantarum*, and *L. delbrueckii* in deep-frozen and freeze-dried form (Anon. 1988). Hill (1987) sprayed concentrated suspensions of *L. brevis* on rye flour, dehydrated the mixture in a fluidized bed, and packaged it under nitrogen. A 70 percent survival rate (5 × 10^{10} viable cells per gram product) and long shelf life was claimed. A mixed starter of *L. brevis* and *Saccharomyces dairensis*, which does not ferment maltose, was proposed by Spiller (1987). A unique dehydrated sourdough starter, developed for household use, featured high concentrates of *L. fermentum* codried with a strain of active dry baker's yeast (Tesch et al. 1976). The nitrogen-protected codried product developed a mature sourdough wheat flour sponge in 20 to 24 h at 30 to 32°C. It resulted in excellent loaves of sourdough white bread. Following a 13-year storage period under refrigeration, one of the nitrogen-packed envelopes of the codried starter was reactivated by the author, using the original sourdough bread recipe. Flavor and texture of the bread scored about 7.0 on the hedonic scale.

San Francisco sourdough bread. Of the white hearth breads made with microbiological sours, the well-known San Francisco sourdough French bread has been studied most intensively (Kline et al. 1970, 1971; Sugihara et al. 1971). Its appealing acid flavor, crisp crust, and chewy crumb have broadened its popularity far beyond its origin, the San Francisco bay area. The natural starter culture favored in its production has been in continuous use for more than 140 years (Sugihara 1985). The formula and process steps leading to San Francisco sourdough bread are shown in Table 2.5. The starter sponge, consisting only of wheat flour, water, and a portion (25 to 40 percent) of the previous batch of natural starter, is fermented for about 8 h at 27°C. The final dough is made up with a portion of the new starter sponge, flour, water, and salt; it is also fermented 8 hours before baking. The starter sponge is easily regenerated, as often as needed, by following the recipe given in Table 4.5. Meigs (1967) has published starter and bread dough formulas suitable for bake shop use.

In their determinative studies, Kline and Sugihara (1971) examined starter sponges from five San Francisco area bakeries. From all of these sponges two heretofore unrecognized microorganisms were isolated: a heterolactic lactobacillus, a new species designated *Lactobacillus sanfrancisco*, and an asporogenous yeast, at first identified as *Torulopsis holmii*, but reclassified by Yarrow (1978) as a new species, which he named *Candida milleri*. These two microorganisms are the sole fermenters in this unusual, stable, noncompetitive ecosystem. *L. sanfrancisco* ferments only maltose, which it requires in relatively high concentrations, and *C. milleri* does not metabolize maltose, but it uses all the other free sugars and some glucofructosans in flour

TABLE 2.5 San Francisco Sourdough Bread Formula

Starter sponge	Dough
100 parts previous starter 100 parts high-gluten flour 46–52 parts water	20 parts starter sponge 100 parts patent flour 60 parts water 2 parts salt
Make up, hold 7–8 h at 27°C; pH: start 4.4–4.5; end 3.8–3.9	Make up, 1 h floor time; Proof 8 h 30°C pH: start 5.2–5.3; end 3.9–4.0

SOURCE: Adapted from Kline et al. (1970), Sugihara (1985).

(Kline and Sugihara 1971; Sugihara et al. 1971; Saunders 1972). *C. milleri* is responsible for the leavening action in the starter and dough preparation stages. In wheat sourdoughs *L. sanfrancisco* produces carbon dioxide and lactic and acetic acids. The acetic acid concentration is higher than that observed in other sourdough systems (Ng 1972). Acetic acid constitutes nearly 50 percent of the total titratable acidity of sourdough bread, a level that is 5 to 10 times that reported for white pan breads (Hunter et al. 1970). In addition to the major acids, Galal et al. (1978) detected six minor acids in commercial San Francisco sourdoughs: propionic, isobutyric, butyric, alpha-methyl-*n*-butyric, isovaleric, and valeric acids. Together these acids amount to less than 1 percent of the total acidity found in the bread, but this small acid fraction is believed to contribute significantly to sourdough flavor since its concentration is greater by 50 percent than the level present in white pan bread.

For its luxuriant growth, *L. sanfrancisco* requires high levels of maltose and a source of organic nitrogen. A peptide chain of five amino acids satisfies this fastidious requirement, and it is secreted by *C. milleri* during sourdough fermentation (Berg et al. 1981).

Following the discovery and elucidation of the role of *Torulopsis holmii* (now *Candida milleri*) in San Francisco sourdough sponges, strains of the same species of yeast were isolated from several microbially leavened products. Among these are certain rye sours commonly made in Germany (Spicher et al. 1979), Finland (Oura et al. 1982), and Russia (Kosmina 1977), and also used in making sourdough starters for Italian cakelike breads and rolls, such as *pannetone* (Milan area), *pandoro* (Verona vicinity), and *colomba* (Sugihara 1977, 1985; Zorzanello and Sugihara 1982). In these Italian products the souring organism is *L. brevis*.

Concentrates of pure starter cultures of *L. sanfrancisco* for use in commercial bakeries have been developed (Sugihara 1985; Kline 1981). Today San Francisco sourdough bread is made in other parts of the world beyond the San Francisco bay area.

Flat breads

Long before bread doughs were baked in ovens, cereal gruels were made into dough, shaped into flat cakes, and cooked or grilled on hot stones or directly in the fireplace. Such products were prepared from different indigenous cereal grasses and starch sources, and they provided early humans with a wide variety of flat breads, including unleavened and fermented products. Many of these ancient and traditional breads are still prepared and consumed as important staple energy foods in large areas of Asia, Africa, and Europe. Faridi and

Rubenthaler (1983) identified 42 main flat breads that are popular in 23 different nations of the Middle East, southern Asia, northern Africa, and Europe. In his global compilation, classification, and description of fermented foods, Campbell-Platt (1987) accounts for 90 major products of fermented cereal flours and starch sources, including a host of flat bread types and closely related pancake-like products.

Two general terms identify flat breads: *Nan* (naan) and *Lavash* (lahvosh). *Nan* is a type of flat bread with a central cavity or pocket; at least 20 closely related varieties are known (Campbell-Platt 1987). *Lavash* is generally associated with several forms of wheat flour flat breads and pancake-like products popular in villages of Middle East countries (Faridi and Rubenthaler 1983). Table 2.6 identifies some of the main varieties. A main feature of these traditional breads is that they are made from almost any kind of cereal flour and starch crop, as indicated in Table 2.2. Such products differ considerably in flavor, texture, and appearance. Typical preparations simply combine flour or meal with water, salt, yeast, and/or leftover fermented dough, or are allowed to ferment naturally by the microflora on the raw ingredients used and the residual populations adhering to process utensils. Flat bread formulations rarely include sugar, shortening, or dry milk. The end products vary from paper-thin to thick pancakes, usually in round or oval form with light, thin crust and mottled-brown spots.

Most flat bread batters and doughs invariably undergo natural acidification due to mixed species of lactic acid bacteria. Unlike the classic studies done on wheat and rye sourdoughs discussed earlier, the microbiology of flat bread batters and doughs is largely unexplored.

TABLE 2.6 Some Main Flat Breads and Areas of Consumption[a]

	Consuming countries
NAN type	
Arabic (*naan*)	Egypt, India, Lebanon, Pakistan
Balady	Egypt, Lebanon
Barbari	Iran
Kamag	Jordan, Syria
Kimaj	Israel
Kubz	Syria
Kulcha	India
Pita	Greece, Cyprus, Saudi-Arabia, USA
Shamsey	Egypt, Sudan
Lavash type	
Kisra	Sudan
Lavash	Afghanistan, Iran, Syria, USA
Markouk	Jordan
Sangak	Iran
Tannouri	Afghanistan, Syria

[a]Compiled from Faridi and Rubenthaler (1983), Campbell-Platt (1987), Steinkraus (1983), and Beuchat (1983).

The sparse data, however, reveal the significant contributions of lactic acid bacteria other than those species associated with the better-known wheat and rye sourdoughs. The principal bacterial species associated with flat breads are *Leuconostoc mesenteroides, Streptococcus faecalis*, and *Pediococcus cerevisiae*. They are involved in the preparation of a closely related group of products: *idli*, its varieties known as *ambali, puto*, and *hopper*; and also some types of *lavash*, mainly *balady* and *sangak* (Campbell-Platt 1987; Azar et al. 1977; Ekmon and Nagodawithana 1983; Mukherjee et al. 1965).

Idli. This pancake-like product, popular in India and Sri Lanka, is made with a thick, salted batter of rice and black gram bean flours incubated overnight. The leavening and acidification is due mainly to *L. mesenteroides*, a heterofermentative rod-shaped bacterium, which initiates the fermentation. In the late stages *S. faecalis* and *P. cerevisiae* are secondary acidifying agents. Both species are homofermentative cocci (Mukherjee et al. 1965).

The mature leavened batter is steamed, forming soft, spongy breadlike cakes with a desirable sour flavor. A thinner batter fried as a pancake is known as *dosai*.

It is a characteristic of *Leuconostoc mesenteroides* to begin growth and fermentation ahead of other lactic species. With its salt tolerance and ability to ferment most common sugars (see Table 2.4), the pH drops rapidly, thereby inhibiting the growth of undesirable microorganisms, especially toxigenic and proteolytic species contaminants. *Streptococcus faecalis*, which also appears early in the fermentation, is a relatively low-acid producer (less than 2 percent) similar to *L. mesenteroides. Pediococcus cerevisiae*, a high-acid species, appears in a minor role, along with yeasts, toward the end of the overnight fermentation (Purushothaman et al. 1983; Ramakrishnan 1983; Mukherjee et al. 1965).

Ambali. The preparation of this steamed, fermented batter made from rice and millet flours follows closely the *idli* procedure. *L. mesenteroides* and *S. faecalis* provide the principal acidification, augmented by *Lactobacillus fermentum* in the later stages of fermentation (Ramakrishnan 1979).

Puto. *Puto*, a steamed rice cake, is a staple food in the Philippines, India, China, and southeast Asia. Formula variations include combinations of wheat or corn (maize) with the glutinous rice flour. The natural starter used is a ripened portion of the previous batter. Initially it contains *L. mesenteroides, S. faecalis* and *P. cerevisiae; Sacch. cerevisiae* participates in the late hours of the 42-h fermentation

(Sanchez 1977). When the holding time prescribed for the traditional process was halved by a mixed starter of *L. mesenteroides, S. faecalis*, and *Sacch. cerevisiae*, the baked product was judged equal in texture, flavor, and volume to *puto* made with natural starter (Sanchez 1977).

***Balady* bread.** This product is one of numerous yeast-leavened, pocket-bearing flat breads popular in eastern Mediterranean countries. It is generally made from doughs of wheat flour (70 to 80 percent extraction), water, baker's yeast, or sourdough from a recent batch of *balady* dough (Faridi and Rubenthaler 1983; Campbell-Platt 1987). Flattened disks of fermented dough are oven-baked rapidly (5 to 10 min, 230 to 260°C), or directly against oven walls (2 to 3 min, 460 to 500°C). The intense heat quickly expands the central dough mass, creating a cavity or pocket. The microbial population of the naturally developed sourdough starter consists mainly of *Lactobacillus* spp. (about 65 percent), yeasts (around 30 percent), and incidental levels of *Leuconostoc* sp., lactic cocci and sporeforming bacilli (Abd-El-Malek et al. 1974). *L. brevis*, the principal fermenter (75 percent) in the dough stage, is assisted by *L. fermentum* (20 percent) in the acidification and leavening. Occasionally *L. plantarum* and *L. casei* are isolated from *balady* doughs.

Hopper. Produced principally in Sri Lanka, *hopper* is a steam-baked, fermented dough consisting of rice and/or wheat flours, coconut water, sugar, and salt and baker's yeast or coconut toddy (fermented coconut milk). The toddy supplies a mixture of unidentified yeasts and lactic acid bacteria. Flavor of the cooked dough varies from coconut milk to sour; longer overnight fermentations yield the most desirable *hopper* taste (Ekmon and Nagodawithana 1983).

Lavash, sangak, taftoon, tannouri. These thin, leavened flat bread varieties are similar in composition and preparation to the *nan*-type pocket breads, such as *balady* bread. The doughs are typically a combination of wheat flour (about 85 percent extraction), water, salt, baker's yeast, and sourdough from a previous batch of dough, or sodium bicarbonate. After a short fermentation (about 2 h), the dough is flattened to form thin sheets and baked on a hot griddle at 300 to 330°C for 0.5 to 1.5 min. At this lower baking temperature there is no formation of the central pocket which is typical of *nan*-type flat breads. Leavening is due chiefly to yeast fermentation and in part to bicarbonate. Lactic bacteria contributing to acidic flavor development include variable and mixed populations of lactobacilli. *Leuconostoc* sp., and lactic cocci are also often present (Campbell-Platt 1987).

Sangak, one of Iran's important flat breads, uses a sourdough starter known as *torsh* (Azar et al. 1977). Its dominant heterolactic, *L. mesenteroides*, constitutes about 80 percent of the bacterial population and accounts for most of the carbon dioxide needed for leavening the dough. The remainder, in varying concentrations, consists of *L. plantarum* and the heterofermentative *L. brevis*, and *P. cerevisiae*, plus incidental yeasts (Azar et al. 1977).

Kisra. This most popular flat bread of the Sudan is made from sorghum flour, alone or mixed with millet or wheat flours for better quality. A portion of a previous batch of vigorously fermenting batter serves as starter for the 18-h dough fermentation. The heavy mature dough is diluted with water to a thin batter suitable for pouring sparingly upon preoiled, hot (150 to 160°C) earthenware or metal plates, and given a short bake (1 to 2 min).

Three main fermentations develop in *kisra* dough: lactic acid by *Lactobacillus* sp., ethanol and carbon dioxide by *Sacch. cerevisiae*, and acetic acid from ethanol in the late stages by the oxidative *Acetobacter* sp. (Abd-El-Gadir and Mohamed 1983). It is reported that *kisra* bakers perpetuate the starter with frequent dough reseedings and infrequent cleaning of the fermentation vessels. Thorough cleaning may occur every 3 to 4 months, after which a fresh starter is borrowed from a friendly neighbor.

Cassava bread. Cassava tubers, peeled and grated, undergo fermentation for several days to soften the starchy pieces and hydrolyze linamarin, the toxic cyanogenic glucoside. The fermented dough is formed into disks and baked or steamed on flat trays. The flat round bread, a staple food in West Africa (Ghana) and West Indies, is often grilled, browned, or toasted to increase flavor (Campbell-Platt 1987).

The necessary retting action which inactivates linamarin is believed to be due to species of *Corynebacterium* and *Bacillus* (Okafor et al. 1984). To achieve the acidic flavor of cassava bread, unidentified species of *Leuconostoc* and *Lactobacillus* are involved. Toward the end of the long fermentation period, some species of *Candida* and *Geotrichum* may appear.

In conclusion, bread in any one of its many forms, whether leavened with yeasts and acid-producing bacteria, or unleavened, has been the human species' basic food for millennia. The mixed flora impart desired textural, flavor, and safety characteristics to the finished products.

References

Anon.: "Cultures for Bread," Technical Bulletin EN 920–588, 8 pp., Chr. Hansen's Laboratorium A/S, Horsholm, Denmark, 1988.

Anon.: "Product Shipments of Bakery and Related Products," *Food Engrg.* 61(6), 86, 1989.

Abd-El-Gadir, A. M., and M. Mohamed: "Kisra: Sudanese Lactic/Acetic Acid Fermented Sorghum Bread," in *Handbook of Indigenous Fermented Foods* (K. H. Steinkraus, ed.), Marcel Dekker, Inc., New York, 1983, pp. 175–179.

Abd-El-Malek, Y., M. A. Leithy, and Y. N. Awad: "Microflora of Balady Bread Starter," *Chem. Mikrobiol. Technol. Lebensm.* 3, 1974, pp. 148–153.

Azar, M., N. Ter-Sarkissian, H. Chavifek, T. Ferguson, and H. Ghassemi: "Microflora of Sangak Bread Starter," *J. Food Sci. Technol.* 14, 1977, pp. 251–254.

Berg, R. W., W. E. Sandine, and A. W. Anderson: "Identification of a Growth Stimulant for *Lactobacillus sanfrancisco*," *Appl. Environ. Microbiol.*, 42, 1981, pp. 786–788.

Beuchat, L. R.: "Indigenous Fermented Foods," in *Biotechnology* (H. J. Rehm and G. Reed eds.), vol. 5, Verlag Chemie, Weinheim, 1983, pp. 477–528.

Boeh-Ocansey, O.: "Developments and Challenges in Africa's Food Industry," *Food Technol.* 43(5), 1989, pp. 84–92.

Brock, T. D.: *Milestones in Microbiology*, Am. Soc. Microbiol., Washington, D.C., 1975.

Campbell-Platt, G.: *Fermented Foods of the World—A Dictionary and Guide*, Butterworths, London, 1987.

Carlin, G. T.: "The Fundamental Chemistry of Breadmaking," *Proc. Am. Soc. Bakery Eng.*, 1958, pp. 55–61.

Darby, W. J.: *Food, The Gift of Osiris*, Nutrition Foundation Press, New York, 1977.

Ekmon, T. D., and T. Nagodawithana: "Fermented Foods of Sri Lanka," in *Handbook of Indigenous Fermented Foods* (K. H. Steinkraus, ed.), Marcel Dekker, Inc., New York, 1983, pp. 172–175.

Faridi, H. A., and G. L. Rubenthaler: "Ancient Breads and a New Science: Understanding Flat Breads," *Cereal Foods World*, 28, 1983, pp. 627–629.

Faridi, H. A., P. L. Finney, G. L. Rubenthaler, and J. D. Hubbard: "Functional (Breadmaking) and Compositional Characteristics of Iranian Flat Breads," *J. Food Sci.*, 47, 1982, pp. 926–929, 932.

Galal, A. M., J. A. Johnson, and E. Varriano-Marston: "Lactic and Volatile (C_2-C_5) Organic Acids of San Francisco Sourdough French Bread," *Cereal Chem.* 55, 1978, pp. 461–468.

Hansen, E. C.: *Practical Studies in Fermentation*, C. Griffin & Co., London, 1896.

Hill, F. F.: "Dry Living Microorganisms, Products for the Food Industry," in *Biochemical Engineering* (H. Chmiel, W. P. Hammes, and J. E. Bailey, eds.), Gustav Fischer Verlag, Stuttgard–New York, 1987.

Hunter, I. R., M. K. Walden, and L. Kline: "The Acetic Acid Content of Sour French Bread and Dough as Determined by Gas Chromatography," *Cereal Chem.* 47, 1970, pp. 189–193.

Jackel, S. S.: "Fermentation Flavors of White Bread," *Bakers Digest*, 43, 1969, pp. 24–26.

Jacob, H. E.: *Six Thousand Years of Bread*, Doubleday Doran and Co., Garden City, New York, 1944.

Johnson, J. A., and C. R. S. Sanchez: "The Nature of Bread Flavor," *Bakers Digest* 47(5) 1973, pp. 48–50.

Kent-Jones, D. W., and A. J. Amos: *Modern Cereal Chemistry*, 6th ed., Northern Publishing Co., Liverpool, 1967, pp. 511–538.

Kline, Leo: "Freeze-Dried Natural Sour Dough Starter," U. S. Patent 4,243,687, Jan. 6, 1981.

Kline, L., T. F. Sugihara, and L. B. McCready: "Nature of the San Francisco Sourdough French Bread Process," *Bakers Digest* 44(2) 1970, pp. 48–50.

Kline, L., and T. F. Sugihara: "Microorganisms of the San Francisco Sourdough Bread Process," *Appl. Microbiol.* 21, 1971, pp. 449–456.

Kosmina, N. P.: *Biochemie der Brotherstellung*, VEB Fachbuchverlag, Leipzig, 1977.

Kreger-van Rij, N. J. W.: *The Yeasts—a Taxonomic Study*, Elsevier Science Publishers B. V., Amsterdam, 1984.

Linko, Y., J. A. Johnson, and B. S. Miller: "The Origin and Fate of Certain Carbonyl Compounds in White Bread," *Cereal Chem.* 39, 1962, pp. 468–476.

Lorenz, K.: "Sourdough Processes—Methodology and Biochemistry," *Bakers Digest* 55(1), 1981, pp. 32–36.

Magoffin, C. D., and R. C. Hoseney: "A Review of Fermentation," *Bakers Digest* 48(6), 1974, pp. 22–23, 26–27.
Meigs, H. T.: *Pacific Slope Sour Dough Hearth Breads*, Am. Soc. Bakery Engrs., Bull. 183, 14 pp., 1967.
Mukherjee, S. K., M. N. Albury, C. S. Pederson, A. G. Van Veen, and K. H. Steinkraus: "Role of *Leuconostoc mesenteroides* in Leavening the Batter of Idli, a Fermented Food of India," *Appl. Microbiol.* 13, 1965, pp. 226–231.
Ng, Henry: "Factors Affecting the Organic Acid Production of Sour Dough (San Francisco) Bacteria," *Appl. Microbiol.*, 23, 1972, pp. 1153–59.
Okafor, N., B. Ijioma, and C. Oyulu:'"Studies in the Microbiology of Cassava Retting for Foo-Foo Production," *J. Appl. Bact.* 56, 1984, pp. 1–13.
Oura, E., H. Suomalainen, and R. Viskari: "Breadmaking" in *Fermented Foods*, vol. 7, Economic Microbiology series (A. H. Rose, ed.), Academic Press, London, 1982, pp. 87–146.
Purushothaman, D., N. Dhanapal, and G. Rangaswami: "Microbiology and Biochemistry of Idli Fermentation," in *Handbook of Indigenous Fermented Foods* (K. H. Steinkraus, ed.) Marcel Dekker, Inc., New York, 1983, pp. 132–146.
Pyler, E. J.: *Baking Science and Technology*, vol. 2, 3d ed., Sosland Publishing Co., Merriam, Kansas, 1988, pp. 776–814.
Ramakrishnan, C. V.: "Studies on Indian Fermented Foods," *Baroda J. Nat.* 6, 1979, pp. 1–57.
Ramakrishnan, C. V.: "Fermented Foods of India," in *Handbook of Indigenous Fermented Foods* (K. H. Steinkraus, ed.), Marcel Dekker, Inc., New York, 1983, pp. 132–146.
Reed, G., and H. J. Peppler: *Yeast Technology*, AVI Publishing Co., Westport, Conn., 1973.
Rehm, H.-J., and G. Reed (eds.): *Biotechnology: Food and Feed Production with Microorganisms*, vol. 5, Verlag Chemie, Weinheim, 1983.
Rogers, R. F., and C. W. Hesseltine: "Microflora of Wheat and Wheat Flour from Six Areas of the United States," *Cereal Chem.*, 55(6), 1978, pp. 889–898.
Rohrlich, M., and J. Stegeman: "Multiplication of Sour Dough Organisms and Acid Formation during Dough Fermentations," *Brot Gebaeck* 12, 1958, pp. 41–63.
Rose, A. H. (ed.): *Economic Microbiology*, vol. 7, *Fermented Foods*, Academic Press, Oxford, 1982.
Sanchez, P. C.: "Starter Cultures for Making Puto," *Philippine Agric.* 61, 1977. pp. 134–137.
Saunders, R. M., H. Ng, and L. Kline: "The Sugars of Flour and Their Involvement in the San Francisco Sour Dough French Bread Process," *Cereal Chem.* 49(1), 1972, pp. 86–91.
Schuldt, E. H.: "Bakers' and Food Yeasts—the Comestible Yeasts," *Dev. Indust. Microbiol.*, 25, 1984, pp. 203–212.
Sneath, P. H. A., (ed.): *Bergey's Manual of Systematic Bacteriology*, vol. 2, Williams and Wilkins, Baltimore, 1986.
Spicher, G.: "Baked Goods" in *Biotechnology* (H.-J. Rehm and G. Reed, eds.), vol. 5, Verlag Chemie, Weinheim, 1983, pp. 1–80.
Spicher, G., and R. Schröder: "Sourdough Starters in Bakeries," *Z. Lebensm. Unters. Forsch.* 167, 1978, pp. 343–354.
Spicher, G., R. Schröder, and K. Schölehammer: "Rye Sourdoughs," *Z. Lebensm. Unters. Forsch.* 169, 1979, pp. 77–81.
Spicher, G., and H. Stephan: *Sourdough Biology, Biochemistry and Technology*, BBV Wirtschaftsinformationen GmbH, Hamburg, 1982.
Spicher, G.: "Sourdough Fermentation," *Chem. Mikrobiol. Technol. Lebensm.* 10(3–4), 1986, pp. 65–77.
Spiller, Monica A.: "Admixture of a *Lactobacillus brevis* and a *Saccharomyces dairensis* for Preparing Leavening Barm." U. S. Patent 4,666,719, May 19, 1987.
Stamer, J. R., B. O. Stoyla, and B. A. Dunckel: "Growth Rates and Fermentation Patterns of Lactic Acid Bacteria Associated with the Sauerkraut Fermentation," *J. Milk Food Technol.*, 34, 1971, pp. 521–525.
Steinkraus, K. H. (ed.): *Handbook of Indigenous Fermented Foods*, Marcel Dekker, New York, 1983.

Steinkraus, K. H.: "Lactic Acid Fermentation in the Production of Foods from Vegetables, Cereals, and Legumes," *Antonie van Leeuwenhoek*, 49, 1983, pp. 337–348.
Sugihara, T. F.: "Non-traditional Fermentations in the Production of Baked Goods," *Bakers Digest*, 51(5), 1977, pp. 76–80, 142.
Sugihara, T. F.: "Microbiology of the Soda Cracker Process," *J. Food Protect.* 41, 1978, pp. 977–982.
Sugihara, T. Frank: "Microbiology of Breadmaking," in *Microbiology of Fermented Foods* (Brian J. B. Wood, ed.), vol. 1, Elsevier Applied Science Publishers, London and New York, 1985, pp. 249–261.
Sugihara, T. Frank: "The Lactobacilli and Streptococci—Bakery Products," in *Bacterial Starter Cultures for Foods* (S. E. Gilliland, ed.) CRC Press, Boca Raton, Florida, 1985, pp. 119–125.
Sugihara, T. F., L. Kline, and M. W. Miller: "Microorganisms of San Francisco Sour Dough Bread Process," *Appl. Microbiol.* 21(3), 1971, pp. 456–465.
Suomalainen, H., J. Dettwiler, and E. Sinda: "Alpha-Glucosidase and Leavening of Baker's Yeast," *Proc. Biochem.* 7(5), 1972, pp. 16–19, 22.
Tang, R. T., R. J. Robinson, and W. C. Hurley: "Qualitative Changes in Various Sugar Concentrations during Breadmaking," *Bakers Digest* 46(4), 1972, pp. 48–55.
Tannahill, Reay: *Food in History*, Stein and Day, New York, 1973.
Tesch, W., R. F. Dale, J. Middleton, G. Reed, and H. J. Peppler: "Sourdough Starter Comprised of Co-dried Bakers Yeast and *Lactobacillus fermentum*," unpublished data, 1976.
Wood, B. J. B.: "The Yeast/Lactobacillus Interaction," in *Mixed Culture Fermentations* (M. E. Bushell and J. H. Slater, eds.), Academic Press, London, 1981, pp. 139–150.
Yarrow, D.: "*Candida milleri*, sp. nov.," *Int. J. System. Bacteriol.* 28, 1978, pp. 608–610.
Zorzanello, D., and T. F. Sugihara: "The Technology of Pandoro Production," *Bakers Digest* 56(6), 1982, pp. 12–15.

Chapter

3

Microbial Interactions during Wine Production

Linda F. Bisson and Ralph E. Kunkee

Department of Viticulture and Enology
University of California
Davis, California

Introduction

Wine is the product of the microbial fermentation of grape juice. The alcoholic fermentation, the conversion of glucose and fructose to ethanol, is principally conducted by the yeast *Saccharomyces cerevisiae* and closely related species. Since "grape must" (the crushed grapes ready for fermentation) is not a sterile medium under current commercial practices, many other microorganisms may be present at different stages of the fermentation or during postfermentation treatments of the wine. These organisms may contribute in either a positive or a negative way to the overall organoleptic quality of the wine and to wine stability. One important class of microorganisms found during vinification are the malolactic bacteria, members of the genera *Leuconostoc, Lactobacillus*, or *Pediococcus*. These bacteria are responsible for the conversion of malate present in the grape juice to lactate, which has been termed the *malolactic fermentation*. The conversion of malate, a dicarboxylic acid, to lactate, a monocarboxylic acid, results in an important deacidification of the wine. In addition, these bacteria may produce minor end products that contribute to the flavor and aroma profiles of the resultant wines. Wines that have undergone a malolactic fermentation are more microbiologically stable and, from a production perspective, if this fermentation is desired, it is important that it be completed prior to bottling of the wine. The interactions between the yeast *Saccharomyces* and the malolactic bacteria during fermentation are complex and influenced by juice composition, temperature of fermentation, postfermentation production practices, as well as by strain-specific factors, which are poorly understood. In an attempt to gain better control over both the alcoholic fermentation and the malolactic fermentation, inoculation with yeast and/or with malolactic bacteria is becoming a common practice in the wine industry, particularly in California.

Other microorganisms may also contribute to the wine-making process by conferring increased perceived complexity as a consequence of the production of specific metabolic end products. Under certain controlled conditions, some of the wild yeasts indigenous to grapes may produce significantly higher amounts of aromatic esters than do *Saccharomyces* species. In addition, many wine spoilage microorganisms exist which can produce high concentrations of acetic acid and other off-flavors that detract from wine quality. Production of a sound wine, therefore, requires proper exploitation of mixed populations of microorganisms, encouraging the growth of desired organisms while simultaneously discouraging that of other microbes.

Numerous researchers dating back to Pasteur (1866) have investigated the microbial flora of wine and have provided a very consistent picture of the microorganisms involved. In general, very few genera and species of microorganisms are capable of growth in grape juice or in wine. Grape juice represents a relatively limited or hostile microbial environment as a consequence of low pH (ideally 3.0 to 3.5) and high sugar [20 to 23 percent (weight/volume, w/v) equimolar mixture of glucose and fructose] content. Fermentations rapidly become anaerobic with a high dissolved CO_2 concentration. Ethanol levels quickly reach concentrations inhibitory to most microorganisms, typically ranging from 11 to 14 percent (volume/volume, v/v). The nutrient content of juice directly affects the interactions and numbers of microorganisms that may be present, as well as the overall rate of the fermentation and length of time required to complete the fermentation. Nitrogen and other nutrients are generally in large enough supply to allow the growth of enough biomass of cells ($\sim 10^8$ cells per milliliter) to carry out the fermentation at a desirable speed. Nevertheless, approximately half of the alcoholic fermentation is conducted by stationary phase, nongrowing yeasts. Occasionally, musts may require nutrient supplementation or aeration in order for the yeasts to be vigorous enough to complete fermentation to dryness. At the end of the fermentation, factors which affect the liberation of material from the spent yeast determine the content of residual nutrients in the wine and therefore the ability of spoilage organisms to grow and multiply.

The purpose of this review is to summarize our current understanding of the different kinds of interactions among microorganisms that occur during the fermentation of grape juice and how population dynamics are affected by factors such as grape flora and nutrient composition. Such interactions entail competition for limiting nutrients; the production of metabolic inhibitors by one group of organisms, such as ethanol by the fermentative yeasts; and the release of nutrients at the end of fermentation by the yeast population that may stimulate subsequent growth of other organisms. In the future the metabolic activities and kinds and numbers of specific microorganisms may be more readily manipulatable to produce a specific desired effect in the finished product.

This review will concern itself with the microbes of "table wines," which are defined as those containing 14% ethanol or less, and to which no wine spirits have been added. The high concentration of ethanol in the "dessert and appetizer" wines precludes the growth of microorganisms (with the notable exception of the very highly ethanol-tolerant dessert wine spoilage bacterium *Lactobacillus fructivorans* (formerly *L. trichodes*) (Fornachon et al. 1949, Radler 1984).

Natural Flora of Grapes

The microorganisms found on the grapes at harvest will then be present in the juice upon crushing of the grapes. Some of these organisms may be very active in the early stages of fermentation, before the inhibiting effect of ethanol from *Saccharomyces* is apparent. The proponents of natural fermentations believe that the early metabolic activities of the wild microbial flora enhance wine complexity. All natural fermentations are eventually dominated by *Saccharomyces cerevisiae*. However, use of natural fermentations to produce wine is high risk, as many spoilage microorganisms may be present on the grapes as well and the winemaker has no control over the kinds and numbers of organisms present. One of our current research activities is to identify those members of the natural flora contributing in a positive way to wine quality so that those organisms may be used as inocula for grape juice in much the same way that *Saccharomyces* and the malolactic bacteria are used as inocula today. Even so, the wild microflora of grapes will be present in the juice in the absence of any antimicrobial treatment. Ideally, one would like to be able to manipulate the microbial population naturally or through inoculation, without the need to resort to use of antimicrobial treatments. Fleet et al. (1989) indicate the temperature of the fermentation has an important effect on the viability of the so-called wild yeast, such as *Kloeckera*, and that at cool temperatures they may be much more ethanol tolerant than previously supposed. That is, although these yeast may not be taking part in the alcoholic fermentation, they may, in fact, be contributing to the final sensory quality of the product.

Many environmental and physical factors affect the natural flora of the grapes, and there appear to be some varietal-specific influences on population size and dynamics. Estimation of the microbial flora on natural substrates such as grapes is difficult and a variety of techniques have been utilized. Enrichment techniques allow the detection of microorganisms present in low number, but it could be argued that any sort of plating technique is an enrichment of some kind, so at best we obtain only a qualitative picture of the major kinds of organisms present. Martini et al. (1980) have recently addressed the question of isolation and enumeration techniques and have found that for many natural substrates including grapes, aggressive techniques such as sonication are essential to dislodge all of the microorganisms associated with the fruit. While they were able to obtain greater quantities of yeasts from grape surfaces using sonication, they did not identify any novel microorganisms (Rosini et al. 1982) and their results confirm qualitatively those of previous investigators using much milder isolation procedures (Davenport 1974, Barnett et al. 1972, Castelli 1957, Minarik 1971, Van Zyl and DuPlessis 1961). Furthermore, it has been our experience that estimation of very low concentrations of the various yeasts, where no dilution of the grape juice is employed, is empirically especially difficult because of the presence of molds, which even at moderate concentration can wreak havoc on the solid medium because of rapid and confluent growth, engulfing the colonies to be enumerated.

Survey of microorganisms found on grapes

Numerous studies of the microbial flora of developing and mature grapes have been conducted (Barnett et al. 1972; Castelli 1955, 1957; Davenport 1974, 1976; Goto and Yokotsuka 1977; Relan and Vyas 1971; Minarik 1971; Van Zyl and DuPlessis 1961; Rosini et al. 1982; Parish and Carroll 1985) and several major reviews have addressed this topic (Amerine and Kunkee 1968; Lafon-Lafourcade and Ribéreau-Gayon 1984; Rankine 1963; Benda 1982; Kunkee and Amerine 1970). Investigations of the microbial flora associated with grape diseases have been undertaken (Guerzoni and Marchetti 1987) as well as of microorganisms found in vineyard soils (Adams 1960, Davenport 1974). As a consequence of these studies, a consistent picture has arisen of the microbial flora of grapes. The two major yeast species found as natural residents of grapes are *Kloeckera apiculata* (perfect form *Hanseniaspora uvarum*) and *Candida pulcherrima* (perfect form *Metschnikowia pulcherrima*). Other yeasts of the genera *Torulopsis, Brettanomyces, Candida, Cryptococcus,* and *Rhodotorula* may also be commonly present. Numerous other yeasts have been identified less frequently and are listed in the review of Kunkee and Amerine (1970). The bacterial and fungal flora have received less attention, and in most studies organisms of these two classes, if enumerated, were not taxonomically identified. In some of the few studies of bacteria on grape berries, Radler (1958) was unable to detect any lactic acid bacteria; however, Balloni et al. (1988) found *Leuconostoc oenos* and two species of lactobacilli on grapes being dried for *vin santo* production in the Chianti region of Italy. Other bacterial species that have been identified as associated with grapes are members of the genera *Pseudomonas, Bacillus, Micrococcus, Bacterium, Lactobacterium* and *Chromobacterium* (Zhuravleva 1963), in addition to molds. Adams (1964) in studies of airborne microorganisms in horticultural areas such as vineyards, found that only 4 to 5 percent of the total eukaryotic microorganisms present were yeast. A similar observation was made of vineyard soils, especially in the absence of decaying fruit (Adams 1960). In general, the natural resident molds and bacteria seem unable to persist in grape must, perhaps due to the low pH and anaerobic conditions of the juice as a consequence of yeast metabolism. Soil and horticultural area molds are commonly found on grapes, members of the genera *Aspergillus, Penicillium, Rhilopus,* and *Mucor* (Amerine et al. 1980), some of which are of enological importance as they may affect grape yield and quality (Martini 1966). One important fungus, *Botrytis cinerea,* will be discussed in detail in a later section as grapes from certain viticultural regions infected with this mold yield a unique and highly prized style of wine.

Interestingly, the yeast *Saccharomyces* is rarely isolated from vineyards (Davenport 1974, Goto and Yokotsuka 1977, Parle and DiMenna 1965), even utilizing the sonication technique of Rosini et al. (1982). *Saccharomyces* may be introduced through the practice of spreading pomace or yeast lees in the vineyard as fertilizer or by virtue of a high airborne population due to colocalization of the vineyard and winery facility. Similarly, the lactic acid bacteria responsible for the malolactic fermentation are not normally found on grapes or in vineyards (see Amerine and Kunkee 1968) and the most likely source of these organisms during natural fermentations is the winery equipment (Peynaud and Domercq 1959, Webb and Ingraham 1960) also the most likely source of *Saccharomyces* and of spoilage microorganisms (Amerine et al. 1980; Parle and DiMenna 1965).

Factors affecting the natural microbial flora of grapes

Numerous environmental factors affect the microbial populations of grapes. As previously mentioned, the practice of disposal of winery wastes on vineyard soils alters the microbial flora. Other parameters also may affect yeast flora. Benda (1982) and Parle and DiMenna (1965) concluded that insect vectors were the primary means by which yeast species are transmitted to mature grapes, principally by members of the genus *Drosophila*, and the data of Stevic (1962) implicate bees and wasps as transmission vectors of normal yeast flora. While viable yeasts can be readily isolated from the digestive tracts of these insects, Shihata and Mrak (1951) pointed out that yeasts do not survive passage through many insect digestive systems. Microorganisms adhering to the external surfaces of insects may be more readily deposited on fruit surfaces. Factors such as available food sources, including decaying fruit material, can dramatically affect the population size of insects such as *Drosophila* that have rapid generation times. For example, the University of California vineyards in Davis are located in an agricultural area rich in tomato fields. The tomato fields support a large population of species of *Drosophila*, which then invade the vineyards as the tomato crop is harvested.

Vineyard practices such as nitrogen fertilization may also perturb the natural microbial flora. Burchill and Cook (1971) investigated the mechanism of inhibition of the causative agent of apple scab, *Venturia inaequalis*, resulting from the foliar application of urea. These authors determined that growth of *V. inaequalis* was suppressed because of the stimulation of growth of yeasts and bacteria as a consequence of the provision of the nitrogen source urea. Presumably application of nitrates and nitrites to soil to stimulate vine growth, a very common viticultural practice worldwide, may selectively stimulate the growth of microorganisms capable of utilizing these substances as nitrogen sources.

Fungicides and other compounds used to control microbial diseases of grapes certainly affect population dynamics and may adversely affect wine quality. Elemental sulfur applied to vines to control powdery mildew can be reduced by yeasts during fermentation to H_2S (Schütz and Kunkee 1977) an undesirable compound in finished wine. Castor et al. (1957) showed that captan residues following fungicidal treatment of grapes were inhibitory to yeast, resulting in prolonged fermentation times. Fungicides in the past were only rarely tested for inhibitory effects against the harmless natural resident microbial flora. Due to the current concern over the use of fungicides and pesticides, control of the microbial flora through more natural means such as nutrient competition may be desirable. We are unaware of any studies to control fungal diseases of grapes by the inoculation of grapes with healthy, harmless natural microbial competitors, but research into this area is clearly needed.

In addition to viticultural practices, the grape microflora may be influenced by physical factors such as climate and altitude. Castelli (1957) in studies of the yeast flora of grapes and musts in Italy, Spain, Israel, and the Bordeaux region of France found that both climate and altitude can influence the microbial flora. In very hot growing regions yeasts of the genus *Hanseniaspora* dominated while *Kloeckera* was present in greater numbers in cool growing regions. In moderately hot growing areas, both yeasts were found in approximately equal numbers. *Kloeckera* was more frequently isolated at high altitude, while *Hanseniaspora* was more prevalent at low altitude (Castelli 1957). Adams (1960), working in Canada, found that *Kloeckera* and *Torulopsis* were the most frequent isolates from the Niagara Peninsula, a cooler growing region which resembles northern Italy. While these studies have not been exhaustive, they do indicate that the physical location of the vineyard may influence the microbial flora of the grapes.

Of particular importance is the effect of climate on the growth of *Botrytis cinerea*. *B. cinerea* is a very common inhabitant of grapes and has been isolated from grapes worldwide (Nelson and Amerine 1957). *Botrytis* infection of grapes results in loss of water from the grapes since the mold mycelia is able to penetrate the grape skin and absorb water. This fungus is able to metabolize grape organic acids and produces high concentrations of end products such as glycerol (Amerine et al. 1980). The Sauterne wines of France owe their special character to the *Botrytis* infection, as do some of the late-harvest German wines (Ribéreau-Gayon et al. 1980, Nelson and Amerine 1957). While *Botrytis* has wide distribution in nature, proper infection of grapes requires highly specific climatic conditions as detailed by Nelson and Amerine (1957). Berries of higher sugar content are also more susceptible to infection (Nelson 1951). Thus, while *Botrytis* itself is not present during wine fermentation, infection of grapes by this fungus yields grape juice with an altered composition which is poorly fermentable by yeasts, yielding a unique style of wine.

Specific varietal factors may also influence grape microflora both quantitatively and qualitatively. Differences in grape skin thickness or composition may affect susceptibility to fungal attack. Rosini et al. (1982) found that the yeast population was greater closer to the peduncle as compared to the central and lower part of the bunch, suggesting that nutrients are more readily available in this region of the cluster. Cluster dynamics can vary dramatically with the variety (Amerine et al. 1980). Also, certain varieties produce tight clusters and are more susceptible to bunch rot due to the physical damage inflicted by berry growth on adjacent berries which would certainly affect nutrient availability and microbial flora. Secretion of antimicrobial substances by damaged grape or vine tissues has not been explored, but may also affect population dynamics.

Thus, many physical and environmental factors affect the population dynamics of grape microorganisms. One class of microorganisms, the yeasts, can be isolated from grape musts early in fermentation and their collective metabolic activities may affect wine quality as well as the speed with which *Saccharomyces* species are able to dominate a natural fermentation. Inoculation of grape musts with *Saccharomyces*, a common practice in the newer wine production regions and becoming more common world wide, allows this microorganism to take over the fermentation more rapidly, potentially minimizing any effects of the population of non-*Saccharomyces* wild yeasts introduced to the musts via their presence on the surface of the intact grapes.

Population Dynamics during Fermentation

Wine production may proceed either by natural (uninoculated) fermentation or by inoculation with a starter culture. The starter culture used may be in the form of rehydrated active dry yeast from a commercial supplier, or must taken from the active or tumultuous stage of another fermentation. In the former case, the commercial yeast inoculum has been prepared with vigorous aeration and ample micronutrients (vitamins and minerals), and with low concentrations of glucose, which promotes formation of the cells' respiratory pathways. Yeast grown under these conditions can readily initiate and complete a wine fermentation provided that the juice is not seriously deficient in macronutrients (nitrogenous materials or phosphate) or contains residual inhibitory compounds such as fungicides. The practice of serial inoculation of tanks has potential problems. First of all, there may be transfer of contaminating or spoilage microorganisms, which may then gain a foothold in a tank. Secondly, this practice will tend to limit the synthesis of the so-called survival factors (sterols and unsaturated long-chain fatty acids) (Lafon-Lafourcade et al. 1979, Traverso-Rueda and Kunkee 1982), materials needed by the yeast for increased tolerance to ethanol. The solubility of molecular oxygen is very low, and the formation of the survival factors is restricted as the must becomes anaerobic. The preformed survival factors must then be distributed to the succeeding yeast populations, often leading to their deficiency and a sluggish or stuck fermentation. The effect of serial transfer on the availability of other micronutrients at the cellular level could affect the fermentation similarly. Obviously, the dynamics of the interactions of microorganisms during fermentation will be dramatically influenced by the method of initiation of fermentation.

Survey of microorganisms present in grape juice

The microbial flora of the grape will generally be introduced to the juice immediately upon crushing of the grapes. In addition, the natural microflora of the winery and winery equipment will also be introduced in the juice during grape processing (Peynaud and Domercq 1959). The numbers of microorganisms present on winery equipment may increase as the harvest season (or crush) progresses. Pardo et al. (1989) investigated both the presence and persistence throughout fermentation of various microorganisms in Spanish wines, comparing musts from four different varieties. In the different musts they obtained mold counts of 2 to 3×10^4 cfu/mL at the start of fermentation. By midfermentation no mold isolates were detectable. Similarly, at the start of fermentation 2×10^2 to 1×10^3 cfu/mL of lactic acid bacteria were present, which decreased by midfermentation. Yeast counts ranged from 4×10^5 to 1×10^6 in fresh musts. Yeast population density increased to 5×10^7 to 1×10^8 cfu/mL at peak of fermentation, followed by loss of viability observed in late stationary phase. The yeasts present in all four musts were members of the genera *Candida, Kloeckera*, and *Torulopsis*. Some musts had initial high cell counts of *Saccharomyces cerevisiae* ($\sim 10^5$ cfu/mL)[1] while others had no detectable yeasts of this genus. There were differences noted in the particular species present and relative numbers of the three main yeast genera, and one must had a relatively high content of *Pichia* species. By midfermentation the *Saccharomyces* yeasts dominated in cell number and by the end of fermentation the only genus of yeasts detectable besides *Saccharomyces* was *Pichia*. This recent study is in excellent agreement with the early and definitive work of Peynaud and Domercq (1959) on the progression of yeast species during fermentation. It is important to note, in comparing recent studies to earlier work, that many of those species of *Saccharomyces* are now considered to be synonymous with *S. cerevisiae* (Kreger-van Rij 1984).

Heard and Fleet (1985, 1986) compared the yeast flora of natural and inoculated juices. In addition to *Saccharomyces cerevisiae*, the major yeast species found were *Kloeckera apiculata, Candida pulcherrima, C. stellata*, and *C. colliculosa*. Other yeasts such as *Hansenula anomala* and *T. delbrueckii* were observed less frequently. There was a greater total number of different yeast species in the natural versus inoculated fermentations, and there were differences in

[1]cfu = colony forming unit

the organisms present in comparing juices from six different varieties. The authors noted that *C. pulcherrima* was observed only in musts of grapes grown in southern Australia. In musts containing *C. pulcherrima*, this species persisted for 6 to 9 days of the natural fermentations and 4 to 9 days for the inoculated fermentations. *Candida pulcherrima* (imperfect form of *Metschnikowia pulcherrima*) is known to be found only in certain growing regions (Benda 1982). This yeast may be an important "positive" contributor to wine flavor and aroma because it produces low levels of volatile acids (acetic acid) but relatively high concentrations of esters and some fusel oils (Benda 1982). *K. apiculata* likewise produces high ester concentrations but also produces high volatile acidity relative to *Saccharomyces* and *Metschnikowia* (Benda 1982). However, Van Zyl et al. (1963) did not find high volatile acidity production by *K. apiculata*. Either variation in environmental conditions or strain differences may explain these apparently conflicting results. The important finding of the work of Heard and Fleet (1985, 1986) is the persistence of some of the wild yeast flora in fermentations initially inoculated with *S. cerevisiae* at approximately 10^7 cells per milliliter. The special persistence at low temperatures has already been mentioned (Fleet et al. 1989). Earlier studies (Rankine and Lloyd 1963, Rosini 1984) also showed that an added yeast strain will affect population dynamics of the natural flora, and while the added strain may dominate at the end of fermentation, *Saccharomyces* species present as normal winery flora will also be found (see Rosini 1984). In spite of the above work and of other promising laboratory results, we have found no confirmed reports of actual flavor enhancement resulting from the presence of any wild species of yeast during fermentation.

Rapid and accurate differentiation of strains of *S. cerevisiae* is difficult. Addition of a strain as inoculum and proving recovery of that *S. cerevisiae* strain at the end of fermentation will be greatly facilitated in the future by the adaptation of pulsed-field gel electrophoresis for the determination of yeast chromosomal karyotypes for the taxonomic identification of yeast isolates. This technique and the various electrophoretic systems that have been designed have been recently reviewed (Lai et al. 1989). The use of karyotype to identify strains of the same species negates the need for genetic markers for differentiation. In the near future more definitive studies of population dynamics of strains of *S. cerevisiae* should be possible.

Nevertheless, a genetically marked wine yeast has been used to confirm the practicality of the use of starter cultures for commercial wine fermentations. This work showed that in a high proportion (34 of 40 trials), the inoculated organism was the dominant one (greater than 95 percent of the total population) at the end of the fermentations. For this study, the inoculated organism had a killer phenotype ("K1") of *S. cerevisiae*, and was resistant to two specific mitochondrial inhibitors (Delteil and Aizac 1989, Vezinhet et al. 1989).

Physiological factors allowing *Saccharomyces* species to dominate wine fermentations

Several biochemical and physiological characteristics of the yeast *Saccharomyces* ensure its dominance of wine fermentations. This yeast is one of the most ethanol-tolerant organisms known, capable of producing high concentrations of ethanol as an end product of fermentation. In fact, *S. cerevisiae* is the organism of choice for the commercial production of ethanol (Stewart 1985). [The bacterium *Zymomonas mobilis* is also highly tolerant to ethanol and is found as the fermentation agent in some natural alcoholic fermentations such as pulque, and as spoilage agents in others, such as cider and perry (Swings and DeLey 1977).] Neither *K. apiculata* nor *M. pulcherrima* are as ethanol tolerant as *S. cerevisiae*, nor do they produce high ethanol concentrations (Benda 1982, Mota et al. 1984). The majority of the other microorganisms found in musts are very ethanol sensitive (Amerine et al. 1980). Only the spoilage yeasts, such as *Brettanomyces* and *Zygosaccharomyces* (see below), are as ethanol tolerant as *Saccharomyces* species. Thus, one strategy utilized by *Saccharomyces* for the domination of fermentations is the rapid production of an end product which is toxic to most other organisms.

Another factor which may contribute to loss of wild yeast flora during fermentation is temperature. Rapid fermentation of sugar by *Saccharomyces* generates waste energy in the form of heat. Amerine et al. (1980) estimate that for each 10 g/L of sugar consumed, an increase of 1.3°C in temperature in a large tank is expected. Strains of *Saccharomyces* grow readily at temperatures up to 38°C (Watson 1987), while many of the wild flora are inhibited at temperatures greater than 25°C (see below). The production of heat by *Saccharomyces* coupled to the ability to withstand warmer temperatures than other organisms in must might also contribute to dominance of the fermentation.

The relative growth rates of the various resident yeasts probably do not play an important role in population dynamics of vinification. *Saccharomyces* strains are widely studied in research laboratories due, in part, to their rapid generation times (70 to 90 min doubling times in rich media, depending upon strain and strain background). However, these rates are considerably slowed to doubling times as high as 12 h under winemaking conditions because of the inhibitory effect of sugars present at high concentrations in grape juice. That is, although the rates of growth of wine yeast, compared with wild yeast, might be greater when observed in laboratory media, all of these rates are curbed in grape juice and become rather similar during the short, early time of the growth phase of wine fermentation (Karuwanna 1976, Kunkee, unpublished observations). Only later, when the inhibitory effects from ethanol and from other end products of the fermentation are observed, or from competition for nutrients (see below), do the differences between growth rates, and final cell concentrations, become noticeable.

A final important factor in population dynamics and dominance of one organism over others is competition for limiting nutrients. While much work has been done on nutrient uptake and utilization parameters in *Saccharomyces* (Suomalainen 1968, Suomalainen and Oura 1969), and some work on the competition between *Saccharomyces* and bacteria (see below, p. 53), very little comparative work has been done with *Saccharomyces* and indigenous wild yeast such as species of *Kloeckera* or *Metschnikowia*.

Factors Affecting Population Dynamics of Yeast Species

The population dynamics during alcoholic fermentation of wine are concerned mainly with the relationship between the major fermentation agents (the wine strains of *Saccharomyces* and the other indigenous yeasts). The populations of bacteria and spoilage yeast are very low at the beginning of the alcoholic fermentation, and their influence at this time can generally be ignored. We begin this subject by looking at the dominance of *Saccharomyces*; but two exceptions, dealing with bacteria, will be brought up later in this section. The physiological factors allowing *Saccharomyces* species to dominate fermentation have been discussed. However, many other parameters such as juice composition, cooling during the fermentation, and the introduction of compounds with antimicrobial activity may also affect interactions among the microorganisms present in grape must.

Nutrient limitation

The nutrient content of juices varies widely and is affected by factors such as nitrogen fertilization of vines (Bell et al. 1979). Fermentation rates have been shown to be directly related to nitrogen content of the juice (Ough and Amerine 1966). Limitation of nitrogenous materials has also been shown to inhibit the activity of sugar carriers in *Saccharomyces* (Lagunas et al. 1982, Busturia and Lagunas 1986,

Salmon 1989), which is believed to be the rate-limiting step of glycolysis under anaerobic conditions. Utilization of available nitrogen sources can be affected by such parameters as pH and ion concentrations as well as by cellular energy levels as most nitrogen-containing compounds are taken up via active transport systems (Cooper 1982). Competition for both macro- and micronutrients among the various microorganisms present in fermenting must directly affects population dynamics. Unfortunately, very little is known about the specific nutrient requirements of the different yeasts present initially in must or their abilities to compete with *Saccharomyces* for limiting nutrients. Our research (Monteiro and Bisson, to be reported) and that of other investigators (Castor and Guymon 1952, Monk et al. 1987, Maule et al. 1966) demonstrated that most amino acids disappear quite quickly from grape juice at the start of fermentation, within the first few Brix drop, indicating that juices quickly become nitrogen depleted. How this affects population dynamics remains to be determined.

In addition to competition for essential nutrients, there may also be competition among species for the above-mentioned survival factors, such as the plasma membrane components ergosterol and unsaturated fatty acids (Lafon-Lafourcade et al. 1979), since in the absence of molecular oxygen, biosynthesis of these two essential classes of molecules cannot occur (Henry 1982). The fatty acid composition of different grape varieties can vary dramatically (Gallander and Peng 1980). Molecular oxygen may also serve the role of survival factor at the end of fermentation (Lafon-Lafourcade et al. 1979, Larue et al. 1980, Traverso-Rueda and Kunkee 1982, Larue and Lafon-Lafourcade 1989). Larue and Lafon-Lafourcade (1989) found that must composition and treatment as well as fermentation temperature affect the yeast sterol requirement. At low cellular sterol content, fermentation is inhibited (Larue et al. 1980). Ethanol tolerance is directly correlated with the ability to synthesize unsaturated fatty acids or their availability in the medium (Thomas et al. 1978). Although much information is available on the requirements of *Saccharomyces* for survival factors, very little is known about the requirements of the rest of the microorganisms comprising the flora of the must. Specific information on vitamin requirements of the different yeast species is also lacking, although as pointed out by Lodder et al. (1958), vitamin requirements appear to be strain specific and not generally useful taxonomic characteristics.

It is becoming more and more common to supplement grape juices prior to fermentation with yeast nutrients (Amerine et al. 1980) including diammonium phosphate and mixtures of vitamins and minerals available in a variety of forms and under a variety of trade names.

These supplements and their appropriate levels have been designed around the requirements of *S. cerevisiae*, but should also support the growth of other microorganisms present in the must. However, the extent to which nutrient supplementation prolongs the activity and survival of the wild microbial flora has not been thoroughly investigated.

Temperature

The temperature of fermentation will dramatically influence the kinds of microorganisms present during must fermentation. In many wine regions, the grapes and the cellars are very cool, usually below 15°C. In other regions, white wine fermentations are cooled by external means and controlled at temperatures from 7 to 15°C (Long 1981). Red wine fermentations are conducted at higher temperature, 18 to 29°C (Ough and Amerine 1966). Since red wines are fermented on the skins for maximal pigment extraction, a cap of skins forms on the top of the tank and may be the site of vigorous microbial activity. Cap temperatures may be as much as 6°C higher than wine temperature (Ough and Amerine 1966). The kinds of microorganisms found in red versus white fermentations may differ greatly. Stokes (1969) reported similar temperature profiles for growth of *K. apiculata* and *Saccharomyces* species. Heard and Fleet (1988), in studying the effect of temperature on the growth of different yeast species in grape juice, also obtained similar results for *S. cerevisiae* and *K. apiculata*. *Candida pulcherrima* did not grow in juice at low temperature (10°C), but in general persisted least well of all species tested with the exception perhaps of *H. anomala*. *C. stellata* and *C. krusei* appeared to survive better at low (10°C) versus high (30°C) temperature. These findings are in contrast with those of Sharf and Margalith (1983) who found that the apiculate yeasts overgrew *Saccharomyces* at 10°C. Strain differences may account for this discrepancy. Tromp (1984) found differences in fermentation ability among four different strains of *S. cerevisiae* tested over the range of 10 to 15°C. Differences in juice composition may also explain differences in growth at low and high temperatures. Ciolfi et al. (1985a, b) have given extensive profiles of dozens of end products formed from fermentations by 15 strains of wine yeast at temperatures from 10 to 40°C.

Ethanol concentration affects temperature tolerance (Loureiro and van Uden 1982, Sá-Correia and van Uden 1983, Sá-Correia 1986). Ethanol concentration decreases the maximal temperature for growth and raises the minimum temperature, in effect narrowing the temperature range supporting the growth of yeast (Sá-Correia and van Uden 1982). The effect of ethanol on growth temperature was much more severe in *Kluyveromyces fragilis* than in *S. cerevisiae* (Sá-Correia and van Uden 1982). The simultaneous effects of ethanol and temperature may exert a more dramatic effect on population dynamics during fermentation than of either parameter alone. In addition, there is evidence that octanoic and decanoic acids which are produced as minor products during fermentation are inhibitory to yeast growth (Viegas et al. 1989) and also affect the kinetics of thermal death in yeast (Sá-Correia 1986). Fatty acids are also believed to inhibit fermentation rate, but can be removed from fermentations by treatment with yeast cell ghosts (Lafon-Lafourcade et al. 1984). Again, little comparative data are available on sensitivity to fatty acids in the wild microbial flora versus *Saccharomyces* species.

pH

The pH of the grape juice may affect the numbers and kinds of yeast species present. Grape juices vary generally over the pH range 3.0 to 3.9 (Amerine et al. 1980). Heard and Fleet (1988) found that *K. apiculata* and *C. stellata* survived better and persisted longer at a higher pH (3.5). *C. pulcherrima* also seemed to do better at higher pH, but the effect was not as dramatic. *S. cerevisiae* seems to be little affected by pH in this range.

Yeast killer factors

Strains of *S. cerevisiae* have been observed to produce low-molecular-weight proteins (11 to 18 kD) that are lethal to sensitive yeasts of the same species. Two recent reviews by Wickner (1986) and Young (1987) have described killer factors to which the reader is referred. Killer factors were first described by Makower and Bevan (1963). A yeast strain may have one of three phenotypes with respect to killer factors. Killer strains both produce a specific killer factor and are resistant to that toxin. Resistant strains are insensitive to one or more killer factors, but do not produce any toxin themselves. Sensitive strains are fully susceptible to killer toxin, which causes inhibition of macromolecular synthesis and cellular leakage resulting in cell death (Bussey and Sherman 1973). At least three different killer factors have been described in strains of *S. cerevisiae*, called K_1, K_2, and K_3, and strains may be resistant to one killer factor, but sensitive to the others (Young 1987). Killer character is conferred by dsRNA molecules that have been likened to virus particles although they are not infectious (Wickner 1986). Killer activity has been observed in other yeast genera, including *Torulopsis, Kluyveromyces, Pichia, Debaryomyces, Hansenula, Candida*, and *Cryptococcus* (Philliskirk and Young 1975, Radler et al. 1985, Stumm et al. 1977, Young and Yagiu 1978). Killer factors are in general active against sensitive members of the same species (Young 1987), but there are reports of killer activity against strains of different genera (Bussey and Skipper 1976, Rosini 1985, Shimizu et al. 1985).

The activity of killer toxin produced during vinification may alter the microbial flora if sensitive yeasts are present. Several investigators have screened wine strains of *S. cerevisiae* and have found that many strains produce killer factors (Barre 1980, 1984; Rosini 1983; Shimizu et al. 1985; van Vuuren and Wingfield 1986; Heard and Fleet 1987). The K_2 toxin appears to be active at wine pH (Barre 1980, van Vuuren and Wingfield 1986) and sensitive strains will be inhibited during white wine vinification (van Vuuren and Wingfield 1986, Heard and Fleet 1987). Heard and Fleet (1987) found no inhibiting effect of K_2 toxin on non-*Saccharomyces* yeast flora, nor did they identify any wild non-*Saccharomyces* must flora producing a killer toxin. Factors such as temperature and vigorous agitation affect killer factor activity (Young 1987), so it is not clear that killer toxin would function under all conditions. Barre (1980) found K_2 toxin to be ineffective in killing sensitive yeast during red wine production. The effectiveness of toxin activity also depends upon the relative population sizes of killer and sensitive yeasts (Heard and Fleet 1987). Killer strains have been suggested as a cause of stuck fermentations in South African wines (van Vuuren and Wingfield 1986), but a healthy killer strain of *Saccharomyces* should itself be able to complete the fermentation. Clearly this subject merits further investigation. However, killer toxin production can influence interactions among different strains of *S. cerevisiae* under vinification conditions. We have already mentioned the use of killer phenotypes as markers for identification of the inoculated yeast strains at the end of wine fermentation.

Factors Affecting Population Dynamics of Bacterial Species

The involvement of bacteria during the alcoholic fermentation of wine is not common, but when it occurs the effects can be significant. One example is given by Vaughn (1955) of the so-called rapid acetification, which was found in wines in some warm regions of California during the post-Prohibition era (Amerine et al. 1972). Whenever this happened, high concentrations of acetic acid bacteria grew in association with wine yeast, the latter presumably arising from premature commencement of the fermentation of broken and badly fungus-infected berries. The yeast then ceased to function because of the high concentration of acetic acid formed, and the bacteria would begin to oxidize the sugars. These musts ended with excessive amounts of volatile (acetic) and fixed (gluconic) acidities, and were given the sensory description, at that time, of "mousey."

Another involvement of bacteria during the alcoholic fermentation can be found in the new practice, in California, of attempting to obtain the malolactic fermentation at the same time as the alcoholic. The malolactic fermentation is the decarboxylation of malic acid to lactic acid by certain lactic acid bacteria, resulting in a deacidification of the wine, a more microbiologically stable wine, and the possibility of a wine with an increased sensory quality (Kunkee 1974, Davis et al. 1985). Traditionally, this fermentation occurs several months after the primary fermentation (Kunkee 1967a). The rational for the desire of an early malolactic fermentation is that this will allow the winemaker the option of more quickly finishing the wine and placing it in a position to protect it from spoilage. Bacterial starters, available commercially, are needed in large amounts for this. Procedures for propagation by the winemaker of large amounts of starter culture have also been given (Kunkee 1981).

The practice of simultaneous alcoholic and malolactic fermentations needs to be applied with caution. The enologists and winemakers in the western and southern regions of France (Lafon-Lafourcade et al. 1980) have cautioned that the presence of substantial amounts of lactic acid bacteria during the alcoholic fermentation will result in formation of large amounts of undesired end products, mainly acetic acid. It was supposed that the acetic acid would be coming from the high concentrations of glucose, serving as excess substrate for the bacteria. It has subsequently been suggested that the acetic acid comes from the wine yeast rather than the bacteria. The bacteria compete with the wine yeast for the available nutrients, placing the yeast under stressful conditions (Dombek and Ingram 1989), which in turn results in the production of the off-flavors (Beelman and Kunkee 1987). On the one hand, we have as yet seen no substantiated evidence for this advisory for California winemaking (Blackburn 1984), and neither has the problem been reported in the eastern regions of France (P. Bréchot, personal communication). On the other hand, the grape berries and musts of newer vineyard regions as found in California would be thought of as being richer in nutrients and not susceptible to this potential problem.

Nonetheless, competition between the yeast and the bacteria in the fermentation exists. Lonvaud-Funel et al. (1988a) outline scenarios regarding variations in the relationships between the alcoholic and malolactic fermentations, depending upon the conditions of the grape must and the natural strains of organisms present. Working with pure cultures of wine yeast and malolactic bacteria, King and Beelman (1986) found complex metabolic interactions. Rapidly growing yeast were definitely antagonistic to bacterial development, especially inhibiting their initial growth, and thus not an indirect inhibitory effect from the yeast end product, ethanol. Furthermore, there was an inhibitory effect of the bacteria on the yeast, whereby in the presence of the bacteria the death rate of the yeast was accelerated. The latter observation agrees with the suggestion given above where simultaneous alcoholic and malolactic fermentations may bring about stress in the metabolism of yeast. Prahl et al. (1988) have suggested the addition of large amounts of lactic acid bacteria to the grape must to bring about the completion of the malolactic fermentation before the commencement of the alcoholic fermentation. This novel procedure is currently being tested in several wine regions (C. Prahl, personal communication).

The inhibitory effect of the yeast on the bacteria in the mixed fermentation may be explained by the results from several laboratories: the production of medium-chain fatty acids (e.g., decanoic acid), which have been shown to be inhibitory to yeast (Sá-Correia 1986) are formed at highest concentrations early during the fermentation (Lee and Kunkee 1989), and also seem to be inhibitory to these bacteria (Edwards and Beelman 1987, Lonvaud-Funel et al. 1988b). However, the inhibitory effect of the bacteria on the yeast is probably more complex than simply competition for nutrients. King and Beelman (1986) found that a dilution of the juice removed the inhibitory effect of the bacteria. Thus the effect might be due to some inhibitory compound formed by the bacteria, such as ornithine, which was not found, or lactic acid. Canel-Llaubéres et al. (1989) and Llaubéres (1988) have discovered the formation of certain complex polysaccharides by some lactic acid bacteria (and yeast), and these might provoke this inhibition (M. Feuillat, personal communication).

The lactic acid bacteria are nutritionally fastidious and their growth is dependent upon nutritional components of the wine. Several studies have been made to pinpoint the nutritional needs of some of the malolactic strains, the results of which might be used for better control (inhibition or stimulation) of the malolactic fermentation (Weiller and Radler 1972, Peynaud et al. 1965, Lafon-Lafourcade 1983). This goal is confounded by the lack of control of the nutritional complement of the wine itself. The initial composition of the grape juice (vitamins, amino acids, and trace metal ions), the extent of uptake of the nutrients by the yeast, and the "release" of the materials from the yeast after the alcoholic fermentation are all variable. The extent of the latter is considerable (Feuillat and Charpentier 1982). It was observed several years ago that the strain of yeast used for the alcoholic fermentation seems to have an influence on the subsequent malolactic fermentation (Fornachon 1968). This is probably because of the differential effects, as just outlined, which the various yeast strains may have on final nutritional composition of the wine. Since these effects will be variable and dependent upon the conditions, regardless of strain, it is not surprising there is no consensus with regard to which yeast strains might stimulate or might inhibit lactic acid bacteria. In this regard, the formation of SO_2 by yeast, depending on strain, should also be mentioned (Eschenbruch 1974).

Effect of Antimicrobial Agents on the Natural Wine Flora

Sulfur dioxide is added to grape must to prevent oxidative browning. This compound is an effective inhibitor of polyphenol oxidase activity, which is responsible for enzymatic browning as well as limiting chemical oxidation (Amerine et al. 1980). In addition, SO_2 has antimicrobial activity. The form of SO_2 that is inhibitory is undissociated H_2SO_3 (King et al. 1981). The amount of H_2SO_3 in solution is a function of the concentration of SO_2 gas in the medium and of pH. At pH 3.0, approximately 5.5 percent of the total free SO_2 present is in the form of H_2SO_3. This value drops to 1.8 percent at pH 3.5 (King et al. 1981). The concentration of H_2SO_3 is also affected by the presence of compounds such as acetaldehyde capable of forming addition compounds with SO_2 and thus reducing the total concentration of this species. Gaseous SO_2 can be lost from the medium and such loss is accelerated by vigorous fermentation. *Saccharomyces* strains can produce SO_2 as a consequence of sulfur metabolism (Amerine et al. 1980), thus added SO_2 does not reflect the potential concentration in the wine. Strains of *S. cerevisiae* are resistant to high concentrations of SO_2 while the non-*Saccharomyces* wild yeast flora appear to be more sensitive to inhibition by this compound (Amerine et al. 1980). In comparing the sensitivity of an SO_2-resistant commercial strain of *S. cerevisiae* to *H. anomala*, King et al. (1981) found only a twofold difference in sensitivity, and calculated that 100 to 200 μg/L of SO_2 was needed to yield enough H_2SO_3 to be inhibitory to *Saccharomyces*.

Goto (1980) examined the effect of SO_2 addition on the microbial flora of natural fermentations. *Kloeckera apiculata* was found to be rapidly killed by SO_2, while *Torulopsis* and *Saccharomyces* species were about equally resistant. Interestingly, *S. cerevisiae* dominated the fermentation not treated with SO_2 more readily than the treated musts. Studies of relative sensitivities of SO_2 of yeasts in pure culture must be treated with caution. Metabolic activities of more resistant yeasts like *S. cerevisiae*, such as the production of acetaldehyde and vigorous fermentation, may reduce SO_2 concentration below that required to inhibit other microorganisms.

Because of their natural low population, the wine bacteria have not been included in the above effects of antimicrobial agents on the natural flora of grapes; most of the studies on wine bacteria have been made after the completion of the alcoholic fermentation (see below). However, the effect on bacteria becomes important when the bacteria are present in large amounts, as when they are added for induction of the malolactic fermentation. In grape juice, in the absence of ethanol, the effects of SO_2, and pH, on the malolactic bacteria are very similar to the effects on the wine yeast (Kunkee 1967). Two studies on the effects of pH and SO_2 on wine lactic acid bacteria in nonalcoholic medium have recently appeared. In one investigation, Henick-Kling (1989) showed that the malolactic activity in *Leuconostoc oenos* is regulated in the whole cell by the rate of transport of malate into the cell. Even though the optimal pH for growth of the organism is much higher than the pH of wine, the malolactic activity of whole cells was highest with cells grown at low pH (3.5). Nonetheless, in terms of the malolactic conversion in wine, the biomass of cells is the operative factor; the velocity of the conversion is directly related to the pH of the medium (Bousbouras and Kunkee 1971, Liu and Gallander 1983). In another study in nonalcoholic medium, Delfini and Morsiani (1989) examined the resistance of malolactic strains to SO_2. They found that the bacteria in the presence of large concentrations of SO_2 could remain inactive for long periods of time without loss of reproductive capacity, and they proved what had long been supposed, that the bacteria can increase their resistance to SO_2 by adaption.

Two materials having high potential as antimicrobial agents are the killer factor of yeast, already mentioned, and the bacteriophages (viruses) of the lactic acid bacteria. Van Vuuren et al. (1986) have given strong evidence of the involvement of killer factor as the agent for some stuck fermentations in wineries in South Africa. Henick-Kling et al. (1986) explore the possibilities of attack of malolactic bacterial starter cultures, and wine undergoing malolactic fermentation, by strain-specific bacteriophages. Winemakers should be alerted to these dangers; however, at this writing we have seen no problem arising globally from either of these potential disasters. The only cure at present, if either problem should occur, would seem to be to have other strains of yeast or bacteria, presumably insensitive to these specific agents, in readiness for reinoculation.

Other natural antimicrobial substances may also be present in grape juice. Cantarelli (1989) recently reported that certain phenolic fractions present in white juices are inhibitory to growth of *S. cerevisiae*. The sensitivity of non-*Saccharomyces* yeasts to these compounds has not been determined. The use of other antimicrobial agents, dimethyl dicarbonate, and fumaric acid will be discussed in the following section.

Postfermentation Flora of Wines and Microbial Spoilage

Microbial contamination of wine can cause a variety of spoilage problems. Off-flavors and aromas can arise as a consequence of microbial metabolism. Structural defects such as ropiness, a problem arising from microbial slime formation, can also occur (Lüthi 1957; Lonvaud-Funel and Joyeux 1989). Once a wine has been bottled, microbial growth leading to turbidity or carbonation is also considered spoilage. In certain wines, a malolactic fermentation is undesirable and growth of the malolactic bacteria is therefore considered a problem in microbial stability. If residual sugar is present in the wine, then even *S. cerevisiae*, the agent of the alcoholic fermentation, may cause a spoilage problem.

Bacteria

The low pH and high alcohol content of wine deter the growth of most prokaryotic microorganisms, and only some acetic and lactic acid bacteria are able to grow under these conditions.

Of the acetic acid bacteria, *Gluconobacter* and *Acetobacter*, only *A. aceti* is of great importance to the winemaker (Drysdale and Fleet 1988). The copious production of acetic acid by this species, and its esterification product, ethyl acetate, which are both organoleptically objectionable in wine, is the principle problem associated with these bacteria. Spoilage may arise in the finished wine or during the course of fermentation, particularly in red wines, which are fermented at warmer temperature. Growth of the acetic acid bacteria requires exposure to oxygen, but only low concentrations are necessary for noticeable spoilage (Lafon-Lafourcade and Ribéreau-Gayon 1984). The avoidance of this spoilage is the main postfermentative activity of the conscientious winemaker, who prevents excess exposure of the wine to oxygen, either by abolition of head spaces in wine storage vessels or by replacement of the air of the head space with inert gas. Legal limits for the maximum amount of acetic acid in wine have been set (Amerine et al. 1980). Joyeux et al. (1984) found that acetic acid bacteria were present in high concentration on grapes infected with *Botrytis cinerea* and Rankine (1963) noted a connection between the growth of acetic acid bacteria and film-forming yeast.

The lactic acid bacteria have been presented in a positive light as agents of the malolactic fermentation, but they may also be the causative organisms of wine spoilage, especially if the malolactic fermentation is not desired. This depends upon the need for deacidification or microbial stabilization of the wine and upon the effect the fermentation might have on the sensory quality of the wine. The winemaker must weigh the effect of the flavor components produced by the bacteria, notably diacetyl, against the inherent delicacy of the wine itself (Kunkee et al. 1965). The control of the malolactic fermentation is made by an interplay, on the one hand, of the inhibitory or stimulatory parameters of the wine, pH, nutrients, and concentrations of SO_2, and ethanol; and, on the other hand, the physical manipulations such as filtration, temperature control, and bacterial inoculation (Kunkee 1967b, Liu and Gallander 1983, Henick-Kling et al. 1989, Lafon-Lafourcade et al. 1983).

Davis et al. (1985) state that it is perhaps risky for the winemaker to consider wines to be bacteriologically stable merely because a malolactic fermentation has taken place. Indeed, it has been shown, at least under Australian winemaking conditions, that the increase in pH, as a consequence of the malolactic fermentations, may allow growth of another resident bacterial strain (Costello et al. 1985, Wibowo et al. 1985). We have, as yet, to notice such an occurrence in other winemaking regions, possibly because the initial pHs of wines are generally lower elsewhere, as compared to those in Australia.

The lactic acid bacteria can also produce acetic acid, as well as a variety of other end products (Pilone et al. 1966, Lafon-Lafourcade 1983, Lafon-Lafourcade et al. 1980). Certain species of *Lactobacillus* are capable of producing metabolites of lysine responsible for the "mousiness" defect (Heresztyn 1986), an off-flavor of wine already mentioned with respect to acetic acid bacteria. If sorbic acid has been used as an antimicrobial agent, some lactic acid bacteria can metabolize this compound, producing what is known as the "geranium tone," an off-odor (Crowell and Guymon 1975, Edinger and Splittstoesser 1986).

In the United States, it is legally permitted to add fumaric acid to wine to prevent the growth of lactic acid bacteria (Amerine et al. 1980, p. 568). Fumaric acid is effective in the presence of SO_2 if the bacterial population is already low and the pH is below 4.0. Its effective concentration is about the same as its solubility in wine; this is a major drawback to its use, especially where large volumes of wine are to be treated.

There is good agreement amongst enologists worldwide as to the kind of lactic acid bacteria found in wine. However, it is interesting to find that the species isolated from the thermovinified red wine, where heat is employed to get better color extraction, are much more thermotolerant (Barre 1978).

Members of the genus *Bacillus*, only rarely considered as wine spoilage organisms (Gini and Vaughn, 1962), have been described as a current spoilage menace in wines of eastern Europe (T. Török, personal communication).

Yeast

Several yeast genera can cause problems in wine stability, particularly in wines with residual sugar. The ability of a given yeast to cause a spoilage problem is often strain specific, so identification of a particular yeast in wine does not necessarily mean that wine will spoil. Thomas and Ackerman (1988) have developed a diagnostic medium useful in distinguishing strains that will cause a problem from those that are harmless. The yeast causing the most common spoilage in bottled wine, at least where wines have been bottled with small (to large) amounts of residual sugar, are wine strains of *S. cerevisiae*. The method of choice to prevent this sort of infection is the use of sterile filtration, followed by sterile bottling (Kunkee and Goswell 1972, pp. 370–371; Amerine et al. 1980, pp. 308–309). Other methods include pasteurization, which means careful heating of the wine followed by sterile bottling or heating the wine in the bottle (Amerine et al. 1980, pp. 310–312; Malletroit et al. 1989); and inhibition of the yeast with potassium sorbate (Amerine et al. 1980, pp. 206–207). A chemical additive ideally suited for control of wine yeast in finished wines with residual sugar is dimethyl dicarbonate (Ough et al. 1988). This compound, which kills the microorganisms and is quickly hydrolyzed, seems also to be effective against malolactic bacteria, especially when present in concert with SO_2.

The two other yeasts that most often cause wine spoilage are *Brettanomyces* and *Zygosaccharomyces*. Species of *Brettanomyces* (or its perfect form *Dekkera*) produce a variety of objectionable end products, variously described as "metallic," "sweaty horse blanket," "Band-Aid," and those responsible for "mousiness" (Kunkee and Goswell 1972, pp. 370; Amerine et al. 1980, p. 175; Hereszty 1986). These organisms are somewhat fastidious and often said to be difficult to culture from spoiled wines (Wright and Parle 1974). The source of these yeasts appears to be contaminated winery equipment and they enter the winery during the crushing season (van der Walt and Van Kerken, 1961).

Zygosaccharomyces bailii is an important spoilage microorganism and is widespread in the food and beverage industry (Thomas and

Davenport 1985). The principal problem associated with growth of this microorganism in finished wine is turbidity. *Z. bailii* is resistant to a variety of antimicrobial agents (Thomas and Davenport 1985), and one of its sources is also likely to be contaminated winery equipment. Recent outbreaks of spoilage by *Z. bailii* of wine not sterilely bottled have been reported in California (Vilas and Kunkee 1986). The yeast seems likely to come from the use of grape juice concentrate, perhaps old and poorly stored, as sweetening agent added at the time of bottling, and where potassium sorbate was used in attempts to obtain microbial stability. *Z. bailii* is resistant, in some cases highly resistant, to the antimicrobial agents used in wine, including sorbate, SO_2, and dimethyl dicarbonate (Thomas and Davenport 1985; Vilas and Kunkee 1986).

Film-forming yeasts such as *Pichia membranefacieus* (imperfect form *Candida valida*, and also synonymous with *C. mycoderma* (Kreger and van Rij 1984) can also cause spoilage problems (Rankine 1966). These yeasts are nonfermentative and thus require that the wine be exposed to oxygen. The main problem with this spoilage is the unsightly film on the top of the new wine, but the growth of these yeasts can result in the production of off-flavors and aromas.

Mold contamination of wine is generally not a problem as these microorganisms have a high oxygen requirement, although moldy cooperage can impart a definite fungal character to the wine. Molds are common inhabitants of cellars due to high humidity and are readily able to utilize wood as a substrate. *Byssochlamys* and other heat-related molds have presented spoilage problems in some grape juice products, but not wine (King et al. 1969). Some moldy flavor characteristics in wine have been attributed to cork itself (Boidron et al. 1984), to mold infections of the cork (Lebrevre et al. 1983), and to the nonmicrobial presence of trichloranisole in faulty corks (Rigaud et al. 1984).

Conditions affecting microbial spoilage of wines

Microbial spoilage requires the presence of suitable residual nutrients in the wine to serve as growth substrates. These nutrients may be present in the wine as a consequence of oversupplementation by the winemaker or more frequently due to the leaching out of *S. cerevisiae* during the decline phases of growth (Ferrari and Feuillat 1988). Lurton et al. (1989), estimated that as much as 30 percent of the nitrogen content of yeast can be liberated during this extraction, which is probably different from the release of materials during autolysis of yeast found in the *méthode champanoise* production of sparkling wine

(Feuillat and Charpentier, Troton et al. 1989). Wine may also contain carbon sources, malate, lactate, citrate, succinate, and sugar alcohols and pentoses not utilized by *S. cerevisiae* or the malolactic bacteria during fermentation. Temperature of storage and exposure to oxygen are also important factors affecting the microbial stability of wines. Perhaps the most important parameter in microbial spoilage of wine is the extent of contamination of winery equipment as this is most often the true source of the spoilage yeasts and bacteria.

In our experience with isolation of microbes from a spoiled wine, we have generally found a preponderance of a single microorganism. In the older literature, other workers have indicated the opposite experience, where a variety of microbes, both yeast and bacteria, were found (Lüthi 1957, Rankine 1963). With mixed cultures, the organisms may be producing kinds of defects different from that found in pure culture (Lüthi 1957). Thomas and Davenport (1985) have suggested that vitamin production by *H. anomala* may foster the growth of *Z. bailii*. Perhaps many such synergistic relationships exist during wine spoilage, indicating a fertile field for further investigations.

Conclusion

The production of wine is a complex microbiological process. Many different kinds of microorganisms are present at different stages of production starting with the natural grape flora and ending with the natural or contaminating flora of the winery and winery equipment. The kinds of microbial interactions observed are intricate and diverse and population dynamics during vinification are influenced by a variety of factors, both natural and artificial. Much more information is needed on the fundamental aspects of microbial interactions during grape must fermentation before our understanding of this complex process, begun over one hundred years ago by Pasteur, is complete.

References

Adams, A. M. 1960. Yeasts in Horticultural Soils. In 1959–60 Report of the Horticultural Experiment Station and Products Laboratory. Vineland, Ontario, Canada, Ontario Department of Agriculture, Toronto. Pages 79–82.

Adams, A. M. 1964. Airborne Yeasts from Horticultural Sites. *Can. J. Microbiol.* 10: 641–646.

Amerine, M. A. and R. E. Kunkee. 1968. Microbiology of Winemaking. *Ann. Rev. Microbiol.* 22:323–358.

Amerine, M. A., H. W. Berg and W. V. Cruess. 1972. *Technology of Wine Making*, 3d edition, pp. 577–578, AVI Publishing Co., Westport, Connecticut.

Amerine, M. A., H. W. Berg, R. E. Kunkee, C. S. Ough, V. L. Singleton and A. D. Webb. 1980. *The Technology of Wine Making*, 4th edition, AVI Publishing Co., Westport, Connecticut.

Balloni, W., L. Granchi, L. Giovannetti and M. Vincenzini. 1988. Evolution of Yeast and

Lactic Acid Bacteria in Chianti Grapes during the Natural Drying for "Vin Santo" Production. *Yeast* 5:S37–S41.

Barnett, J. A., M. A. Delaney, E. Jones, B. Magson and B. Winch. 1972. The Number of Yeasts Associated with Winegrapes of Bordeaux. *Arch. Mikrobiol.* 23:52–55.

Barre, P. 1978. Identification of Thermobacteria and Homofermentative, Thermophilic, Pentose-utilizing Lactobacilli from High Temperature Fermenting Grape Musts. *J. Appl. Bacteriol.* 44:125–129.

Barre, P. 1980. Role of Facteur "Killer" dans la Concurrence entre Souches de Levures. *Bull. O.I.V.* 593–594:560–567.

Barre, P. 1984. Le Mechanisme Killer dans la Concurrence entre Souches de Levures. Evaluation et Prise en Compte. *Bull. O.I.V.* 641–642:635–643.

Beelman, R. B. and R. E. Kunkee. 1987. Inducing Simultaneous Malolactic/Alcoholic Fermentations. *Pract. Winery Vyd.* July/Aug. pp. 44–56.

Bell, A. A., C. S. Ough and W. M. Kliewer. 1979. Effects on Must and Wine Composition, Rates of Fermentation, and Wine Quality of Nitrogen Fertilization of *Vitis vinifera* var. Thompson Seedless Grape Vines. *Am. J. Enol. Vitic.* 30:124–129.

Benda, I. 1982. Wine and Brandy. Pages 293–402 in Reed, G. (ed). *Prescott and Dunn's Industrial Microbiology*. AVI Publishing Co., Inc., Westport, Connecticut.

Bousbouras, G. E. and R. E. Kunkee. 1971. Effect of pH on Malo-lactic Fermentation in Wine. *Am. J. Enol. Vitic.* 22:121–126.

Blackburn, D. 1984. Malolactic Bacteria. Research and Applications. *Pract. Winery Vyd.* 5(3):44–48.

Boidron, J. N., A. Lefebvre, J. M. Bibovlet and P. Ribéreau-Gayon. 1984. Les Substances Volatiles Susceptibles d'être Cédées au Vin par le Bouchon de ége. *Sci. Ailments* 4:609–616.

Burchill, R. T. and R. T. A. Cook. 1971. The Interaction of Urea and Micro-organisms Suppressing the Development of Perithecia of *Venturia inaequalis* (Cke). Wint. Pages 471–483 in Preece, T. F. and C. H. Dickinson (eds.). *Ecology of Leaf Surface Micro-Organisms*. Academic Press, New York.

Bussey, H. and D. Sherman. 1973. Yeast Killer Factor: ATP Leakage and Coordinate Inhibition of Macromolecular Synthesis in Sensitive Cells. *Biochem. Biophys. Acta* 298:868–875.

Bussey, H. and N. Skipper. 1976. Killing of *Torulopsis glabrata* by *Saccharomyces cerevisiae* Killer Factor. *Antimicrob. Agents Chemother.* 9:352–354.

Busturia, A. and R. Lagunas. 1986. Catabolite Inactivation of the Glucose Transport System in *Saccharomyces Cerevisiae. J. Gen. Microbiol.* 132:379–385.

Canal-Llaubéres, R.-M., D. Dubourdieu, B. Richard and A. Lonvaud-Funel. 1989. Structure Moléculaire β-D-Glucane Exocullulaire de *Pediococcus* sp. *Conn. Vigne Vin* 23: 49–52.

Cantarelli, C. 1989. Phenolics and Yeast: Remarks Concerning Fermented Beverages. Proc. 7th Intl. Symp. Yeasts. *Yeast* 5:S53–S61.

Castelli, T. 1955. Yeasts of Wine Fermentations from Various Regions of Italy. *Am. J. Enol. Vitic.* 6:18–19.

Castelli, T. 1957. Climate and Agents of Wine Fermentation. *Am. J. Enol. Vitic.* 8:149–156.

Castor, J. G. B. and S. F. Guymon. 1952. On the Mechanism of Formation of Higher Alcohols during Alcoholic Fermentation. *Science* 115:147–149.

Castor, J. G. B., K. E. Nelson and J. M. Harvey. 1957. Effect of Captan Residues on Fermentation of Grapes. *Am. J. Enol. Vitic.* 8:50–57.

Ciolfi, C., M. Castino and R. di Stefano. 1985a,b. Studio Sulla Risposta Metabolics di Lieviti di Specie Diverse Fermentanti in Unico Mosto a Temparature Comprese fra 10 e 40° C. *Riv. Viticol. Enol.* 38:447–470, Nota I; 489–506, Nota II.

Cooper, T. G. 1982. Transport in *Saccharomyces cervisiae*. Pages 389–461 in Strathern, J. N., E. W. Jones, and J. R. Broach (eds.) *The Molecular Biology of the Yeast Saccharomyces: Metabolism and Gene Expression*. Cold Spring Harbor Laboratory, New York.

Costello, P. J., G. J. Morrison, T. H. Lee and G. H. Fleet. 1983. Numbers and Species of Lactic Acid Bacteria in Wine during Vinification. *Fd. Technol. Australia* 35:14–18.

Crowell, E. A. and J. F. Guymon. 1975. Wine Constituents Arising from Sorbic Acid

Addition and Identification of 2-Ethoxyhexa-3,5-diene as Source of Geranium-like Off-odor. *Am. J. Enol. Vitic.* 26:97–102.

Daudt, C. E. and C. S. Ough. 1980. Action of Dimethyldicarbonate on Various Yeasts. *Am. J. Enol. Vitic.* 31:21–23.

Davenport, R. R. 1974. Microecology of Yeasts and Yeast-like Organisms Associated with an English Vineyard. *Vitis* 13:123–130.

Davenport, R. R. 1976. Distribution of Yeasts and Yeast-like Organisms from Aerial Surfaces of Developing Apples and Grapes. Pages 325–359 in Dickson, C. H. and T. F. Preece (eds.) *Microbiology of Aerial Plant Surfaces*. Academic Press, New York.

Davis, C. R., D. Wibowo, R. Eschenbruch, T. H. Lee and G. H. Fleet. 1985. Practical Implications of Malolactic Fermentation: A Review. *Am. J. Enol. Vitic.* 36:290–301.

Delfini, D. and M. G. Morsiani. 1989. Study on the Resistance of Sulfur Dioxide of Malolactic Strains of Leuconostoc oenos and of Lactobacillus Isolated from Wines. VI Symposium international d'oenologie, Bordeaux, June n.p.

Delteil, D. and T. Aizac. 1989. Yeast Inoculation Techniques with a "Marked" Yeast Strain. *Pract. Winery Vyd.* May/June :43–47.

Dombek, K. M. and L. O. Ingram. 1986. Nutrient Limitation as a Basis for the Apparent Toxicity of Low Levels of Ethanol during Fermentation. *J. Ind. Microbiol.* 1:219–225.

Drysdale, G. S. and G. H. Fleet. 1988. Acetic Acid Bacteria in Winemaking: A Review. *Am. J. Enol. Vitic.* 39:143–154.

Edinger, W. D. and D. F. Splittstoesser. 1986. Production by Lactic Acid Bacteria of Sorbic Alcohol, the Precursor of the Geranium Odor Compound. *Am. J. Enol. Vitic.* 327:34–38.

Edwards, C. G. and R. B. Beelman. 1987. Inhibition of the Malolactic Bacterium, *Leuconostoc oenos* (PSU-1), by Decanoic Acid and Subsequent Removal of the Inhibition by Yeast Ghosts. *Am. J. Enol. Vitic.* 38:239–242.

Ferrari, G. and M. Feuillat. 1988. l'Elevage sur Lie des Vins Blancs de Bourgogne. I. Etude des Composés Azotés, des Acides Gras et Analyse Sensorielle des Vins. *Vitis* 27:183–197.

Feuillat, M. and C. Charpentier. 1982. Autolysis of Yeasts in Champagne. *Am. J. Enol. Vitic.* 33:6–13.

Fleet, G. H., G. M. Heard and C. Gao. 1988. The Effect of Temperature on the Growth and Ethanol Tolerance of Yeast During Fermentation. *Yeast* 5:543–546.

Fornachon, J. C. M., H. C. Douglas and R. H. Vaughn. 1949. *Lactobacillus trichodes*. nov. spec., a Bacterium Causing Spoilage in Appetizer and Dessert Wine. *Hilgardia* 19:129–132.

Gallander, J. F. and A. C. Peng. 1980. Lipid and Fatty Acid Composition of Different Grape Types. *Am. J. Enol. Vitic.* 31:24–27.

Gini, B. and R. H. Vaughn. 1962. Bacilli in Wine. *Am. J. Enol. Vitic.* 13:20–31.

Goto, S. 1980. Changes in the Wild Yeast Flora of Sulfited Grape Musts. *J. Inst. Enol. Vitic. Yamanashi Univ.* 15:29–32.

Goto, S. and I. Yokotsuka. 1977. Wild Yeast Populations in Fresh Grape Musts of Different Harvest Times. *J. Ferment. Technol.* 55:417–422.

Guerzoni, E. and R. Marchetti. 1987. Analysis of Yeast Flora Associated with Grape Sour Rot and of the Chemical Disease Markers. *Appl. Environ. Microbiol.* 53:571–576.

Heard, G. M. and G. H. Fleet. 1985. Growth of Natural Yeast Flora during the Fermentation of Inoculated Wines. *Appl. Environ. Microbiol.* 50:727–728.

Heard, G. M. and G. H. Fleet. 1986. Occurrence and Growth of Yeast Species during the Fermentation of Some Australian Wines. *Food Tech. Austrail.* 38:22–25.

Heard, G. M. and G. H. Fleet. 1987. Occurrence and Growth of Killer Yeasts during Wine Fermentation. *Appl. Environ. Microbiol.* 53:2171–2174.

Heard, G. M. and G. H. Fleet. 1988. The Effects of Temperature and pH on the Growth of Yeast Species during the Fermentation of Grape Juice. *J. Appl. Bacteriol.* 65:23–28.

Henick-Kling, T. 1989. pH and Regulation of Malolactic Activity in *Leuconostoc oenos*. VI Symposium international d'oenologie, Bordeaux, June, n.p.

Henick-Kling, T., T. H. Lee and D. J. D. Nicholas. 1986. Inhibition of Bacterial Growth and Malolactic Fermentation in Wine by Bacteriophage. *J. Appl. Bacteriol.* 61:287–293.

Henick-Kling, T., W. E. Sandine and D. A. Heatherbell. 1989. Evaluation of Malolactic Bacteria Isolated from Oregon Wines. *Appl. Environ. Microbiol.* 55:2010–2016.

Henry, S. A. 1982. Membrane Lipids of Yeast. Biochemical and Genetic Studies. In J. N. Stratern, E. N. Jones and J. R. Broach, (eds.). In "The Molecular Biology of the Yeast Saccharomyces Metabolism and Gene Expression," pp. 101–158. Cold Spring Harbor Laboratory, New York.

Heresztyn, T. 1986. Formation of Substituted Tetrahydropyridines by Species of *Brettanomyces* and *Lactobacillus* Isolated from Mousy Wines. *Am. J. Enol. Vitic.* 37: 127–132.

Joyeux, A., S. Lafon-Lafourcade and P. Ribéreau-Gayon. 1984. Evolution of Acetic Acid Bacteria during Fermentation and Storage of Wine. *Appl. Environ. Microbiol.* 48: 153–156.

Karuwanna, P. 1976. Wine Fermentations with Mixed Yeast Cultures. MS thesis, University of California, Davis.

King, S. W. and R. B. Beelman. 1986. Metabolic Interactions Between *Saccharomyces cerevisiae* and *Leuconostoc oenos* in a Model Grape Juice/Wine System. *Am. J. Enol. Vitic.* 37:53–60.

King, A. D., Jr., H. D. Michener and K. A. Ito. 1969. Control of *Byssochlamys* and Related Heat-resistant Fungi in Grape Products. *Appl. Microbiol.* 18:166–173.

King, A. D., J. D. Ponting, D. W. Sanshuck, R. Jackson and K. Mihara. 1981. Factors Affecting Death of Yeast by Sulfur Dioxide. *J. Food Protect.* 44:92–97.

Kreger-van Rij, N. J. W. 1984. *The Yeasts: A Taxonomic Study*, 3d ed. Elsevier Science Publishers, B.V., Amsterdam, p. 1082.

Kunkee, R. E. 1967a. Malolactic Fermentation. *Adv. Appl. Microbiol.* 9:235–279.

Kunkee, R. E. 1967b. Control of Malolactic Fermentation Induced by *Leuconostoc citrovorum*. *Am. J. Enol. Vitic.* 18:71–77.

Kunkee, R. E. 1974. Malolactic Fermentation and Winemaking. Pages 151–170 in Webb, A. D., (ed.) *Chemistry of Winemaking* Adv. Chem. Ser. 137, American Chemical Society, Washington, D.C.

Kunkee, R. E. 1981. Rapid Malolactic Fermentation. *East. Grape Grow. Winery News* 7(5):64–65.

Kunkee, R. E. and M. A. Amerine. 1970. Yeasts in Wine-Making, Pages 50–71 in Rose, A. M. and J. S. Harrison (eds.) *The Yeasts*, vol. 3, Academic Press, London.

Kunkee, R. E. and R. Goswell. 1977. Table Wines. Pages 315–386 in Rose, A. H. (ed.), *Economic Microbiology*, vol. 1, chap. 4, Academic Press, New York.

Kunkee, R. E., G. J. Pilone, and R. C. Combs. 1965. The Occurrence of Malolactic Fermentation in Southern California Wines. *Am. J. Enol. Vitic.* 16:219–223.

Lafon-Lafourcade, S. 1983. Wine and Brandy. Pages 81–163 in Rehn, H. J. and G. Reed (eds.), *Biotechnology: A Comprehensive Treatise in 8 Volumes*, vol. 5 of *Food and Feed Production with Microorganisms*, Verlag Chemie, Deerfield Beach, Florida.

Lafon-Lafourcade, S. and P. Ribéreau-Gayon. 1984. Developments in the Microbiology of Wine Production. *Prog. Indust. Microbiol.* 19:1–45.

Lafon-Lafourcade, S., F. Larue and P. Ribéreau-Gayon. 1979. Evidence for the Existence of "Survival Factors" as an Explanation for Some Particularities of Yeast Growth Especially in Grape Must of High Sugar Concentration. *Appl. Environ. Microbiol.* 38:1069–1073.

Lafon-Lafourcade, S., C. Geneix and P. Ribéreau-Gayon. 1984. Inhibition of Alcoholic Fermentation of Grape Must by Fatty Acids Produced by Yeasts and Their Elimination by Yeast Ghosts. *Appl. Environ. Microbiol.* 47:1246–1249.

Lagunas, R., C. Dominguez, A. Busturia and M. J. Saez. 1982. Mechanisms of Appearance of the Pasteur Effect in *Saccharomyces cerevisiae.*: Inactivation of Sugar Transport Systems. *J. Bacteriol.* 152:19–25.

Lai, E., B. W. Birren, S. M. Clark, M. I. Simon and L. Hood. 1989. Pulsed Field Gel Electrophoresis. *BioTech.* 7:34–42.

Larue, F. and S. Lafon-Lafourcade. 1989. Survival Factors in Wine Fermentation. Pages 193–215 in van Uden, N. (ed.), *Alcohol Toxicity in Yeasts and Bacteria*. CRC Press, Boca Raton, Florida.

Larue, F., S. Lafon-Lafourcade and P. Ribéreau-Gayon. 1980. Relationship Between the Sterol Content of Yeast Cells and Their Fermentation Activity in Grape Must. *Appl. Environ. Microbiol.* 39:808–811.

Lee, S. O. and R. E. Kunkee. 1988. Relationship between Yeast Strain and Production or Uptake of Medium Chain Fatty Acids during Fermentation. Technical Abstracts, 39th Annual Meeting, *Am. Soc. Enol. Vitic.* p. 10.

Lefebvre, A., J. M. Riboulet, J. N. Boidron, P. Ribéreau-Gayon. 1983. Incidence des Micro-organismes du Liège sur les Altératons Olfactives du Vin. *Sci. Ailments* 8:265–278.

Llaubéres, R.-M. 1988. Les Polysaccharides Sécrétés dans les Vins par *Saccharomyces cerevisiae* et *Pediococcus* sp. Ph.D. Thesis. Université de Bordeaux II, France.

Llaubéres, R.-M., D. Dubourdie, and J. C. Villettaz. 1987. Exocellular Poly-saccharides from *Saccharomyces* in wine. *J. Sci. Fd. Agric.* 41:27.

Lodder, J., W. C. Slooff and N. J. W. Kreger-van Rij. 1958. The Classification of Yeasts. Pages 1–62 in Cook, A.H. (ed.), *The Chemistry and Biology of Yeasts.* Academic Press, New York.

Long, Z. R. 1981. White Table Wine Production in California's North Coast Region. Pages 29–57 *In* Amerine, M.A. (ed) Wine Production Technology in the United States. *Am. Chem. Soc.*, Washington, D.C.

Lonvaud-Funel, A. and A. Joyeux. 1989. Etude d'Alternations des Vins per des Bactéries Lactiques. IV Symposium international d'oenologie, Bordeaux, June, n.p.

Lonvaud-Funel, A., J.-P. Masclef, A. Joyeux and Y. Paraskevolpoulos. 1988a. Etude des Interactions entre Levures et Bactéries Lactiques dans de Mout de Raisin. *Conn. Vigne Vin* 22:11–24.

Lonvaud-Funel, A., A. Joyeux and C. Desens. 1988b. Inhibition of Malolactic Fermentation of Wines by Products of Yeast Metabolism. *J. Sci. Fd. Agric.* 44:183–191.

Loureiro, V. and N. van Uden. 1982. Effects of Ethanol on the Maximum Temperature for Growth of *Saccharomyces cerevisiae*: A Model. *Biotech. Bioeng.* 24:1881–1884.

Lurton, L., J. P. Segain and M. Feuillat. 1989. Étude de la Protéolyse au Cours de l'Autolyse de Levures en Milieu Acide. *Sci. Ailments* 9:111–124.

Lüthi, H. 1957. Symbiotic Problems Relating to the Bacterial Deterioration of Wines. *Am. J. Enol. Vitic.* 8:176–181.

Makower, M. and E. A. Bevan. 1963. The Inheritance of Killer Character in Yeast. *Proc. 11th Intl. Cong. Genetics* 1:202.

Martini, L. P. 1966. The Mold Complex of Napa Valley Grapes. *Am. J. Enol. Vitic.* 17: 87–94.

Martini, A., F. Federici and G. Rosini. 1980. A New Approach to the Study of Yeast Ecology of Natural Substrates. *Can. J. Microbiol.* 26:856–859.

Maule, D. R., M. A. Pinnegar, A. D. Portno and A. L. Whitear. 1966. Simultaneous Assessment of Changes Occurring during Batch Fermentation. *J. Inst. Brew.* 72:488–494.

Minarik, E. 1971. Étude de la Flore Levurienne des Regions Viticoles Peripheriques en Tchecoslovaquie. *Conn. Vigne Vin* 5:185–197.

Monk, P. R., D. Hook and B. M. Freeman. 1987. Amino Acid Metabolism by Yeasts. Pages 129–133 in (Lee, T. (ed.), *Proceedings of the Sixth Australian Wine Industry Technical Conference.* Australian Industrial Publishing, Adelaide.

Mota, M., P. Strehaiano and G. Goma. 1984. Studies on Conjugate Effects of Substrate (Glucose) and Product (Ethanol) on Cell growth Kinetics during Fermentation of Different Yeast Strains. *J. Inst. Brew.* 90:359–362.

Nelson, K. E. 1951. Factors Influencing the Infection of Table Grapes by *Botrytis cinerea.* Phytopath. 41:319–326.

Nelson, K. E. and M. A. Amerine. 1957. Use of *Botrytis cinerea* for the Production of Sweet Table Wines. *Am. J. Enol. Vitic.* 7:131–136.

Ough, C. S. and M. A. Amerine. 1966. The Effects of Temperature on Wine-Making. *Calif. Agric. Exp. Stat. Bull.* 827:1–36.

Ough, C. S., R. E. Kunkee, M. R. Vilas, E. Bordeu and M.-C. Huang. 1988. The Interaction of Sulfur Dioxide, pH, and Dimethyl Dicarbonate on the Growth of *Saccharomyces cerevisiae* Montrachet and *Leuconostoc oenos* MCW. *Am. J. Enol. Vitic.* 38:279–282.

Pardo, I., M. J. Garcia, M. Zungia and F. Uruburu. 1989. Dynamics of Microbiol Populations during Fermentation of Wines from the Utiel-Requena Region of Spain. *Appl. Environ. Microbiol.* 55:539–541.
Parle, J. N. and M. E. DiMenna. 1965. The Source of Yeasts in New Zealand Wines. *N. Zeal. J. Agric. Res.* 9:98–107.
Pasteur, L. 1866. *Études sur le vin*, Victor Masson, Paris.
Parish, M. E. and D. E. Carroll. 1985. Indigenous Yeasts Associated with Muscadine (*Vitis rotundifolia*) Grapes and Musts. *Am. J. Enol. Vitic.* 36:165–169.
Peynaud, E. and S. Domercq. 1959. A Review of Microbiological Problems in Wine-Making in France. *Am. J. Enol. Vitic.* 10:69–77.
Philliskirk, G. and T. W. Young. 1975. The Occurrence of Killer Character in Yeasts of Various Genera. *Ant. van Leeuwen.* 41:147–151.
Prahl, C., A. Lonvaud-Funel, S. Korsbaard, E. Morrison and A. Joyeux. 1988. Etude d'un Nouveau Procédeé de Déclenchement de la Fermentation Malolactique. *Conn. Vigne Vin* 22:197–207.
Radler, F. 1958. Untersuchung des Biologischen Säureabbaus im Wein. Isolierung und Charakterisierung von Äpfelsäure-abbauenden Bakterien. *Arch. Bikrobiol.* 30:64–72.
Radler, F. and S. Hartel. 1984. *Lactobacillus trichodes*, ein alkoholabhängiges Milchsäurebakterium. *Wein-Wissen.* 39:106–112.
Radler, F., P. Pfeiffer and M. Dennert. 1985. Killer Toxins in New Isolates of the Yeasts *Hanseniaspora Uvarum* and *Pichia Kluyveri. FEMS Microbiol. Lett.* 29:269–272.
Rankine, B. C. 1963. The Microbiology of Winemaking. *Aust. Wine Brew. Spirit Rev.* 81:11–16.
Rankine, B. C. 1966. *Pichia membranefaciens*, A Yeast Causing Film Formation and Off-flavor in Table Wine. *Am. J. Enol. Vitic.* 17:82–86.
Rankine, B. C. and B. Lloyd. 1963. Quantitative Assessment of Dominance of Added Yeast in Wine Fermentations. *J. Sci. Food Agric.* 14:793–798.
Rankine, B. C. and D. A. Pilone. 1974. Yeast Spoilage of Bottled Table Wine and its Prevention. *Aust. Wine Brew. Spirit Rev.* 92:36–40.
Relan, S. and S. R. Vyas. 1971. Nature and Occurrence of Yeasts in Haryana Grapes and Wines. *Vitis* 10:131–135.
Ribéreau-Gayon, J. P. Ribéreau-Gayon and G. Seguin. 1980. *Botrytis cinerea* in Enology. In, (Coley-Smith, J. R., K. Verhoeff, and W. R. Jarris, (eds.). *The Biology of Botrytis*. Academic Press, New York.
Rigaud, J., S. Issanchou, J. Sarris and D. Langlois. 1984. Incidence des Composés Volatils Issues du Liège sur le "Goût de Bouchon" des Vins. *Sci. Ailments* 4:81–93.
Rosini, G. 1983. The Occurrence of Killer Characters in Yeasts. *Can. J. Microbiol.* 29: 1462–1464.
Rosini, G. 1984. Assessment of Dominance of Added Yeast in Wine Fermentation and Origin of *Saccharomyces Cerevisiae* in Wine-making. *J. Gen. Appl. Microbiol.* 30: 249–256.
Rosini, G. 1985. Interaction Between Killer Strains of *Hansenula anomala* var. *anamala* and *Saccharomyces cerevisiae* Yeast Species. *Can J. Microbiol.* 31:300–302.
Rosini, G., F. Federici and A. Martini. 1982. Yeast Flora of Grape Berries during Ripening. *Microb. Ecol.* 8:83–39.
Sá-Correia, I. 1986. Synergistic Effects of Ethanol, Octanoic, and Decanoic Acids on the Kinetics and the Activation Parameters of Thermal Death in *Saccharomyces bayanus. Biotech. Bioeng.* 28:761–763.
Sá-Correia, I. and N. van Uden. 1983. Temperature Profiles of Ethanol Tolerance: Effects of Ethanol on the Minimum and the Maximum Temperatures for Growth of the Yeasts *Saccharomyces cerevisiae* and *Kluyveromyces fragilis. Biotech. Bioeng.* 25: 1665–1667.
Salmon, J. M. 1989. Effect of Sugar Transport Inactivation in *Saccharomyces cerevisiae* on Sluggish and Stuck Enological Fermentations. *Appl. Environ. Microbiol.* 55:953–958.
Schütz, M. and R. E. Kunkee. 1977. Formation of Hydrogen Sulfide from Elemental Sulfur during Fermentation by Wine Yeast. *Am. J. Enol. Vitic.* 28:137–144.
Sharf, R. and P. Margalith. 1983. The Effect of Temperature on Spontaneous Wine Fermentation. *Eur. J. Appl. Microbiol. Biotech.* 17:311–313.

Shihata, A. M. E-T. A. and E. M. Mrak. 1951. The Fate of Yeast in the Digestive Tract of *Drosophila*. *Am. Natural.* 85:381–383.
Shimizu, K., T. Adachi, K. Kitano, T. Shimazaki, A. Totsuka, S. Hara and H. H. Dittrich. 1985. Killer Properties of Wine Yeasts and Characterization of Killer Wine Yeasts. *J. Ferment. Technol.* 63:421–429.
Stevic, B. 1962. The Importance of Bees (*Apis* sp) and Wasps (*Vespa* sp) as Carriers of Yeasts for the Microflora of Grapes and the Quality of Wines. *Archiv. Poljoprivredre Nauke, Beograd.* 15:80–91.
Stewart, G. G. 1985. New Developments in Ethanol Fermentation. *J. Am. Soc. Brew. Chem.* 43:61–65.
Stokes, J. L. 1969. Influence of Temperature on the Growth and Metabolism of Yeasts. Pages 118–134 in Rose, A. H. and J. S. Harrison (eds.), *The Yeasts*, vol. 2. Academic Press, New York.
Stumm, C., J. M. H. Hermans, E. J. Middelbeek, A. F. Croes and G. J. M. L. DeVries. 1977. Killer-Sensitive Relationships in Yeasts from Natural Habitats. *Ant. van Leeuwen.* 43:125–128.
Suomalainen, H. 1968. Penetration of Some Metabolic Compounds into and from the Yeast Cell. *Suomen Kemistilehti A* 41:239–254.
Suomalainen, H. and E. Oura. 1969. Yeast Nutrition and Solute Uptake. Pages 1–74 in Rose, A. M. and J. S. Harrison (eds.) *The Yeasts*, vol. 2. Academic Press, London.
Swings, J. and J. DeLey. 1977. The Biology of *Zymomonas*. Bacteriol. Rev. 41:1–46.
Thomas, D. S. and R. R. Davenport. 1985. *Zygosaccharomyces bailii*—a Profile of Characteristics and Spoilage Activities. *Food Microbiol.* 2:157–169.
Thomas, D. S., J. A. Hossack and A. H. Rose. 1978. Plasma Membrane Lipid Composition and Ethanol Tolerance in *Saccharomyces cerevisiae. Arch. Microbiol.* 117:239–245.
Thomas, D. S. and J. C. Ackerman. 1988. A Selective Medium for Detecting Yeasts Capable of Spoiling Wine. *J. Appl. Bacteriol.* 65:299–308.
Traverso-Rueda, S. and R. E. Kunkee. 1982. The Role of Sterols on Growth and Fermentation of Wine Yeasts Under Vinification Conditions. *Dev. Indust. Microbiol.* 23: 131–143.
Tromp, A. 1984. The Effect of Yeast Strain, Grape Solids, Nitrogen and Temperature on Fermentation Rate and Wine Quality. *S. Afr. J. Enol. Vitic.* 5:1–6.
Troton, D., M. Charpentier, B. Robillard, R. Calvayrac, and B. Duteurtre. 1989. Evolution of the Lipid Contents of Champagne Wine during the Second Fermentation of *Saccharomyces cerevisiae. Am. J. Enol. Vitic.* 40:175–182.
van Vuuren, H. J. J. and B. D. Wingfield. 1986. Killer Yeasts Cause of Stuck Fermentation in a Wine Cellar. *S. Afr. J. Enol. Vitic.* 7:113–118.
van der Walt, J. P. and A. E. van Kerken. 1961. The Wine Yeasts of the Cape. Part V. Studies on the Occurrence of *Brettanomyces intermedius* and *Brettanomyces schanderlii. Ant. van Leeuwen.* 27:81–90.
Van Zyl, J. A. and L. de W. DuPlessis. 1961. The Microbiology of South African Winemaking. I. The Yeasts Occurring in S. African Vineyards, Musts and Wines. *S. Afr. J. Agric. Sci.* 4:393–403.
Van Zyl, J. A., M. J. DeVries and A. S. Zeeman. 1963. The Microbiology of South African Winemaking. III. The Effect of Different Yeasts on the Composition of Fermented Musts. *S. Afr. Agric. Sci.* 6:165–180.
Vaughn, R. H. 1955. Bacterial Spoilage of Wines with Special Reference to California Conditions. *Adv. Food Res.* 6:67–108.
Vezinhet, F., D. Delteil and M. Valde. 1989. Les Apports du Marquage Gentique de Souches de Levures Oenologiques pour le Suivi des Populations Levuriennes en oenologie. VI. Symposium international d'oenologie, Bordeaux, June, n.p.
Viegas, C. A., M. F. Rosa, I. Sá-Correia and J. M. Novais. 1989. Inhibition of Yeast Growth by Octanoic and Decanoic Acids Produced During Ethanolic Fermentation. *Appl. Environ. Microbiol.* 55:21–28.
Vilas, M. R. and R. E. Kunkee. 1986. Isolation, Identification and Characterization of *Zygosaccharomyces bailii* from a California Wine. *Technical Abstracts. 39th Annual Meeting, Am. Soc. Enol. Vitic.* p. 3.
Watson, K. G. 1987. Temperature Relations. Pages 41–71 in Rose, A. H. and J. S. Harrison (eds.). *The Yeasts*, vol. 2, 2d edition. Academic Press, N.Y.

Webb, R. B. and J. L. Ingraham. 1960. Induced Malo-lactic Fermentations. *Am. J. Enol. Vitic.* 11:59–63.
Weiller, H. G. and F. Radler. 1972. Vitamin- und Aminosäurebakterien aus Wein und von Rebenblättern. *Mitt. Rebe Wein Obstbau Fruechteverwert (Klosterneuburg).* 22: 4–18.
Wibowo. D., R. Eschenbruch, C. R. Davis, G. H. Fleet and T. H. Lee. 1985. Occurrence and Growth of Lactic Acid Bacteria in Wine: A Review. *Am. J. Enol. Vitic.* 36:313–320.
Wickner, R. B. 1986. Double-Stranded RNA Replication in Yeast: The Killer System. *Ann. Rev. Biochem.* 55:373–395.
Wright, J. M. and I. N. Parle. 1974. *Brettanomyces* in the New Zealand Wine Industry. *N.Z. J. Agric. Res.* 17:273–278.
Young, T. W. 1987. Killer Yeasts. Pages 131–164 in Rose, A. H. and J. S. Harrison (eds). *The Yeasts*, vol. 2. 2d edition. Academic Press, New York.
Young, T. W. and M. Yagiu. 1978. A Comparison of the Killer Character of Yeasts of Various Genera. *Ant. van Leeuwen.* 44:59–77.
Zhuravleva, V. P. 1963. The Microflora of Grape Must and its Change during the Process of Fermentation. Izvest. Akad. Nauk. Turkmen. *SSR Ser. Biol. Nauk.*, 1963(2): 19–24 (English translation of Russian title).

Chapter

4

Mixed Cultures in Vegetable Fermentations

H. P. Fleming[1]

Food Fermentation Laboratory
U. S. Department of Agriculture
Agricultural Research Service

North Carolina Agricultural Research Service
Department of Food Science
North Carolina State University
Raleigh, North Carolina

Introduction

Many vegetables have been preserved by salting, with various degrees of fermentation, depending upon the salt concentration. Cucumbers, cabbage (for sauerkraut) and olives account for the largest volume of fermented vegetables in the western hemisphere, but smaller quantities of peppers, carrots, cauliflower, and okra are preserved by brining.[2] The fermentation of vegetables involves a complexity of physical, chemical, and microbiological factors that have been well characterized over the past several decades. The reader is encouraged to consider previous reviews on this subject[3] in addition to the original research cited.

---

[1]This chapter was prepared by a U.S. government employee as a part of his official duties and legally cannot be copyrighted.

[2]Mention of a trademark or proprietary product does not constitute a guarantee or warranty of the product by the U. S. Department of Agriculture or North Carolina Agricultural Research Service, nor does it imply approval to the exclusion of other products that may be suitable.

[3]Andersson et al. 1988; Etchells et al. 1975; Fernandez Diez 1983; Fleming 1982; Pederson and Albury 1969; Stamer 1983, 1988; Vaughn 1954, 1982.

Microbial activities during the natural fermentation and storage of vegetables have been divided into four stages: initiation of fermentation, primary fermentation, secondary fermentation, and postfermentation (Table 4.1). This classification was based upon characterizations of the fermentations of cucumbers, cabbage, and olives by various researchers during this century. This review is organized according to this classification to emphasize similarities and dissimilarities in the fermentation of these commodities. Although some overlapping of activities among stages occurs, criteria for distinguishing the stages are sufficiently distinct to serve as a guide in developing controlled fermentation methods.

During the past 30 years, efforts have been made to develop pure-culture and controlled fermentation methods, many of which are discussed herein. Some novel control procedures have been employed commercially, such as the purging of CO_2 from fermenting cucumbers to prevent bloater formation. However, comprehensive controlled fermentation methods such as those used in the production of alcoholic beverages and fermented dairy products have not been used commercially on a large scale. Economic and technical reasons are responsible for this state of affairs. It is conceivable that technological advancements will mandate commercial acceptance of improved methods for fermentation control. One such advancement could be the development of microbial cultures with novel and valuable properties. The likelihood of developing such cultures during the next decade seems reasonable, considering the recent progress in genetic technology of microorganisms. To take advantage of such microbial technology, however, other technological advances also must occur (e.g., improved

TABLE 4.1 Stages of Microbial Activities During the Natural Fermentation of Vegetables

Stage	Prevalent microorganisms (conditions)
Initiation of fermentation	Various gram-positive and -negative bacteria
Primary fermentation	Lactic acid bacteria, yeasts (sufficient acid has been produced to inhibit most bacteria)
Secondary fermentation	Fermentative yeasts (when residual sugars remain and LAB have been inhibited by low pH)
	Spoilage bacteria (degradation of lactic acid when pH is too high and/or salt/acid concentration is too low; e.g., propionic acid bacteria, clostridia)
Postfermentation	Open tanks: surface growth of oxidative yeasts, molds, and bacteria
	Anaerobic tanks: none (provided the pH is sufficiently low and salt or acid concentrations are sufficiently high)

SOURCE: Modified from Fleming, 1982.

methods for sanitization and containment of the vegetables). It is the intent of this review to provide a better understanding of the principles that govern successful fermentation of vegetables, with the hope of encouraging the development of improved controlled fermentation methods that may eventually become commercially acceptable.

Initiation of Fermentation

Many vegetables, in the presence of appropriate concentrations of salt (NaCl) and under suitable environmental conditions, will undergo fermentation by lactic acid bacteria (LAB). Although the number of LAB is usually quite low compared to the total number of microorganisms (Table 4.2), the LAB eventually predominate due to the production of acids and other products which restrict growth and survival of other groups of microorganisms. The rate and consistency of LAB gaining predominance also is a reflection of many factors, including initial populations of all microorganisms, physical and chemical properties of the vegetables, and the environment (physical and chemical) in which the vegetables are held. Each vegetable may reflect unique responses during initiation of fermentation as exemplified with cucumbers, cabbage, and olives. Salt is known to exert two important effects in its preservative role: It directs the course of microbial activities and it prevents softening of the vegetable tissue. Although salt is commonly added in vegetable fermentations, the concentration used (Table 4.3) and its effects on the fermentation and product quality vary widely.

Cucumbers

Pickling varieties of cucumbers are used in various methods for commercial processing, including brine fermentation. Fermentation accounts for about 40 percent of commercially processed cucumbers, with pasteurization (40 percent) and refrigeration (20 percent) ac-

TABLE 4.2 Microorganisms on Raw Vegetables Used for Fermentation

	Number/g fresh weight		
	Cucumber [a]		
Microorganism	Fruit	Flower	Cabbage[b]
Total aerobes	1.6×10^4	1.8×10^7	1.3×10^5
Enterobacteriaceae	3.9×10^3	6.4×10^6	3.9×10^3
Lactic acid bacteria	5×10^0	2.6×10^4	4.2×10^1
Yeasts	1.6×10^0	3×10^3	<10

[a]From Etchells et al. 1975.
[b]After trimming outer leaves. From Fleming et al. 1988a.

TABLE 4.3 Salt Concentrations Used for the Fermentation and Storage of Vegetables

Method of salting	Salt concentration, %[a]		Vegetable
	Fermentation	Storage	
Dry salting	1.5–2.5	1.5–2.5	Cabbage
Brining	5–8	5–16	Cucumbers
Brining	4–8	4–8	Green olives

[a]The concentrations of salt indicated generally are used by commercial firms in the United States. Modified from Fleming, 1982.

counting for the remainder (Fleming and Moore, 1983). The primary purpose of fermentation is to serve as a means for temporary preservation of the cucumbers in bulk tanks until they are needed for processing into various types of sweet and sour pickles. The fermented cucumber products may or may not be pasteurized, depending upon the type of product and its intended use.

The cucumbers may be harvested by hand or mechanically, and usually are graded by size before being placed in bulk tanks containing a salt solution where they are allowed to ferment. The cucumbers usually are not intentionally washed before brining, so the initial microbial load reflects the soil type and weather conditions to which the fruit were exposed, as well as harvesting and handling methods. The fruit may be rinsed if they are received in a water-cushioned hopper or pit or conveyed in liquid. Most companies do not brush-wash the fruit before brining, although washing has been suggested for use in controlled fermentation procedures (Etchells et al. 1973, Fleming et al. 1988a). The microbial load on the raw fruit may vary widely, particularly if the flower remains attached, as is more common with smaller fruit (Fig. 4.1).

The flower is of particular significance also because it may harbor pectinolytic enzymes from mold growth which may result in softening of the cucumbers. In the southeastern United States where softening was a severe problem with small cucumbers for many years, pectinolytic enzymes resulting from mold growth within the flowers were found to be the causative agents of softening (Bell 1951, Etchells et al. 1958). Previously, aerobic, spore-forming bacilli had been thought to be responsible for softening (Vaughn et al. 1954). The softening problem with small fruit was greatly reduced by removal of the flowers before brining or by draining of the cover brines from the cucumbers after about 37 h (Etchells et al. 1955). In recent years the practice of draining brine from cucumbers to prevent softening has been curtailed due to environmental problems associated with salt disposal. Attempts now are made to remove the flowers from fruit be-

Figure 4.1 Flowers and immature fruit of the cucumber plant. (*From Etchells et al., 1953.*)

fore brining. Also, the addition of calcium salts to the cover brine has been shown to reduce softening of brined cucumbers (Fleming et al. 1978, 1987), even in the presence of mold polygalacturonase (Buescher et al. 1979, 1981).

The microflora are located mostly on or in the outer surface of raw cucumbers (Samish et al. 1963). For many years some scientists in the field considered the interior of healthy, fresh cucumbers to be sterile. However, it has been reported that the interior of fresh cucumbers may contain a latent population of bacteria (Samish and Dimant 1959, Meneley and Stanghellini 1974), probably of the family Enterobacteriaceae. Furthermore, it has been shown that bacteria may grow within the cucumbers during fermentation (Samish et al. 1963), and that the relative amount of fermentation by lactic acid bacteria within the cucumbers and in the brine surrounding the cucumbers can be manipulated by modifying the internal gas atmosphere of the fresh fruit before brining (Daeschel and Fleming 1981). Oxygen within the fresh cucumbers is quickly metabolized after the fruit are brined, and CO_2 is formed. The CO_2 is much more soluble than oxygen, thus a vacuum is formed within the fruit (Corey et al. 1983, Fleming et al. 1980). Bacteria may enter the fruit through stomata located in the skin (Fig. 4.2). Yeasts do not enter the fruit because of their larger size (Daeschel et al. 1985).

Cucumbers, graded by size, are conveyed into bulk tanks of 7500- to 75,000-L capacity containing brine (6 to 12% NaCl). The tanks are

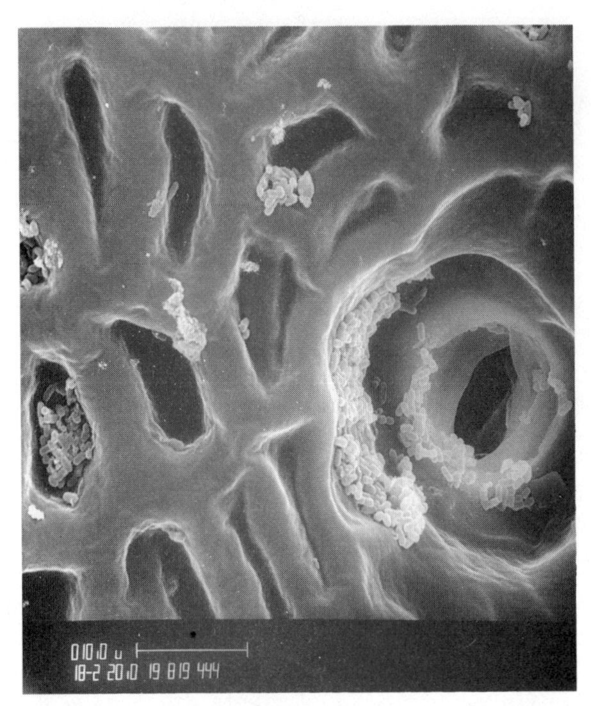

Figure 4.2 Bacteria associated with a stomate on the surface of a cucumber fermented by *L. plantarum*. Bar, 10 μm. (*From Daeschel and Fleming, 1981.*)

made of wood, fiberglass, or polyethylene. After the tanks are filled with cucumbers, wooden headboards are mounted over the cucumbers for their restraint after further brine addition. Additional brine is added to cover the cucumbers and to a level of 6 to 12 inches above the headboards, which are spaced or perforated to allow movement of liquid above and below the boards. The tanks (Fig. 4.3) are held outdoors to allow sunlight to strike the brine surface and thereby prevent growth of oxidative microorganisms (e.g., molds and film yeasts) at the brine surface. Dry salt is added onto the headboards as needed to attain the final equalized concentration for fermentation (5 to 8 percent), or to compensate for rainwater. Although open-top tanks have been used commercially for many years, recent efforts have been made to develop a closed-top (anaerobic) tanking system (Fleming et al. 1983a, 1988a), as is discussed later.

After brining, salt from the cover solution diffuses into the cucumbers and soluble substrates diffuse from the cucumber into the brine, the rate of diffusion being dependent upon the cucumber size and other variables (Potts et al. 1986). The cucumbers become flaccid to

Figure 4.3 A tank yard for the fermentation and storage of pickling cucumbers. The tops of the tanks are open to the atmosphere, necessitating exposure of the brine surface to sunlight to prevent surface growth of yeasts and molds. Note the tall, white tank in the center background which is used for liquid nitrogen storage. Nitrogen gas is piped to all tanks for use in purging of CO_2 from fermenting brines to prevent bloater damage. (*From Fleming, 1984.*)

varying degrees, depending upon fruit size and concentration of salt, due to osmotic differential. Bacteria may be drawn into the fruit due to the vacuum created therein (Daeschel and Fleming 1981).

Cabbage

The microflora of cabbage are more numerous on outer leaves and diminish in number within the head and the relative number of LAB is higher within the head (Keipper et al. 1932). Cabbage heads are trimmed of green, outer leaves, which amounts to about 30 percent weight loss, before being further processed. The cabbage then may be washed, which reduces the total microflora in relation to the LAB (Keipper et al. 1932). The heads are then cored (the core is either removed or diced mechanically while within the head) and sliced (Stamer 1983). Efforts are made to avoid inclusion of discolored or diseased cabbage and foreign material before shredding, since it is practically impossible to do so later. Dry salt is sprinkled onto the shredded cabbage before it is conveyed into the fermentation vessel.

Fermentation vessels may be constructed of reinforced concrete, wood, or fiberglass and hold 20 to 180 tons of shredded cabbage (Stamer 1983). Brine is quickly generated due to osmotic conditions created by the added salt. The salt added typically constitutes 2 to

2.25 percent (w/w) of the shredded cabbage in the United States. Kandler (1981) reported use of 0.75 to 1.5 percent salt in German kraut manufacture. The amount of salt influences the amount of brine generated and the rate and type of microbial action. The concentration of salt necessary to ensure acceptable texture of sauerkraut is much lower than that of cucumbers, although softening has been attributed to low salt concentration (Pederson and Albury 1969). Neither enzymatic activity nor microorganisms responsible for such softening have been well characterized. The relatively low salt concentration and more immediate availability of nutrients from the sliced cabbage result in more rapid initial growth of microorganisms than in the case of cucumbers.

Olives

While differences in the fermentation of cucumbers and cabbage can be attributed largely to differences in salt concentration used and physical handling of the product, the fermentation of olives is unique partly because of the presence of the bitter phenolic glucoside, oleuropein (Fig. 4.4). In the Spanish-style process for production of fermented green olives, the olives are exposed to alkali to degrade the oleuropein (thereby eliminating bitterness) and then washed to remove excess alkali. Then the olives are placed in brine (to equalize with the olives at about 7% NaCl) where they are allowed to ferment. The initial microbial load is influenced by the alkali treatment, in addition to other variables.

The possible presence of a microbial inhibitor(s) in olives has been conjectured for many years, and oleuropein was suggested as an inhibitor of the LAB (Vaughn 1954). It has been shown, however, that oleuropein itself is not very inhibitory to the lactic acid bacteria, but certain of its hydrolysis products are (Fleming et al. 1973c). The aglycone of oleuropein and elenolic acid (Fig. 4.4) both have been shown to inhibit LAB. Further evidence that oleuropein is not appreciably inhibitory is the fact that it has been shown to serve as a sub-

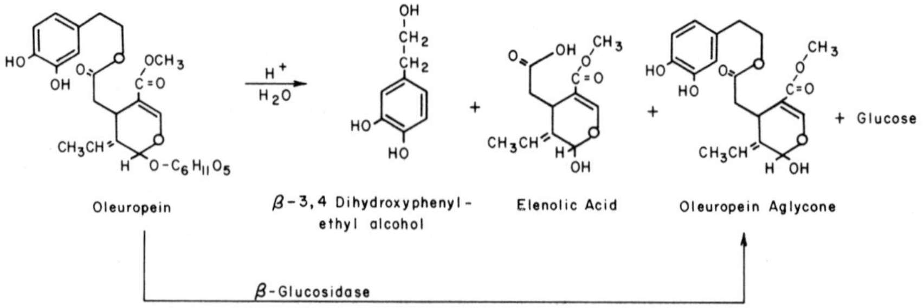

Figure 4.4 Oleuropein and its hydrolysis products. (*From Walter et al., 1973.*)

strate for the LAB (Garrido Fernandez and Vaughn 1978). If green olive varieties containing appreciable levels of oleuropein (e.g., Manzanillo, Mission) are brined without (or inadequate) lye treatment, they will undergo fermentation by yeasts and not LAB (Etchells et al. 1966). Neither the hydrolysis products of oleuropein nor oleuropein itself inhibit yeasts species tested (Fleming et al., 1973c).

It has been speculated that inhibitory hydrolysis products of oleuropein are formed when olives are brined (Fleming et al. 1973c). However, the mechanism by which these products are formed has not been established. It is thought that beta-glucosidase or a similar enzyme hydrolyzes oleuropein to the inhibitory compounds after the olives are brined. In fact, exposure of pure oleuropein to beta-glucosidase has been shown to produce the aglycone, which is inhibitory (Walter et al. 1973).

More recent studies have shown that phenolic compounds from olives previously shown to be noninhibitory, when tested in complex media, were inhibitory when *Lactobacillus plantarum* was exposed to the purified compounds in water (Ruiz Barba et al. 1990). These authors suggested that the presence of organic nitrogenous compounds in complex media can mask the bactericidal effects because of binding with the phenolic compounds. Thus, while the mechanism of inhibition of LAB in green olive fermentation has not been fully established, continuing interest in the subject promises to reveal new insights.

Some types of olives are not treated with alkali before fermentation either because they contain low levels of oleuropein and, thus, are low in bitterness (e.g., Sicilian type); or they are allowed to ripen on the tree before harvest, which apparently reduces the bitterness (Greek-type ripe olives); or they are allowed to ferment by yeasts (for processing into ripe olives). More detailed accounts on various types of olive fermentations are available (Fernandez Diez 1983; Fernandez Diez et al. 1985; Vaughn 1954, 1982). The present review will emphasize the Spanish-type process, since it accounts for the largest quantity of fermented olives and the yeast fermentation of non-alkali-treated olives. The salt-free storage of olives without fermentation for processing into "green-ripe" olives has become widely used in the United States in the last 20 years due to salt disposal problems, but will not be discussed herein (see Vaughn et al. 1969b, Vaughn 1982).

Primary Fermentation

Lactic acid bacteria

Four genera of lactic acid–producing bacteria have been reported in fermenting vegetables (Table 4.4). *Streptococcus* species occur on the fresh vegetables (Mundt et al. 1967) and may be present in the initi-

TABLE 4.4 Lactic Acid–Producing Bacteria Involved in Vegetable Fermentations

Genus and species	Fermentation type[a]	Main product (molar ratio)	Configuration of lactate
Streptococcus faecalis	Homofermentative	Lactate	L (+)
Streptococcus lactis	Homofermentative	Lactate	L (+)
Leuconostoc mesenteroides	Heterofermentative	Lactate:acetate: CO_2 (1:1:1)	D (−)
Pediococcus pentosaceus	Homofermentative	Lactate	DL, L (+)
Lactobacillus brevis	Heterofermentative	Lactate:acetate: CO_2 (1:1:1)	DL
Lactobacillus plantarum	Homofermentative	Lactate	D (−), L (+), DL
	Heterofermentative[b]	Lactate:acetate (1:1)	D (−), L (+), DL
Lactobacillus bavaricus	Homofermentative	Lactate	L (+)

[a]With respect to hexose fermentation.
[b]Heterofermentative with respect to pentoses (facultatively heterofermentative).
SOURCE: Adapted from Kandler, 1983.

ation stage of fermentation. Their significance in the primary fermentation and upon product quality has not been established. The other three genera (*Leuconostoc, Pediococcus*, and *Lactobacillus*) characteristically appear in vegetable fermentations. The relative numbers, time of occurrence, and significance of each vary with the vegetable commodity, salt concentration, and temperature, as is more fully discussed later in reference to specific vegetables.

Fructose and glucose are the principal fermentable carbohydrates in vegetables, although small quantities of sucrose and other sugars may be present. Fermentation of fructose and glucose yields mainly lactic acid by homofermentative and facultatively heterofermentative species, but obligately heterofermentative species produce CO_2, ethanol, acetic acid, and mannitol in addition to lactic acid. Organic acids of the vegetables also may be metabolized. Malic acid is degraded to lactic acid and CO_2 (McFeeters et al. 1982b). Metabolism of other organic acids such as citric (to acetic, pyruvate, and CO_2) by LAB has been reported (DuPlessis 1964). Similar reactions might be expected in vegetable fermentations, depending upon presence of the acids and appropriate species of bacteria. Amino acids may be decarboxylated to yield CO_2 and amines (Rodwell 1953).

Yeasts

Yeasts may or may not be involved in the primary fermentation of vegetables, depending upon their occurrence on the raw product and the chemical and physical environments in which the product is fermented. Species of yeasts associated with the fermentation of cucumbers and olives are summarized in Table 4.5. Little information has been published on the involvement of yeasts in the fermentation of sauerkraut.

While the presence of oxygen has little influence on the fermentation products of LAB, it dictates metabolic end products of yeasts involved in vegetable fermentations. Under anaerobic conditions, yeasts produce mainly ethanol and CO_2 from hexoses. Aerobically, they produce mainly CO_2. Although anaerobic conditions normally prevail during vegetable fermentations, the use of air for purging of cucumber brines has been shown to encourage aerobic metabolism by yeasts present (Potts and Fleming 1979). Many species of yeasts can oxidize

TABLE 4.5 Yeasts Associated with the Fermentation of Cucumbers and Olives

Yeast[a]	Cucumbers[b]	Olives[c]
Candida diddensii		D
Candida krusei	B	C
Candida rugosa		C
Candida solani		C
Candida tenuis		C
Candida valida (C. mycoderma)		C
Debaromyces hansenii (*D. membranaefaciens* var. Holl.)	B	
Hansenula anomala	A	D
Hansenula subpelliculosa	A	C
Pichia membranaefaciens		C,D
Pichia ohmeri (Endomycopsis ohmeri)	B	
Rhodotorula sp.	B	
Rhodotorula glutinus var. *glutinus*		E
Rhodotorula minuta var. *minuta*		E
Rhodotorula rubra		E
Saccharomyces baillii (S. elegans)	A	
Saccharomyces delbrueckii	A	
Saccharomyces oleaginosus		D
Saccharomyces rosei	A	
Saccharomyces rouxii (Zygosaccharomyces halomembranis)	B	
Torulopsis candida		D
Torulopsis holmii	A	C
Torulopsis lactis-condensii (T. caroliniana)	A	
Torulopsis sphaerica (imperfect form of *Kluyveromyces lactis*)		C
Torulopsis versatilis (Brettanomyces versatilis)	A	

[a]Names consistent with Lodder (1970). Names in parentheses are synonyms used by the authors indicated.

[b]The letter "A" indicates Etchells et al. (1961) as the source, and they classified these as fermentative species. "B" indicates Etchells and Bell (1950) as the source of occurring in cucumber fermentations, and they classified these as oxidative species.

[c]The letter "C" indicates Mrak et al. (1956) as the source of occurring in green olive fermentations. The letter "D" indicates Duran-Quintana and Gonzalez-Cancho (1977) as the source of occurring in natural black olive fermentations. The letter "E" indicates Vaughn et al. (1969b) as the source occurring in green olive fermentations.

lactic and other organic acids aerobically, which can lead to spoilage if insufficient acids remain to inhibit growth of undesirable bacteria.

Until recently, yeasts have been considered to be incidental during primary fermentation of vegetables at best, and spoilage microorganisms at worst. Purging of brines to remove CO_2 and, thereby, preventing the formation of gas pockets within the vegetable has opened the possibility for use of yeasts and other gas-forming microorganisms in the fermentation, as is discussed below.

Cucumbers

Natural fermentation. Salt concentration can exert a great influence on the fermentation of cucumbers. In the early decades of this century, it was common commercial practice to use high concentrations of salt to prevent softening and other spoilage problems. Perhaps the best illustration of the effects of salt concentration on relative activities of LAB, yeasts, and Enterobacteriaceae during cucumber fermentation is in the work of Etchells and Jones (1943; Fig. 4.5). A companion paper summarizes the corresponding chemical changes at the three salt concentrations tested (Jones and Etchells, 1943; Fig. 4.6). These papers showed that a high salt concentration (15 percent) resulted in slow and limited growth by LAB and low acid production with vigorous yeast growth after about 14 days and high concentrations of CO_2. A low salt concentration (5 percent) resulted in rapid growth by LAB and high acid production, with limited yeast growth and low concentrations of CO_2.

Maintenance of structural integrity is highly important in the fermentation of whole vegetables such as cucumbers, which adds complications not encountered in the fermentation of liquid products such as wine and beer. Bloater damage (Fig. 4.7) results from gas accumulation inside the cucumbers during fermentation. Tissue softening may be caused by pectinolytic enzymes of either microbial (primarily fungal) origin (Etchells et al. 1958) or of the cucumber fruit itself (McFeeters et al. 1980). Off-flavors and off-colors also may result from improper fermentation. Bloater damage has been attributed to CO_2 production by yeasts (Etchells et al. 1953), Enterobacteriaceae (Etchells et al. 1945), heterofermentative LAB (Etchells et al. 1968b), and even homofermentative LAB (Fleming et al. 1973a). Sorbic acid has been used to reduce the yeast growth and bloater damage resulting therefrom (Costilow et al. 1957).

Pure culture fermentation. To ensure only growth of the added LAB culture, the natural flora must be inactivated or prevented from growing. Etchells et al. (1964) accomplished this by use of ionizing irradi-

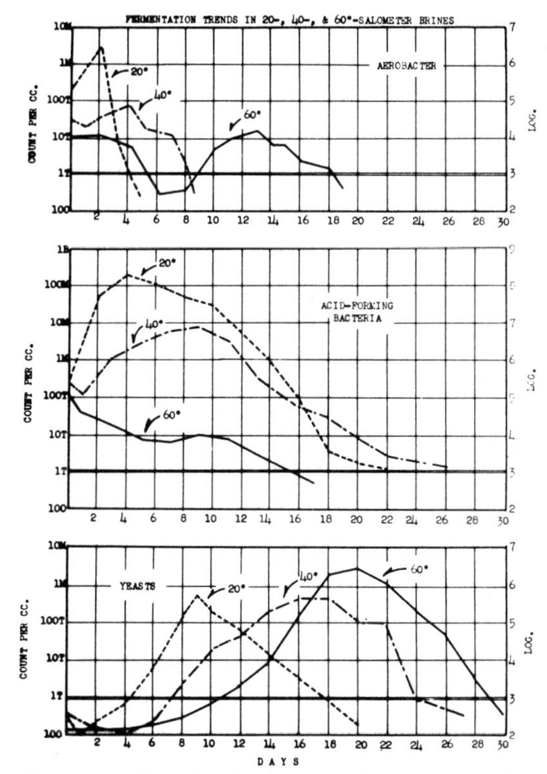

Figure 4.5 Growth of microorganisms in cucumber fermentations at initial brine concentrations of 20, 40, and 60°S (percent saturation with respect to salt = 5.3, 10.6, and 15.8%, w/w, respectively). (*From Etchells and Jones, 1943.*)

ation or heat shocking of the raw fruit and mild acidification of the cover brine. Fermentable sugars were not reported in this study, but, based on the author's experience, sugars undoubtedly remained in the fermentations with homofermentative LAB. Although the pure-culture procedure has been suggested for commercial production in small containers (Etchells et al. 1968*a*), it has not been considered practical for bulk fermentation due to energy requirements to inactivate the natural microflora and the aseptic conditions required for its success.

Controlled fermentation. Various efforts have been made to influence the fermentation of cucumbers by addition of cultures to the unpasteurized raw product. Such procedures are considered to be controlled rather than pure-culture fermentations, since the natural

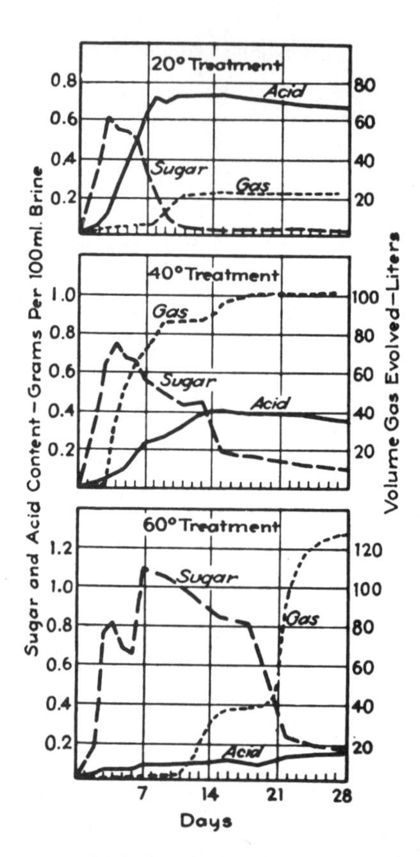

Figure 4.6 Chemical changes in cucumber fermentations at initial brine concentrations of 20, 40, and 60°S. (*From Jones and Etchells, 1943.*)

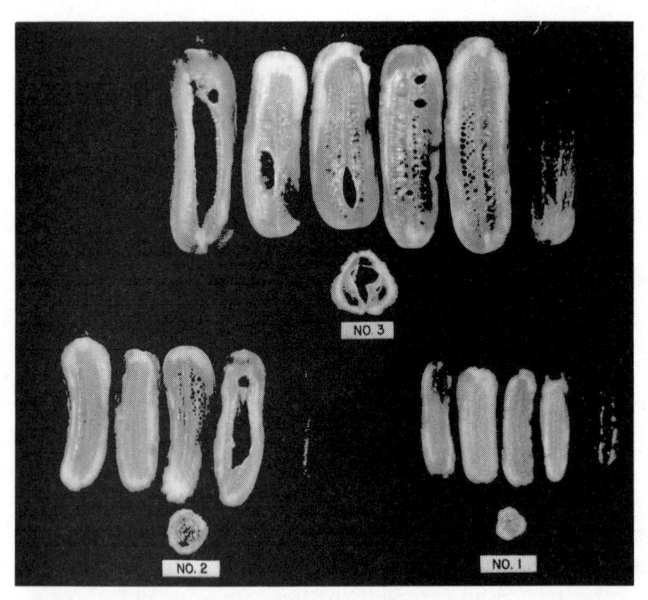

Figure 4.7 Bloater damage in fermented cucumbers. Note the more serious damage in larger fruit.

microflora may play a significant role. For example, Pederson and Albury (1961) found that *Lactobacillus* plantarum eventually predominated fermentations regardless of the species of LAB added initially. Apparently, the greater acid tolerance of *L. plantarum* permitted its growth after less-acid-tolerant species were inhibited by acid produced during fermentation.

The controlled fermentation procedure of Etchells et al. (1973) included the use of an acid-tolerant culture of *L. plantarum*. In addition, other control measures helped to assure success of the procedure. These measures included washing of the fruit before brining, chlorination and mild acidification of the cover brine, addition of sodium acetate buffer to assure complete sugar utilization, and nitrogen purging of CO_2 from the fermenting brine to prevent bloater formation. Certain features of this procedure have been successfully applied on a commercial scale, but the overall procedure has not been accepted commercially.

Purging of CO_2 from brines to prevent bloater formation represents an evolving story that continues today. Further research is needed and a historical perspective is warranted. The need for purging was recognized when cucumbers fermented by *L. plantarum* were observed to have bloater damage, even though there was no evidence of heterofermentative LAB or yeasts (Etchells et al. 1969, unpublished observation). This was surprising since *L. plantarum* does not produce CO_2 from glucose or fructose, the fermentable sugars of cucumbers. A subsequent study revealed that *L. plantarum* produced sufficient CO_2 from cucumber fermentation, which when added to that produced by the cucumber tissue, was sufficient to cause bloater damage (Fleming et al. 1973a). A nondestructive method for monitoring bloater formation, based on brine volume increase due to gas formation within the tissue, was used. Nitrogen purging was shown to relieve the expansion volume after bloater formation had occurred, or to prevent bloater formation altogether if CO_2 was continuously purged from the brine. The CO_2 produced during fermentation was shown to be derived about equally from the cucumber tissue and the *L. plantarum* culture (Fleming et al. 1973b). Respiratory CO_2 from the cucumbers was expected, but *L. plantarum* was not expected to produce CO_2. Later, malic acid was found to be the principal organic acid of cucumbers (McFeeters et al. 1982a) and was shown to be the major source of CO_2 produced by the *L. plantarum* starter culture (McFeeters et al. 1982b). A procedure was developed for mutation and selection of *L. plantarum* strains that do not produce CO_2 from malic acid (Daeschel et al. 1984; Fig. 4.8).

The pickle industry has adopted purging as a general practice, particularly for larger sizes of fruit that are more susceptible to bloater

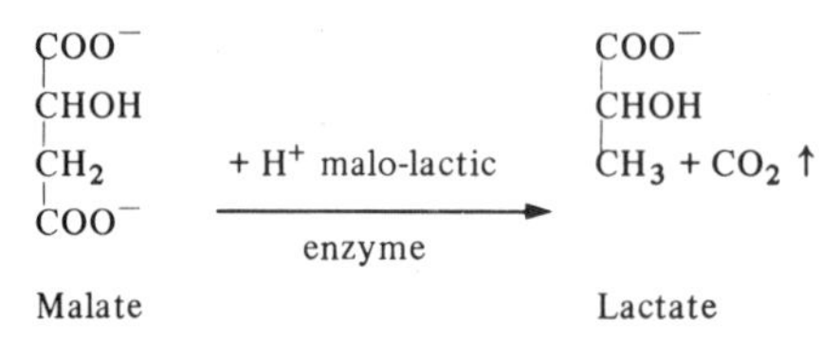

Figure 4.8 Proposed malolactic reaction in cucumber fermentation. (*From McFeeters et al., 1982b.*)

damage. Both nitrogen and air have been shown to effectively prevent bloater formation. Nitrogen has been recommended to avoid growth of undesired aerobic microorganisms and potential oxidative changes in flavor and color components (Etchells et al. 1973). However, air is used by many companies using a side-arm purging device described by Costilow et al. (1977) or modifications because it is less expensive.

Cucumbers fermented solely by *L. plantarum* may be too acidic in flavor without removal of part of the acid before final processing. Thus, use of microorganisms that produce neutral end products may be desirable. Heterofermentative LAB have been shown to ferment all of the fermentable carbohydrates in green beans with the production of a mild flavor (Chen et al. 1983a). *Lactobacillus cellobiosus* was found to be particularly efficacious in this regard and has shown promise for cucumber fermentations. Fermentative yeasts, historically considered to be spoilage agents because of bloater damage (Etchells et al. 1953), also have been considered as a means of producing more mildly acidic cucumber products. Fermentation of sterile cucumber juice could be manipulated to produce varying amounts of lactic acid and ethanol by mixed cultures of *L. plantarum* and the yeast *Saccharomyces cerevisiae* (Daeschel et al. 1988).

Before any procedure is accepted for commercial fermentation and storage of cucumbers, it must be economically justifiable. Furthermore, in order for controlled fermentation to occur, the environment in which the product is held must be controlled. Since the pickle industry has become accustomed to use of open tanks (see Fig. 4.3) for various reasons, the problem of control is exacerbated. However, the pickle industry has supported research to develop a closed-top tanking system for the anaerobic fermentation and storage of cucumbers. Pilot tanks and the overall handling system have been described (Fleming et al. 1983a). A simplified brining procedure was developed for use in closed tanks, which included addition of calcium acetate buffer to assure complete utilization of fermentable sugars (Fleming et al. 1988a). The calcium served to improve firmness retention in the cucumbers. Using the procedure in pilot tanks, approximately 95 percent of the fermentable sugars were converted to lactic acid, which is consistent with the *L. plantarum* starter culture that was added. However, the starter culture did not predominate the entire fermentation, as is indicated by recovery of the marked culture in Fig. 4.9. Apparently, naturally occurring homofermentative LAB overcame the added culture after a day or so.

Sauerkraut

Natural fermentation. The natural fermentation of sauerkraut is characterized by a progression in the growth of LAB species, typically starting with the acid-sensitive *Leuconostoc mesenteroides* and terminating with the acid-tolerant *L. plantarum* (Pederson and Albury 1969). *Pediococcus cerevisiae* and *Lactobacillus brevis* species have been indicated to grow between these two extremes. Temperature and salt concentration are extremely important factors which regulate the type of fermentation. Optimum conditions to produce high-quality sauerkraut are temperatures of 13 to 24°C and salt concentrations of 1.8 to 2.25 percent (Pederson and Albury 1969). Various spoilage problems have been reported for sauerkraut including softening, discoloration, and off-flavors, some of which occur during primary fermentation due to low or uneven salt concentrations. The aggravating

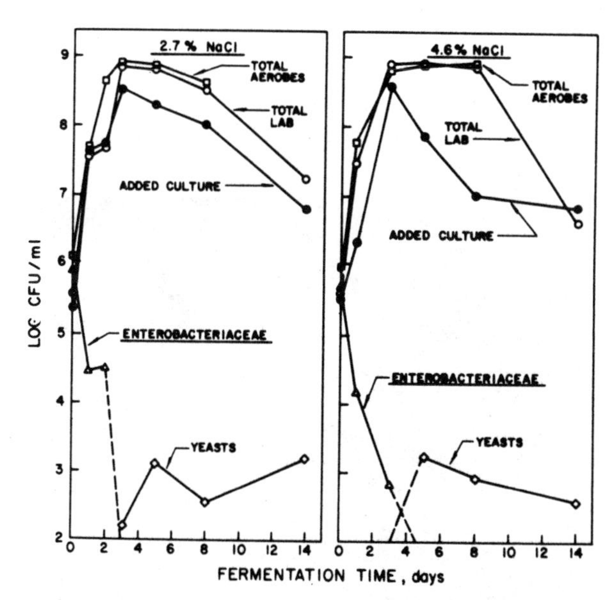

Figure 4.9 Microbial growth in brined cucumbers inoculated with an *L. plantarum* starter culture. Note that the added culture (marked with streptomycin resistance for detection) did not predominate the lactic acid fermentation at either 2.7 or 4.6% NaCl. (*From Fleming et al., 1988a.*)

problem of heaving occurs during the first day or so after shredded cabbage and salt are placed into the fermentation tank. This problem is due to the high concentration of CO_2 produced by the cabbage and by *L. mesenteroides* and other gas-forming bacteria early in the fermentation. After heaving, the tank must be reheaded. Tanks today typically are headed with a plastic cover weighted down by water, but the primitive method of weighting down the fermenting cabbage with stones is still practiced to a limited extent (Fig. 4.10).

Considerable differences exist between methods for sauerkraut production in the United States and certain other parts of the world. The above conditions cited by Pederson and Albury (1969) refer principally to sauerkraut produced in the United States. The sauerkraut is fermented and held in bulk tanks until it is needed for further processing. Thus, the acidity varies greatly, depending upon the initial concentration of fermentable sugars in the raw cabbage. In Europe, however, the sauerkraut is held only until the desired acidity is reached (e.g., 1 percent as lactic acid). The product is then canned, heat processing being used to stabilize the product to further fermentation. Thus, American processors retain the economic advantage of bulk storage but compromise on product uniformity in comparison to their European counterparts. There are additional economic reasons for the differences in methods of sauerkraut production beyond the scope of this review.

Controlled fermentation. Various attempts have been made to influence the fermentation of sauerkraut by addition of bacterial cultures

Figure 4.10a Heading of sauerkraut tanks with either water (*a*) or stones (*b*).

Figure 4.10b *(Continued)*

(Pederson and Albury 1969). However, some have concluded that culture additions are not needed to yield desirable sauerkraut, provided optimum temperature and salt concentrations are maintained (Pederson and Albury 1969, Stamer 1983). Stetter and Stetter (1980) isolated *Lactobacillus bavaricus* from sauerkraut, which produces exclusively L(+) lactic acid. A German patent has been granted for the production of sauerkraut with this culture (Eden-Waren 1976). There has been some health concern by the presence of high concentration of D(−) lactic acid (FAO/WHO 1966), thus the above culture may provide a significant advantage in such products. More recent indications are that the presence of D(−) lactic acid in the diet may only be important in infants (WHO 1974).

A significant factor influencing the uniformity of quality of sauerkraut, or lack thereof, is the distribution of salt within the product. Noel et al. (1979) have suggested a brine circulation system to ensure uniformity of salt and acid throughout the product. A fermenter has been designed which permits creation of a vacuum to prevent the heaving problem (Christ et al. 1981).

The sauerkraut fermentation is highly complex and involves physical, chemical, and microbiological factors which influence quality of the product. The fermentation has been broadly categorized into two stages, gaseous and nongaseous, based on studies using a specially designed laboratory fermentor (Fleming et al. 1988b). The physical, chemical, and microbiological changes that occurred during fermentation used to characterize these two stages are given in Figs. 4.11 to

4.13. During the first 2 days, CO_2 rapidly increased, and the brine level increased due to gaseous expansion within the shreds of cabbage (Fig. 4.11). Purging of CO_2 from the brine relieved the gas pressure, as evidenced by a decrease in the brine level (Fig. 4.11*a*). Heterofermentative LAB, presumably *L. mesenteroides*, predominated for approximately 8 days, after which homofermentative LAB predominated and continued to ferment until all sugar was fermented or fermentation ceased due to low pH. This conclusion is based on the evidence of rapid fructose depletion and rapid formation of mannitol, acetic acid, and ethanol, which are products of heterofermentative metabolism (Fig. 4.12). Also, the microbial growth profile indicated two stages of LAB growth, which included a rapid increase in numbers of LAB during the first 4 days (which were determined to be all gas-formers), followed by a decrease thereafter and a subsequent slight increase in numbers (all non-gas-formers) after 8 days (Fig. 4.13).

Kimchi

The early phase of *kimchi* fermentation is similar to sauerkraut fermentation. However, *kimchi* typically includes several vegetables (e.g., Korean cabbage, garlic, green onion, red pepper powder) that are allowed to ferment to only a slightly acidic product. In this regard, *kimchi* is highly perishable and must be consumed within a few days, depending upon temperature, before excessive acidity results. Like

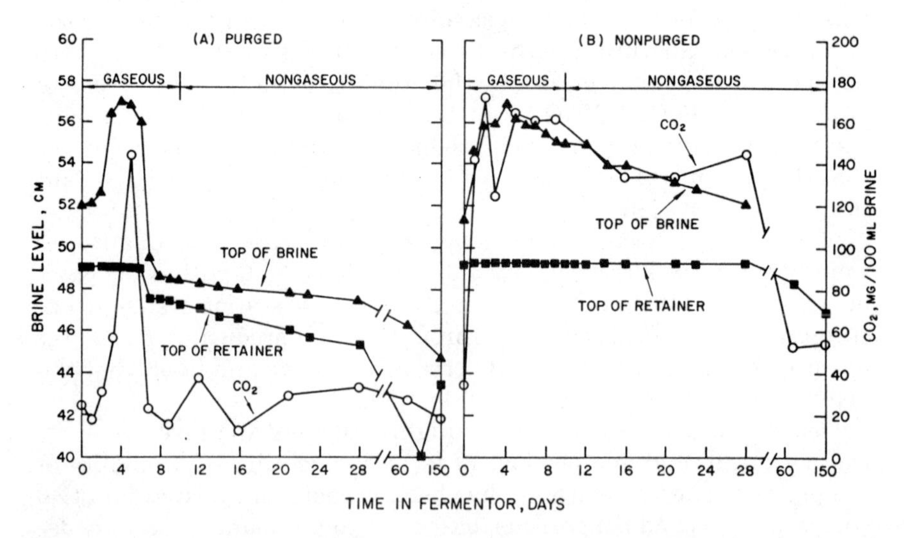

Figure 4.11 Changes in liquid and kraut bed volumes and in CO_2 concentration of purged and nonpurged sauerkraut fermentations. Note that purging of CO_2 from the brine (*a*) resulted in CO_2 reduction and reduced internal gas pressure, as evidenced by lower brine levels. (*From Fleming et al., 1988b.*)

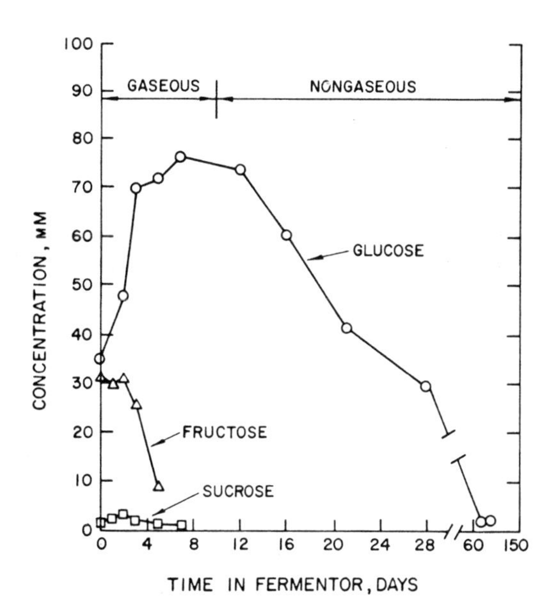

Figure 4.12a Substrate depletion in sauerkraut fermentation. (*From Fleming et al., 1988b.*)

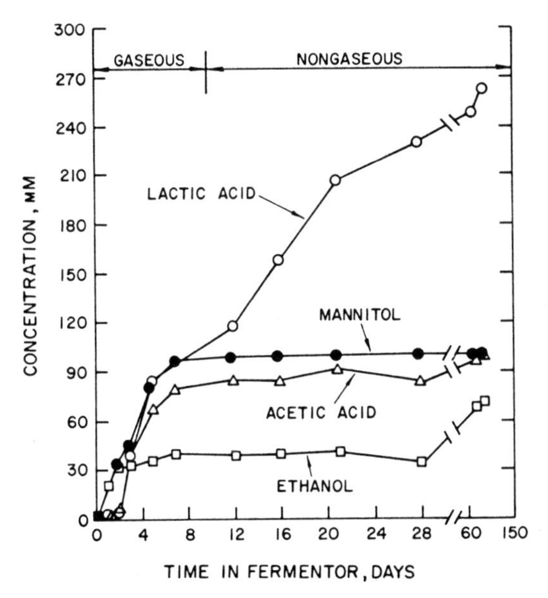

Figure 4.12b Product formation in sauerkraut fermentation. (*From Fleming et al., 1988b.*)

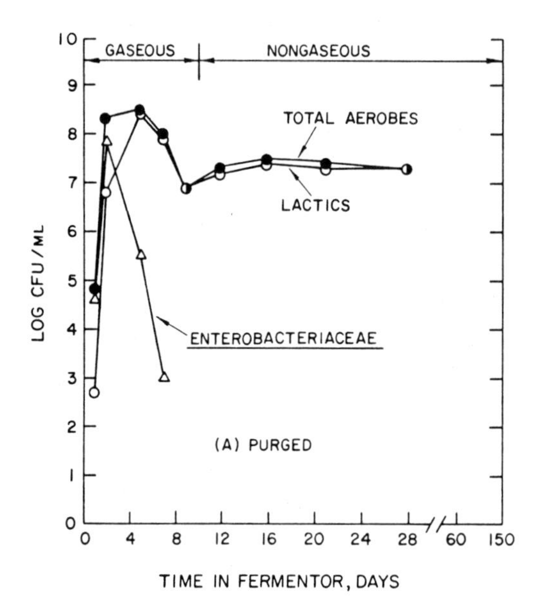

Figure 4.13 Microbial growth in sauerkraut fermentation. (*From Fleming et al., 1988b.*)

sauerkraut, *L. mesenteroides* is involved in the early phase of fermentation. *L. plantarum*, which grows later, is considered to be a spoilage microbe when it produces excessive amounts of acid (Mheen and Kwon 1984). The optimum pH, acidity (as lactic acid), and salt of *kimchi* was found to be 4.2, 0.6 to 0.8 percent, and 3.0 percent, respectively, and the temperature for achieving the best flavor to be 5 to 14°C (Mheen and Kwon 1984).

Olives

Natural fermentation. After alkali treatment and washing to remove excess alkali according to the Spanish-type process previously mentioned, the olives undergo fermentation by naturally occurring LAB. The species of LAB involved in the fermentation vary with the salt concentration used. Some olive varieties such as Sevillano (which are more susceptible to shrivel) are brined at 4 to 5% NaCl, while most varieties today are brined at 7 to 8% NaCl (Vaughn 1982). *L. plantarum* is consistently found in all fermentations, while other LAB may occur, depending on olive variety and salt concentration used (Vaughn 1982). *L. mesenteroides* and *Streptococcus faecalis* have been reported in olive fermentations at low salt concentrations, while *L. brevis* and *P. cerevisiae* may or may not occur at various periods during the fermentation (Vaughn et al. 1943, Vaughn 1982).

L. plantarum has been shown to be primarily responsible for the occurrence of "yeast spots" on the surface of fermented olives (Vaughn et al. 1953). This blemish, which appears as raised white spots on the olive surface, may be considered by some to be unsightly, but the olives are not otherwise affected adversely. Gas pocket formation within the flesh of fermenting olives, however, is a serious defect. This problem is analagous to bloater formation in cucumbers and has been shown to be associated with the growth of gas-forming bacteria and yeasts (Vaughn 1982). Some of the gas-forming bacteria such as Enterobacteriaceae (Vaughn et al. 1943, West et al. 1941) or butyric acid bacteria (Gilliland and Vaughn, 1943) may grow during the initiation stage of fermentation, before primary fermentation by lactic acid bacteria becomes established. These problems likely are associated with the improper removal of alkali and resultant initial high pH. Most processors add sufficient acid to the brine today to lower the pH to 4.5 to 5.0 (Vaughn 1982), which should reduce these problems. Tissue softening and malodorous fermentations also result from growth of these same types of bacteria (Vaughn 1982).

Yeasts also may be active during fermentation of Spanish-type olives (Table 4.5). Mrak et al. (1956) reported predominantly fermentative yeasts during the first 7 weeks of olive fermentation and a more aerobic flora during weeks 9 to 16. Some yeasts have been associated with the softening and gas-pocket formation of brined olives (Vaughn et al. 1969a, 1972). In non-alkali-treated olives, yeasts are primarily responsible for the fermentation (Duran Quintana and Gonzalez Cancho 1977; Table 4.5).

Spanish researchers have begun to explore ways to take advantage of yeasts for the fermentation of non-alkali-treated olives for the production of ripe olives. Two relatively recent areas of study involve naturally ripe black olives (i.e., olives that are allowed to ripen on the tree before harvest) and post-harvest-ripened black olives (i.e., olives that are harvested green and subsequently debittered and blackened by the use of NaOH and air).

The fermentation of naturally ripe black olives of four varieties was shown to be due principally to yeasts (Duran Quintana et al. 1971). Naturally occurring yeasts from the fermentation have been identified (Duran Quintana and Gonzalez Cancho 1977, Gonzalez Cancho et al. 1975; Table 4.5). A serious problem with this type of fermented olive is the formation of gas pockets within the flesh, referred to as *alambrado* by Spanish researchers. The problem is analogous to bloater formation in brined cucumbers. The problem is more severe in tree-ripened olives than in green olives, due apparently to the softer flesh of the tree-ripened fruit. Purging of CO_2 from the brine with air or nitrogen prevented *alambrado* (Garcia Garcia et al. 1982). Procedures for air purging of brines to prevent the problem have been described (Duran Quintana et al. 1986, Garcia Garcia et al. 1985). More recently, air purging has been shown to be useful in the fermentation of green olives for postharvest ripening (Brenes Balbuena et al. 1986a,b).

Controlled fermentation. Inoculation of alkali-treated green olives has been done with pure cultures of LAB (Borbolla y Alcala et al. 1952, 1964; Cruess 1937; Etchells et al. 1966) and with actively fermenting brine (Cruess 1930). Commercially, however, inoculation has not been practiced extensively. Although various spoilage problems occur, apparently none has justified the use of cultures. Stuck fermentations (failure to develop acidity) is one such problem that occurs occasionally. The causes of stuck fermentations are not always understood. Lack of sufficient sugar and the presence of compounds inhibitory to LAB have been suggested as possible reasons (Vaughn 1954). The inhibitory compound(s) of olives is inactivated by alkali (Fleming and Etchells 1967, Juven et al. 1968), which is coincidental to the use of alkali to remove bitterness. The presence of such inhibitors complicates the use of LAB cultures. When cultures were added to brined olives that had been neither alkali-treated nor heated, only yeasts grew (Etchells et al. 1966). The LAB culture was inhibited. When heat-shocked (74°C, 3 min), however, the olives underwent rapid fermentation (Fig. 4.14). *L. mesenteroides* was found to be more inhibited than

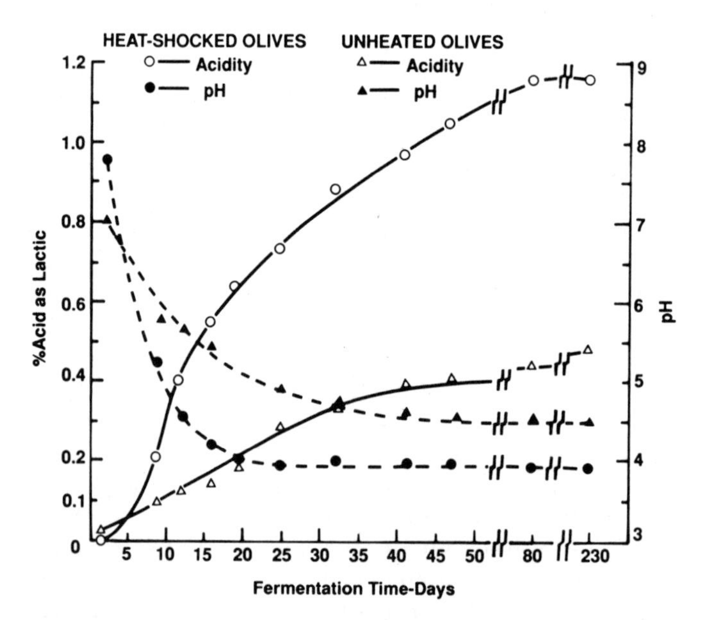

Figure 4.14 Effect of heat shocking Manzanillo olives on their subsequent acid fermentation by *L. plantarum*. Lye-treated, washed olives were either heated (74°C, 3 min) or unheated before inoculation. (*From Etchells et al., 1966.*)

either *L. plantarum* or *P. cerevisiae* (Fleming and Etchells 1967).

Although heating renders green olives fermentable by LAB even if the olives are not treated with alkali, this phenomenon has not been exploited on a large commercial basis. If a practical method for debittering olives other than alkali treatment could be developed, heat-shocking and inoculation with LAB could prove advantageous.

Secondary Fermentation

Yeasts

The problem of secondary fermentation by fermentative yeasts is particularly important with cucumbers in bulk vessels because of the bloater problem. Sorbic acid or its potassium salt may be added to suppress the growth of fermentative yeasts during primary fermentation (Costilow et al. 1957). However, the presence of residual fermentable sugars may allow eventual growth by yeasts if the sorbic acid is degraded or diluted. Thus, it is preferable for all fermentable sugars to be utilized before purging is discontinued. A buffering agent such as sodium or calcium acetate may be added to the cover brine (Etchells et al. 1973, Fleming et al. 1988a), or the acid may be partially neutralized with base during fermentation to ensure complete sugar utilization. Secondary yeast fermentation also is important in various fermented vegetables that are not heat-processed after packing for consumer use. Gas pressure buildup in the jar and cloudy brine in packaged products is undesirable.

Cucumbers, green beans, tomatoes, and peppers that were fermented with pH control to remove all fermentable carbohydrates were microbiologically stable (Fleming et al. 1983b). Carrots and red beets, however, could not be fermented by LAB to complete sugar utilization; these products underwent fermentation by yeasts after growth of LAB ceased. These observations led to the conclusion that the fermented vegetables that were microbiologically stable achieved stability due to: (1) the absence of fermentable carbohydrates, (2) the pH being 3.8 or lower, and (3) exclusion of oxygen. The maximum pH for assuring inhibition of spoilage bacteria is not known. Although pH 3.8 resulted in stability of several vegetables (Fleming et al. 1983b), in another instance butyric acid spoilage of cucumbers resulted when the lactic fermentation terminated at pH 3.7 (Fleming et al. 1989), as is discussed below.

Bacteria

Spoilage of Spanish-style green olives may result from two types of clostridial action, both of which result in malodorous products. In one type, *Clostridium butyricum* and closely related species produced butyric acid from sugars during the primary fermentation stage (Gilliland and Vaughn 1943). A concentration of 7 to 8 percent salt in brines was reported sufficient to prevent this type of spoilage. In a second type of malodorous fermentation, *zapatera* spoilage resulted from decomposition of lactic acid when little or no sugar was present and before the pH had decreased below 4.5 (Kawatomari and Vaughn 1956). These workers stated that *zapatera* spoilage is prevented if the brine is pH 3.8 or below.

Plastourgos and Vaughn (1957) isolated propionic acid bacteria from commercial olives with *zapatera* spoilage. All isolates grew at 7 percent salt concentration in lactate medium between pH 4.8 and 8.0. They conjectured that propionic acid bacteria grew first in the olives and caused a rise in pH which permitted *Clostridium* species to grow. Borbolla y Alcala et al. (1975) and Rejano Navarro et al. (1978) reported the formation of propionic acid in Sevillian olives and suggested that acidity, salt concentration, and pH can be controlled to prevent its formation. These workers observed that the degradation of lactic acid to acetic and propionic acids occurs in the late stages of fermentation and concluded that this action could be prevented by sufficiently high salt concentration and sufficiently low pH.

Instability of fully fermented (no residual sugars) cucumbers due to bacterial action has only recently been documented. Fleming et al. (1989) observed butyric acid spoilage of cucumbers that had undergone a normal primary fermentation which resulted in > 1 percent titratable acidity (as lactic) and pH 3.7. The cucumbers were fermented and stored anaerobically. Products formed during secondary fermentation in relative order of concentration were acetic acid > butyric acid > *n*-propanol > propionic acid. Fermentation balances after primary fermentation and after spoilage are shown in Table 4.6. No botulinal toxin was detected. *Clostridium tertium* was identified as contributing to the spoilage but did not produce propionic acid or *n*-propanol under test conditions. Similarities exist between instability problems identified above for olives and the spoiled cucumbers. It was hypothesized that lactic acid was degraded to propionic and acetic acids by unidentified bacteria (possibly propionic acid bacteria), with resultant rise in pH. This rise in pH eventually allowed growth of *C. tertium* and perhaps other clostridia. The fact that the fermented cucumbers were pH 3.7 prior to subsequent microbial spoilage is an important distinction from the problem noted above with olives (where pH values in the 4.8 to 8.0 range were found necessary for growth of propionic acid bacteria). Bacterial instability of fermented cucumbers under anaerobic conditions at pH 3.7 was not expected, especially since *zapatera* spoilage in brined olives does not occur at pH 3.8 or below (Kawatomari and Vaughn 1956). However, the cucumbers were

TABLE 4.6 Substrates and Products of a Cucumber Fermentation that Underwent a Secondary Butyric Acid Fermentation[a]

Compound	Before fermentation	After primary fermentation	After spoilage
	Concentration of compounds, mM		
Glucose	27.1	ND	ND
Fructose	34.3	ND	ND
Malic acid	12.6	ND	ND
Acetic acid	53.2 (0.0)[b]	63.7 (12.5)[b]	105.3 (39.6)[b]
Lactic acid	ND[c]	140.1	ND
Ethanol	ND	7.3	1.7
Propionic acid	ND	ND	8.1
Propanol	ND	ND	34.5
Butyric acid	ND	ND	38.7
	Elemental recoveries, %		
Carbon		109.8	79.4
Hydrogen (H)		115.0	85.9
Oxygen (O)		104.9	46.2
	Hydrogen/oxygen ratio of compounds, mM		
	1.88	2.06	3.84

[a]After primary fermentation (30 days), the brine (2.3% NaCl) contained 1.09% acidity (as lactic) and was pH 3.7. After spoilage, the brine contained 0.59% acidity and was pH 4.8. From Fleming et al., 1989.

[b]Values in parentheses are those used in calculation of elemental recoveries and reflect the assumption that acetic acid was not a substrate for products measured.

[c]ND = none detected.

stored at only 2.3% NaCl, which is much lower than typical concentrations used commercially (5 to 12 percent).

Mannitol is formed from fructose during the heterolactic fermentation of vegetables, which may or may not be susceptible to metabolism, depending upon pH. Chen et al. (1983b) found that mannitol was not metabolized by *L. plantarum* at pH 3.5 and below, but was when the pH was raised to 3.9.

Postfermentation

Normally it is desirable for microbial activity of brined vegetables to cease after primary fermentation. Secondary fermentation can be prevented by suitable pH, acidity, and salt concentration; the absence of fermentable carbohydrates; and the exclusion of oxygen. The exclusion of oxygen is dictated by design of the fermentation and storage vessel, while the other parameters can be controlled by brine additions.

The open-top tanks used for brined cucumber storage (refer to Fig. 4.3)

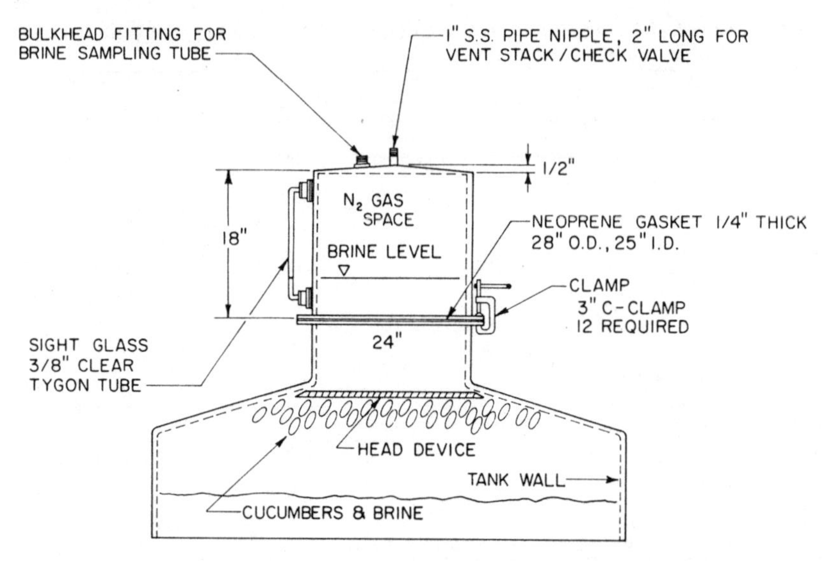

Figure 4.15 Cap assembly for anaerobic storage of brined cucumbers. (*From Fleming et al., 1983a.*)

Figure 4.16 Experimental anaerobic tanks for the fermentation and storage of brined cucumbers. The tanks are made of reinforced fiberglass, coated on the interior with a food-grade resin, and contain about 30,000 L.

create multiple problems. As noted earlier, open tanks must be left outdoors to allow exposure of the brine surface to sunlight and resultant inhibition of aerobic microorganisms. If allowed to grow, film yeasts and molds will oxidize lactic acid, resulting in a rise in pH and growth of other spoilage microorganisms. Growth of molds and production of pectinolytic enzymes may result in softening of the cucumbers. Extra salt must be added to compensate for rainwater and to prevent freezing damage to the cucumbers in colder regions during winter. An anaerobic tanking system has been designed and is being tested commercially. An important feature of the tank design is the anaerobic headspace which is maintained with nitrogen (Fig. 4.15). This principle has been applied to commercial size tanks (Fig. 4.16).

Sauerkraut is covered with plastic sheeting, weighted down with water (Fig. 4.10). After the heaving problem that occurs in initiation and primary fermentation, the cover provides an excellent anaerobic seal. Aerobic spoilage microorganisms may grow, however, if the cover becomes dislodged. Also, if the cover is ruptured, the salt and acid of the sauerkraut will be diluted, which can result in spoilage.

Anaerobic tanks for olive fermentation and storage have been in use for over 20 years. An example of such tanks in southern California is shown in Fig. 4.17*a,b*. These tanks are being buried to provide a cooler temperature and better texture retention of the olives. The Spanish commonly use spherically shaped tanks constructed of fiberglass, which are buried, for olive fermentation (Fernandez Diez et al. 1985).

Figure 4.17a Anaerobic tanks for fermentation of Spanish-type green olives.

Figure 4.17b (*Continued*)

Only the manway for filling and emptying extends above a concrete pad. The tanks have a typical capacity of 16,000 L. The Spanish claim that burying provides moderate temperatures that are suitable for fermentation and storage, thus avoiding extremes due to seasonal temperature changes.

Summary

General principles involved in the fermentations of cucumbers, cabbage, and olives are summarized in this review. Traditionally, vegetables have been allowed to undergo spontaneous fermentation with salt concentration and temperature largely dictating the fermentation type. Recent studies have suggested additional controls which may enhance the potential of fermentation as an economical means for preservation and bulk storage of vegetables. These controls include purging of CO_2 from fermenting brine, use of anaerobic vessels, pH control, Ca^{2+} addition, and use of selected and genetically modified microbial cultures.

Acknowledgment

This chapter is based on a presentation at the Z. John Ordal Memorial Symposium held at the Annual Meeting of the Institute of Food Technologists, New Orleans, Louisiana, June 21, 1988.

References

Andersson, R. E., M. A. Daeschel, and C. E. Ericksson: "Controlled Lactic Acid Fermentation of Vegetables," in G. Durand, L. Bobichon, and J. Florent (eds.), *8th International Biotechnology Symposium Proceedings*, vol. II, Societe Française de Microbiologie, Paris, France, 1988, pp. 855–868.

Bell, T. A.: "Pectolytic Enzyme Activity in Various Parts of the Cucumber Plant and Fruit," *Bot. Gaz.* 113:216–221 (1951).

Borbolla y Alcala, J. M. R. de la, M. J. Fernandez Diez, and F. Gonzalez Cancho: "Use of Pure Cultures of *Lactobacillus* in the Preparation of Green Olives," *Grasas y Aceites* 15:6–11 (1964).

Borbolla y Alcala, J. M. R. de la, C. Gomez Herrera, and A. I. Tamayo: "The Use of Pure Cultures of *Lactobacillus* in the Pickling of Green Olives," *Grasas y Aceites* 3:91–94 (1952).

Borbolla y Alcala, J. M. R. de la, L. Rejano Navarro, and M. Nosti Vega: "Propionic Acid Formation During the Conservation of Table Green Olives," *Grasas y Aceites* 26:153–160 (1975).

Brenes Balbuena, M., P. Garcia Garcia, M. C. Duran Quintana, and A. Garrido Fernandez: "Comparative Study of Several Methods for the Conservation of the Raw Material for the Elaboration of Ripe Olives," *Grasas y Aceites* 37:123–128 (1986a).

Brenes Balbuena, M., P. Garcia Garcia, and A. Garrido Fernandez: "Comparative Study of Several Methods for the Storage of the Raw Material for the Elaboration of Ripe Olives. II. Effects on Colour and Texture of the Final Product," *Grasas y Aceites* 37: 301–306 (1986b).

Buescher, R. W., J. M. Hudson, and J. R. Adams: "Inhibition of Polygalacturonase Softening of Cucumber Pickles by Calcium Chloride," *J. Food Sci.* 44:1786–1787 (1979).

Buescher, R. W., J. M. Hudson, and J. R. Adams: "Utilization of Calcium to Reduce Pectinolytic Softening of Cucumber Pickles in Low Salt Conditions," *Lebensm. Wiss. Technol.* 14:65–69 (1981).

Chen, K. H., R. F. McFeeters, and H. P. Fleming: "Fermentation Characteristics of Heterolactic Acid Bacteria in Green Bean Juice," *J. Food Sci.* 48:962–966 (1983a).

Chen, K. H., R. F. McFeeters, and H. P. Fleming: "Stability of Mannitol to *Lactobacillus plantarum* Degradation in Green Beans Fermented with *Lactobacillus cellobiosus*," *J. Food Sci.* 48:972–974, 981 (1983b).

Christ, C., J. M. Lebeault, and C. Noel: Apparatus for Ensilaging and Fermenting, U. S. Patent 4,293,655, 1981.

Corey, K. A., D. M. Pharr, and H. P. Fleming: "Pressure Changes in Oxygen-Exchanged, Brined Cucumbers," *J. Am. Soc. Hort. Sci.* 108:61–65 (1983).

Costilow, R. N., C. L. Bedford, D. Mingus, and D. Black: "Purging of Natural Salt-Stock Pickle Fermentations to Reduce Bloater Damage," *J. Food Sci.* 42:234–240 (1977).

Costilow, R. N., F. M. Coughlin, E. K. Robbins, and W. T. Hsu: "Sorbic Acid as a Selective Agent in Cucumber Fermentations. II. Effect of Sorbic Acid on the Yeast and Lactic Acid Fermentations in Brined Cucumbers," *Appl. Microbiol.* 5:373–379 (1957).

Cruess, W. V.: *Pickling Green Olives*, California Agricultural Experiment Station Bulletin 498, 1930.

Cruess, W. V.: "Use of Starters for Green Olive Fermentations," *Fruit Prods. J.* 17:12 (1937).

Daeschel, M. A., and H. P. Fleming: "Entrance and Growth of Lactic Acid Bacteria in Gas-Exchanged, Brined Cucumbers," *Appl. Environ. Microbiol.* 42:1111–1118 (1981).

Daeschel, M. A., H. P. Fleming, and R. F. McFeeters: "Mixed Culture Fermentation of Cucumber Juice with *Lactobacillus plantarum* and Yeasts," *J. Food Sci.* 53:862–864, 888 (1988).

Daeschel, M. A., H. P. Fleming, and E. A. Potts: "Compartmentalization of Lactic Acid Bacteria and Yeasts in the Fermentation of Brined Cucumbers," *Food Microbiol.* 2: 77–84 (1985).

Daeschel, M. A., R. F. McFeeters, H. P. Fleming, T. R. Klaenhammer, and R. B. Sanozky: "Mutation and Selection of *Lactobacillus plantarum* Strains That Do Not Produce Carbon Dioxide from Malate," *Appl. Environ. Microbiol.* 47:419–420 (1984).

DuPlessis, L. de W.: "The Microbiology of South African Winemaking. VII. Degradation of Citric Acid and L-Malic Acid by Lactic Acid Bacteria from Dry Wines," *S. Afr. J. Agric. Sci.* 7:31–42 (1964).

Duran Quintana, M. C., P. Garcia Garcia, and A. Garrido Fernandez: "Fermentation in an Aerobic Medium of Natural Black Olives with Alternating Air Injection. Study of the Effect of Calcium Chloride on Texture of Fruits," *Grasas y Aceites* 37:242–249 (1986).

Duran Quintana, M. C., A. Garrido Fernandez, F. Gonzalez Cancho, and M. J. Fernandez Diez: "Ripe Black Olives in Brine. I. Physical Chemical and Microbiological Study of Fermentation," *Grasas y Aceites* 22:167–177 (1971).

Duran Quintana, M. C., and F. Gonzalez Cancho: "Responsible Yeasts for the Fermentation Process of Naturally Black Olives in Brine," *Grasas y Aceites* 28:181–187 (1977).

Eden-Waren, GmbH.: *Lactobacillus bavaricus*, Its Production in Pure Culture, and Use in Lactic Acid Fermentation of Plant Material, German Patent 2,440,516, 1976.

Etchells, J. L., and T. A. Bell: "Classification of Yeasts from the Fermentation of Commercially Brined Cucumbers," *Farlowia* 4:87–112 (1950).

Etchells, J. L., T. A. Bell, and R. N. Costilow: Pure Culture Fermentation Process for Pickled Cucumbers, U. S. Patent 3,403,032, 1968a.

Etchells, J. L., T. A. Bell, H. P. Fleming, R. E. Kelling, and R. L. Thompson: "Suggested Procedure for the Controlled Fermentation of Commercially Brined Pickling Cucumbers—the Use of Starter Cultures and Reduction of Carbon Dioxide Accumulation," *Pickle Pak Sci.* 3:4–14 (1973).

Etchells, J. L., T. A. Bell, and I. D. Jones: "Morphology and Pigmentation of Certain Yeasts from Brines and the Cucumber Plant," *Farlowia* 4:265–304 (1953).

Etchells, J. L., T. A. Bell, and I. D. Jones: "Cucumber Blossoms in Salt Stock Mean Soft Pickles," *Res. and Farm.* 13:14–15 (1955).

Etchells, J. L., T. A. Bell, R. J. Monroe, P. M. Masley, and A. L. Demain: "Populations and Softening Enzyme Activity of Filamentous Fungi on Flowers, Ovaries, and Fruit of Pickling Cucumbers," *Appl. Microbiol.* 6:427–440 (1958).

Etchells, J. L., A. F. Borg, and T. A. Bell: "Influence of Sorbic Acid on Populations and Species of Yeasts Occurring in Cucumber Fermentations," *Appl. Microbiol.* 9:139–144 (1961).

Etchells, J. L., A. F. Borg, and T. A. Bell: "Bloater Formation by Gas-Forming Lactic Acid Bacteria in Cucumber Fermentations," *Appl. Microbiol.* 16:1029–1035 (1968b).

Etchells, J. L., A. F. Borg, I. D. Kittel, T. A. Bell, and H. P. Fleming: "Pure Culture Fermentation of Green Olives," *Appl. Microbiol.* 14:1027–1041 (1966).

Etchells, J. L., R. N. Costilow, T. E. Anderson, and T. A. Bell: "Pure Culture Fermentation of Brined Cucumbers," *Appl. Microbiol.* 12:523–535 (1964).

Etchells, J. L., F. W. Fabian, and I. D. Jones: *The Aerobacter Fermentation of Cucumbers During Salting*, Michigan State Agricultural Experiment Station Technical Bulletin 200, 1945.

Etchells, J. L., H. P. Fleming, and T. A. Bell: "Factors Influencing the Growth of Lactic Acid Bacteria During Brine Fermentation of Cucumbers," in J. G. Carr, C. V. Cutting, and G. C. Whiting (eds.), *Lactic Acid Bacteria in Beverages and Food*, Academic Press, New York, 1975, pp. 281–305.

Etchells, J. L., and I. D. Jones: "Bacteriological Changes in Cucumber Fermentation," *Food Ind.* 15:54–56 (1943).

FAO/WHO: *Specifications for the Identity and Purity of Food Additives and Their Toxicological Evaluation: Some Antimicrobials, Antioxidants, Emulsifiers, Stabilizers, Flour-Treatment Agents, Acids, and Bases*, 9th Report Joint FAO/WHO Expert Committee, Food Additives, Rome, Italy, 1966.

Fernandez Diez, M. J.: "Olives," in G. Reed (ed.), *Biotechnology. A Comprehensive Treatise in 8 Volumes*, vol. 5, Verlag Chemie, Deerfield Beach, Florida, 1983, pp. 379–397.

Fernandez Diez, M. J., R. De Castro y Ramos, A. Garrido Fernandez, F. Gonzalez Cancho, F. Gonzalez Pelliso, M. Nosti Vega, A. Heredia Moreno, M. I. Minguez Mosquera, L. Rejano Navarro, M. Del C. Duran Quintana, D. F. Sanchez Roldan, P. Garcia Garcia, and A. C. Gomez-Millan: *Biotecnologia De La Aceituna De Mesa*, Instituto De La Grasa Y Sus Derivados, Consejo Superior De Investigaciones Cientificas, Seville, Spain, 1985, 475 pp.

Fleming, H. P.: "Fermented Vegetables," in A. H. Rose (ed.), *Economic Microbiology. Fermented Foods*, vol. 7, Academic Press, New York, 1982, pp. 227–258.
Fleming, H. P.: "Developments in Cucumber Fermentation," *J. Chem. Tech. Biotechnol.* 34B:241–252 (1984).
Fleming, H. P., M. A. Daeschel, R. F. McFeeters, and M. D. Pierson: "Butyric Acid Spoilage of Fermented Cucumbers," *J. Food Sci.* 54:636–639 (1989).
Fleming, H. P., and J. L. Etchells: "Occurrence of an Inhibitor of Lactic Acid Bacteria in Green Olives," *Appl. Microbiol.* 15:1178–1184 (1967).
Fleming, H. P., E. G. Humphries, and J. A. Macon: "Progress on development of an anaerobic tank for brining of cucumbers. *Pickle Pak Sci.* VII:3–15 (1983a).
Fleming, H. P., R. F. McFeeters, M. A. Daeschel, E. G. Humphries, and R. L. Thompson: "Fermentation of Cucumbers in Anaerobic Tanks," *J. Food Sci.* 53:127–133 (1988a).
Fleming, H. P., R. F. McFeeters, and E. G. Humphries: "A Fermentor for Study of Sauerkraut Fermentation," *Biotech. Bioeng.* 31:189–197 (1988*b*).
Fleming, H. P., R. F. McFeeters, and R. L. Thompson: "Effects of Sodium Chloride Concentration on Firmness Retention of Cucumbers Fermented and Stored with Calcium Chloride," *J. Food Sci.* 52:653–657 (1987).
Fleming, H. P., R. F. McFeeters, R. L. Thompson, and D. C. Sanders: "Storage Stability of Vegetables Fermented with pH Control," *J. Food Sci.* 48:975–981 (1983*b*).
Fleming, H. P., and W. R. Moore, Jr.: "Pickling," in G. Fuller and G. G. Dull, "Processing of Horticultural Crops in the United States," I. A. Wolff (ed.), *CRC Handbook of Processing and Utilization in Agriculture*, vol. II, part 2: *Plant Products*, CRC Press, Boca Raton, Florida, 1983, pp. 397–463.
Fleming, H. P., D. M. Pharr, and R. L. Thompson: "Brining Properties of Cucumbers Exposed to Pure Oxygen Before Brining," *J. Food Sci.* 45:1578–1582 (1980).
Fleming, H. P., R. L. Thompson, T. A. Bell, and L. H. Hontz: "Controlled Fermentation of Sliced Cucumbers," *J. Food Sci.* 43:888–891 (1978).
Fleming, H. P., R. L. Thompson, J. L. Etchells, R. E. Kelling, and T. A. Bell: "Bloater Formation in Brined Cucumbers Fermented by *Lactobacillus plantarum*," *J. Food Sci.* 38:499–503 (1973a).
Fleming, H. P., R. L. Thompson, J. L. Etchells, R. E. Kelling, and T. A. Bell: "Carbon Dioxide Production in the Fermentation of Brined Cucumbers," *J. Food Sci.* 38:504–506 (1973b).
Fleming, H. P., W. M. Walter, Jr., and J. L. Etchells: "Antimicrobial Properties of Oleuropein and Products of Its Hydrolysis from Green Olives," *Appl. Microbiol.* 26: 777–782 (1973*c*).
Garcia Garcia, P., M. C. Duran Quintana, and A. Garrido: "Modifications in the Natural Fermentation Process of Black Olives in Order to Avoid Spoilage," *Grasas y Aceites* 33:9–17 (1982).
Garcia Garcia, P., M. C. Duran Quintana, and A. Garrido Fernandez: "Aerobic Fermentation of Natural Black Olives in Brine," *Grasas y Aceites* 36:14–20 (1985).
Garrido Fernandez, A., and R. H. Vaughn: "Utilization of Oleuropein by Microorganisms Associated with Olive Fermentations," *Can. J. Microbiol.* 24:680–684 (1978).
Gililland, J. R., and R. H. Vaughn: "Characteristics of Butyric Acid Bacteria from Olives," *J. Bact.* 46:315–322 (1943).
Gonzalez Cancho, F., M. Nosti Vega, M. C. Duran Quintana, A. Garrido Fernandez, and M. J. Fernandez Diez: "The Fermentation Process of Ripe Black Olives in Brine," *Grasas y Aceites* 26:297–309 (1975).
Jones, I. D., and J. L. Etchells: "Physical and Chemical Changes in Cucumber Fermentations," *Food Indus.* 15:62–64 (1943).
Juven, B., Z. Samish, Y. Henis, and B. Jacoby: "Mechanism of Enhancement of Lactic Acid Fermentation of Green Olives by Alkali and Heat Treatments," *J. Appl. Bact.* 31:200–207 (1968).
Kandler, O.: "Microbiology of Fermented Vegetables," *Ferm. Lebensm.* 23:3–19 (1981).
Kandler, O.: "Carbohydrate Metabolism in Lactic Acid Bacteria," *Antonie van Leeuwenhoek* 49:209–224 (1983).
Kawatomari, T., and R. H. Vaughn: "Species of *Clostridium* Associated with Zapatera Spoilage of Olives," *Food Res.* 21:481–490 (1956).
Keipper, C. H., E. B. Fred, and W. H. Peterson: "Microorganisms on Cabbage and Their

Partial Removal by Water for the Making of Sauerkraut," *Zent. fur Bakteriol., Parasit. und Infek.* II., 1932.
Lodder, J.: *The Yeasts, A Taxonomic Study*. North-Holland Publishing Co., London, England, 1970.
McFeeters, R. F., T. A. Bell, and H. P. Fleming: "An Endo-polygalacturonase in Cucumber Fruit," *J. Food Biochem.* 4:1–16 (1980).
McFeeters, R. F., H. P. Fleming, and R. L. Thompson: "Malic and Citric Acids in Pickling Cucumbers," *J. Food Sci.* 47:1859–1861, 1865 (1982a).
McFeeters, R. F., H. P. Fleming, and R. L. Thompson: "Malic Acid as a Source of Carbon Dioxide in Cucumber Juice Fermentations," *J. Food Sci.* 47:1862–1865 (1982b).
Meneley, J. C., and M. E. Stanghellini: "Detection of Enteric Bacteria Within Locular Tissue of Healthy Cucumbers," *J. Food Sci.* 39:1267–1268 (1974).
Mheen, Tae-Ick, and Tai-Wan Kwon: "Effect of Temperature and Salt Concentration on Kimchi Fermentation," *Kor. J. Food Sci. Technol.* 16:443–450 (1984).
Mrak, E. M., R. H. Vaughn, M. W. Miller, and H. J. Phaff: "Yeasts Occurring in Brines During the Fermentation and Storage of Green Olives," *Food Technol.* 10:416–419 (1956).
Mundt, J. O., W. F. Graham, and I. E. McCarty: "Spherical Lactic Acid-Producing Bacteria of Southern-Grown Raw and Processed Vegetables," *Appl. Microbiol.* 15:1303–1308 (1967).
Noel, C., A. M. Deschamps, and J. M. Lebeault: "Sauerkraut Fermentation: New Fermentation Vat," *Biotech. Let.* 1:321–326 (1979).
Pederson, C. S., and M. N. Albury: "The Effect of Pure Culture Inoculation on Fermentation of Cucumbers," *Food Technol.* 15:351–354 (1961).
Pederson, C. S., and M. N. Albury: *The Sauerkraut Fermentation*, New York State Agricultural Experiment Station Bulletin 824, 1969.
Plastourgos, S., and R. H. Vaughn: "Species of Propionibacterium Associated with Zapatera Spoilage of Olives," *Appl. Microbiol.* 5:267–271 (1957).
Potts, E. A. and H. P. Fleming: "Changes in Dissolved Oxygen and Microflora During Fermentation of Aerated, Brined Cucumbers," *J. Food Sci.* 44:429–434 (1979).
Potts, E. A., H. P. Fleming, R. F. McFeeters, and D. E. Guinnup: "Equilibration of Solutes in Nonfermenting, Brined Pickling Cucumbers," *J. Food Sci.* 51:434–439 (1986).
Rejano Navarro, L., F. Gonzalez Cancho, and J. M. R. de la Borbolla y Alcala: "Formation of Propionic Acid During the Conservation of Table Green Olives. II.," *Grasas y Aceites* 29:203–210 (1978).
Rodwell, A. W.: "The Occurrence and Distribution of Amino Acid Decarboxylases Within the Genus *Lactobacillus*," *J. Gen. Microbiol.* 8:224–232 (1953).
Ruiz-Barba, J. L., R. M. Rios-Sanchez, C. Fedriani-Iriso, J. M. Olias, J. L. Rios, and R. Jimenez-Diaz: "Bactericidal Effect of Phenolic Compounds from Green Olives on *Lactobacillus plantarum*," *System. Appl. Microbiol.* 13:199–205 (1990).
Samish, Z. and D. Dimant: "Bacterial Population in Fresh, Health Cucumbers," *Food Manuf.* 34:17–20 (1959).
Samish, Z., R. Etinger-Tulczynskza, and M. Bick: "The Microflora Within the Tissue of Fruits and Vegetables," *J. Food Sci.* 28:259–266 (1963).
Stamer, J. R.: "Lactic Acid Fermentation of Cabbage and Cucumbers," in H. J. Rehm and G. Reed (eds.), *Biotechnology*, vol. 5, Verlag Chemie, Weinheim, 1983, pp. 365–378.
Stamer, J. R.: "Lactic Acid Bacteria in Fermented Vegetables," in R. K. Robinson (ed.), *Developments in Food Microbiology*, vol. 3, Elsevier Applied Science Publishers, New York, 1988, pp. 67–85.
Stetter, H., and K. O. Stetter: "*Lactobacillus bavaricus* Sp. Nov., a New Species of the Subgenus *Streptobacterium*," *Zentralbl. Bakteriol. Parasitenkd. Infektionkr. Hyg. Abt.* 1:Org. Reihe C; 70, 1980.
Vaughn, R. H.: "Lactic Acid Fermentation of Cucumbers, Sauerkraut, and Olives," in *Industrial Fermentations*, vol. II, Chemical Publishing, New York, 1954, pp. 417–479.
Vaughn, R. H.: "Lactic Acid Fermentation of Cabbage, Cucumbers, Olives, and Other Produce," in G. Reed (ed.), *Industrial Microbiology*, 4th ed., Avi Publishing, Westport, Connecticut 1982, pp. 185–236.

Vaughn, R. H., H. C. Douglas, and J. R. Gililland: *Production of Spanish-Type Green Olives*, University of California Agricultural Experiment Station Bulletin 678, 1943.
Vaughn, R. H., T. Jakubczyk, J. D. MacMillan, T. E. Higgins, V. A. Dave, and V. M. Crampton: "Some Pink Yeasts Associated with Softening of Olives," *Appl. Microbiol.* 18:771–775 (1969a).
Vaughn, R. H., J. H. Levinson, C. W. Nagel, and P. H. Krumperman: "Sources and Types of Aerobic Microorganisms Associated with the Softening of Fermenting Cucumbers," *Food Res.* 19:494–502 (1954).
Vaughn, R. H., M. H. Martin, K. E. Stevenson, M. G. Johnson, and V. M. Crampton: "Salt-Free Storage of Olives and Other Produce for Future Processing," *Food Technol.* 23:124–126 (1969b).
Vaughn, R. H., K. E. Stevenson, B. A. Dave, and H. C. Park: "Fermenting Yeasts Associated with Softening and Gas-Pocket Formation in Olives," *Appl. Microbiol.* 23: 316–320 (1972).
Vaughn, R. H., W. D. Won, F. B. Spencer, D. Pappagianis, I. O. Foda, and P. H. Krumperman: "*Lactobacillus plantarum*, the Cause of 'Yeast Spots' on Olives," *Appl. Microbiol.* 1:82–85 (1953).
Walter, W. M., Jr., H. P. Fleming, and J. L. Etchells: "Preparation of Antimicrobial Compounds by Hydrolysis of Oleuropein from Green Olives," *Appl. Microbiol.* 26: 773–776 (1973).
West, N. S., J. R. Gililland, and R. H. Vaughn: "Characteristic of Coliform Bacteria from Olives," *J. Baceteriol.* 41:341–352 (1941).
WHO: *Toxicological Evaluation of Some Food Additives Including Anticaking Agents, Antimicrobials, Antioxidants, Emulsifiers, and Thickening Agents*, WHO Food Addit. Ser. 5, 1974.

Chapter

5

Mixed Cultures in Dairy Fermentations

Mary Ellen Sanders

Consultant
7119 S. Glencoe Ct.
Littleton, Colorado

Historical Perspective

Dairy fermentation microbiology

Since ancient times, undefined mixtures of genera, species, strains, and variants have been integral to the fermentation of milk. The growth of this mixed flora not only serves to preserve the milk, primarily by the production of lactic acid from lactose, but also results in a variety of metabolic end products and enzyme activities which give distinction to the vast assortment of fermented milks and cheeses available today.

Historically, these mixed flora developed from growth of natural contaminants of the milk. Today, genera and species of the microbial components of starters used for specific fermented milk products are mostly known, although many starters remain as undefined mixtures of strains of these bacteria. The genera and species used in dairy fermentations are listed in Table 5.1. Metabolic end products which contribute to flavor (or off-flavors) are also listed. Combinations of these microbes used to produce many traditional fermented milks and cheeses are shown in Table 5.2.

Although mixed cultures are truly the backbone of dairy fermentation microbiology, research on mixed and/or undefined cultures is difficult to conduct and interpret. This review will focus on interactions of mixed cultures.

TABLE 5.1 Microorganisms Used in Dairy Fermentations and Their End Products

Culture[a]	End products
Lactococcus lactis subsp. *lactis*	Lactic acid
Lactococcus lactis subsp. *cremoris*	Lactic acid
Lactococcus lactis subsp. *lactis* biovar. *diacetylactis*	Diacetyl, CO_2, acetaldehyde
Streptococcus salivarius subsp. *thermophilus*	Lactic acid, acetaldehyde
Lactobacillus	Lactic acid, acetic acid, fatty acids
Leuconostoc	CO_2, diacetyl, ethanol
Propionibacterium	CO_2, propionic acid
Yeasts, molds	Deacidifaction, lipolysis, CO_2, proteolysis

[a]Culture designations reflect recent taxonomy changes. For convenience, throughout this text cultures will be referred to as *Lactococcus lactis, Lactococcus cremoris, Lactococcus diacetylactis* and *Streptococcus thermophilus.*

TABLE 5.2 Mixed Cultures Used in the Production of Fermented Milks and Cheeses

	Culture for	
Product	Acid	Flavor, ripening, gas
	Cheeses	
Cheddar	*L. lactis, L. cremoris*	
Gouda, fontina	*L. lactis, L. cremoris*	*S. diacetylactis* *Leuconostoc cremoris*
Emmental, gruyere	*L. bulgaricus, S. thermophilus*	*Propionibacterium shermanii*
Roquefort, blue, gorgonzola, stilton	*L. lactis, L. cremoris*	*L. diacetylactis* *Leuconostoc cremoris*
Camembert, brie	*L. lactis, L. cremoris*	*Penicillium* sp. *L. diacetylactis*
Brick, limburger, bel paese	*L. lactis, L. cremoris, S. thermophilus*	*Brevibacterium linens*
	Fermented milks	
Yogurt	*L. bulgaricus, S. thermophilus*	
Acidophilus/bifidus milk	*L. acidophilus, Bifidobacterium* sp.	
Viili	*L. cremoris*[a]	
Buttermilk	*L. cremoris, L. lactis*	*L. diacetylactis*
Kefir, koumis, yakult	*Lactobacillus* sp.	*Saccharomyces* sp.

[a]Special ropy strains used for thickening.
SOURCE: Adapted from Roginski 1988, and Vedamuthu and Washam, 1983.

TABLE 5.3 Genetic Factors in Evolution of Lactic Acid Bacteria

Process	Description
Mutation	Small genetic change in DNA: inversion, deletion, addition of a base.
Transposons	Addition or deletion of transposon can lead to gene interruption; imprecise excision leaves anomalies in remaining DNA.
Transformation	Uptake of naked DNA; can be small or large changes due to such acquisition.
Transduction	Phage-mediated transfer of DNA.
Conjugation	Cell-to-cell transfer of genes.

Strain evolution

Dairy lactic acid bacteria probably adapted for growth in milk sometime after the domestication of mammals. The fact that wild-type strains of *Lactococcus cremoris* have never been found outside subcultured starter cultures is evidence for their specific adaptation to the milk environment. This adaptation was likely promoted by the many mechanisms which exist in dairy bacteria that mediate rapid genetic change (Table 5.3). These mechanisms, combined with unique selective pressures of the new environment, likely led to rapid evolution of these bacteria. Factors present in the dairy environment that promote genetic change include: milk as a novel substrate for growth, nutrient availability, phage infections, inhibitors in milk, competing flora, and temperature, salt, and moisture levels. Genetic systems in lactic acid bacteria have been recently reviewed.[1]

Conjugation has been reported for *Lactococcus* (McKay et al. 1980, Neve et al. 1981), *Leuconostoc* (Pucci et al. 1988), *Streptococcus* (Romero et al. 1987) and *Lactobacillus* (Gibson et al. 1979, Tannock 1987, and Thompson and Collins 1988) strains. Although many conjugation systems for lactic acid bacteria have been elucidated using antibiotic-resistant plasmids originating from non-lactic streptococci, conjugal transfer of traits inherent to lactic culture functionality has also been described, albeit primarily for the lactococci. These traits include lactose utilization (McKay et al. 1980), bacteriophage resistance (Klaenhammer 1984), and bacteriocin production and resistance (Klaenhammer 1988). Conjugation, which transfers DNA from a live donor cell to a live recipient cell, is likely responsible for both intra- and intergenus mixing of genes among the lactic acid bacteria. Genes have been conjugally passed between *Lactococcus* and *Lactobacillus helveticus* (Thompson and Collins, 1988); *Lactococcus* and *Lacto-*

[1] Davies and Gasson 1981, de Vos 1986, Fitzgerald and Gasson 1988, and Kondo and McKay 1985.

bacillus species *reuteri, fermentum*, and *murinus* (Tannock 1987); *Lactococcus* and *Leuconostoc* spp. (Pucci et al. 1988), *Pediococcus* to *Lactococcus* (Gonzales and Kunka 1983), as well as among different species of the same genus. The power of such cross-species gene transfer is theorized to promote rapid bursts in evolution (Syvanen 1985). Although gene transfer is known to occur in the laboratory on solid surfaces, it is possible that over time, fluid or coagulated milk could serve as a menstrum for natural conjugation.

The plethora of phages found associated with lactic acid bacteria suggest that phage-mediated transduction is also a mechanism for rapid genetic evolution. Transduction has been demonstrated in some lactic acid bacteria under laboratory conditions. A specialized transducing phage was shown to transfer the lactose-metabolism genes (Lac) or both Lac and proteolytic genes (Prt) in *Lactococcus lactis* C2 (McKay and Baldwin 1974). Transduction has also been demonstrated in *Lactobacillus* (Raya et al. 1989), and *S. thermophilus* (Mercenier et al. 1988).

The recent identification of insertion sequences in lactic acid bacteria has made evident yet another mechanism for rapid adaptation of these microbes in mixed-culture habitats. Shimizu-Kadota et al. (1985) isolated and sequenced ISL*1*, a 1256-bp insertion sequence from *Lactobacillus casei*. This element mediated the transition from the temperate to the virulent state of a prophage harbored by *L. casei*. A second insertion sequence ISSIS (808 bp) was found to mediate cointegration of the lactose plasmid and a conjugal plasmid in *Lactococcus lactis* ML3. Romero and Klaenhammer (1989) described the presence of a putative insertion sequence which resides on the conjugal phage-resistance plasmid pTR2030 and mediates conjugal mobilization of coresident plasmids. These examples suggest that illegitimate recombination mediated by transposition of insertion sequences can contribute to gene mixing among strains of lactic acid bacteria present in mixed-culture habitats.

Biological evidence exists today which suggests that the associations of naturally mixed starter cultures is a result of continued evolution since these microbes first invaded our milk supply. This evidence includes:

1. Demonstration of inter- and intraspecies gene transfer
2. Plasmid linkage of many genes of importance to growth and survival of lactic bacteria in milk (Kondo and McKay 1985)
3. A continuous spectrum of genetic variation among strains, so much that definitive distinctions among some species is difficult
4. No wild-type sources known for some lactic acid bacteria, e.g., *Lactococcus lactis* subsp. *cremoris*.

Dairy starter culture development

The need for mixed starter cultures to consistently perform in fermented dairy products is perhaps more necessary today than ever. Rigorous standards of quality exist for products sold in today's market, and the microbiological components of the production contribute greatly to final quality. Much progress has been made in recent years in understanding and controlling the microflora in dairy fermentations. In some cases it is quite simple. For example, the starter culture contribution to cottage cheese is primarily lactic acid. In fact, direct acidification is widely used in the United States for cottage cheese manufacture to eliminate microbial inconsistencies. On the other hand, the production of aged, full-flavored cheeses requires an intricate balance of microflora to achieve a consistent, high-quality product. A general list of criteria for the selection or development of strains for today's market is shown in Table 5.4.

Modern strain development is approached by a variety of strategies:

1. Isolation of strains from environmental sources: traditional, undefined starter cultures; natural flora of milk; mammals; green plant material
2. Isolation of mutants from currently available strains
3. Directed genetic development of strains using recently elucidated technologies of conjugation, transduction, and transformation to introduce new genes or gene systems.

Programs to isolate new strains from the field have been the backbone of strain development efforts for years. To combat the constant threat of phage infections, reliance has also been placed on the isolation of phage-resistant mutants from otherwise well-performing cultures. This can be done in an undefined manner where phage contamination of mixed, undefined starters is promoted, as in the Dutch cheese industry (Stadhouders and Leenders 1984), or in a defined manner where single strains are challenged with phage, and phage-resistant survivors are isolated and characterized (Limsowtin and Terzaghi 1974, Thunell et al. 1981). However, metabolic deficiencies of many phage-resistant mutants and the narrow spectrum of resistance often conferred by mutation have limited the usefulness of this approach. The directed genetic development of strains using powerful systems of gene transfer of naturally occurring plasmids or plasmids engineered through recombinant DNA work is an attractive possibility. Today massive efforts worldwide are being directed toward understanding the genetics, biochemistry, and molecular biology of dairy starter cultures. Applied on a commercial level, these efforts have already resulted in the use of conjugally constructed phage-resistant

TABLE 5.4 Characteristics of Dairy Starter Cultures

Trait	Comments
Acid-producing activity	Usually, the more acid, the better. But acid must be controllable. Some starters are not profuse acid producers (*Propionibacteria, Leuconostoc*). Best evaluation is under conditions (temp, salt, substrate) of final use.
Bacteriophage resistance	The more, the better. Perhaps the most limiting trait of cultures.
Temperature sensitivity	Depends on product and make procedure. Can be used to control culture activity; temperature of make is important to cheese properties, therefore limited ranges are often used. Cultures should not produce acid under refrigeration conditions (prevents postacidification of yogurt, cottage cheese, etc.)
Salt tolerance	Significant NaCl levels in cheese requires salt-tolerant organisms for ripening. Salt-sensitive strains may enable better control of the final pH of some cheeses.
Consistency in make	Essential that culture perform consistently day to day and not be overly sensitive to minor fluctuations in make procedure.
Flavor	Strains should produce desired flavor compounds in proper ratios to other metabolites (diacetyl, acetaldehyde, ethanol, CO_2, lactic acid, etc.)
Off-flavor	Bitter, green, malty, defects in cheese are strain-related.
Texture/Body	Production of exopolysaccharides can impart a thickness or ropiness to fermented milks and yogurt, a desirable trait in some applications. Gas production and proteolytic ability can also contribute to the openness or texture of cheese.
Proteolysis	Related to bitterness and ability of strains to grow in milk.
Compatibility	Lack of inhibitor production is preferred for strains used in blends. Bacteriocin or inhibitor production is desirable for cultures applied to extend product shelf life after manufacture.
Sensitivity to phosphates	Phosphates are widely used to buffer and chelate calcium in bulk starter media. This helps guarantee an active, phage-free bulk starter. Strains sensitive to phosphates cannot be used in this application.
Agglutination	Important in cultures used in nonrennet applications (e.g., cottage cheese). Tendency for culture to agglutinate should be low.
Strain uniqueness	Important, primarily for phage resistance and to prevent duplication of efforts during strain characterization. DNA (plasmid or chromosomal) analyses give best strain differentiation, especially in the absence of any distinctive phenotype.

strains (Sanders et al. 1986) and hold promise for development of improved or novel strains for a variety of other applications. Application of certain of these technologies (e.g., recombinant DNA techniques), however, will require government approval. This is a time-consuming, costly effort with no guarantee of approval and ultimately of consumer acceptance. Certainly, scientific advances have provided an abundance of genetic tools and information on genes of interest that make strides in strain development possible.

Mixed Dairy Cultures

Genera

As was shown in Table 5.1, a variety of genera are commonly used in the manufacture of cheese and fermented milk. As a rule, the more important aging and ripening is to a cheese, the greater the complexity of flora associated with that cheese. Choisy et al. (1987) discussed the microbiological aspects of cheese ripening. Different genera involved in cheesemaking include *Lactococcus, Lactobacillus, Streptococcus, Leuconostoc, Propionibacterium, Micrococcus, Brevibacterium, Penicillium, Geotrichum*, and a variety of yeasts. Some of these are added deliberately to cheese during manufacture; others are indirect contaminants. *Pediococcus, Acetobacter*, and yeasts are also used in some fermented milk production (Rozinski 1988). In products such as fermented milk where processing steps and aids (enzymes, salting, cooking, whey expulsion) are minimal, the starter culture flora is the most important factor dictating the type and character of the final product. Each of these genera offers its own contributions to the character of the final product. In addition to the metabolic end products listed in Table 5.1 for each genus, the action of these microbes in milk-cheese systems is dependent on salt tolerance, ranges of growth temperature, requirement for or stimulation by peptides from proteolysis, oxygen availability, availability of substrates for growth, pH tolerance, and compatibility with other strains.

In general, the genera of dairy starter cultures are distinct, and their role is suitably specific that in practice alternative genera are not interchanged. One exception to this may be the choice of *Leuconostoc mesenteroides* subsp. *cremoris (Leuconostoc cremoris)* or *Lactococcus lactis* subsp. *lactis* biovar. *diacetylactis (L. diacetylactis)*. The predominant function of both of these organisms is the production of diacetyl and CO_2. Usually, one or both are combined with suitable fast acid-producing mesophilic strains of *Lactococcus lactis*. *Leuconostoc cremoris* cannot acidify milk in pure culture (Cogan 1985) and frequently isolates of *L. diacetylactis* are proteolytically deficient

(M. E. Sanders, unpublished observations). Both of these microbes produce diacetyl and CO_2 from citrate, and this activity is pH-dependent (Cogan 1985, Sandine 1985). In addition, acetaldehyde can be produced by lactic acid bacteria (Lees and Jago 1978a, b). Accumulation of acetaldehyde is dependent primarily on the enzymatic capability of the cells to reduce acetaldehyde to other compounds (Lees and Jago 1978a, b). Although acetaldehyde is an important flavor compound in some products, it can also result in a "green" off-flavor if present in unbalanced quantities (Lees and Jago 1978b). *L. cremoris* more readily converts acetaldehyde to ethanol than *L. diacetylactis*, and its use can avoid a flavor defect resulting from too much acetaldehyde (Sharpe 1979). *L. diacetylactis* produces copious CO_2 and may be the choice for certain applications. Although seemingly quite similar, subtleties may dictate the choice of culture containing one or the other of these genera. It should be mentioned however, that different strains within each genus may have quite different metabolic patterns, and strain selection may be more important than genus selection.

Species, subspecies

Different species and subspecies are often found together in mixed starter cultures, and are often deliberately blended to provide subtle flavor qualities or to provide consistent acid production. For example, *Lactococcus lactis* subsp. *cremoris* is not likely to produce a bitter defect in aged cheese. However, *Lactococcus lactis* subsp. *lactis* is on the whole a faster acid-producing subspecies. Therefore, sometimes blends of these subspecies can be used to balance good flavor with rate of acid production. Other examples are found with different species of *Lactobacillus*. As a genus, *Lactobacillus* is quite diverse. It includes obligate homofermentatives (*L. delbrueckii* subsp. *bulgaricus* and subsp *lactis, L. helveticus*, and *L. acidophilus*), facultative, heterofermentative (*L. casei*) and obligate heterofermentative (*L. brevis, L. fermentum*, and *L. kefir*) members. Each of these can be used in the manufacture of some fermented milks and cheeses. Most of these are thermophilic but some can grow in mesophilic ranges (Rozinski 1988). The choice among these species of *Lactobacillus* is based on the desirability of specific metabolic end products and the temperature at which milk processing is conducted. For example, *L. acidophilus* may be used by those who believe it enhances favorable gut interactions; however, this species is considered too acidogenic to be used for a flavorable yogurt in most countries. The ability of *L. helveticus* to metabolize galactose and the failure of *L. bulgaricus* to do this (Hickey et al. 1986) may influence the choice of *L. helveticus* strains for incorporation into mozarella cheese where browning can be a defect upon cooking.

Strains

Strain-to-strain variabilities within species can be remarkably broad. Occhino et al. (1986) examined six strains of *Streptococcus thermophilus* for β-galactosidase (β-gal) activity and found a range of 0 to 58 units per milligram of protein. Of nine strains of *L. lactis*, β-gal activity ranged from 1 to 73 units per milligram of dry cell weight. The range for *L. cremoris* strains was 1 to 9, and was 1 to 13 for *L. diacetylactis* strains (McKay et al. 1980). Phospo-β-galactosidase activity was also determined for these same strains. Ranges of 5 to 172, 25 to 246, and 71 to 269 units per milligram of dry cell weight were found (McKay et al. 1980). Five strains of *L. casei* differed in their relative activities of β-galactosidase and P-β-galactosidase (ratios ranged from 1.4 to 6) (Jimeno et al. 1984). Galactose released into the growth medium from lactose varied from 26.5 to 33.6 μmole/μL for four *L. bulgaricus* strains, and 0.3 to 1.8 μmole/μL for six *L. helveticus* strains (Hickey et al. 1986). Autolytic activity of group N streptococci was assayed by comparison of maximum to minimum absorbances when cells were grown and stored in M17 broth for up to 2 weeks. Six *S. cremoris* strains lysed 16 to 79 percent, four *S. lactis* 24 to 87 percent, and four *S. diacetylactis* 24 to 89 percent (Langsrud et al. 1987). Thirty-six *S. thermophilus* strains showed temperature maxima that varied from 39.3 to 46.1°C (Martley 1983). Strain-to-strain variability among lactococci also exists in plasmid patterns (Fujita et al. 1984, Pechmann and Teuber 1980), chromosomal DNA digest patterns (Le Bourgeois et al. 1989), plasmid linkage of traits (Kondo and McKay 1985), phage susceptibility,[2] resistance to physiological stress (Yang and Sandine 1979), ability to be concentrated by centrifugation, receptiveness to conjugation, transduction and transformation (Kondo and McKay 1985), and many other properties.

It is evident that a broad range can exist among strains of the same species with regard to many different properties. These differences can be exploited in making proper strain selections, and may be important in the evolution of undefined mixed-strain cultures.

Mutants

Perhaps the most widespread use of mutants in dairy microbiology is for bacteriophage resistance (or insensitivity). Culture systems have

[2] Accolas and Spillmann 1979, Daniell and Sandine 1981, Krusch et al. 1987, Mata et al. 1986, Terzaghi 1976.

been based on application of such mutants (Hoier and Jorck-Ramberg 1988, Thunell et al. 1981), although phage-resistant mutants may lose parental fermentative properties (Hull 1977, Limsowtin and Terzaghi 1977, Marshall and Berridge 1976). It is interesting that King et al. (1983) showed that mutation to phage resistance and slow acid production were independent genetic events. Although the above approaches describe the use of selected and purified phage-resistant mutants, an alternative approach relies on the outgrowth of phage-resistant strains and mutants from a mixed undefined culture challenged with phage-containing whey from current manufacture (Stadhouders and Leenders 1984). In natural, undefined cultures, mutants are likely to play a significant role, as a variety of selective pressures and genetic processes serve to favor growth of resistant variants over others.

Less effort has been placed on isolation of mutants for properties other than bacteriophage resistance. Perhaps this is because it is easier to screen for desired strains than to deliberately mutagenize to obtain them. Again, concern over disruption of other desirable traits probably contributes to this practice.

Prt^- (Richardson et al. 1983) or mixtures of Prt^+ and Prt^- (Stadhouders et al. 1988) variants of lactococci have been used in the manufacture of cheddar cheese. Benefits from this practice were reported to be decreased phage sensitivity, decreased bitter flavor, and increased yield of the cheese. Some natural, undefined cultures contain a high proportion of Prt^- strains, and cultures containing nearly 2 to 20 percent Prt^+ are often indistinguishable in acid-producing capacity from the 100 percent Prt^+ cultures (Stadhouders et al. 1988).

Lac^- mutants were used for accelerating cheddar cheese ripening (Grieve and Dulley 1983, Grieve et al. 1983). Enhanced levels of intracellular enzymes could be delivered to the cheese by the Lac^- biomass, and the acid profiles of the cheesemake would not be disturbed. Although an interesting concept, the data were not statistically evaluated, and therefore no conclusion could be made regarding effects on flavor development.

Types of Interactions

It is clear that dairy starter cultures are central to the manufacture of fermented milks and cheeses. The following discussion will focus in more detail on the biological nature of these interactions. Hugenholtz (1986) presented an excellent review of this subject specific to dairy fermentations, and articles by Slater (1981), Steinkraus (1982) and Harrison (1978) discuss the topic more generally.

Microbial interactions can be defined (Table 5.5). Examples of competition, parasitism, amensalism, commensalism, and mutualism exist among dairy microbes, and are discussed in the following sections.

Competition

Competition is a prevalent microbial interaction in dairy cultures. Competition exists among two or more populations that are limited by a common factor. The individual with the fastest specific growth rate can often outcompete other individuals in the population. However, the presence of a variety of mutants, strains, species, and genera that coexist in starters maintained by traditional ongoing transfer (Hugenholtz 1986) suggest that competition among these mixed dairy starters does not result in the domination of one homogeneous population. During growth in milk, conditions change (temperature, concentration of organic acids, pH, amino acid availability) such that spe-

TABLE 5.5 Microbial Interactions

Type	Description	Stability	Example
Competition	Two or more populations limited by a common factor (eg. nutrient, environmental conditions)	Often the fastest grower takes over the population	Lac^- variants in Lac^+ population in milk
Parasitism	One population whose survival is at the expense of a host population	Survival of parasite is dependent on survival of a population of host, therefore can be quite stable	Phage or bacteria (lysogenic or lytic)
Amensalism	Growth of one microbe interferes with and inhibits growth of another (e.g., toxic metabolites, bacteriocins, etc.)	"Aggressor" normally predominates population	Nisin-producing *S. lactis*; lactic acid bacteria vs. pathogens
Commensalism	One population benefits from growth of another population	Stable	Prt^- population can feed on amino acids generated by Prt^+ growing in milk
Mutualism	Two populations benefit from growth of each other	Stable	*Lactobacillus bulgaricus*/ *S. thermophilus* growth in milk

cific growth rates of different populations change. The result is a mixed population in which one organism does not completely dominate. Some populations, however, unable to compete under most circumstances, would be dominated by other populations, and would never attain detectable levels. The fact that mixed cultures remain heterogeneous in dairy fermentations may also be a result of the diverse mechanisms of adaptation present in these microbes. Conjugal transfer of plasmids or chromosomal DNA, transposition of genes, phage-mediated gene transfer, and plasmid curing all may provide constant change and adaptation among these mixed cultures in response to changes in the competitive and environmental conditions. Although this may promote mixed populations, the specific composition of the mixture may shift as internal populations undergo genetic change. Cultures carried successively through many fermentations may not show extreme strain dominance, whereas cultures prepared in the lab from a mixture of defined single strains have been demonstrated to be dominated by one strain (Collins 1955).

Centralized production of starter cultures by culture houses emphasizes the importance of preservation of cultures and its effect on population balance. Prior to centralized preparation of starters, cultures were propagated from fresh ferment from the previous day's production. Methods to store and preserve culture integrity were developed, and focused primarily on dried, liquid, or frozen states. Developments were required to alleviate the stress of storage, to provide media, and to define growth conditions to maintain strain balance yet promote high population levels during production. These mixed cultures then faced new physiological stresses that changed the competitive picture. Populations more resistant to injury from freezing or drying gained a competitive advantage. Shifts in strain dominance during frozen storage of multiple-strain concentrated cultures was demonstrated. Also, resistance to low pH became less of a selective factor in light of the development of pH-controlled fermentors.

Little is published on true competitive interactions among dairy cultures. However, the occurrence of Lac$^-$ variants in Lac$^+$ populations is a good example of competition. Lac$^-$ derivatives are often found naturally in dairy cultures (McKay 1978). However, in a medium where lactose is essentially the sole carbohydrate (e.g., milk and certain laboratory media), Lac$^-$ derivatives never dominate the population. If transferred to a medium providing glucose, Lac$^-$ variants can compete with the Lac$^+$ population. In fact, it is even speculated that Lac$^-$ may out compete Lac$^+$ in glucose-containing media due to a decreased energy requirement for the maintenance of plasmids that encode lactose metabolism.

Parasitism

Perhaps no other biological factor plays such a dominant role in the dynamics of mixed-strain dairy culture populations as parasitism. Parasitism occurs when a predator population survives at the expense of a host population. This frequently occurs in the case of bacteriophage infection of starters. Although some genera or species may be less vulnerable than others [*Lactobacillus* is thought to be less frequently attacked and devastated than *S. thermophilus* by phage (Rajagopal and Sandine 1989)], dairy fermentations are plagued by many forms of phage attack. Phage can persist as lytic, lysogenic, or phage-carrying states (Sanders 1987). In each case, their presence is viewed as a threat to culture maintenance and performance.

The effects of phage on mixed-strain cultures has been examined in detail by Collins (1952a,b; 1955a–c). This series of papers elucidated several aspects of phage-host interaction for lactic streptococci as described below:

1. Strains not normally sensitive to a particular phage showed retarded acid development in the presence of large numbers of nonhomologous phage (Collins 1952a). No increase in viable phage was seen. Later, a mechanism for this action was proposed to be "lysis-from-without," whereby phages that adsorb but do not replicate on a specific host can cause cell death (Oram and Reiter 1968). This provided early evidence that the use of undefined, mixed-strain cultures may not be the best solution to phage attack. Collins (1952b) later said about the use of mixed cultures: "their use also has been found to complicate diagnosis of starter difficulty and to make the effect of bacteriophage action less predictable."
2. The relative contribution of phage-sensitive and phage-resistant strains to acid production in the presence of phage (Collins 1952b) was examined. With the assay strains used, Collins found that when the susceptible strain made up less than 50 percent of the inoculum, acid production was affected only slightly. But even in mixtures containing 90 to 95 percent sensitive bacteria, appreciable acid production could be obtained in the presence of phage (assuming the absence of the "nascent phenomenon"). These conclusions, however, are influenced by the lag phase and burst size of the phage and the acid-producing capabilities of the remaining resistant strains.
3. Relatedness of strains composing several commercial mixed starter cultures (Collins 1955c) was determined. Collins established phage relationship tables and a whey activity test which distinguished

> strains with similar phage-sensitivity patterns. He suggested that this information be used to establish culture rotation programs. Such strategies are still in use today.

Many other investigators worldwide have addressed the phenomenon of phage infection of dairy starters. These efforts have focused on (1) description of the problem and (2) strategies for coping with the problem. The importance of the phage-starter strain interaction is evident from the many recent reviews that have addressed this topic.[3] Simply stated, the problem is that phage infection interferes with the function of a starter culture, be it acid production, production of flavor compounds, achieving sufficient populations to provide enzymes important for ripening, or inhibiting spoilage of milk. This problem is especially complex because the phage is not a static entity. Phage are capable of complex genetic and phenotypic modifications. Therefore, development of a strain resistant to the effects of one phage strain does not necessarily render it immune to another phage type. Furthermore, phage have evolved a variety of mechanisms to persist in the presence of their hosts: lysogeny, phage carrier state, or lytic state. Therefore, the management of such potentially destructive parasites is a continued concern for the starter culture technologist.

Phage control methods have recently been reviewed (Klaenhammer 1984, Sanders 1987) and therefore will not be discussed fully here. Focus of control primarily relies on:

1. Prevention of phage contamination by the use of well-designed cheese plants, aseptic bulk starter tanks, and phage-inhibitory bulk starter media. It should be noted, however, that some starter systems, especially those used in The Netherlands, allow deliberate contamination of the starter with phage with the idea that whatever grows in the starter will be resistant to current phages (Stadhouders and Leenders 1984).
2. Starter strain rotation. The use of phage-unrelated strains in sequence provides fresh strains on at least a daily basis that should be resistant to whatever phages have developed against preceding strains.
3. Use of strains selected or developed to have broad resistance to phage.

---

[3]Klaenhammer 1984, Lahbib-Mansais et al. 1988, Mata and Ritzenthaler 1988, Sanders 1988, Sechaud et al. 1988.

Amensalism

Amensalism refers to the inhibition of the growth of one population by a second population. This is usually mediated by the production of toxic metabolites or specific antagonistic compounds, such as bacteriocins. A result of amensalism is that one population can dominate or exclude other populations. Lactic acid bacteria have come into prominence with regard to food preservation effects largely because of their amensalistic action on other microbial populations. This interaction, therefore, is very important in the action and understanding of lactic acid bacteria in food systems.

Prior to refrigeration, the principle means to preserve milk as a valuable food source was through the lactic fermentation. Evidence suggests that this method of preservation of milk has been in practice for 8000 to 9000 years. Antagonistic substances or conditions resulting from growth of the lactic acid bacteria leading to enhanced preservation of milk include: organic acids, primarily lactic, but also acetic and propionic; H_2O_2; low pH; reduced water activity (in the case of cheese, not fermented milks); lowered redox potential; diacetyl; and bacteriocins. Organic acids, low pH, and reduced Aw and redox potential are the principle inhibitory factors. In fact, the long history of safety of fermented dairy products is due to the inability of pathogens to grow in, or often to even survive, the environment of fermented milks and most cheeses.

Aside from the long history of the safety and preservation of fermented dairy products, many specific studies have demonstrated the antagonistic action of lactic acid bacteria against pathogens and spoilage organisms. Schaack and Marth (1988) and Speck (1972) have concisely reviewed the subject of pathogen inhibition. Babel (1977) reviewed more generally antibiosis of lactic culture bacteria. Some of the studies specifically directed toward pathogens and spoilage bacteria are summarized in Table 5.6. It is interesting that Frank and Marth (1977) in their study on inhibition of enteropathogenic *E. coli* by *Lactococcus lactis* and *cremoris* showed that at 21°C, an undefined mixed lactic culture was more inhibitory than the defined *lactis* and *cremoris* cultures. However, since the individual strains were not isolated from the mixed culture, direct comparisons could not be made.

It is important to note that even in light of the effectiveness of lactic culture inhibition of pathogens and spoilage of microbes, fermented dairy products do spoil and can serve as vehicles for foodborne disease. Pasteurization of milk has done much to reduce these outbreaks and enhance shelf stability of fermented dairy products. However, the continuing practice of raw milk use for cheesemaking in some countries continues to put consumers at risk (Brackett 1988). The recent linkage of *Listeria* to foodborne outbreaks from cheese consumption has caused a reexamination of the safety of cheese prepared from raw milk

TABLE 5.6 Examples of Inhibition of Pathogens and Spoilage Bacteria by Dairy Lactic Acid Bacteria

Inhibition culture	Inhibition demonstrated against	Test system	Mechanism of inhibition	Reference
Lactococcus lactis *Lactococcus cremoris*	EEC[a]	Skim milk, 21 + 32°C; 7°C storage	Lactic acid	Frank and Marth 1977
Leuconostoc citrovorum	*Salmonella gallinarum*	TGY medium	Acetic acid Lactic acid	Sorrells and Speck 1970
S. thermophilus *L. bulgaricus*	*Salmonella typhimurium*	Yogurt	Lactic acid	Rubin and Vaughan 1979
Lactic streptococci	*S. aureus*	Milk	Lactic acid —	Gilliland and Speck 1974, 1977
L. acidophilus	*Staph.* aureus *Salmonella typhimurium* EEC *Clostridium perfringens*	Broth	H_2O_2	Gilliland and Speck 1974, 1977
Lactococcus cremoris *Lactococcus lactis*	*Enterobacter cloacae* *Enterobacter aenogenes* *Hafnia* sp.	Skim milk	Lactic acid	Rutzinski and Marth 1980
Leuconostoc cremoris	*Pseudomonas fragi* *Pseudomonas putrefaciens* Coliforms	Cottage cheese Dressing		Babel 1977
Lactococcus diacetylactis *Leuconostoc cremoris*	*Pseudomonas fragi*	Disk assay on plates	Acetic acid, glyoxylic acid, malonic acid, α-ketoglutaric acid	Pinheiro et al. 1968
Lactococcus diacetylactis	*Pseudomonas fluorescens* *P. fragi* *P viscosa* *Alcaligenes metalcaligenes* *P. aeruginosa* *Clostridium perfringens* *E. coli* *Salmonella tennessee* *Vibrio parahemolyticus*	Skim milk, lactic broth	Acids —	Daly et al. 1970
L. cremoris *L. lactis*	*Listeria monocytogenes*[b]	Skim milk	—	Schaack and Marth 1988
Lactococcus sp.	*Campylobacter jejuni*	Cheddar and cottage cheese	Lactic acid	Ehlers et al. 1982

[a]EEC, enteropathogenic *E. coli*
[b]*Listeria* increased in numbers during fermentation under some inoculation and strain conditions; growth was pH-dependent.

and aged 60 days or more. *Listeria* has been shown to withstand this aging process in laboratory-prepared cheeses (Doyle 1988), and can be an especially significant threat in semisoft, surface-ripened cheese where the mold surface flora neutralize acids and bring the surface pH close to 6.0 (Doyle 1988).

In addition to the general antimicrobial nature of organic acids and H_2O_2 produced by lactic cultures in milk, more specific antimicrobials, bacteriocins, are also produced. Bacteriocins produced by lactic acid bacteria have been thoroughly reviewed by Klaenhammer (1988) and are summarized in Table 5.7. Daeschel (1989) reviewed this subject relative to food applications. These proteinacious substances are chemically diverse, but normally exhibit inhibition against a narrow range of species closely related to the producer (Tagg 1976). This type of activity can be quite effective in establishing dominance among strains within mixed cultures, making the criterion of "compatibility" quite important to starter culture technologists when defining or designing new mixed cultures.

However, in some cases, bacteriocins have been shown to have broader ranges of activity and these activities are of great interest and show promise for commercial applications. The best studied and one that has become a commercial reality is nisin. Produced by some strains of *Lactococcus lactis*, the five forms of nisin show bactericidal activity against gram-positive bacteria in general (Hurst 1983), but not against yeast, fungi, or gram-negative organisms. *Lactococcus cremoris*, however, is sensitive to nisin, while some fecal streptococci are resistant (Hurst 1983). Nisin is useful commercially in acidic foods where gram-positive microbes need to be controlled, yet government approval has been slow in some countries [although Great Britain permitted its use in 1959 (Hurst 1983)]. Recently, nisin was approved for certain uses throughout Europe, India, and some African, Asian, and Latin American countries (Hurst 1983). Nisin was also approved for use in pasteurized processed cheese spreads in the United States (FDA 1988). The recent report of the gene sequence for nisin (Buchmann et al. 1988) has made possible the development through site-directed mutagenesis of a family of peptide antibiotics, potentially with different spectra of action or physical properties. This promising line of research may expand the numbers and functionality of nisin-derived peptide antibiotics.

Other bacteriocins which seem especially useful include those that have activity against *Listeria*. This pathogen's presence in milk and dairy products has made its control a foremost concern. Recently, bacteriocins from *Pediococcus*[4] and nisin (Benkerroum and Sandine 1988) have been shown to have antilisterial activity. These sub-

[4]Bhunia et al. 1988, Carminatl et al. 1989, Hoover et al. 1988, Pucci et al. 1988.

TABLE 5.7 Bacteriocins Isolated from Lactic Acid Bacteria[a]

Bacteriocin	Producer organism	Spectrum of activity	Reference
Diplococcin	*Lactococcus cremoris*	*Lactococcus*	Davey and Richardson 1981
Nisin	*Lactococcus lactis*	Gram$^+$ microbes	Hurst 1983
Pediocin A	*Pediococcus pentosaceus*	Gram$^+$ microbes	Daeschel and Klaenhammer 1985
PA-1	*Pediococcus acidilactici*	*Listeria monocytogenes* *Pediococcus* *Lactobacillus*	Pucci et al. 1988
Pediocin AcH	*Pediococcus acidilactici*	*Clostridium* *Lactobacillus* *Leuconostoc* *Listeria monocytogenes* *Staphylococcus aureus* *Aeromonas hydrophilia* *Bacillus cereus* *Pseudomonas putida*	Bhunia 1988
Lactocin 27	*Lactobacillus helveticus*	*L. acidophilus* *L. helveticus*	Upreti and Hindsdill 1973
Helveticin J	*Lactobacillus helveticus*	*L. helveticus* *Lactobacillus lactis*	Joerger and Klaenhammer 1986
Lactocin B	*L. acidophilus*	*L. leichmannii* *L. helveticus* *L. bulgaricus* *L. lactis*	Barefoot and Klaenhammer 1983
Lactocin F	*L. acidophilus*	*L. fermentum* *Streptococcus faecalis*	Muriana and Klaenhammer 1987
Plantaricin A	*L. plantarum*	*L. plantarum* *P. pentosaceus* *Leuconostoc paramasenteroides*	Daeschel et al. 1986
Plantacin B	*L. plantarum*	*L. plantarum* *Leuconostoc mesenteroides* *Pediococcus damnosus*	West and Warner 1988
Lactostrepcins	*L. lactis* *L. diacetylactis*	*Lactococcus* sp. gps. A, C, G, streptococci *Bacillus cereus* *Lactobacillus helveticus*	Kozak et al.

[a]Reports of unnamed and not fully characterized antagonistic activities from lactococci have indicated the potential for bacteriocins in addition to those listed above from these microbes. See Geis et al. 1983, Neve et al. 1984, Scherwitz et al. 1983.

stances, or the organisms producing them, are likely to gain acceptance for commercial application.

Preservative activities of dairy lactic cultures are also known, and have found use as cultures applied to refrigerated products. Raw milk inoculants are sold by several commercial companies to inhibit psychrotrophic growth. Microgard, a product of Wesman Foods, is marketed as an "all-natural" inhibitor of gram-negative bacteria, yeasts, and molds for products such as cottage cheese, sour cream, yogurt, and salad dressings. It is a pasteurized product of skim milk, presumably fermented by *Propionibacterium shermani* (Daeschel 1989). A similar application for Lac^- *Lactococcus diacetylactis* strains was proposed by Gonzales (1983) in a European patent application.

Cultured dairy products have been consumed not only for their taste and nutritional value but also for supposed secondary beneficial effects of some dairy cultures in vivo. These activities have been associated with putative antipathogenic action of certain *Lactobacillus* strains in the human intestine (Fernandes et al. 1988, Truslow 1986). Although some research, including treatment of patients suffering from irritable bowel syndrome with unfermented acidophilus milk, showed no improvement of symptoms (Newcomer 1983), other positive results (Fernandes et al. 1988) have spurred continued investigation. Therefore, the amensalistic behavior of the microbes used in some fermented dairy products may also be useful for some intestinal disease management in humans.

In summary, amensalism is a dominant interaction in dairy lactic cultures. It plays a key role in lactic culture dominance over pathogenic and spoilage organisms in cultured dairy products and milk and in strain dominance within lactic cultures. Of all the types of mixed-culture interactions, amensalism may be the most important.

Commensalism

In commensalism, one population benefits from the growth of a second population. In mixed dairy cultures, there is evidence of commensalism among Prt^+ and Prt^- members of a population (Hugenholtz 1986), among different genera[5] and between species (Dahiya and Speck 1963). As with all studies of mixed populations, the complexity of the interactions makes it difficult to pinpoint specific factors involved in interactive effects. Furthermore, effects may change depending on strains, specific growth conditions (medium, temperature, pH, pH-control, etc.) or parameters examined (populations, metabolites,

[5]Liu and Moon 1982, Parker and Moon 1982, Subramanian and Shankar 1985, and Wood 1981.

chemical changes vs. growth rate, acidification rate, etc.). Studies most meaningful to dairy fermentations are ones conducted in milk and under conditions which mimic industrial processes. However, these experiments are often more difficult to do and provide less clear results. However, some studies done to date do suggest commensalistic potential of some mixed-strain dairy cultures.

A fairly clear case of commensalism exists among frequently occurring Prt^- variants within normal Prt^+ populations of *Lactococcus* strains. Prt^- individuals occur at high frequencies. The mechanism of instability has been shown to be due to the plasmid linkage of proteinase genes (McKay and Baldwin 1975). In a milk environment, where free-amino-acid levels are quite low, Prt^- populations cannot achieve populations comparable to Prt^+ strains. However, Prt^- strains can persist and even dominate cultures that contain Prt^+ populations (Stadhouders et al. 1988). This is due to crossfeeding of Prt^- strains with peptides and amino acids produced by breakdown of casein by the Prt^+ strains. Clearly, Prt^- strains benefit from this interaction.

Yeast and *Lactobacillus* exist in mixed culture in the production of certain fermented milks, such as kefir and koumiss. Subramanian and Shankar (1985) showed that growth of *Lactobacillus acidophilus* was stimulated in the presence of yeast, although a mechanism for this stimulation was not discussed. Furthermore, cultures of *L. acidophilus* and yeast coagulated the milk in about half the time required by *L. acidophilus* alone. Amensalism was also noted in this study with populations of both *Escherichia coli* and *Bacillus cereus* declining to negligible levels after four days in the fermented milk product.

In a study of mixed cultures of *Lactococcus cremoris* and *Leuconostoc lactis*, it was found that a mixed culture improved the maximum growth rate of the *Leuconostoc* by 50 percent, but did not affect the *Lactococcus* strain. However, total bacterial populations were lower by 39 percent than the sum of each pure culture (Boquien et al. 1988).

Mutualism

Mutualism, or symbiosis, a state where both populations in a mixed culture benefit from the mixed growth, is perhaps best exemplified in dairy fermentations by the *Streptococcus thermophilus* and *Lactobacillus bulgaricus* interaction. These two organisms are used in the manufacture of many fermented milks and cheeses where processing temperatures are above 40°C, e.g., yogurt and Swiss and Italian cheeses. This interaction was reviewed by Radke-Mitchell and

Sandine (1984) and Matalon and Sandine (1986). Mutualism between these two microorganisms is useful in that it helps to assure that the mixed culture will not be dominated by one or the other genus. An approximate rod to coccus ratio of 1:1 is considered to be important to flavor and texture of the final product as well as the rate of acid production throughout the fermentation. It is important to note, however, that not all strains of *S. thermophilus* and *L. bulgaricus* achieve the mutualistic interaction described and strains of these genera must be paired with caution to avoid incompatibility (1974).

The conventional belief is that the nature of symbiosis between *S. thermophilus* and *L. bulgaricus* is due to formate production by *S. thermophilus* and amino acid release by *L. bulgaricus*. Formate is stimulatory to *L. bulgaricus* and amino acids stimulate *S. thermophilus*. This associative growth results in levels of acid production and cell growth greater than would occur in pure cultures (Radke-Mitchell and Sandine 1984). However, there is argument in the literature regarding the exact amino acids responsible, with reports claiming valine (Pette and Lolkema 1950); glycine and histidine (Bautista et al. 1966); glutamine, methionine, histidine, arginine, phenylalanine, and tryptophan (Bracquart et al. 1978); or peptides (Shankar and Davies 1977) as the stimulating factors. Shanker and Davies (1977) commented that stimulation of *S. thermophilus* by these amino acids was poor, and stimulatory fractions containing no amino acids were obtained from milk-grown *L. bulgaricus* filtrates. Possibly some variability in results may be due to seasonal variation or heat treatment of the milk used for growth. Formate produced by *S. thermophilus* appears to consistently stimulate the growth of *L. bulgaricus* (Varinga et al. 1968). However, in work describing the behavior of *S. thermophilus* and *L. bulgaricus* in chemostat cultures, Driessen (1981) concluded that formic acid is not the only limiting substrate for *L. bulgaricus*. It was interesting that *L. bulgaricus* and *S. thermophilus* were found to reach steady state in chemostats (Driessen 1981). The fact that neither strain was washed out attested to the mutualistic nature of these strains.

Radke-Mitchell and Sandine (1986) examined the influence of temperature on associative growth of nine strains of *S. thermophilus* and ten strains of *L. bulgaricus*. It was found that *S. thermophilus* strains always reached greater numbers than *L. bulgaricus* in mixed cultures with average ratios of rod to coccus of 1:8. This was found even at temperatures at or near the optimum for the rod, and up to 7°C off the optimum for the coccus. It was hypothesized that increases in levels of *S. thermophilus* stimulants offset the decrease in growth rate due to temperature. It was also interesting that in single-strain milk cultures, optimum growth rates were not the same as optimum acid production rates, implying an uncoupling of growth and acid production for both the rod and coccus.

Types of Mixed Starter Systems

Different types of starter systems have been developed to cope with the demands of modern dairy fermentations. The main challenge is to develop a starter system that provides uniform, high-quality product with the desired flavor and textural complexity, day in and day out. Those inclined toward more traditional fermentations, desiring a unique and complex final product on a small scale, will have different needs than those operating modern, large-scale cheese factories in which the product is earmarked for processed cheese. One main hindrance to consistency of starter system operation is phage (discussed in the previous section under "Parasitism"). Many strategies to deal with phage have been proposed, and current starter systems reflect them. Starter systems consist of both a source of culture and, in some cases, a medium in which to grow the culture. In modern years, a wide selection of both media and cultures have been available from commercial sources.

Culture systems can often accommodate either undefined or defined cultures. Until the 1970s, most commercial programs offered undefined blends of cultures with a known history of performance. These undefined cultures were suitable for many applications, but they suffered certain drawbacks. For one, their undefined nature made it difficult to reliably detect phage appearance, especially at low levels or against only one component of the culture. This could mask an impending problem. Secondly, they were subject to shifts in strain dominance due to seemingly insignificant shifts in environmental parameters (temperature, milk supply, media formulation, phage attack). Therefore, although the culture bore the same name, its composition could be significantly altered. This led to the development of defined-strain systems. The defined-strain system, applied in New Zealand in 1977 (Limsowtin et al. 1977), allowed careful and precise monitoring of phage development and determination of phage relatedness due to the homogeneity of each culture being analyzed. In defined systems, phage-sensitive components of blends could be removed and replaced with a phage-resistant mutant or phage-unrelated strain without disruption of the other strains in the blend. This system has worked quite well in the United States, New Zealand, Canada, and Ireland, especially in applications where acid production is the predominant starter function. In cases where more complex flavor development is desired in a cheese, or where the cheesemaker wants assurance that the final product is distinct from a competitor's, the defined-strain approach has been more slowly accepted. Certainly, the complexity of strain interactions in traditional, undefined starters may not be duplicated by simply reformulating the blend with a few dominant isolates. Qualities including gas formation, flavor compounds, metabolites, and intracellular enzymes all may have achieved a unique balance by a mixed-strain culture that may be difficult to mimic. However, from a microbiologist's perspective, working with pure, defined cultures provides process control which is more elusive with undefined cultures.

Whether the choice is undefined or defined, culture systems can be used either in rotation or on a repeated basis. Rotations are generally conducted on a time schedule and are not based on the appearance of a specific phage. Schedules are daily or perhaps more often with large-capacity plants. A series of cultures are identified and these are changed or rotated on a regular schedule. When undefined cultures are used in rotation, phage monitoring is often not part of the regular program. Theoretically, the sequence of phage-unrelated cultures is sufficient to prevent any problematic phage buildup. If rotations are conducted with defined strains, however, regular phage monitoring may be done: results are more sensitive and reliable, and may dictate a culture change before a problem occurs.

Alternatively, a single defined or undefined blend can be used repeatedly until some problem is detected. This is widely practiced with "back-slopping," where the previous day's ferment is used to inoculate the current production run. When a problem arises, the cheesemaker simply resorts to using a backup culture provided by a neighbor, a competitor, or a culture house. Repeated use of undefined cultures is also practiced in Holland. Increasing concern, however, is emerging in The Netherlands over the loss or irreversible shift in the priceless cultures that have worked over the decades (Hugenholtz 1986). Portions of these cultures have been frozen back, but these are not in unlimited supply.

Repeated use of defined blends has also been proposed (Daniell and Sandine 1981, Thunell et al. 1981, Timmons et al. 1988). Phage monitoring is integral to this approach. When a component to the blend is shown to be susceptible to phage, it is removed and replaced by a suitable substitute. This system has shown great success as measured by consistent cheesemakes and high-quality products in some well-documented studies (Thunell et al. 1981, Timmons et al. 1988). The drawback to its use is that the cheesemaker must be committed to monitoring phage, and if a problem is detected, no backup starter tank is available as a replacement. Therefore, some risk still remains.

Summary

In conclusion, dairy starter cultures are dynamic mixtures of different genera, species, strains, and mutants. The interaction of these populations constitutes the challenge of and provides the tools for production of a vast array of fermented dairy products. The nature of these interactions can be inhibitory, stimulatory, or neutral, and can be directed at other dairy starter cultures or inadvertent milk contaminants. Opportunity exists for development of new blends or new types of culture for unique dairy products, or the use of these diverse cultures for applications outside of dairy.

References

Accolas, J.-P., and H. Spillmann, 1979. Morphology of bacteriophages of *Lactobacillus bulgaricus, L. lactis*, and *L. helveticus. J. Appl. Bacteriol. 47*:309–319.

Babel, F. J., 1977. Antibiosis by lactic culture bacteria. *J. Dairy Sci. 60*:815–821.

Barefoot, S. F., and T. R. Klaenhammer, 1983. Detection and activity of lactacin B, a bacteriocin produced by *L. acidophilus. Appl. Environ. Microbiol. 45*:1801–1815.

Bautista, E. S., R. S. Dahiya, and M. L. Speck, 1966. Identification of compounds causing symbiotic growth of *Streptococcus thermophilus* and *Lactococcus bulgaricus* in milk. *J. Dairy Res. 33*:299–307.

Benkerroum, N., and W. E. Sandine, 1988. Inhibitory action of nisin against *Listeria monocytogenes. J. Dairy Sci. 71*:3237–3245.

Bhunia, A. K., M. C. Johnson, and B. Ray, 1988. Purification, characterization, and antimicrobial spectrum of a bacteriocin produced by *Pediococcus acidilactici. J. Appl. Bacteriol. 65*:261–268.

Boquien, C.-Y., G. Corrieu, and M. J. Desmazeaud, 1988. Effect of fermentation conditions on growth of *Streptococcus cremoris* AM2 and *Leuconostoc lactis* CNRZ 1091 in pure and mixed cultures. *Appl. Environ. Microbiol. 54*:2527–2531.

Brackett, R. E., 1988. Presence and persistence of *Listeria monocytogenes* in food and water. *Food Technol. 42*:162–178.

Bracquart, P., D. Lorient, and C. Alias, 1978. Effects of amino acids and small peptides on the acidification of milk by *Streptococcus thermophilus*. 20th International Dairy Congress, Paris. E:512–513.

Buchmann, G. W., S. Banerjee, and J. N. Hansen, 1988. Structure, expression and evolution of a gene encoding the precursor of nisin, a small peptide antibiotic. *J. Biological Chem. 263*:16260–16266.

Carminati, D., G. Giraffa, and M. G. Bossi, 1989. Bacteriocin-like inhibitors of *Streptococcus lactis* against *Listeria monocytogenes. J. Food Protection. 52*:614–617.

Choisy, C., M. Gueguen, J. Lenoir, J. L. Schmidt, and C. Tourneur, 1987. The ripening of cheese. Microbiological aspects. In A. Eck. *Cheesemaking Science and Technology*, Lavoisier Publishing, New York, pp. 259–292.

Cogan, T. M., 1985. The leuconostocs: milk products. In S. E. Gilliland *Bacterial Starter Cultures for Foods*, CRC Press, Boca Raton, Florida, pp. 25–40.

Collins, E. B., 1952a. Action of bacteriophage on mixed strain starter cultures. I. Nature and characteristics of the "nascent phenomenon." *J. Dairy Sci. 35*:371–380.

Collins, E. B., 1952b. Action of bacteriophage on mixed strain starter cultures. II. Relation to acid production of the proportion of resistant bacteria. *J. Dairy Sci. 35*:381–387.

Collins, E. B., 1955a. Action of bacteriophage on mixed strain cultures. III. Strain dominance due to the action of bacteriophage and variations in the acid production of secondary growth bacteria. *Appl. Microbiol. 3*:137–140.

Collins, E. B., 1955b. Action of bacteriophage on mixed strain cultures. IV. Domination among strains of lactic streptococci. *Appl. Microbiol. 3*:141–144.
Collins, E. B., 1955c. Action of bacteriophage on mixed strain cultures. V. Similarities among strains of lactic streptococci in commercially used cultures and use of a whey activity test for culture selection and rotation. *Appl. Microbiol. 3*:145–148.
Daeschel, M. A., 1989. Antimicrobial substances from lactic acid bacteria for use as food preservatives. *Food Technol. 43*:164–167.
Daeschel, M. A., and T. R. Klaenhammer, 1985. Association of a 13.6 Mdal plasmid in *Pediococcus pentosaceus* with bacteriocin activity. *Appl. Environ. Microbiol. 50*:1538–1541.
Daeschel, M. A., M. C. McKenney, and L. C. McDonald, 1986. *ASM Bacteriol. Proc. Abstract.* pp. 13, 277.
Dahiya, R. S., and M. L. Speck, 1963. Identification of stimulatory factor involved in symbiotic growth of *Streptococcus lactis* and *Streptococcus cremoris. J. Bacteriol. 85*: 585–589.
Daly, C., W. E. Sandine, and P. R. Elliker, 1970. Associative growth and inhibitory properties of *Streptococcus diacetylactis. J. Dairy Sci. 53*:637–638.
Daniell, S. D., and W. E. Sandine, 1981. Development and commercial use of a multiple strain starter. *J. Dairy Sci. 64*:407–415.
Davey, G. P., and B. C. Richardson, 1981. Purification and some properties of diplococcin in *Streptococcus cremoris* 346. *Appl. Environ. Microbiol. 41*:84–89.
Davies, F. L., and M. J. Gasson, 1981. Reviews of the progress of dairy science: genetics of lactic acid bacteria. *J. Dairy Res. 48*:363–376.
Davies, F. L., and M. J. Gasson, 1983. Genetics of dairy cultures. *Irish J. Food Sci. Technol. 7*:49–60.
deVos, W. M., 1986. Gene cloning in lactic streptococci. *Neth. Milk Dairy J. 40*:141–154.
Doyle, M. P., 1988. Effect of environmental and processing conditions on *Listeria monocytogenes. Food Technol. 42*:169–171.
Driessen, F. M., 1981. Protocooperation of yogurt bacteria in continuous cultures. In M. E. Bushall, and J. H. Slater *Mixed Culture Fermentations*. Academic Press, New York, p. 1–24.
Ehlers, J. G., M. Chapparo-Serrano, R. Richter, and C. Vanderzant, 1982. Survival of *Campylobacter fetus* subsp. *jejuni* in cheddar and cottage cheese. *J. Food Protection. 45*:1018–1021.
FDA, 1988. Nisin preparation: affirmation of GRAS status as a direct human food ingredient. Food and Drug Administration, *Federal Register* 53:11247.
Fernandes, C. F., K. M. Shahani, and M. A. Amer, 1988. Control of diarrhea by lactobacilli. *J. Appl. Nutrition. 40*:32–43.
Fitzgerald, G. F., and M. J. Gasson, 1988. In vivo gene transfer systems and transposons. Biochimie *70*:489–502.
Frank, J. F., and E. H. Marth, 1977. Inhibition of enteropathogenic *Escherichia coli* by homofermentative lactic acid bacteria in skim milk. 1. Comparison of strains of *Escherichia coli. J. Food Protection 40*:749–753.
Fujita, Y., T. Okamoto, and R. Irie, 1984. Plasmid distribution in lactic streptococci. *Agric. Biol. Chem. 48*:1895–1898.
Geis, A., J. Singh, and M. Teuber, 1983. Potential of lactic streptococci to produce bacteriocin. *Appl. Environ. Microbiol. 45*:205–211.
Gibson, E. M., N. M. Chace, S. B. London, and J. London, 1979. Transfer of plasmid-mediated antibiotic resistance from streptococci to lactobacilli. *J. Bacteriol. 137*:614–619.
Gilliland, S. E., and M. L. Speck, 1974. Antagonism of lactic streptococci toward *Staphylococcus aureus* in associative milk cultures. *Appl. Microbiol. 28*:1090–1093.
Gilliland, S. E., and M. L. Speck, 1977. Antagonistic action of *Lactobacillus acidophilus* toward intestinal and foodborne pathogens in associative cultures. *J. Food Protection. 40*:820–823.
Gonzales, C. F., 1983. Preservation of foods with non-lactose fermenting *Streptococcus lactis* subspecies *diacetilactis*. European Patent Application #0092183, filed April 14, 1983.

Gonzales, C. F., and B. S. Kunka, 1983. Plasmid transfer in *Pediococcus* spp.: intergeneric and intrageneric transfer of pIP501. *Appl. Environ. Microbiol. 46*:81–89.
Grieve, P. A., and J. R. Dulley, 1983. Use of *Streptococcus lactis* Lac$^-$ mutants for accelerating cheddar cheese ripening. 2. Their effect on the rate of proteolysis and flavor development. *Aust. J. Dairy Technol. 38*:49–54.
Grieve, P. A., B. A. Lockie, and J. R. Dulley, 1983. Use of *Streptococcus lactis* Lac$^-$ mutants for accelerating cheddar cheese ripening. 1. Isolation, growth, and properties of a C2 Lac$^-$ variant. *Aust. J. Dairy Technol. 38*:10–13.
Harrison, D. E. F., 1978. Mixed cultures in industrial fermentation processes. *Adv. Appl. Microbiol. 24*:129–164.
Hickey, M. W., A. J. Hillier, and G. R. Jago, 1986. Transport and metabolism of lactose, glucose, and galactose in homofermentative lactobacilli. *Appl. Environ. Microbiol. 51*:825–831.
Hoier, E., and D. Jorck-Ramberg, 1988. A direct vat starter system based on phage-insensitive strains. *Irish J. Food Sci. Technol. 12*:97.
Hoover, D. G., P. M. Walsh, K. M. Kolaetis, and M. M. Daly, 1988. A bacteriocin produced by *Pediococcus* species associated with a 5.5-megadalton plasmid. *J. Food. Protection 51*:29–31.
Hugenholtz, J., 1986. Population dynamics of mixed starter cultures. *Neth. Milk Dairy J. 40*:129–140.
Hull, R. R., 1977. Control of bacteriophage in cheese factories. *Aust. J. Dairy Technol. 32*:65–66.
Hurst, A., 1983. Nisin and other inhibitory substances from lactic acid bacteria. In A. L. Branen, P. M. Davidson, *Antimicrobials in Foods*, Marcel Dekker, New York.
Jimeno, J., M. Casey, F. Hofer, 1984. The occurrence of β-phosphogalactosidase in *Lactobacillus casei* strains. *FEMS Microbiol Lett. 25*:275–278.
Joerger, M. C., and T. R. Klaenhammer, 1986. Characterization and purification of Helveticin J and evidence for a chromosomally determined bacterioicin produced by *Lactobacillus helveticus* 481. *J. Bacteriol. 167*:439–446.
King, W. R., E. B. Collins, and E. L. Barrett, 1983. Frequencies of bacteriophage-resistant and slow acid-producing variants of *Streptococcus cremoris. Appl. Environ. Microbiol. 45*:1481–1485.
Klaenhammer, T. R., 1988. Bacteriocins of lactic acid bacteria. *Biochimie 70*:337–349.
Klaenhammer, T. R., 1984. Interactions of bacteriophages with lactic streptococci. *Adv. Appl. Microbiol. 30*:1–29.
Kondo, J. K., and L. L. McKay, 1985. Gene transfer systems and molecular cloning in group N streptococci: a review. *J. Dairy Sci. 68*:2143–2159.
Kozak, W., J. Bardowski, and W. T. Dobrzanski, 1978. Lactostrepcins-acid bacteriocins produced by lactic streptococci. *J. Dairy Res. 45*:247–257.
Krusch, U., H. Neve, B. Luschei, and M. Teuber, 1987. Characterization of virulent bacteriophages of *Streptococcus salivarius* subsp. *thermophilus* by host specificity and electron microscopy. *Kieler Milchwirtschaftliche Forschungsberichte. 39*:155–167.
Lahbib-Mansais, Y., M. Mata, and P. Ritzenthaler, 1988. Molecular taxonomy of *Lactobacillus* phages. *Biochimie 70*:429–436.
Langsrud, T., A. Landaas, and H. B. Castberg, 1987. Autolytic properties of different strains of group N streptococci. *Milchwissenschaft 42*:556–560.
Le Bourgeois, P., M. Mata, and P. Ritzenthaler, 1989. Genome comparison of *Lactococcus* strains by pulsed-field gel electrophoresis. *FEMS Microbiol. Lett. 59*:65–70.
Lees, G. J., and G. R. Jago, 1978a. Role of acetaldehyde in metabolism: a review. 1. Enzymes catalyzing reactions involving acetaldehyde. *J. Dairy Sci. 61*:1216–1224.
Lees, G. J., and G. R. Jago, 1978b. Role of acetaldehyde in metabolism: a review. 2. The metabolism of acetaldehyde in cultured dairy products. *J. Dairy Sci. 61*:1216–1224.
Limsowtin, G. K. Y., and B. E. Terzaghi, 1976. Phage resistant mutants: their selection and use in cheese factories. *New Zealand J. Dairy Sci. Technol. 11*:251–256.
Limsowtin, G. K. Y., H. A. Heap, and R. C. Lawrence, 1977. A multiple starter concept for cheesemaking. *New Zealand J. Dairy Sci. Technol. 12*:101–106.

Liu, J. A. P., and N. J. Moon, 1982. Commensalistic interaction between *Lactobacillus acidophilus* and *Propionibacterium shermanii*. *Appl. Environ. Microbiol.* *44*:715–722.
Marshall, R. J., and N. J. Berridge, 1976. Selection and some properties of phage-resistant starters for cheese-making. *J. Dairy Res.* *43*:449–458.
Martley, F. G., 1983. Temperature sensitivities of thermophilic starter strains. *New Zealand J. Dairy Sci. Technol.* *18*:191–196.
Mata, M., A. Trautwetter, G. Luthaud, and P. Ritzenthaler, 1986. Thirteen virulent and temperate bacteriophages of *Lactobacillus bulgaricus* and *Lactobacillus lactis* belong to a single DNA homology group. *Appl. Environ. Microbiol.* *52*:812–818.
Mata, M., and P. Ritzenthaler, 1988. Present state of lactic acid bacteria phage taxonomy. *Biochimie* *70*:395–400.
Matalon, M. E., and W. E. Sandine, 1986. *Lactobacillus bulgaricus, Streptococcus thermophilus* and yogurt: a review. *Cultured Dairy Products J.* *21*:6–12.
McKay, L. L., 1978. Microorganisms and their instability in milk and milk products. *Food Technol.* *32*:181–185.
McKay, L. L., and K. A. Baldwin, 1974. Simultaneous loss of proteinase- and lactose-utilizing enzyme activities in *Streptococcus lactis* and reversal of loss by transduction. *Appl. Microbiol.* *28*:342–346.
McKay, L. L., and K. A. Baldwin, 1975. Plasmid distribution and evidence for a proteinase plasmid in *Streptococcus lactis* C2. *Appl. Microbiol.* *29*:546–548.
McKay, L. L., K. A. Baldwin, and P. M. Walsh, 1980. Conjugal transfer of genetic information in group N streptococci. *Appl. Environ. Microbiol.* *40*:84–91.
McKay, L. L., A. Miller, III, W. E. Sandine, and P. R. Elliker, 1970. Mechanisms of lactose utilization by lactic acid streptococci: enzymatic and genetic analyses. *J. Bacteriol.* *102*:804–809.
Mercenier, A., P. Slos, M. Faelen, and J. P. Lecocq, 1988. Plasmid transduction in *S. thermophilus*. *Mol. Gen. Genet.* *312*:386–389.
Moon, N. J., and G. W. Reinbold, 1974. Selection of active and compatible starters for yogurt. *Cultured Dairy Prod. J.* *9*:10–12.
Muriana, P. M., and T. R. Klaenhammer, 1987. Conjugal transfer of plasmid-encoded determinants for bacteriocin production and immunity in *L. acidophilus* 88. *Appl. Environ. Microbiol.* *53*:553–560.
Neve, H., A. Geis, and M. Teuber, 1984. Conjugal transfer and characterization of bacteriocin plasmids in group N (lactic acid) streptococci. *J. Bacteriol.* *157*:833–838.
Neve, H., A. Geis, and M. Teuber, 1987. Conjugation, a common plasmid transfer mechanism in lactic acid streptococci of dairy starter cultures. *System. Appl. Microbiol.* *9*:151–157.
Newcomer, A. D., H. S. Park, P. C. O'Brien, and D. B. McGill, 1983. Response of patients with irritable bowel syndrome and lactose deficiency using unfermented acidophilus milk. *Am. J. Clin. Nutr.* *38*:257–263.
Occhino, L. A., H. A. Morris, and D. A. Savaiano, 1986. A comparison of beta-galactosidase specific activities in strains of *Streptococcus thermophilus*. *J. Dairy Sci.* *69*:2583–2588.
Oram, J. D., and B. Reiter, 1968. The adsorption of phage to group N streptococci. The specificity of adsorption and the location of phage receptor substances in cell-wall and plasma-membrane fractions. *J. Gen. Virol.* *3*:103–119.
Parker, J. A., and N. J. Moon, 1982. Interactions of *Lactobacillus* and *Propionibacterium* in mixed culture. *J. Food Protection* *45*:326–330.
Pechmann, H., and M. Teuber, 1980. Plasmid pattern of group N (lactic) streptococci. *Zbl. Bakt. Hyg., I Abt. Orig.* *C1*:133–136.
Pette, J. W., and H. Lolkema, 1950. Yogurt II. Growth stimulating factors for *S. thermophilus*. *Neth. Milk Dairy J.* *4*:209–224.
Pinheiro, A. J. R., B. J. Liska, and C. E. Parmilee, 1968. Properties of substances inhibitory to *Pseudomonas fragi* produced by *Streptococcus citrovorus* and *Streptococcus diacetilactis*. *J. Dairy Sci.* *51*:183–187.
Pucci, M. J., M. E. Monteschio, and C. L. Kemker, 1988. Intergeneric and intrageneric

conjugal transfer of plasmid-encoded antibiotic resistance determinants in *Leuconostoc* spp. *Appl. Environ. Microbiol. 54*:281–287.

Pucci, M. J., E. R. Vedamuthu, B. S. Kunka, and P. A. Vandenberg, 1988. Inhibition of *Listeria monocytogenes* by using bacteriocin PA-1 produced by *Pediococcus acidilactici* PAC1.0. *Appl. Environ. Microbiol. 54*:2349–2353.

Radke-Mitchell, L., and W. E. Sandine, 1984. Associative growth and differential enumeration of *Streptococcus thermophilus* and *Lactobacillus bulgaricus*: a review. *J. Food Protection 47*:245–248.

Radke-Mitchell, L. C., and W. E. Sandine, 1986. Influence of temperature on associative growth of *Streptococcus thermophilus* and *Lactobacillus bulgaricus. J. Dairy Sci. 69*: 2558–2568.

Rajagopal, S. N., and W. E. Sandine, 1989. Isolation of *Streptococcus thermophilus* and *Lactobacillus bulgaricus* bacteriophages from Italian cheese whey. *Cultured Dairy Prod. J. 24*:18–21.

Raya, R. R., E. G. Kleeman, J. B. Luchansky, and T. R. Klaenhammer, 1989. Characterization of the temperate bacteriophage Øadh and plasmid transduction in *Lactobacillus acidophilus* ADH. *Appl. Environ. Microbiol. 55*:2206–2213.

Richardson, G. H., C. A. Ernstrom, J. M., Kim, and C. Daly, 1983. Proteinase negative variants of *Streptococcus cremoris* for cheese starters. *J. Dairy Sci. 66*:2278–2286.

Roginski, H., 1988. Fermented milks. *Aust. J. Dairy Technol. 43*:37–46.

Romero, D. A., and T. R. Klaenhammer, 1990. Characterization of insertion sequence IS *946*, an ISO-ISS*1* element, isolated from the conjugative lactococcal plasmid pTR2030. *J. Bacteriol. 172*:4151–4160.
342, Institute of Food Technologists Annual Meeting, Chicago.

Romero, D. A., P. Slos, C. Robert, I. Castellino, and A. Mercenier, 1987. Conjugative mobilization as an alternative vector delivery system for lactic streptococci. *Appl. Environ. Microbiol. 53*:2405–2413.

Rubin, H. E., and F. Vaughan, 1979. Elucidation of the inhibitory factors of yogurt against *Salmonella typhimurium. J. Dairy Sci. 62*:1873–1879.

Rutzinski, J. L., and E. H. Marth, 1980. Behavior of *Enterobacter* species and *Hafnia* species in skim milk during fermentation by lactic acid bacteria. *J. Food Protection. 43*:720–728.

Sanders, M. E., 1987. Bacteriophages of industrial importance. In S. M. Goyal, C. P. Gerba, and G. Bitton *Phage Ecology* John Wiley and Sons, New York.

Sanders, M. E., 1988. Phage resistance in lactic acid bacteria. *Biochimie 70*:411–421.

Sanders, M. E., P. J. Leonhard, W. D. Sing, and T. R. Klaenhammer, 1986. Conjugal strategy for construction of fast acid-producing, bacteriophage-resistant lactic streptococci for use in dairy fermentations. *Appl. Environ. Microbiol. 52*:1001–1007.

Sandine, W. E., 1985. The streptococci: milk products. In S. E. Gilliland *Bacterial Starter Cultures for Foods*, CRC Press, Boca Raton, Florida, pp. 5–23.

Schaack, M. M., and E. H. Marth, 1988. Interaction between lactic acid bacteria and some foodborne pathogens: a review. *Cultured Dairy Prod. J. 23*:14–19.

Scherwitz, K. M., K. A. Baldwin, and L. L. McKay, 1983. Plasmid linkage of a bacteriocin-like substance in *Streptococcus lactis* subsp. *diacetylactis* strain WM4: transferability to *Streptococcus lactis. Appl. Environ. Microbiol. 45*:1506–1512.

Sechaud, L., P. J. Cluzel, M. Rousseau, A. Baumgartner, and J. P. Accolas, 1988. Bacteriophages of lactobacilli. *Biochimie 70*:401–410.

Shankar, P. A., and F. L. Davies, 1977. Associative bacterial growth in yogurt starters; initial observations on stimulatory factors. *J. Soc. Dairy Technol. 30*:31–32.

Sharpe, M. E., 1979. Lactic acid bacteria in the dairy industry. *J. Soc. Dairy Technol. 32*:9–18.

Shimizu-Kadota, M., M. Kiwaki, H. Hirokawa, and N. Tsuchida, 1985. ISL*1*: a new transposable element in *Lactobacillus casei. Mol. Gen. Genet. 200*:193–198.

Slater, J. H., 1981. Mixed cultures and microbial communities. In M. E. Bushell, T. N. Slater, *Mixed Culture Fermentations*, Academic Press, New York, pp. 1–24.

Sorrells, K. M., and M. L. Speck, 1970. Inhibition of *Salmonella gallinarum* by culture filtrates of *Leuconostoc citrovorum. J. Dairy Sci. 53*:239–241.

Speck, M. L., 1972. Control of food-borne pathogens by starter cultures. *J. Dairy Sci. 55*:1019–1022.
Stadhouders, J., and G. J. M. Leenders, 1984. Spontaneously developed mixed-strain cheese starters. Their behavior towards phages and their use in the Dutch cheese industry. *Neth. Milk Dairy J. 38*:157–189.
Stadhouders, J., L. Toepoel, and J. T. M. Wouters, 1988. Cheese making with Prt^- and Prt^+ variants of N-streptococci and their mixtures. Phage sensitivity, proteolysis and flavour development during ripening. *Neth. Milk Dairy J. 42*:183–193.
Steinkraus, K. H., 1982. Fermented foods and beverages: the role of mixed cultures. *Microb. Interact. Communities. 1*:407–442.
Subramanian, P., and P. A. Shankar, 1985. Commensalistic interaction between *Lactobacillus acidophilus* and lactose-fermenting yeasts in the preparation of acidophilus-yeast milk. *Cultured Dairy Prod. J. 20*:19–26.
Syvanen, M., 1985. Cross-species gene transfer; implications for a new theory of evolution. *J. Theor. Biol. 112*:333–343.
Tagg, J. R., A. S. Dajani, and L. W. Wannamaker, 1976. Bacteriocins of gram-positive bacteria. *Bacteriological Rev. 40*:722–756.
Tannock, G. W., 1987. Conjugal transfer of plasmid pAMβ1 in *Lactobacillus reuteri* and between *Lactobacilli* and *Enterococcus faecalis. Appl. Environ. Microbiol. 53*:2693–2695.
Terzaghi, B. E., 1976. Morphologies and host sensitivities of lactic streptococcal phages from cheese factories. *New Zealand J. Dairy Sci. Technol. 11*:155–163.
Thompson, J. K., and M. A. Collins, 1988. Evidence for the conjugal transfer of the broad host range of plasmid pIP501 into strains of *Lactobacillus helveticus. J. Appl. Bacteriol. 65*:309–319.
Thunell, R. K., W. E. Sandine, and F. R. Bodyfelt, 1981. Phage-sensitive, multiple-strain starter approach to cheddar cheese making. *J. Dairy Sci. 64*:2270–2277.
Timmons, P., M. Hurley, F. Drinan, C. Daly, and T. M. Cogan, 1988. Development and use of a defined strain starter system for cheddar cheese. *J. Soc. Dairy Technol. 41*: 49–53.
Truslow, W. W., 1986. *Lactobacillus* and fermented foods. *J. Holistic Med. 8*:36–46.
Upreti, G. C., and R. D. Hindsdill, 1973. Isolation and characterization of a bacteriocin from a homofermentative *Lactobacillus. Antimicrob. Agent Chemother. 4*:487–494.
Vedamuthu, E. R., and C. Washam, 1983. Cheese. *Biotechnology 5*:231–313.
Veringa, H. A., T. E. Galeshoot, and H. Davelaar, 1968. Symbiosis in yogurt. *Neth. Milk Dairy J. 22*:114–120.
West, C. A., and P. J. Warner, 1988. Plantacin B, a bacteriocin produced by *Lactobacillus plantarum* NCDO 1193. *FEMS Microbiol. Lett. 49*:163–165.
Wood, B. J. B., 1981. The yeast/*Lactobacillus* interaction; a study in stability. In M. E. Bushell and J. H. Slater, *Mixed Culture Fermentations*. Academic Press, New York, p. 1–24.
Yang, N. L., and W. E. Sandine, 1979. Acid-producing activity of lyophilized lactic streptococcal cheese starter concentrates. *J. Dairy Sci. 62*:908–915.

Chapter

6

Microbiological Safety of Fermented Foods

Eric A. Johnson

Department of Food Microbiology and Toxicology
University of Wisconsin
Madison, Wisconsin

Introduction

Fermentation has been used for several thousand years as an effective and low-cost means to preserve the quality and safety of foods. All fermented foods depend on the metabolic activities of beneficial microorganisms, particularly molds, yeasts, and lactic acid bacteria. Food fermentations involve mixed cultures of microorganisms that grow simultaneously or in succession, and many employ starter cultures of complex yet stable associations (Hesseltine 1965). Fermented foods of the world have been reviewed (e.g., Campbell-Platt 1987, Hesseltine and Wang 1986, Pederson 1979, Wood 1985). The present chapter addresses the characteristics of food fermentations that contribute to the microbiological safety of the fermented products. Fermented foods have been reported to have anticarcinogenic and antimutagenic qualities (e.g., Hirayama, 1982), but these aspects of food safety are not covered in this chapter.

The safety of fermented foods is evident from the long history of production and consumption. Fermented foods provide a source of safe food in several areas of the world where unsanitary conditions are found. Many fermented foods are shelf-stable and do not require refrigeration. Preservation by fermentation has been promoted in tropical countries that have high ambient temperatures and low refrigeration capacities (Cooke et al. 1987). The safety of fermented foods depends on the activities of mixed cultures that convert raw food substrates to a form that cannot support the growth of pathogens. Many of these beneficial organisms thrive at low pHs and in high concentrations of salt. Spoilage and pathogenic organisms are inhibited or destroyed by unfavorable conditions that develop during fermentations, including increased acidity, lowered water activity and oxidation-reduction potential, competition for nutrients, formation of antibiotic substances, and spatial exclusion. Starter cultures are now routinely used in many fermentations worldwide to achieve consistency in the products (Hesseltine 1985). Curing processes involving secondary microorganisms also contribute to product safety by elimination of pathogens and toxins from the food, as occurs in the ripening of cheese.

Importance of Foodborne Disease

Foodborne disease has a large social impact and disastrous results on national and international economies (Todd 1987, 1989; Archer and Kvenberg 1985). Evidence indicates that the incidence of acute foodborne diseases, unlike most noncommunicable diseases, is increasing globally (Todd 1987). Consumption of contaminated foods leads to millions of diarrheal cases in humans each year. Enteric diseases rank second in incidence to respiratory diseases in the United States (Archer and Young 1988). Newer foodborne pathogens including *Campylobacter jejuni, Escherichia coli* O157:H7, *Listeria monocytogenes*, as well as the Norwalk agent and other human enteric viruses have been recognized in the past decade. Traditional pathogens such as *Salmonella* are also becoming more troublesome with changes in food production practices and social patterns. The long-range significance of acute foodborne disease is underscored by the realization that such disease can trigger chronic disease sequelae that appear later in the lives of humans (Archer and Young 1988). It has been estimated that a significant proportion of acute foodborne illness cases may develop into chronic problems including joint diseases, immune diseases, afflictions of heart and vascular system, renal diseases, and nervous system and neuromuscular disorders (Archer and Young 1988).

Fermentation is an effective method of improving the safety of the world's food supply. Fermented foods have an impressive record of safety compared to many other foods (National Academy of Sciences 1985, Bryan 1988). Certain fermented products, however, have been found to periodically harbor pathogens and toxins and to transmit disease. Concerns have been raised recently that newly recognized pathogens such as *L. monocytogenes* could survive during some food fermentations. Molds associated with fermented foods have been suspected to produce mycotoxins. Despite these concerns, fermented foods are comparatively safe, and examination of their microbiology provides an opportunity to understand the biological mechanisms of pathogen exclusion and to develop new technologies for improving food safety.

Diseases Transmitted in Fermented Foods

Considerable attention has been given to acute foodborne diseases and many pathogens have been recognized (Table 6.1). Most of these are infectious pathogens that colonize the human gastrointestinal tract and inflict damage to intestinal cells. In addition to the infectious pathogens, certain bacteria and fungi produce toxins (Table 6.2), including botulinum toxin, which can migrate through intestinal epithelia to peripheral nerves and cause permanent damage. Toxins are introduced to the food in raw materials or may be synthesized during improperly conducted fermentations.

Bacterial pathogens known to cause foodborne disease can be taxonomically classified into a few phylogenetic lineages (Johnson and Pariza 1989). Newly recognized pathogens usually derive from one of these taxonomic groups. Studies of the ecology and genetics of pathogens have indicated that they evolve along distinct clonal lines (Selander and Musser 1990). It is expected that new pathogens will be recognized and they will share genetic homology with related pathogenic strains within the lineage. Periodically, at an unknown frequency, virulence properties are naturally transferred between species and genera. For example, an organism was isolated from an infant with botulism that had acquired the ability to produce a toxin that was neutralized by antitoxin to type E botulinum neurotoxin. Characterization of the organism showed that it belonged to the species *Clostridium butyricum* (McCroskey et al. 1986), an organism commonly associated with dairy products and other fermented foods. Promiscuous plasmids capable of transferring virulence traits have been found in foodborne pathogens, particularly within the Enterobacteriaceae. Therefore, the food processor should be aware that organisms associated with fermented foods may acquire pathogenic characteristics by interspecies gene transfer.

The impact of viruses on foodborne disease is not well known. Over 100 enteric viruses have been detected in human feces (Melnick et al. 1978). Families of viruses that have caused foodborne illness include hepatitis type A, poliovirus, calciviruses, astroviruses, Norwalk virus, Snow Mountain agent, and non-A non-B hepatitis viruses (Blackwell et al. 1985). Viruses usually contaminate foods by human fecal pollution, especially in polluted water. Low numbers of many viruses can cause severe infections. Hepatitis A is endemic in countries that are

TABLE 6.1 Known and Suspected Causative Agents of Foodborne Disease

Microbial group	Organism	Reservoirs	Known food (particularly fermented & related)
Bacteria			
Enterobacteriaceae			
	Salmonella typhi	Infected humans	Raw milk, dairy products meat products, vegetables
	Salmonella paratyphi	Infected humans	
	Salmonella sp.	Animals & animal	Meats, eggs, raw milk products
	Pathogenic		
	Escherichia coli	Infected animals, dairy cattle	Raw milk, Brie cheese
	Shigella sp.	Feces from infected humans, water	Vegetables, salads, water
	Yersinia enterocolitica	Animals, especially swine	Raw milk, tofu
Vibrionaceae	*Vibrio cholerae*	Feces, water	Raw shellfish, raw vegetables
	Vibrio parahaemolyticus	Sea water	Marine fish and crustacea
	Aeromonas hydrophila	Aquatic organisms	Shellfish
	Plesiomonas shigelloides	Sea water	Salted fish, shellfish
Spirillaceae	*Campylobacter* sp.	Infected animals	Raw milk, poultry
Nonfermentative gram-negative bacilli	*Pseudomonas cocovenenans*		Poorly fermented *tempe bongkrek*
	Flavobacterium farinofermentans		Poorly fermented cornmeal
Miscellaneous gram-negative bacteria	*Brucella* sp.	Infected animals	Raw, milk, dairy products, cheese from raw milk, goat's milk cheese
	Mycobacterium bovis	Infected cows	Raw milk, cheese from nonpasteurized milk
Gram-positive cocci	*Staphylococcus aureus*	Infected humans and animals	Improperly fermented dairy products and sausages

TABLE 6.1 Known and Suspected Causative Agents of Foodborne Disease (*Continued*)

Microbial group	Organism	Reservoirs	Known food (particularly fermented & related)
Bacteria			
Gram-positive nonsporeforming rods	*Erysipelothrix rhysiopathiae*	Swine, fish	Infected animals
	Listeria monocytogenes	Soil, animals, humans	Raw milk, soft-ripened cheeses, poorly fermented meats, vegetables
	Corynebacterium diphtheriae	Infected animals	Raw milk
	Streptococcus pyogenes	Infected humans, mastitic cows	Raw milk
Gram-positive sporeformers	*Clostridium botulinum*	Soil	Vegetables, fermented fish
	Clostridium perfringens	Soil, feces	Meats, vegetables
	Bacillus cereus	Soil	Rice, vegetables
	Bacillus anthracis	Infected animals	Undercooked sausage
Viruses and other agents			
	Rotavirus	Infected humans	Various foods, water
	Astrovirus	Infected humans	Water and shellfish
	Calciviruses (Norwalk)	Infected humans	Shellfish, salads, water
	Echovirus	Infected humans	Coleslaw
	Enteroviruses, reoviruses, poliovirus, etc.	Infected humans	Shellfish, raw-contaminated foods
	Hepatitis A	Infected humans	Shellfish, raw milk
	Reoviruses	Infected humans	
	Coxiella burnetii	Infected animals	Raw milk and products

TABLE 6.2 Microbial Toxins Found in Foods

Toxin	Producing organisms	Potentially contaminated fermented foods
Bacterial toxins		
Botulinum toxin	*Clostridium botulinum*	Fermented meats, especially fish; fermented bean curd; other poorly fermented foods
Staphylococcal enterotoxin	*Staphylococcus aureus*	Poorly fermented meats and dairy products
Bacillus cereus emetic and diarrheal toxin	*Bacillus cereus*	Rice-based foods; sausage
Bongkrek toxin	*Flavobacterium farinofermentans*	Poorly fermented cornmeal
	Pseudomonas cocovenenans	Poorly fermented tempe bongkrek
Mycotoxins		
Aflatoxins	*Aspergillus flavus* group	Milk, cheese, yogurt, fermented grain products, breads, sausage
Cyclopiazanic acid	*Aspergillus* sp.	Corn, cheese
Ochratoxin	*Aspergillus ochraceus*	Grains, pork, poultry, sausage, cheese
Patulin	*Penicillium* sp.	Fruits and fruit juices, apple cider; sausages
Penicillic acid	*Penicillium* sp.	Meats, sausages
Penetrem	*Penicillium* sp.	Cheese, beer
Sterigmatocystin	*Penicillium* sp.	Rice, corn, cheese, bread, sausages
Tricothecenes	*Fusarium* sp.	Grain-based foods
Seafood toxins		
Ciguatera toxin	*Gambierdiscus toxicus* and possibly other dinoflagellates	Various fishes
Paralytic shellfish toxin (saxitoxin)	*Gonyaulax catenella*, *Gonyaulax tamarensis*, and possibly marine bacteria	Mussels, clams
Okadaic acid	Dinoflagellate	Mussels, clams
Biogenic amines		
Tyramine	Fecal streptococci, proteolytic clostridia	Aged cheeses, other fermented foods
Histamine	*Morganella morganii*, *Klebsiella pneumoniae*, coliforms	Fish, usually of Scombridae; aged cheeses

overcrowded and have poor sanitation and hygiene. Viral diseases are difficult to diagnose, and their magnitude is probably much higher than currently reported (Blackwell et al. 1985).

Certain microbial toxins may contaminate food and cause foodborne intoxications (Table 6.2). Several fermented foods are suspected to transmit mycotoxins, since mycotoxigenic-related fungi are used in certain fermentations. Aflatoxins are produced by *Aspergillus flavus* and *Aspergillus parasiticus*, fungi which are closely related to organisms used for production of various Oriental-style fermented foods such as *shoyu* (Hesseltine 1985). Bacteria, including *Clostridium botulinum, Flavobacterium farinofermentans*, and *Pseudomonas coconvenenans*, have also been found to produce potent toxins during certain food fermentations. Recognized pathogen or intoxicant risks in fermented foods and processing inadequacies enabling contamination are summarized in Table 6.3.

Incidence of Foodborne Disease Transmitted in Fermented Foods

In countries that have conducted surveillance for foodborne disease, fermented foods have a good safety record compared to many other foods. Fermented foods are often much safer than the raw substrates from which they are prepared. For example, raw milk consumption was responsible for transmittal of several diseases with disastrous consequences in the United States in the early 1900s (Bryan 1983). In 1985, the largest *Salmonella* outbreak on record in the United States, affecting more than 16,000 humans, was caused by consumption of contaminated milk (Ryan et al. 1987). Fermented milk products have been responsible for considerably less disease transmission than fluid milk (Bryan 1983, Bryan 1988). The implementation of various technologies in the dairy industry including raw milk management, milk pasteurization and heat treatment, starter culture management, and curing processes have significantly reduced foodborne disease transmitted by fermented dairy products (Johnson et al. 1990).

Other fermented foods, including fermented vegetables and sausages, have infrequently transmitted foodborne disease. Bryan (1988) estimated that of 766 reported outbreaks of foodborne disease in the United States during 1977 to 1982, only 12 (1.6 percent) were caused by inadequate fermentation. During 1961 to 1982, only 25 of 1918 outbreaks (1.3 percent) were caused by improper fermentations (Table 6.4). In other countries fermented products are also a relatively safe class of foods (Hesseltine 1985). The safety record is particularly significant when it is considered that fermented foods comprise 25 percent or more of the diet of humans in many parts of the world.

TABLE 6.3 Recognized Pathogen and Intoxicant Risks in Certain United States Fermented Foods

Food product	Pathogen or intoxicant risks	Source of pathogen	Probable processing inadequacies or conditions causing risk
Fermented meat or poultry	*Salmonella*	Infected animal; environmental contamination	Poor acid formation; improper A_w; growth during processing
	Staphylococcus aureus	Infected animal or human; soil contamination	Poor acid formation; improper A_w; growth during processing
	Listeria monocytogenes	Infected animal; environmental contamination	Poor acid formation; improper A_w; growth during processing
Fermented fish	*Clostridium botulinum*	Water contamination	Improper acid formation; improper A_w; survival of toxin during process; growth during process
Cheese	*Salmonella*	Infected animal; environmental contamination	Poor acid formation; improper A_w: organism survives processing; growth during processing; insufficient duration of aging; postprocessing contamination
	Staphylococcus aureus	Infected animal or human	Poor acid formation; Improper A_w; organism survives processing; growth during processing; insufficient duration of aging
	Brucella	Infected animal or worker	Poor acid formation; improper A_w; organism survives processing; growth during processing; insufficient duration of aging
	Enteropathogenic *Escherichia coli*	Infected animal or worker	Poor acid formation; improper A_w; organism survives processing; growth during processing; insufficient duration of aging; postprocessing contamination

TABLE 6.3 Recognized Pathogen and Intoxicant Risks in Certain United States Fermented Foods (*Continued*)

Food product	Pathogen or intoxicant risks	Source of pathogen	Probable processing inadequacies or conditions causing risk
Cheese (*Cont.*)	*Listeria monocytogenes*	Infected animal; environmental contamination	Poor acid formation; improper A_w; organism survives processing; growth during processing; insufficient duration of aging; postprocessing contamination
	Histamine		Poor acid formation; improper A_w; multiplication of producing organism during processing

TABLE 6.4 Outbreaks of Foodborne Disease Caused by Faulty Fermentations, United States, 1961–1982

Disease[a]	No. of outbreaks due to improper fermentation[b]	Percent of total outbreaks
Salmonellosis	2	0.6
Staphylococcal food poisoning	2	0.7
Botulism	18	12.2

[a]No outbreaks of food poisoning by improper fermentations were attributed to *C. perfringens, Shigella* sp., *Vibrio* sp., *Salmonella typhi, B. cereus*, or viruses during 1961–1982.
[b]Overall, Bryan (1988) reported that improper fermentation resulted in 16 outbreaks of foodborne disease in homes from 1973–1982 (4.6 percent of total), and 8 occurred in commercial food processing plants (10.7 percent of total).
SOURCE: Adapted from Bryan, 1988.

Microbiological safety of fermented foods depends on: (1) raw materials that are free of toxic substances that are stable during fermentation, and (2) prevention of pathogen growth and survival during fermentation and subsequent storage of the processed food. Representative outbreaks that have occurred in fermented and related foods are presented in Table 6.5. Fermented foods prepared from animal-derived substrates, such as meat and dairy products, have caused a higher incidence of foodborne disease in the United States than those of botanic origin such as sauerkraut, breads, and fermented cucumbers. The majority of the outbreaks listed in Table 6.5 have occurred in foods that were prepared in the home where starter cultures were not used.

TABLE 6.5 Representative Outbreaks of Foodpoisonings in Fermented and Related Foods

Etiologic agent	Foods	Comments	Country	Reference
Clostridium botulinum	Blood and liver sausages	Mortality ~50%	Germany	Kerner, 1817
Clostridium botulinum	Fermented bean curd from 1958–1983	705 outbreaks	China	Ying and Shuyan, 1986
Clostridium botulinum	Fermented marine fish and mammals	Home prepared	Alaska, Canada	Hauschild, 1989
Clostridium botulinum	"Izushi," "kirikomi"		Japan	Nakano and Kodama, 1970
Clostridium botulinum	Home-preserved vegetables	Low-acid foods	Various	Hauschild, 1989
Clostridium botulinum	Fermented trout		Scandinavia	Hauschild, 1989
Clostridium botulinum	Soft cheese		France Switzerland	Johnson et al., 1990
Salmonella sp.	Cheddar		United States	
	Cheddar Vacherin cheese		Canada Switzerland, France, United States	Johnson et al., 1990
Salmonella typhimurium	Salami sticks		England	Cowden et al., 1989
Brucella	Dairy products	Brucellosis	Various, esp. Mexico	Bryan, 1983.
Yersinia enterocolitica	Tofu	Isolated from water in manufacturing plant		Aulisio et al., 1983
Pseudomonas cocovenenans	Tempeh bongkrek	Spasms, death		van Veen, 1967
Flavobacterium farinofermentans	Fermented corn meal	Severe intoxication		Hu et al., 1984; Anonymous, 1980
Staphylococcus aureus	Sausages	Staphylococcal food poisoning		Bacus and Brown, 1981
	Genoa salami			Bergdoll, 1989

TABLE 6.5 **Representative Outbreaks of Foodpoisonings in Fermented and Related Foods** *(Continued)*

Etiologic agent	Foods	Comments	Country	Reference
Staphylococcus aureus (*Cont.*)	Various cheeses	Slow or failed starters	United States	Johnson et al., 1990
	Swiss cheese	Contaminated starters	Canada	Johnson et al., 1990
Listeria monocytogenes	Soft cheese Vacherin cheese		United States Switzerland	Johnson et al., 1990
Shigella sonnei	Brie cheese		Scandinavia	Johnson et al., 1990
Enteropathogenic *Escherichia coli*	Brie cheese		France, consumed in United States	Johnson et al., 1990

Control of Pathogens and Toxins in Specific Fermented Foods

Fermented meats

Freshly slaughtered meats are an excellent growth substrate for many spoilage and pathogenic microorganisms (Mossel and Ingraham 1953). Inherent defenses in meats are lost on slaughter, and colonization by pathogenic organisms can occur readily. It was recognized many years ago that addition of salt and sugar to raw meats followed by a suitable incubation period prevented spoilage by various microorganisms, particularly gram-negative enteric bacteria (Bacus and Brown 1981). Later it was found that salt-resistant gram-positive bacteria converted the sugar to lactic acid and used the available oxygen. The resulting high acidity and anaerobic conditions prevented growth of pathogens and spoilage organisms. Prior to 1940, meat fermentations were usually carried out by flora naturally present in the meats or by backslopping from a previous fermentation. The success with starter cultures in dairy fermentations led to the eventual adoption of starters in fermented meats (Bacus and Brown 1981, Bacus 1984, Nurmi 1966). Starter cultures are now routinely employed for many sausage fermentations and typically contain homofermentative lactobacilli and pediococci (*Lactobacillus plantarum, Lactobacillus curvatus, Lactobacillus sake, Pediococcus acidilactici, Pediococcus pentosaceus*) and Micrococcaceae (*Micrococcus varians, Staphylococcus xylosus*, and *Staphylococcus carnosus*) (Bacus and Brown 1981, Hammes 1990). Since meat is usually fermented raw, the starter organisms must compete effectively with other microorganisms present in the meat substrate. This differs from dairy fermentations, in which pasteurization or thermization of milk is usually carried out, destroying much of the microflora and providing the starters with a competitive advantage. Controlled studies have clearly shown that starter cultures lower the pH more rapidly than the inherent microflora in a variety of food systems (Hammes 1990). In meats, the American Meat Institute recommends that a suitable temperature and time of fermentation be employed to rapidly lower the pH to ≤ 5.3 (American Meat Institute 1982). The addition of chemical acidulants together with starters rapidly lowers the pH and has been recommended for sausage fermentation (Bacus 1984, Metaxopoulos et al. 1981). At least four groups of microorganisms are involved in the ripening process of dry sausages, including lactic acid bacteria, yeasts, molds, and Micrococcaceae (Bacus and Brown 1981, Leistner, 1990). These secondary organisms help to prevent pathogen contamination by lowering water activity, spatially excluding pathogens, and activating curing agents.

Pathogens of known concern in dry and semidry fermented sausages include *Staphylococcus aureus, Salmonella* sp., *Listeria monocytogenes*, and *Clostridium botulinum* (Genigeorgis 1976, Hauschild, 1989, J. L. Johnson et al. 1990). Prevention of pathogen contamination in fermented sausages depends on the actions of competitive flora (lactic acid starter and secondary cultures), which lower the redox potential and water activity, and produce nitrite and acidity (Barber and Deibel 1972, Daly et al. 1973, Hammes 1990, Meisel et al. 1989). Hygiene during processing and control of the levels of contaminants in raw meats and ingredients also contributes to the safety of fermented meats.

Staphylococcus aureus has caused outbreaks of food poisoning in fermented sausages by production of stable enterotoxins early in the fermentations (Bryan 1980). The primary reservoirs of *S. aureus* are the human nose and other areas of the skin, and infected animal tissues. Staphylococci were detected in over 50 percent of raw pork sausages for retail sale, and 95 percent of sampled sausages had log counts of 2.4 to 2.7 colony-forming units (cfu) per gram (Genigeorgis 1976). Barber and Deibel (1972) reported that staphylococcal counts of 4×10^7 cfu/g of sausage were necessary to produce detectable enterotoxin. Factors that may increase the initial numbers or competitive ability of *S. aureus* include: (1) use of raw meat from animals infected with staphylococcal lesions, (2) freezing of substrates prior to fermentation, and (3) bacteriophage infection or chemical inhibition of the starter cultures (Genigeorgis 1976). Bryan (1980) attributed an incident of staphylococcal poisoning in Genoa salami to high initial numbers of *S.*

aureus in the raw meat and a favorable temperature for staphylococcal growth. Several investigators have studied growth and survival of *S. aureus* during meat fermentations. *S. aureus* competes poorly with lactic acid–forming organisms in most fermented meats (Genigeorgis 1989). However, if acid formation is slow, then *S. aureus* can grow to numbers sufficient to produce enterotoxin. The primary attribute of *S. aureus* that enables potential growth in dry and semidry sausages is the ability of the pathogen to grow at low water activity. The minimum A_w for growth under anaerobic conditions was 0.90 (Genigeorgis 1976), which decreased to 0.83 to 0.86 in aerobic conditions (Genigeorgis 1976). An aerobic environment favors development of *S. aureus*. Barber and Deibel (1972) demonstrated that staphylococcal growth was localized in outermost areas of sausages where oxygen was available. Aerobic and anaerobic staphylococcal growth could be prevented by chemical acidulation with 1.5% glucono-delta-lactone. Biological acidification with a high inoculum of *Pediococcus cerevisiae* only partially prevented aerobic growth. Increasing the incubation temperature was found to favor starter activity and prevent staphylococcal growth (Peterson et al. 1964). Inhibitory substances produced by competitive bacteria include hydrogen peroxide and antibiotic-like substances (Troller and Frazier 1963, Haines and Harmon 1973, Hurst 1973). Competition for essential nutrients may also contribute to control of *S. aureus* (Smith and Palumbo 1981). Micrococci involved in sausage fermentations produce antibiotic-like substances (micrococcins) (Heatley and Doery, 1952) that may help to exclude staphylococci.

Salmonella contamination of fermented sausages has resulted in foodborne illness outbreaks (Cowden et al. 1989, Smith et al. 1975). Generally, salmonellae are poor competitors during acidogenic meat fermentations, but slow acid production promotes their survival. Salmonellae present in beef-pork mixtures were reduced by lactic starter cultures (Goepfert and Chung 1970, Masters et al. 1981, Smith et al. 1975). Low pH in conjunction with salt was the most important factor in decreasing the numbers of salmonellae. It was observed in some experiments that low numbers of salmonellae persisted during drying in fermented sausages. The surviving salmonellae were not characterized, but it is possible that they adapted to the acid-salt conditions. Salmonellae have been demonstrated to induce protective proteins and metabolic pathways when stressed by salt or acid (Foster and Hall 1990, Ingraham, 1987). It would be of interest to determine if acid-salt-adapted salmonellae survive better than unadapted cultures during food fermentations that select for nonpathogens by acid and salt.

Clostridium botulinum has long been recognized as a potential hazard in fermented meat products (Hauschild 1989, Meyer and Eddie 1965, Tompkin 1980). The incidence of botulism outbreaks in meat products (excluding fermented fish, see below) has been higher in European countries than in the United States and Canada (Tompkin 1980), probably due to increased numbers of spores associated with the raw meats in Europe. Growth of *C. botulinum* is inhibited during sausage fermentations by sodium chloride, acidity, and "curing salts" (nitrite) (Roberts and Ingram 1973). Carbohydrate addition to the fermentation promotes inhibition of *C. botulinum* by increased acid formation (Christiansen et al. 1975, Tanaka et al. 1980).

Concern has recently been raised that *Listeria monocytogenes* could be a hazard in fermented sausages. *L. monocytogenes* is salt-tolerant and can grow at an A_w of 0.92 (John Troller, personal communication). *L. monocytogenes* is also relatively resistant to other environmental stresses, including acid. The pathogen is frequently associated with raw meats and has been detected in cured and fermented meat products (Buchanan et al. 1989, Farber et al. 1988, J. L. Johnson et al. 1990). No outbreaks of listeriosis have been attributed to fermented meats.

Challenge studies indicated that *L. monocytogenes* survived but did not grow during sausage fermentations, and that cell viability decreased during the drying period (Farber et al. 1988). J. L. Johnson et al. (1988) found that during the preparation of hard salami, *L. monocytogenes* viable counts decreased by about 1 $\log_{10}$ cfu/g during the fermentation, and also decreased during the 9-day drying period. Similar results have been found in other laboratory-fermented sausages (Johnson et al. 1990). *L. monocytogenes* survived in hard salami (A_w = 0.79) for 12 weeks. To eliminate *L. monocytogenes* from fermented meats, it is probably necessary to promote a vigorous acidogenic fermentation. Certain pediococci and lactobacilli produce bacteriocins which inactivate *L. monocytogenes* (Berry et al. 1990, Pucci et al. 1988, Schillinger and Lucke 1989). Berry et al. (1990) reported that a bacteriocin-forming culture of *Pediococcus* inactivated a proportion of *L. monocytogenes* in semidry fermented sausage but did not eliminate the pathogen if large numbers were initially present. The bacteriocin appeared to inhibit independently from acid formation. Bacteriocins may have potential applications for the control of *L. monocytogenes* and other pathogens in fermented foods. More research is needed to evaluate their effectiveness.

Several studies have demonstrated that properly conducted sausage fermentations successfully prevent pathogen growth and survival. A guideline for control of bacterial pathogens in shelf-stable fermented meats specifies the following: (1) pH < 5.0, (2) $A_w < 0.91$, or (3) pH ≤ 5.2 and $A_w \leq 0.95$ (Hauschild 1989).

During the drying of fermented sausages, mycotoxigenic molds can develop on the surface of the sausages (Leistner 1990). Aged salamis and sausages may potentially become contaminated with aflatoxins, ochratoxin A, patulin, penicillic acid, sterigmatocystin, and penitrem (Bullerman 1986). Attempts have been made to prevent mold growth with preservatives, including pimaricin and sorbates, or by inoculation with mold spores from known nontoxigenic strains of fungi (Leistner 1990). Use of mold starter cultures could provide product consistency in the surface mold content and increase the microbiological safety of the products. Molds have also been found to be inhibitory to certain bacterial pathogens, including *L. monocytogenes* (Leistner 1990).

Little is known of the potential for transmittal of pathogenic viruses in fermented meats. Asian swine fever virus survived 15 days of the curing period in dried pepperoni and salami (Blackwell et al. 1985). Swine vesicular disease virus was reported to survive for at least 400 days in pepperoni and dried salami that underwent a controlled acid fermentation (Blackwell et al. 1985). Several mammalian viruses, including Newcastle disease virus, canine hepatitis virus, a porcine picornavirus, and diarrheal bovine viruses, were inactivated to various extents in food wastes inoculated with *Lactobacillus acidophilus* or *Saccharomyces cerevisiae* (Gilbert et al. 1983). Virus inactivation was attributed to acid and heat.

Fermented fish products

Finfish and shellfish are usually extremely perishable commodities that undergo rapid spoilage and can also support growth of pathogens. Fish tissue has a high protein content, relatively high water activity and pH, and low carbohydrate content compared to some other foods. Fish muscle usually has a pH of 6.2 to 6.5 and little sugar or glycogen is present. The absence of available carbohydrate in fish flesh prevents rapid acidification by the natural flora or by added cultures. During decomposition ammonia and amines are released causing the pH to remain near neutrality. This provides an environment that can potentially support growth of several pathogens, including *Clostridium botulinum*. Traditionally, preservation was achieved by heavy salting and drying to lower the water activity, and by smoking (Campbell-Platt 1987).

Pathogenic microorganisms associated with fish and crustacean products include the primary hazards *C. botulinum* (type E, nonproteolytic B and F) and *Vibrio* sp. Other potential bacterial pathogens include *L. monocytogenes, Clostridium perfringens, Erysipelothrix rhusiopathiae, Edwardsiella tarda, Salmonella* sp.,

Shigella, and *S. aureus* (Bryan 1980, Liston 1980). Incidental contamination by *Shigella*, virulent salmonellae, and viruses is a high risk in seafoods harvested from polluted waters. Oysters, clams, and other shellfish have been demonstrated to transmit viral diseases when eaten raw or undercooked, but viruses have not been incriminated as vehicles in fermented fish products. Seafoods may also become contaminated with potent toxins, including saxitoxin and tetrodotoxin produced by dinoflagellates during blooms in the oceans. Transmission of these toxins in fermented fish has not been reported (Bryan 1980, Liston 1980). Botulism has been periodically transmitted in fermented fish products, usually in home-prepared items including *pindang* (fermented fish product of Indonesia), Japanese *izushi*, and fermented products of native Americans in Alaska and Canada (Bryan 1980, Campbell-Platt 1987, Hauschild 1989, Meyer 1956, Wainwright et al. 1988).

In Japan, fermented fish products are often prepared by mixing small fish, chopped fish, and sometimes plankton, and salt is added to about 20% (w/w). Fermentation is carried out at 25 to 35°C for several months. The clear liquid obtained is sold as well as the undigested residue (*bagoong*). The high concentration of salt added to the fish substrate prevents growth and survival of pathogens. Other fermented fish products are potentially more susceptible to pathogen colonization. *Izushi*, a fermented product prepared from raw fish, rice, and vegetables in Japan, has been responsible for several botulism outbreaks (Nakano and Kodama 1970). From 1951 to 1968 in Japan there were 57 outbreaks with 83 deaths of botulinal poisoning, most caused by *izushi*. Although *izushi* is produced in various regions of Japan, nearly all botulinal outbreaks have occurred in the Hokkaido and northeastern regions of Japan, probably because salting is less often practiced in these areas. Starter cultures are usually not used, although some preliminary studies on their use have been carried out with encouraging results (Schubring and Kuhlmann 1978). Carbohydrates such as rice or cassava may be used as adjuncts in fish fermentations in tropical countries to promote acidification (Cooke et al. 1987). Twiddy et al. (1987) found that the addition of sugar to fermentations in a fish-salt-starchy substrate model system promoted a rapid drop in pH and resulted in the inhibition of several bacteria including *S. aureus, Salmonella typhimurium, Clostridium sporogenes*, and *E. coli*.

Native American and Eskimo populations in Canada and Alaska have experienced relatively frequent outbreaks of botulism from fermented salmon eggs, whale meat (*muktuk*), seal meat, beaver tail, and other substrates (Dolman 1960, Wainwright, 1988). Since 1947, 59 confirmed or suspected outbreaks of botulism with 156 cases were reported in Alaska (Wainwright 1988). The overall case-fatality rate was 11 percent (17 deaths). The majority of the cases associated with fish were attributed to type E botulinum toxin, while those associated with marine or terrestrial animals were usually types A and B. Botulism was mostly transmitted in fresh fermented products (62 percent of the outbreaks), although significant numbers were also associated with dried and frozen fermented foods. Botulinum toxin is readily destroyed by heat, but it maintains toxicity when dried or frozen, particularly at low pH. Traditional Eskimo fermented foods are usually prepared by placing the food in a pit lined with wood or animal skin, which is covered with moss and allowed to incubate (Wainwright 1988). Recently, fermentation in plastic bags (which provides an anaerobic environment favorable for growth of *C. botulinum*) has resulted in several botulinogenic foods (Eisenberg and Bender 1976). The modern practice of incubating the containers near the home stove also provides a favorable growth temperature for *C. botulinum*. These fermentations are not usually controlled by salting or acid production. Putrefaction occurs during the fermentation and *C. botulinum* develops and produces its potent toxin. The foods are eaten without heating, which would destroy the toxin present.

The tendency of fermenting fish and mammals to become botulinogenic compared to other food substrates is interesting and suggests that the composition of the substrate, for example, its protein and carbohydrate contents, associated microflora and other characteristics promote toxin formation by *C. botulinum*. The major factors encouraging toxin formation include the initial load of spores in the substrate, nutritious and anaerobic growth conditions, and the lack of fermentable carbohydrate in the fish. *C. botulinum* spores are widely distributed in soil and fresh and brackish waters, and most raw foods that contact soil or dust are contaminated. *C. botulinum* has been demonstrated to grow best in complex substrates that satisfy the fastidious nutritional requirements of the organism. Most foods contain sufficient nutrients for growth, and therefore growth must be prevented by processing, including destroying the spores by heating, addition of preservatives or acidification, or storing the foods at refrigeration temperatures. Currently there is concern in the food industry that minimally processed refrigerated foods and low-acid foods (i.e., pH > 4.6) can support botulinogenesis (Conner et al. 1989).

Fermented Dairy Products

Milk is a nutritious, perishable commodity and has been responsible for transmission of severe foodborne diseases (Bryan 1983). Prior to promulgation of pasteurization and widespread use of refrigeration, milk and milk products were the vehicle of thousands of cases of typhoid fever, tuberculosis, scarlet fever, brucellosis, and various diarrheal diseases (Bryan 1983). Currently, dairy products are among the United States' safest foods. Mixed-culture fermentations are extremely important for preserving the quality and safety of fermented dairy products. The safety of dairy products has been thoroughly reviewed (Bryan 1983, D'Aoust 1989, Johnson et al. 1990), and is not discussed in detail here. Pathogens of current concern include *Salmonella* sp., *L. monocytogenes*, and enteropathogenic *E. coli* (D'Aoust, 1989; E. A. Johnson et al., 1990).

Pathogen contamination in fermented dairy products is prevented by a combination of practices, including hygienic bulk milk handling, heat treatments of the milk, controlled lactic fermentations, and ripening conditions. Much work has been done on the factors that contribute to the safety of milk fermentations (Johnson et al. 1990). A major contributing factor is rapid acid production by lactic acid bacteria. Other inhibitory compounds produced by starters include acetate, hydrogen peroxide, and natural antibiotics. The requirement for starter cultures in the dairy industry has increased in recent years with widespread use of heat treatment, which eliminates the natural bacterial flora. Occasionally acid formation is slowed due to starter inhibition by residual antibiotics from mastitic cows undergoing antibiotic therapy, by faulty starter culture management, and by trace quantities of sanitizers. The most frequent cause of poor starter development is infection of the starter cultures by bacteriophages through aerosols and whey contamination (see Chap. 5). Various dairy products involve different rates of acid formation by the starter cultures, which is influenced by temperature, salt concentration, and pH (Lawrence and Thomas 1979). Acid production may continue in the presence of high salt concentration. End-product fermentation by lactococci is influenced by the availability of a carbon source. When glucose or lactose is present in excess, then lactic acid formation predominates, but when sugar is restricted, then heterofermentative metabolism is prevalent. At very low sugar concentrations the principal products are acetate, ethanol, and formate and little lactate is made (Lawrence and Thomas 1979).

Certain dairy products appear to be more likely to transmit microbial disease. Semisoft surface-ripened cheeses may be particularly prone to pathogen infection because of their relatively low salt content, and the increase in pH that occurs on the surface by mold metabolism during ripening. Mold growth on cheddar cheese enabled colonization by staphylococci (Duitschaever and Irvine 1971). Postfermentation contamination appears to be the most common means by which pathogens get into dairy products. Secondary preservation systems, including bacteriocins and antimicrobial enzymes such as lysozyme, potentially could help to eliminate pathogens that enter into products after the lactic fermentation is completed.

Fermented fruits and vegetables

In several countries throughout the world mixed-culture fermentations are used for the preservation of vegetables, including cabbage, carrots, cauliflower, cucumbers, radishes, beans, beets, chard, turnips, and peppers. Fermentation is an excellent means of preservation and there are no documented cases of foodborne disease from commercially pickled vegetables in the United States (Fleming and McFeeters 1981). The ferment of acid and salt prevents growth and survival of pathogens. Also, plant tissue is low in protein but high in carbohydrate, which promotes rapid acid production. During the spontaneous fermentation of plant tissues, the small numbers of lactic acid bacteria and yeasts initially present become the dominant population excluding various other microorganisms (Daeschel et al. 1987). Salting is important to the success of the sauerkraut fermentation. Not only does salting select for gram-positive bacteria, but it also plasmolyzes the cabbage cells and releases nutrients. Several bacterial species use the nutrients and also hydrolyze polymeric components, producing assimilable nutrients required by lactic acid bacteria. Eventually the fermentation is taken over by *Leuconostoc mesenteroides*. The reasons that *L. mesenteroides* dominates at this stage are not clear. In turn, acid-requiring lactobacilli ferment the remaining carbohydrate substrate and lower the pH to about 3.5. The sauerkraut fermentation illustrates how a succession of organisms changes the environment chemically and physically and sets the stage for dominance by a new species. A key to sauerkraut preservation is the final homolactic fermentation by *Lactobacillus*, an organism which requires low pH conditions in order to develop maximally.

Although commercially pickled vegetables have an excellent safety record, home-fermented vegetables and fruits have caused outbreaks of botulism. Botulism outbreaks have occurred in high-acid foods prepared in the United States that underwent fermentation: 14 outbreaks in fruit juices, 18 in tomato products, 3 in pickles, and several in various high-acid fruits and vegetables (Meyer and Eddie 1965). Certain studies have indicated that *C. botulinum* can produce toxin in acid foods, including fruits and vegetables (Odlaug and Pflug 1978). The minimum pH at which *C. botulinum* can grow depends on the initial pH, temperature and time of incubation, the characteristics of the food substrate, initial load of *C. botulinum* spores, and the presence of associated microflora. Incidences of botulism in acid foods often involve a metabiosis (Frazier 1967), in which bacteria and fungi in the food metabolize organic acids present and raise the pH (Meyer and Gunnison 1929, Odlaug and Pflug 1978). This occurs infrequently, since spoilage organisms associated with most foods tend to lower the pH by acid formation. Meyer and Gunnison (1929) found that inoculation into an acid medium of a *Lactobacillus* sp. with *C. botulinum* prevented toxin production, whereas toxin and growth resulted when a yeast and *C. botulinum* were cultured in the acid medium. The pHs of the media ranged from 3.33 to 4.22. Tanner et al. (1940) detected botulinum toxin in fruits inoculated with *C. botulinum* spores when *Penicillium* sp. or *Mycoderma* sp. were also present. The fungi raised the pH to ~ 5 from the initial pHs of 3.25 to 3.75. Other fungi and bacteria have been demonstrated to create a favorable environment for *C. botulinum* growth. Bachmann (1924) reported that juice expressed from string beans containing 2.5% salt was botulinogenic after a two-month fermentation. Salted cabbage juice prepared in the same manner and fermented did not support toxin formation. Bachmann (1924) proposed that the higher content of protein in the beans promoted toxin production, whereas lactic fermentation prevented *C. botulinum* development in the cabbage juice. In addition to preventing *C. botulinum* growth by lowering the pH or scavenging essential nutrients, lactic acid bacteria, lactococci, and certain other bacteria have been reported to destroy preformed botulinum toxin (Dack 1926, Stark et al. 1929).

Evidence has been presented that in special circumstances *C. botulinum* can grow at pH less than 4.6. In 1979, Raatjes and Smelt showed that types A and B of toxin were produced in media containing 3% protein at pH 4.0 if *Bacillus subtilis* or *Bacillus lichinoformis* spores (100 per gram) were added. Apparently utilization of the protein with consequent production of ammonia enabled toxin formation. Tanaka (1982) showed that toxin was produced by mixtures of spore types A and B at pH values from 4.24 to 4.40 in the absence of *Bacillus*. Conditions and events are not completely understood of the factors that allow *C. botulinum* growth in high-acid foods since attempted repetition of some of the experiments have proved negative. In nearly all foods it is unlikely that *C. botulinum* can grow below pH 4.6 unless unpredicted pH changes occur by associated fungal or bacterial activities or unique microenvironments are established.

Fermented Legume and Cereal Products

Fermentations of legumes and cereals often involve mixed cultures of molds, yeasts, and bacteria (Hesseltine 1965, Hesseltine and Wang 1986). Potential microbiological safety hazards include transmission of bacterial pathogens, botulism, poisonings mediated by interesting low-molecular-weight bacterial toxins, and mycotoxin poisonings.

In China, several outbreaks of botulism have occurred in home-fermented bean curd and broad bean sauce (986 outbreaks from 1958 to 1983) (Ying and Shuyan 1986). Epidemiology of botulism is unique in China: 824 outbreaks (86 percent of total recognized) have been caused by bean products, most (74 percent) by improperly fermented bean curd. Botulism occurs mainly in the winter and spring when fresh vegetables are scarce and homemade fermented foods are frequently prepared. Types A, B, and E botulism have been detected in China; type A has accounted for 658 (93.4 percent) of the cases, type B for 37 cases (5.0 percent), mixed types A and B for 3 cases (0.4 percent); and type E for 8 (1.0 percent) (Ying and Shuyan 1986). Specific regions have been linked to toxin types. Botulism incidence in China has decreased dramatically since 1974 because of education of households concerning the hazard, adequate supervision of food factories, inclusion of 10% salt during fermentations, and heating of foods prior to consumption (Ying and Shuyan 1986).

The microbiological safety of *tempeh* prepared from unacidified soybeans was examined by Tanaka et al. (1985b). During laboratory preparation of *tempeh* several pathogens grew, including *Salmonella typhimurium, Staphylococcus aureus, Yersinia enterocolitica*, and *Clostridium botulinum*. Botulinum toxin was produced within 2 days of the soybean fermentation when 139 botulinum spores per gram of *tempeh* were initially present. *Salmonella, Staphylococcus*, and *Yersinia* also survived during *tempeh* preparation. Since all four pathogens survived or multiplied during the fermentation, it is possible that *tempeh* could present a foodborne hazard. Measures should be taken to prevent contamination of the raw product and fermentations should be conducted under hygienic conditions. Samson et al. (1987) reported that the microbiological quality of commercial *tempeh* in the Netherlands was poor. In 110 samples examined, Enterobacteriaceae numbers exceeded 10^5 cfu/g, and *Staphylococcus aureus* and *Bacillus cereus* exceeded 10^5 cfu/g in 13 and 11 percent of the samples, respectively. *Yersinia enterocolitica* was detected in six samples, whereas *Salmonella* was not detected in any of the 25-g samples. To produce a safe *tempeh*, the authors' recommended acidification of the beans, adequate cooking, maintaining hygienic conditions during processing, and storage of the product below 7°C. Hesseltine (1985) pointed out that consumption of *tempeh* is believed to help prevent bacterial infections in humans, and may have beneficial effects for people suffering from dysentery. Some molds used in Oriental food fermentations form antibiotic-like compounds that inhibit bacteria and may be beneficial for human health (Wang et al. 1969).

Tofu, or soybean curd, prepared from soybean mash was implicated in an outbreak of yersiniosis affecting 87 persons in the United States (Aulisio et al. 1983). It is likely that postprocessing handling errors resulted in product contamination, since the manufacture of *tofu* involves two heating steps that are sufficient to kill vegetative bacteria. Szabo et al. (1989) examined 153 samples of *tofu* from 14 manufacturers in Canada for microbiological quality. *S. aureus* counts were generally less than 250 cells per gram, *Y. enterocolitica* was detected in four samples, and *Salmonella* was not detected. The levels of psychrotrophs exceeded 10^6 per gram in more than 45 percent and levels of coliforms exceeded 1000 per gram in more than 35 percent of the *tofu* samples. The high numbers of psychrotrophs detected suggested a potentially unsafe product and need for consistent sanitation and good manufacturing practices.

Another traditional oriental fermented food whose microbiological safety has been questioned is *miso*. Traditionally, *miso* has been prepared since 1600 A.D. in Japan (Shurtleff and Aoyagi 1976). Currently, *miso* is produced by first preparing *koji* by growing *Aspergillus oryzae* on moistened soybeans mixed with rice or barley. Following growth of the mold, the *koji* is salted and fermented. The principal microorganism necessary for the fermentation is *Saccharomyces rouxii*, an osmotolerant yeast, which begins to grow when the pH has dropped below 5.0 (Hesseltine and Wang 1986). Traditional miso is a relatively high salt product, containing 5.5 to 13% (w/w) in fresh miso (Shurtleff and Aoyagi 1976). Tanaka et al. (1985a) evaluated the bacteriological safety of low-salt *misos* (2.36 to 5.79% NaCl) obtained from Japan, and found that several pathogens did not grow or produce toxin including *C. botulinum, Staph. aureus, S. typhimurium*, and *Y. enterocolitica*. Inoculated pathogens progressively died during holding. The *misos* were not susceptible to pathogen infection, probably owing to the combination of relatively high NaCl (2.36 to 5.79%), low pH values of 5.26 to 4.73, and unidentified inhibitory factors in the fermented product.

Traditional fermented vegetable protein foods produced in nonaseptic conditions generally depend on low water activity and pH to prevent pathogen development. In *moromi* mash used in soy sauce fermentation, salt-tolerant organisms dominate including *Pediococcus halophilus, Saccharomyces rouxii*, and some species of *Torulopsis*. At the start of the fermentation, the pH of the *moromi* mash is around 6.5 and the salt content is over 18% (w/v) (Fukushima 1985). *P. halophilus* dominates early and produces lactic acid, lowering the pH over one month to about 5.0. Yeasts begin to grow as the pH decreases and the fermentation becomes alcoholic, principally involving *S. rouxii*. In the last stage *Torulopsis* species produce alkylphenols and aromatic alcohols which contribute to the microbiological safety of the food. The minimum Aw for growth of *P. halophilus* is 0.81 and is 0.79 to 0.81 for *S. rouxii* and *Torulopsis*, corresponding to 24 to 26% salt. *S. rouxii* and *Torulopsis* produce polyols during fermentation in high salt concentrations which promote osmotic tolerance in the yeasts. In the high-salt (> 18%) environment, *S. rouxii* grows only in the pH range 4 to 5, which is a special attribute of the yeast. It requires biotin, thiamine, pantothenic acid, and inositol for growth, and the inositol requirement increases at high salt concentrations. Enzymes formed by molds in the *koji* production supply inositol through digestion of phytin and phosphatidyl inositol in the soybeans. The soy sauce fermentation illustrates the specialized attributes of individual organisms and environmental conditions that select for their growth.

In certain regions of China severe outbreaks of food poisoning and numerous deaths have occurred from consumption of fermented corn meal (anonymous, 1980, Hu et al. 1984). During toxigenic corn flour fermentations a bacterium, *Flavobacterium farinofermentans*, contaminates the desired white fungus (wood ear or silver ear) and produces a potent toxin (Anonymous, 1980). The toxin is of low molecular weight and apparently causes neurological damage and cardiovascular blockage (Hu et al. 1984). Presently there is much concern in China about corn meal poisonings (F. S. Chu, personal communication). In 1984 to 1985 there were 7 outbreaks involving 330 people, of which 140 developed severe symptoms and 48 patients died. Levels of toxin contamination of 2 to 144 ppm have been detected in the fermented corn flour. The consumption of as little as 30 g of silver ear may have caused the death of a 6-year-old child. The LD_{50} of the toxin has been reported to be about 3 mg/kg for mice (F. S. Chu, personal communication).

Food poisonings have also occurred from the consumption of fermented *tempe bongkrek*, which uses defatted coconut as a raw material. In 1977 an outbreak affected more than 400 persons and caused 70 deaths (Gandjar 1986), and in 1980 an outbreak involved 113 people with 20 deaths (Gandjar 1986). Since 1951 nearly 1000 fatalities have been reported (Buckle and Kartadarma 1990). In the preparation of *tempe bongkrek*, coconut presscake is crushed, washed, and steamed; soybeans are added; the whole mixture is steamed and inoculated with a traditional starter that contains spores of *Rhizopus oligosporus* (Rusmin and Ko 1974). The inoculated cake is packed in banana leaves and fermented for 2 days at ambient temperature. In normal fermentations *Rhizopus* grows rapidly and the white mycelium completely permeates the cake, excluding growth of other organisms. In abnormal fermentations toxin-producing organisms successfully compete with *Rhizopus*. Early bacteriological investigations indicated that a gram-negative bacterium, *Pseudomonas cocovenenans* ("the pseudomonad that produces poison from coconut") formed the toxin (Arbianto 1971). Coconut was found to promote formation by *P. cocovenenans* of bongkrek acid and the less toxic yellow antibiotic toxoflavin (van Damme et al. 1960). Ko et al. (1979) reported that addition of 1 to 2% salt to the raw fermentation substrate reduced the production of bongkrek acid and toxoflavin. This method has not been adopted in the preparation of *tempe bongkrek*, which is still produced in homes in Indonesia (Gandjar 1986). Reduction of the pH to 4.5 with acetic acid and addition of 2% sodium chloride was reported to prevent toxin formation by three strains of *P. cocovenenans* (Buckle and Kartadarma 1990). *Rh. oligosporus* has been reported to produce antibacterial substances active against a wide range of bacterial species (Wang et al. 1969). The causative organism of bongkrek poisoning may be identical to *Flavobacterium farinofermentans*, responsible for the corn meal poisonings described earlier.

Mycotoxin Formation in Fermented Foods

Since the discovery of aflatoxin in 1958, concerns have often been raised that fermented foods could become contaminated with mycotoxins by incidental contamination or by production of mycotoxins during fermentation. The potential for mycotoxin formation by fungi associated with food fermentations has received considerable attention (Leistner 1990, Wang and Hesseltine 1981). Several of the molds involved in food fermentations, e.g., species of *Aspergillus, Penicillium, Neurospora,* and *Monascus*, and others, are closely related to certain mycotoxin-producing fungi. In an interesting study, Wicklow and Kurtzmann (1988) found by DNA hybridization that the industrial *koji* molds *Aspergillus oryzae* and *A. sojae* are domesticated forms of *A. flavus* and *A. parasiticus*, which are mycotoxin producers. However, extensive surveys of fungal cultures and of certain fermented foods have indicated that the cultures used in food fermentations do not produce mycotoxins (Engel and Teuber 1989, Leistner 1990, Wang and Hesseltine 1981). Mycotoxin contamination usually occurs after fermentation. Available information suggests that mycotoxin formation has no biological advantage for molds involved in food fermentations and that nontoxigenic strains may have a selective advantage over toxin producers. Since mycotoxins are mostly secondary metabolites and unnecessary for cell growth, they may not be induced during food fermentations. Environmental conditions and nutrients released during fermentation may repress mycotoxin formation. Although mycotoxins are not formed in properly conducted fermentations, they may be formed in moldy-spoiled products, such as moldy cheeses (Scott, 1989).

Aflatoxin was originally discovered in moldy peanuts, and the Indonesian food *oncom* which contains peanuts was suspected some time ago to contain aflatoxin. Research has indicated, however, that growth of *Neurospora* reduced the aflatoxin levels substantially (van Veen et al. 1968). *Neurospora* also inhibited the growth of toxigenic strains of *Aspergillus flavus* (Ko et al. 1974).

The potential for mycotoxin formation has been examined in mold-ripened cheeses (Engel and Teuber, 1989). Several strains of *Penicillium camemberti* and *Penicillium roqueforti* were examined for production of mycotoxins. No mycotoxins were detected during in vitro culture of cheeses produced with the fungi. Olivigni and Bullerman (1977) showed that certain strains of *P. roqueforti* produced patulin in laboratory culture media, but patulin was not detected during experimental production of cheddar or Swiss cheeses. Similarly, patulin was not detected in cheeses prepared with patulin-forming strains of *P. roqueforti* (Engel and Teuber 1989). Stott and Bullerman (1975) proposed that cheese is a poor substrate for patulin formation, probably owing to the low-carbohydrate and protein-rich qualities. Furthermore, Stott and Bullerman (1976) showed that patulin content decreased when it was added to cheese followed by incubation at 5 or 25°C.

Surveys have not provided evidence that mycotoxins are present at detectable levels in beers, wines, and fruit juices (except apple juice and apple cider) (Jelinek et al. 1989). Surveys have shown that mold species used for fermentation of Oriental foods do not produce mycotoxins, but, in fact, inhibit toxin formation by contaminant molds (Wang and Hesseltine 1981).

Survival of Viruses in Fermented Foods

In contrast to many of the bacterial pathogens, relatively little is known concerning the survival of viruses in foods. Enteroviruses have been shown to endure in a variety of foods (Blackwell et al. 1985). Panini et al. (1989) found that foot-and-mouth virus did not survive during preparation of Italian salami. Cliver (1973) reported that poliovirus and influenza virus decreased by 98 and 100 percent during cheddar cheesemaking. Natural inhibitors of viruses have been detected in milk. Panon et al. (1987) found virus inhibitors in raw but not in heated milk. Because of the long incubation times for viral illness to occur, and the difficulties in detecting and culturing many pathogenic viruses, their persistence in fermented foods and their interactions with microflora of fermented foods require further study. The available evidence suggests that viruses do not survive well in fermented foods.

Mechanisms of Exclusion of Pathogens from Fermented Foods

Microbial interactions in fermented foods usually favor development of beneficial organisms and exclusion of human pathogens or spoilage organisms. Pathogen exclusion is accomplished by deliberate alteration of the food environment (especially by salting) to select for nonpathogens, and use of beneficial organisms that outcompete pathogens for essential nutrients and space and elicit specific antagonisms toward pathogens. Starter organisms have evolved specialized phenotypic traits that favor their dominance, particularly acid and osmotic tolerance and possibly the formation of specific toxins. Known exclusion mechanisms are described in Table 6.6. The most common and important means in food fermentations for the exclusion of pathogens is acid formation by lactic cultures, usually in conjunction with salt stress.

Survival and growth of a pathogen in foods depends on the organism's tolerance to the environment and nutrient requirements. Tolerance limits to chemical and biological factors are interactive, and in most foods absolute limits cannot be defined. Tolerance limits are also often complicated in foods owing to the presence of microhabitats (Troller 1986). The current methods of describing microbial environments (e.g., temperature, pH redox potential) have little meaning when considered on the scale of individual bacterial cells. Despite these limitations, approximate limits have been determined for most pathogens that are applicable to foods (Table 6.7). If these limits are exceeded, then the organism is excluded from the environment. An additional exclusion mechanism is limitation of nutrient availability. Essential nutrients may be limited because of inherent binding substances in the food substrates (e.g. avidin, conalbumin, lactoferrin) or because of scavenging by starter organisms. Starter organisms have specialized nutrient requirements that contribute to their dominance of fermentations. For example, lactic acid bacteria require very low concentrations of iron for growth and are not inhibited by its exclusion from a food. Pathogens and spoilage organisms generally have high

TABLE 6.6 Factors Responsible for Pathogen Exclusion in Fermented Foods

Reduction in water activity (A_w)
Increase in acidity
Reduction of available oxygen
Formation of hydrogen peroxide
Formation of diacetyl
Production of antibiotic-like substances
Depletion of microbial nutrients
Spatial exclusion of pathogens

TABLE 6.7 General Growth Characteristics of Foodborne Pathogens at Risk in Fermented Foods

Pathogen	Growth temp. range (°C)	Lowest pH allowing growth	Maximum NaCl (%) allowing growth	Minimum A_w allowing growth
Clostridium botulinum (group 1)	10–48	4.7	8–10	0.94
Clostridium botulinum (group 2)	3.3–45	5.0	5	0.97
Listeria monocytogenes	1–45	5.0	10	0.92
Salmonella sp.	5.2–45	4.05	8	0.94
Staphylococcus aureus	7–45	4.0 ($+O_2$) 4.6 ($-O_2$)	16–18 ($+O_2$) 14–16 ($-O_2$)	0.83 ($+O_2$) 0.90 ($-O_2$)
Enterotoxin production by *S. aureus*	10–45	4.0 ($+O_2$) 5.3 ($-O_2$)	10 ($+O_2$) 9.5 ($-O_2$)	0.90 ($+O_2$) 0.94 ($-O_2$)
Enteropathogenic *Escherichia coli*	4	4.4	8.0	0.94–0.97

iron requirements. Certain yeasts and molds grow only in high-salt and low-pH environments, as described earlier for the soy sauce fermentation.

Successions of microorganisms often occur during the preparation of fermented foods. Eventually, a stable community may be obtained, which is the final stage of succession, called a *climax community* in ecological terms. Oriental food fermentations usually involve complex yet stable associations of microorganisms (Hesseltine 1985). *Miso* is an example of a solid-state fermentation using a mixed pure culture inoculum (Hesseltine 1985). More than 700,000 metric tons are produced annually in Japan. Polished rice is soaked, steamed to reach a moisture content that will support growth of *A. oryzae* but not bacteria. The mold rice or *koji* provides enzymes for digestion of soybeans to sugars. Sodium chloride is added to the steamed soybeans and mold rice, and pure starter cultures are added. The mixture is packed tightly into a container and an anaerobic fermentation proceeds, which prevents mold growth but allows osmophilic yeasts and bacteria to develop. The addition of salt selects for desirable salt-tolerant organisms including *Saccharomyces rouxii* and *Pediococcus halophilus* (Hesseltine 1985). Because of the high salt content, some *misos* can be kept for long periods without refrigeration. Hesseltine (1985) has pointed out the advantages of solid-state fermentations for production of safe and high-quality fermented foods (see also Chap. 1). Solid-state systems impart spatial distribution of microorganisms and their products, which promotes the formation of microenvironments. Cooperative interactions contribute to successful solid-state fermentations.

Response of Starter Organisms and Pathogens to Osmotic and Acid Stresses

Adaptation to osmotic stress is one of the most important mechanisms by which starter organisms dominate food fermentations. Fungi and bacteria respond to external osmotic stress by accumulating a few intracellular solutes including potassium, amino acids and related compounds, and polyols. This increase in cellular osmolarity enables the flow of water back into the cells. This phenomenon was clearly demonstrated in *Salmonella oranienburg* by Christian (1955), who found that certain nutrients promoted growth in osmotically stressed cells. The compatible organic solutes accumulated by fungi are often carbohydrates and polyols (Rose 1989), whereas in bacteria they are often amino and imino acids (Ingraham 1987). Molds, yeasts, and lactic acid bacteria are uniquely suited to thrive in salted food substrates since they are able to continue metabolism and growth even when subjected to the severe osmotic stress.

A second important adaptation mechanism in starter organisms is acid tolerance. Organic acids produced during mixed fermentations are often more toxic to pathogens than to the starter organisms. Organic acids dissolve in the membranes of organisms and disrupt physiological processes. Eventually acidification of the cell contents and death will result. As fermented foods mature, they usually increase in acidity and have less available water for metabolism. The lactic acid bacteria have evolved mechanisms to cope with these stresses (Kashket 1987). Similar to many microorganisms, they maintain a cytoplasm that is more alkaline than the medium. In contrast to most bacteria, they continue to survive and metabolize when the internal pH decreases to low values (Kashket 1987; McDonald et al. 1990). In lactic acid bacteria, intolerable hydrogen ion concentrations are minimized by H^+ efflux by the membrane ATPase and by an electrogenic H + /lactate symporter that is effective in removing protons from the cell during anaerobic fermentation (Konings 1985). pH homeostasis requires respiratory activity in many bacteria including salmonellae (Booth 1985), and therefore in fermented foods where anaerobic conditions prevail enteric organisms will have difficulty maintaining the internal pH required for survival, while lactics will remain viable. Streptococci are also able to adapt for utilization of alternative carbohydrates and to use certain amino acids, particularly arginine, as a source of energy (Thomas and Batt 1969). Arginine was catabolized by *E. lactis* and certain other streptococci even under strong acidic conditions ($pH < 4$), and the degradation products raised the internal pH of the cells (Kashket 1987), enabling survival in the harsh environment of cheese (Thomas and Batt 1969). Acid resistance in certain streptococci has been attributed in part to arginine catabolism (Marquis et al. 1987). The ability to change catabolic pathways enables adaptation and continued metabolism in the face of changing osmotic and acidic conditions. Bacterial cells have evolved chemotaxis responses to external pH and may move to more compatible locations. This illustrates the importance of microenvironments in food safety. In conclusion, lactic acid bacteria and fungi that are beneficial in food fermentations appear better adapted than many pathogenic organisms to tolerate salty and acidic environments, providing a critical advantage for dominance and survival in fermented foods.

Future Research

The cooperative processes among molds, yeasts, and lactic acid bacteria in food fermentations are incompletely understood. Only a few interactions are recognized. Molds release sugars and nutrients (e.g., inositol) which promote development of yeasts during certain fermentations. Yeasts produce vitamins and growth factors that stimulate the growth of lactic acid bacteria. Physical affinities between yeasts and lactic acid bacteria have been observed. The mold-yeast-lactic bacteria connections operative in food fermentations require further study.

Recent efforts in developing countries have emphasized the utility of consistent starter activity in traditional fermented foods (Stanton 1985). Dried starters are being used in Indonesia for the production of *ragi*, and research is being carried out on starters for West African fermented maize doughs, such as *ogi* (Stanton 1985). Starters are being developed for the production of *gari, fufu*, and other cassava fermentation products, which provide a staple food for about 2 billion Africans (Ofuya and Nnajiofor 1989). Further research would contribute to the development of reliable starter cultures for traditional fermentations.

There is presently much interest in using lactic cultures to preserve perishable low-acid foods (Gombas 1989). In most systems, sugar and lactic cultures are added to the food. If the temperature is abused, the lactic organisms develop and prevent the growth of pathogens. Acid production by lactic cultures has been shown to prevent growth of *C. botulinum* in several foods (Gombas 1989). This technology will likely be applied to many low-acid foods that could potentially support growth and toxin formation by *C. botulinum*. By addition of cultures to food systems, many genetic traits can be introduced to a food system without using genetically engineered organisms. This capacity should enable the development of improved fermentation processes and promote the introduction of novel fermentation processes to food preservation.

References

American Meat Institute. 1982. Good manufacturing practices, fermented dry and semidry sausages. American Meat Institute, Washington, D.C.

Anonymous. 1980. A new species of food poisoning bacteria—*Flavobacterium farinofermentans* sp. nov. *Acta Acad. Med. Sin.* 2:77–82.

Arbianto, P. 1971. Studies on bongkrek acid. Ph.D. thesis, University of Wisconsin, Madison, Wisconsin.

Archer, D. L., and J. E. Kvenberg. 1985. Incidence and cost of foodborne diarrheal disease in the United States. *J. Food Prot.* 48:887–894.

Archer, D. L., and F. E. Young. 1988. Contemporary issues: diseases with a food vector. *Clin. Microbiol. Rev.* 1:377–398.

Aulisio, C. C. G., J. T. Stanfield, S. D. Weagant, and J. A. Morris. 1983. Yersiniosis associated with tofu consumption. Serological, biochemical, and pathogenicity studies of *Yersinia enterocolitica* isolates. *J. Food Prot.* 46:226–230, 234.

Bachmann, F. M. 1924. Growth of *Clostridium botulinum* in fermented vegetables. *J. Infect. Dis.* 34:129–131.

Bacus, J. N. 1984. *Utilization of Microorganisms in Meat Processing: A Handbook for Meat Plant Operators.* John Wiley, New York.

Bacus, J. N., and W. L. Brown. 1981. Use of microbial cultures: meat products. *Food Technol.* 35:74–78.

Barber, L. E., and R. H. Deibel. 1972. Effect of pH and oxygen tension on staphylococcal growth and enterotoxin formation in fermented sausage. *Appl. Microbiol.* 24:891–898.

Bergdoll, M. S. 1989. *Staphylococcus aureus*, pp. 463–523. In M. P. Doyle (ed.), *Foodborne Bacterial Pathogens.* Marcel Dekker, New York.

Berry, E. D., M. B. Liewen, R. W. Mandigo, and R. W. Hutkins. 1990. Inhibition of *Listeria monocytogenes* by bacteriocin-producing *Pediococcus* during the manufacture of fermented semidry sausage. *J. Food Prot.* 53:194–197.

Blackwell, J. H., D. O. Cliver, J. J. Callis, N. D. Heidelbaugh, E. P. Larkin, P. D. McKercher, and D. W. Thayer. 1985. Foodborne viruses: their importance and need for research. *J. Food Prot.* 48:717–723.

Booth, I. R. 1985. Regulation of cytoplasmic pH in bacteria. *Microbiol. Rev.* 49:359–378.

Bryan, F. L. 1980a. Foodborne diseases in the United States associated with meat and poultry. *J. Food Prot.* 43:140–150.

Bryan, F. L. 1980b. Epidemiology of foodborne diseases transmitted by fish, shellfish and marine crustaceans in the United States, 1970–1978. *J. Food Prot.* 43:859–876.

Bryan, F. L. 1983. Epidemiology of milk-borne diseases. *J. Food Prot.* 46:637–649.

Bryan, F. L. 1988. Risks of practices, procedures, and processes that lead to outbreaks of foodborne diseases. *J. Food Prot.* 51:663–673.

Buchanan, R. L., H. G. Stahl, M. M. Bencivengo, and F. Del Corral. 1989. Comparison of lithium chloride-phenylethanol-moxolactam and modified Vogel agars for detection of *Listeria monocytogenes* spp. in retail-level meats, poultry, and seafood. *Appl. Environ. Microbiol.* 55:599–603.

Buckle, K. A., and E. Kartadarma. 1990. Inhibition of bongkrek acid and toxoflavin production in tempe bongkrek containing *Pseudomonas cocovenenans*. *J. Appl. Bacteriol.* 68:571–576.

Bullerman, L. B. 1981. Public health significance of molds and mycotoxins in fermented dairy products. *J. Dairy Sci.* 64:2439–2452.

Bullerman, L. B. 1986. Mycotoxins and food safety. A scientific status summary by the Institute of Food Technologist's expert panel on food safety and nutrition. Institute of Food Technologists, Chicago, Illinois.

Campbell-Platt, G. 1987. *Fermented Foods of the World: A Dictionary and Guide.* Butterworths, London.

Christian, J. H. B. 1955. The influence of nutrition on the water relations of *Salmonella oranienburg*. *Aust. J. Biol. Sci.* 8:75–82.

Christiansen, L. N., R. B. Tompkin, A. B. Shaparis, R. W. Johnston, and D. A. Kautter. 1975. Effect of sodium nitrite and nitrate on *Clostridium botulinum* growth and toxin production in a summer style sausage. *J. Food Sci.* 40:488–490.

Cliver, D. O. 1973. Cheddar cheese as a vehicle for viruses. *J. Dairy Sci.* 56:1329–1331.

Cooke, R. D., D. R. Twiddy, and P. J. A. Reilly. 1987. Lactic acid fermentation as a low cost means of food preservation in tropical countries. *FEMS Microbiol. Rev.* 46:369–379.

Conner, D. E., V. N. Scott, and D. A. Kautter. 1989. Potential *Clostridium botulinum* hazards associated with extended shelf-life refrigerated foods: A review. *J. Food Safety* 10:131–153.

Cowden, J. M., M. O'Mahony, C. L. R. Bartlett, B. Rana, B. Smyth, D. Lynch, H. Tillett, L. Ward, D. Roberts, R. J. Gilbert, A. C. Baird-Parker, and D. C. Kilsby. 1989. A national outbreak of *Salmonella typhimurium* DT 124 caused by contaminated salami sticks. *Epidem. Inf.* 103:219–225.

D'Aoust, J.-Y. 1989. Manufacture of dairy products from unpasteurized milk: A safety assessment. *J. Food Prot.* 12:906–914.

Dack, G. M. 1926. Influence of some anaerobic species on toxin of *Clostridium botulinum* with special reference to *Clostridium sporogenes. J. Infect. Dis.* 38:165–173.

Daeschel, M. A., R. E. Andersson, and H. P. Fleming. 1987. Microbial ecology of fermenting plant materials. *FEMS Microbiol. Rev.* 46:357–367.

Daly, C., M. La Chance, W. E. Sandine, and P. R. Elliker. 1973. Control of *Staphylococcus aureus* in sausage by starter cultures and chemical acidulation. *J. Food Sci.* 38:426–430.

Dolman, C. E. 1960. Type E botulism, a hazard of the north. *Arctic* 13:230–256.

Duitschaever, C. L., and D. M. Irvine. 1971. A case study: effect of mold on growth of coagulase positive staphylococci in Cheddar cheese. *J. Milk Food Technol.* 34:583.

Eisenberg, M. S., and T. R. Bender. 1976. Plastic bags and botulism: a new twist to an old hazard of the North. *Alaska Med.* 18:47–49.

Engel, G., and M. Teuber. 1989. Toxic metabolites from fungal cheese starter cultures (*Penicillium camemberti* and *Penicillium roqueforti*), p. 163–192. In H. P. Egmond (ed.), *Mycotoxins in Dairy Products*. Elsevier Applied Science, London and New York.

Farber, J. M., F. Tittiger, and L. Gour. 1988. Surveillance of raw-fermented (dry-cured) sausages for the presence of *Listeria* spp. *Can. Inst. Food Sci. Technol. J.* 21:430–434.

Fleming, H. P., and R. F. McFeeters. 1981. Use of microbial cultures: vegetable products. *Food Technol.* 35:84–88.

Foster, J. W., and H. K. Hall. 1990. Adaptive acidification tolerance response of *Salmonella typhimurium. J. Bacteriol.* 172:771–778.

Frazier, W. C. 1967. *Food Microbiology*. 2d edition. McGraw-Hill, New York.

Fukushima, D. 1985. Fermented vegetable protein and related foods of Japan and China. *Food Rev. Int.* 1:149–209.

Gandjar, I. 1986. Soybean fermentation and other tempe products in Indonesia, pp. 55–84. In C. W. Hesseltine and H. L. Wang (eds.), *Indigenous Fermented Food of Non-Western Origin*. Mycologia memoir No. 11, J. Cramer, Berlin.

Genigeorgis, C. A. 1976. Quality control for fermented meats. *J. Am. Vet. Med. Assoc.* 169:1220–1228.

Genigeorgis, C. A. 1989. Present state of knowledge on staphylococcal intoxication. *Int. J. Food Microbiol.* 9:327–360.

Gilbert, J. P., R. E. Wooley, E. B. Shotts, Jr., and J. A. Dickens. 1983. Viricidal effects of *Lactobacillus* and yeast fermentation. *Appl. Environ. Microbiol.* 46:452–458.

Goepfert, J. M., and K. C. Chung. 1970. Behavior of *Salmonella* during the manufacture and storage of a fermented sausage product. *J. Milk Food Technol.* 33:185–191.

Gombas, D. E. 1989. Biological competition as a preserving mechanism. *J. Food Safety* 10:107–117.

Haines, W. C., and L. G. Harmon. 1973. Effect of selected lactic acid bacteria on growth of *Staphylococcus aureus* and production of enterotoxin. *Appl. Microbiol.* 25:436–441.

Hammes, W. P. 1990. Bacterial starter cultures in food production. *Food Biotechnol.* 4: 383–397.

Hauschild, A. H. W. 1989. *Clostridium botulinum*, p. 111–189. In M. P. Doyle (ed.), *Foodborne bacterial pathogens*. Marcel Dekker, New York.

Heatley, N. G., and H. M. Doery. 1952. The preparation and some properties of purified Micrococcin. *Biochem.* J. 50:247–253.

Hesseltine, C. W. 1965. A millenium of fungi, food, and fermentation. *Mycologia* 57: 149–197.
Hesseltine, C. W. 1985. Fungi, people, and soybeans. *Mycologia* 77:505–525.
Hesseltine, C. W., and H. L. Wang (eds.) 1986. *Indigenous Fermented Food of Non-Western Origin*. Mycologia memoir no. 11. J. Cramer, Berlin.
Hirayama, T. 1982. Relationship of soybean paste soup intake to gastric cancer risk. *Nutr. Cancer* 3:223–233.
Hu. W. J., G. S. Guang, F. S. Chu, H. D. Meng, and Z. H. Meng. 1984. Purification and partial characterization of flavotoxin A. *Appl. Environ. Microbiol.* 48:690–693.
Hurst, A. 1973. Microbial antagonism in foods. *Can. Inst. Food Sci. Technol.* J. 6:80–90.
Ingraham, J. 1987. Effect of temperature, pH, water activity, and pressure on growth, pp. 1543–1554. In F. C. Neidhardt (ed.), *Escherichia coli and Salmonella typhimurium. Cellular and Molecular Biology*. American Society for Microbiology, Washington, D.C.
Jelinek, C. F., A. E. Pohland, and G. E. Wood. 1989. Worldwide occurrence of mycotoxins in foods and feeds-an update. *J. Assoc. Off. Anal. Chem.* 72:223–230.
Johnson, E. A., J. H. Nelson, and M. Johnson. 1990. Microbiological safety of cheese made from heat-treated milk, Part II. Microbiology. *J. Food Prot.* 53:519–540.
Johnson. E. A., and M. W. Pariza. 1989. Microbiological principles for the safety of foods, pp. 135–174. In R. D. Middlekauff and P. Shubik (eds.), *International Food Regulation Handbook*. Marcel Dekker, New York.
Johnson, J. L., M. P. Doyle, and R. G. Cassens. 1990. *Listeria monocytogenes* and other *Listeria* spp. in meat and meat products. A review. *J. Food Prot.* 53:81–91.
Johnson, J. L., M. P. Doyle, R. G. Cassens, and J. L. Schoeni. 1988. Fate of *Listeria monocytogenes* in tissues of experimentally infected cattle and in hard salami. *Appl. Environ. Microbiol.* 54:497–501.
Kashket, E. R. 1987. Bioenergetics of lactic acid bacteria: cytoplasmic pH and osmotolerance. *FEMS Microbiol. Rev.* 46:233–244.
Kerner, J. 1817. Tubinger BlatterNaturwissenchaft. Arzneik 3:1–25.
Ko. S. D., and C. W. Hesseltine. 1979. Tempe and related foods, pp. 115–140. In A. H. Rose (ed.), *Economic Microbiology*, vol. 4., Academic Press, London.
Konings, W. N. 1985. Generation of metabolic energy by end-product efflux. *Trends Biochem. Sci.* 10:317–319.
Lawrence, R. C., and T. D. Thomas. 1979. The fermentation of milk by lactic acid bacteria, pp. 187–219. In A. T. Bull, D. C. Ellwood, and C. Ratledge (eds.), *Microbial Technology: Current state, Future Prospects*. Society of General Microbiology Symposium 29, Cambridge University Press, Cambridge.
Leistner, L. 1990. Mould-fermented foods: recent developments. *Food Biotechnol.* 4:433–441.
Liston, J. 1980. Fish and shellfish and their products, pp. 567–605. In *Microbial Ecology of Foods*, vol. 2, *Food Commodities*. The International Commission on Microbiological Specifications for Foods. Academic Press, New York.
Marquis, R. E., G. R. Bender, D. R. Murray, and A. Wong. 1987. Arginine deiminase system and bacterial adaptation to acid environments. *Appl. Environ. Microbiol.* 53: 198–200.
Masters, B. A., J. L. Oblinger, S. J. Goodfellow, J. N. Bacus, and W. L. Brown. 1981. Fate of *Salmonella newport* and *Salmonella typhimurium* inoculated into summer sausage. *J. Food Prot.* 44:527–530.
McCroskey, L. M., C. L. Hatheway, L. Fenicia, B. Pasolina, and P. Aureli. 1986. Characterization of an organism that produces type E botulinal toxin but which resembles *Clostridium butyricum* from the feces of an infant with type E botulism. *J. Clin. Microbiol.* 23:201–202.
McDonald, L. C., H. P. Fleming, and H. M. Hassan. 1990. Acid tolerance of *Leuconostoc mesenteroides* and *Lactobacillus plantarum*. *Appl. Environ. Microbiol.* 56:2120–2124.
Meisel, C., K. H. Gehlen, A. Fischer, and W. P. Hammes. 1989. Inhibition of the growth of *Staphylococcus aureus* in dry sausages by *Lactobacillus curvatus, Micrococcus varians*, and *Debaryomyces hansenii*. *Food Biotechnol.* 3:145–168.
Melnick, J. L., C. P. Gerba, and C. Wallis. Viruses in water. *Bull. WHO* 56:499–504.
Metaxopoulos, J., C. Genigeorgis, M. J. Fanelli, C. Franti, and E. Cosma. 1981. Produc-

tion of Italian dry salami. II. Effect of starter culture and chemical acidulation on staphylococcal growth in salami under commercial conditions. *Appl. Environ. Microbiol.* 42:863–871.

Meyer, K. F. 1956. The status of botulism as a world health problem. *Bull. WHO* 15: 281–298.

Meyer, K. F., and B. Eddie. 1965. Sixty-five years of human botulism in the United States and Canada. Epidemiology and tabulations of reported cases 1899–1964. George Williams Hooper Foundation, University of California, San Francisco Medical Center.

Meyer, K. F., and J. B. Gunnison. 1929. Botulism due to home canned Bartlett pears. *J. Infect. Dis.* 45:135–147.

Mossel, D. A. A., and I. Ingram. 1955. The physiology of the microbial spoilage of foods. *J. Appl. Bacteriol.* 18:232–268.

Nakano, W., and E. Kodama. 1970. On the reality of "Izushi," the causal food of botulism, and its folkloric meaning, pp. 388–392. In M. Herzberg (ed.), *Toxic Microorganisms*. Proceedings of the first U.S.-Japan Conference, U.S. Department of the Interior, Washington, D.C.

National Academy of Sciences. 1988. An evaluation of the role of microbiological criteria for foods and food ingredients. National Academy Press, Washington, D.C.

Nurmi, E. 1966. Effect of bacterial inoculations on characteristics and microbial flora of dry sausage. *Acta Agralia Fennica* 108:1–77.

Odlaug, T. E., and I. J. Pflug. 1978. *Clostridium botulinum* and acid foods. *J. Food Prot.* 41:566–573.

Odlaug, T. E., and I. J. Pflug. 1979. *Clostridium botulinum* growth and toxin production in tomato juice containing *Aspergillus gracilis. Appl. Environ. Microbiol.* 37: 496–504.

Ofuya, C. O., and C. Nnajiofor. 1989. Development of a starter culture for the industrial production of gari. *J. Appl. Bacteriol.* 66:37–42.

Olivigni, F. J., and L. B. Bullerman. 1977. Simultaneous production of penicillic acid and patulin by a *Penicillium* species isolated from cheddar cheese. *J. Food Sci.* 42: 1654–1657.

Panina, G. F., A. Civardi, I. Massirio, F. Scatozza, P. Paldini, and F. Palmia. 1989. Survival of foot-and-mouth disease virus in sausage meat products (Italian salami). *Int. J. Food Microbiol.* 8:141–148.

Panon, G., S. Tache, and C. Labie. 1987. Antiviral substances in raw bovine milk active against bovine rotavirus and coronavirus. *J. Food Prot.* 50:862–866.

Pederson, C. S. 1979. Microbiology of food fermentations, 2d ed. AVI, Westport, Conn.

Peterson, A. C., J. J. Black, and M. F. Gunderson. 1964. Staphylococci in competition. III. Influence of pH and salt on staphylococcal growth in mixed populations. *Appl. Microbiol.* 12:70–76.

Pucci, M. J., E. R. Vedamuthu, B. S. Kunka, and P. A. Vandenbergh. 1988. Inhibition of *Listeria monocytogenes* by using bacteriocin PA 1 produced by *Pediococcus acidilacti* PAC 1.0. *Appl. Environ. Microbiol.* 54:2349–2353.

Raatjes, G. J. M., and J. P. P. M. Smelt. 1979. *Clostridium botulinum* can grow and form toxin at pH values lower than 4.6. *Nature* 281:398–399.

Roberts, T. A., and M. Ingram. 1973. Inhibition of growth of *C. botulinum* at different pH values by sodium chloride and sodium nitrite. *J. Food Technol.* 8:467–475.

Rose, A. H. 1987. Responses to the chemical environment, pp. 5–40. In A. H. Rose and J. H. Harrison (eds.), *The Yeasts*, vol. 2, 2d ed. Academic Press, London.

Rusmin, S., and S. D. Ko. 1974. Rice-grown *Rhizopus oligosporus* inoculum for tempeh fermentation. *Appl. Microbiol.* 28:347–350.

Ryan, C. A., M. K. Nickels, N. T. Hargrett-Bean, M. E. Potter, T. Endo, L. Mayer, C. W. Langkop, C. Gibson, R. C. McDonald, R. T. Kenney, N. D. Puhr, P. J. McDonnell, R. J. Martin, M. L. Cohen, and P. A. Blake. 1987. Massive outbreak of antimicrobial-resistant salmonellosis traced to pasteurized milk. *J. Amer. Med. Assoc.* 258:3269–3274.

Samson, R. A., J. A. Van Kooij, and E. De Boer. 1987. Microbiological quality of commercial tempeh in the Netherlands. *J. Food Prot.* 50:92–94.

Schillinger, U., and F.-K. Lucke. 1989. Antibacterial activity of *Lactobacillus sake* isolated from meat. *Appl. Environ. Microbiol.* 55:1901–1906.

Schubring, R., and W. Kuhlmann. 1978. Preliminary studies on application of starter cultures in manufacture of fish products. *Lebensmittel Industrie* 25:455–457.

Scott, P. M. 1989. Mycotoxigenic fungal contaminants of cheese and other dairy products, pp. 193–259. In H. P. Egmond (ed.), *Mycotoxins in Dairy Products*. Elsevier Applied Science, London and New York.

Selander, R. K., and J. M. Musser. 1990. Population genetics of bacterial pathogenesis, pp. 11–36. In B. H. Iglewski and V. L. Clark (eds.), *Molecular Basis of Bacterial Pathogenesis, The Bacteria*. vol. 11, Academic Press, New York.

Shurtleff, W., and W. Aoyagi. 1979. *The Book of Tempeh*. Harper & Row, New York.

Smith, J. L., C. N. Huhtanen, J. C. Kissinger, S. A. Palumbo. 1975. Survival of salmonellae during pepperoni manufacture. *Appl. Microbiol.* 30:759–763.

Smith, J. L., and S. A. Palumbo. 1981. Microorganisms as food additives. *J. Food Prot.* 44:936–955.

Stanton, W. R. 1985. Food fermentations in the tropics, pp. 193–211. In B. J. B. Wood (ed.), *Microbiology of Fermented Foods*. Elsevier, London.

Stark, C. N., J. M. Sherman, and P. Stark. 1929. Destruction of botulinum toxin by *Bacillus subtilis*. *Proc. Soc. Exp. Biol. Med.* 26:343–344.

Stott, W. T., and L. B. Bullerman. 1975. Influence of carbohydrate and nitrogen source on patulin in Cheddar cheese. *Appl. Microbiol.* 30:850–854.

Stott, W. T., and L. B. Bullerman. 1976. Instability of patulin in Cheddar cheese. *J. Food Sci.* 41:201–203.

Szabo, R. A., G. A. Jarvis, K. F. Weiss, K. Rayman, G. Lachapelle, and A. Jean. 1989. Microbiological quality of tofu and related products in Canada. *J. Food Prot.* 52:727–730.

Tanaka, N. 1982. Toxin production by *Clostridium botulinum* in media at pH lower than 4.6. *J. Food Prot.* 45:234–237.

Tanaka, N., S. K. Kovats, J. A. Guggisberg, L. M. Meske, and M. P. Doyle. 1985a. Evaluation of the bacteriological safety of low-salt miso. *J. Food Prot.* 48:435–437.

Tanaka, N., S. K. Kovats, J. A. Guggisberg, L. M. Meske, and M. P. Doyle. 1985b. Evaluation of the microbiological safety of tempeh made from unacidified soybeans. *J. Food Prot.* 48:438–441.

Tanaka, N., E. Traisman, M. H. Lee, R. G. Cassens, and E. M. Foster. 1980. Inhibition of botulinum toxin formation in bacon by acid development. *J. Food Prot.* 43:450–457.

Tanner, F., P. R. Beamer, and C. J. Rickher. 1940. Further studies on development of *Clostridium botulinum* in refrigerated foods. *Food Res.* 5:323–333.

Thomas, T. D., and R. D. Batt. 1969. Metabolism of exogenous arginine and glucose by starved *Streptococcus lactis* in relation to survival. *J. Gen. Microbiol.* 58:371–380.

Todd, E. C. D. 1987. Impact of spoilage and foodborne diseases on national and international economies. *Int. J. Food. Microbiol.* 4:83–100.

Todd, E. C. D. 1989. Costs of acute bacterial foodborne disease in Canada and the United States. *Int. J. Food Microbiol.* 9:313–326.

Tompkin, R. B. 1980. Botulism from meat and poultry products—A historical perspective. *Food Technol.* 34:229–236, 259.

Troller, J. A., and W. C. Frazier. 1963. Repression of *Staphylococcus aureus* by food bacteria. *Appl. Microbiol.* 11:163–165.

Troller, J. 1986. Water relations of foodborne bacterial pathogens—An updated review. *J. Food Prot.* 49:656–670.

Twiddy, D. R., S. J. Cross, and R. D. Cooke. 1987. A study of the parameters involved in the production of lactic preserved fish-starchy substrate combinations. *Int. J. Food Sci. Technol.* 22:115–121.

van Damme, P. A., A. G. Johannes, H. C. Cox, and W. Berends. 1960. On toxoflavin, the yellow poison of *Pseudomonas cocovenenans*. *Rec. Trav. Chim. Pays-Bas* 79:255–267.

Van Veen, A. G., D. C. W. Graham, and K. H. Steinkraus. 1968. Fermented peanut presscake. *Cereal Science Today* 13:96–99.

Wainwright, R. B., W. L. Heyward, J. P. Middaugh, C. L. Hatheway, A. P. Harpster, and T. R. Bender. 1988. Foodborne botulism in Alaska, 1947–1985: epidemiology and clinical findings. *J. Infect. Dis.* 157:1158–1162.

Wang, H. L., and C. W. Hesseltine. 1981. Use of microbial cultures: legume and cereal products. *Food Technol.* 35:79–83.
Wang, H. L., D. I. Ruttle, and C. W. Hesseltine. 1969. Antibacterial compound from a soybean product fermented by *Rhizopus oligosporus*. *Proc. Soc. Exp. Biol. Med.* 131: 579–583.
Wicklow, D. T., and C. P. Kurtzman. 1989. Nucleic acid relatedness of mycotoxin-producing fungi, p. 1–6. In S. Natori, K. Hashimoto, and Y. Ueno (eds.), *Mycotoxins and Phycotoxins '88*. Elsevier, Amsterdam.
Wood, B. J. B. 1985. *Microbiology of Fermented Foods*, vols. 1 & 2. Elsevier, London.
Ying, S., and C. Shuyan. 1986. Botulism in China. *Rev. Infect. Dis.* 8:984–990.

Chapter

7

Silage Fermentation

Richard E. Muck

USDA—Agricultural Research Service
U.S. Dairy Forage Research Center
Madison, Wisconsin

Introduction

For centuries farmers have used two basic strategies for preserving crops. The first method is to dry the crop so that both plant and microbial activity essentially ceases. If the stored crop is not rewetted, drying assures that losses during storage are kept to a minimum (81). This method of preservation has worked well for storing both grains and forages. However, the attainment of a stable crop moisture level, especially in humid climates, increases harvest losses from mechanical and weather factors (81, 51) and may require additional energy for artificial drying.

The alternate approach to crop preservation is ensiling. This method has been used for over 3000 years. Egyptian murals show grain being preserved by ensiling as long ago as 1000 to 1500 B.C. (132). In ensiling, a crop is harvested directly or with minimal field wilting, placed in a pile or storage structure (silo) and sealed to create anaerobic conditions. During storage, sugars and some organic acids in the crop are fermented by lactic acid bacteria, producing lactic acid as well as acetic acid, ethanol, and other minor compounds. This fermentation typically reduces crop pH to between 3.8 and 5.

The preservation of the crop in ensiling results from two factors: low pH and anaerobic environment. The low pH decreases plant enzyme activity and prevents the proliferation of detrimental anaerobic microorganisms, especially clostridia and enterobacteria.

However, neither the low pH nor the products of a normal fermentation are correlated with preventing aerobic microbial activity (135). Aerobic spoilage and heating of the crop are prevented by maintaining an anaerobic environment. The anaerobic environment also ensures that the lactic acid–producing bacteria efficiently make lactic acid so that crop pH is quickly lowered. Consequently, the sealing of the silo or pile is essential to the preservation of the crop by ensiling.

A farmer can maintain the anaerobic environment in the silo by good management. However, the farmer generally has little or no information regarding the populations of various microorganisms on the crop, the amount of substrate available for fermentation, or other plant factors affecting fermentation. As a result, the outcome of ensiling a crop, even with good ensiling technique, is not certain. Even so, considerable progress in understanding the ensiling process has been made over the past 3000 years, so that a good farmer usually makes high-quality silage.

The focus of this chapter will be on the interaction and development of the various microbial groups involved in ensiling and on the environmental and plant factors that affect the dominance of particular microbial groups and species. Emphasis will be placed on the ensiling of forage crops, including whole-plant corn, because these crops have been studied the most extensively. However, much of the ensuing discussion will apply to other crops as well.

Important Microbial Species

Lactic acid bacteria

The bacteria on the crop responsible for the fermentation of sugar to lactic acid are termed lactic acid bacteria. They consist of four genera: *Lactobacillus, Pediococcus, Leuconostoc*, and *Streptococcus*. These bacteria are all microaerophilic, gram-positive, non-spore-forming microorganisms. The earliest classification of these organisms (99) divided them into two groups: homofermenters, which produce lactic acid almost exclusively from hexoses, and heterofermenters, which produce substantial quantities of acetic acid or ethanol in addition to lactic acid. These two terms are still widely used in describing lactic acid bacteria, even though some homofermenters under certain environmental conditions will act quite heterofermentatively (12, 110).

Table 7.1 lists many of the lactic acid bacteria that have been reported in silages. Of these, the most commonly reported species are *Lactobacillus plantarum, Lactobacillus casei, Lactobacillus buchneri, Lactobacillus brevis, Pediococcus pentosaceus, Pediococcus acidilactici, Streptococcus faecium*, and *Leuconostoc mesenteroides*. Despite the

TABLE 7.1 **Lactic Acid Bacterial Species Reported in Silages**

Species	Source
Homofermentative rods	
Lactobacillus acidophilus	28, 29
Lactobacillus alimentarius	103
Lactobacillus casei	4, 18, 19, 28, 29, 32
Lactobacillus coryniformis	19, 103
Lactobacillus curvatus	4, 103
Lactobacillus delbrueckii	20
Lactobacillus farciminis	103
Lactobacillus leichmannii	20
Lactobacillus pentosus	103
Lactobacillus plantarum	4, 18, 19, 21, 28, 29, 32, 91
Lactobacillus sake	103
Lactobacillus salivarius	20, 103
Lactobacillus sharpeae	103
Homofermentative cocci	
Pediococcus acidilactici	18, 19, 20
Pediococcus dextrinicus	4
Pediococcus halophilus	124
Pediococcus parvulus	103
Pediococcus pentosaceus	18, 19, 20, 21, 103
Streptococcus bovis	19
Streptococcus faecalis	28
Streptococcus faecium	18, 90
Streptococcus lactis	19
Heterofermentative rods	
Lactobacillus brevis	4, 19, 21, 32, 91
Lactobacillus buchneri	4, 19, 20, 28, 32
Lactobacillus cellobiosus	32, 91
Lactobacillus coprophilus	32
Lactobacillus corynoides	4
Lactobacillus divergens	103
Lactobacillus fermentum	19, 28, 29
Lactobacillus reuteri	103
Lactobacillus viridescens	91
Heterofermentative cocci	
Leuconostoc mesenteroides	4, 32, 90
Leuconostoc paramesenteroides	18, 19, 21

predominance of these organisms in the literature, the most recent studies (4, 19, 20, 32, 103) indicate that other species may play important roles in specific situations.

Most of the lactic acid bacteria are mesophilic microorganisms with optimum growth temperatures between 25 and 40°C and extremes of 5 and 55°C (115). Optimum pH is 5.5 to 5.8. *Lactobacillus* and *Pediococcus* species are generally more tolerant of low pH, and many species will grow at pH 4.0 or less (115). In contrast, *Streptococcus* and

Leuconostoc species do not or very slowly grow below pH 4.5 (115). This difference, as will be seen later, helps explain some of the variation in species numbers with time during fermentation.

The important silage species of all four genera have complex nutritional requirements. Most have requirements for various amino acids and vitamins. Pediococci require most of the amino acids, whereas *Leuconostoc mesenteroides* subsp. *mesenteroides* only requires glutamic acid and valine (115). There is also a wide variation in vitamin requirements, especially the B vitamins (115).

Carbon and energy sources for these bacteria are largely limited to simple sugars and a few organic acids. Glucose is fermented by virtually all species; however, the fermentation of other mono- and disaccharides varies between and often within species. The sugars fermented by a strain are used to presumptively identify the microorganism.

Some species of lactic acid bacteria will ferment organic acids in the plant. The two most important of these acids are citric and malic. *Streptococcus faecalis* and *S. faecium* can ferment citrate; *Lactobacillus brevis* and some *Leuconostoc mesenteroides* strains can use citrate when a fermentable carbohydrate is present (115, 68, p. 71). Malic acid fermentation has been observed in *S. faecalis, S. faecium, Pediococcus* species, *Leuconostoc mesenteroides* and *Lactobacillus plantarum* (68, p. 71). Under aerobic conditions, many species of lactic acid bacteria will oxidize lactic acid to acetic acid and carbon dioxide (13).

The lactic acid bacteria involved in silage fermentation are generally nonproteolytic. Serine and arginine are the only amino acids that have been reported to be fermented by important silage lactic acid bacteria. Serine is fermented to acetoin, ammonia, and CO_2, whereas ornithine, ammonia, and CO_2 are the products of arginine fermentation (97).

Products of fermentation for typical sugars and organic acids are shown in Table 7.2. The species that homofermentatively ferment glucose and fructose produce lactate via glycolysis (43). This results in no dry matter (DM) loss from the crop and only a slight energy loss (0.7 percent) (68, p. 174). In heterofermentation, glucose is converted to gluconate 6-phosphate. Decarboxylation produces xylulose 5-phosphate and CO_2. Xylulose 5-phosphate is split to produce lactate and either acetate or ethanol (43). This results in a 24 percent DM loss, but only a 1.1 percent energy loss (68, p. 174).

Fermentation of pentoses is similar for all lactic acid bacteria, producing lactate and acetate by a similar pathway to heterofermentation of glucose. Citrate and malate are converted to oxalocitrate and then to pyruvate. From pyruvate, a number of products may result: lactate, ac-

TABLE 7.2 Products of Sugar and Organic Acid Fermentation by Lactic Acid Bacteria, Presented as Molar Ratios

Substrate	Products	Comments
1 glucose	2 lactic acid	Homofermentation
1 fructose	2 lactic acid	Homofermentation
1 glucose	1 lactic acid, 1 acetic acid, 1 carbon dioxide	Heterofermentation
1 glucose	1 lactic acid, 1 ethanol, 1 carbon dioxide	Heterofermentation
3 fructose	1 lactic acid, 1 acetic acid, 2 mannitol, 1 carbon dioxide	Heterofermentation
1 xylose	1 lactic acid, 1 acetic acid	Homo and Hetero
2 citrate	3 acetic acid, 1 lactic acid, 3 carbon dioxide	Homo and Hetero
2 citrate	2 acetic acid, 4 carbon dioxide, 1 2,3-butanediol	Homo and Hetero
1 malate	1 lactate and 1 carbon dioxide	Homo and Hetero

etate, ethanol, and 2,3-butanediol. Fermentation of these organic acids result in substantial DM losses (30 to 33 percent); however, there are slight gains in energy (1.5 to 1.8 percent) (68, p. 174).

Consequently, fermentation of sugar and organic acids by lactic acid bacteria results in little energy loss from the crop no matter what species dominate fermentation. However, the dominant lactic acid bacterial species do affect DM losses if the predominant substrates are hexoses.

Clostridia

Clostridia are the anaerobic microorganisms most detrimental to the ensiling process. Some clostridia will ferment lactate to butyric acid, whereas others will ferment amino acids to amines and ammonia. The presence of butyric acid and amines make silages less palatable to animals, reducing feed value.

Clostridia are gram-positive, spore-forming, usually motile bacteria that are far less tolerant of aerobic and acid conditions than lactic acid bacteria (115). Gibson (26) found seven species of clostridia important in silage (Table 7.3). Reports by others (1, 29, 126) confirm that these species are the most significant. The seven species are divided into three groups based on the substrates which can be fermented. *C. butyricum, C. paraputrificum* and *C. tyrobutyricum* ferment lactic acid and various sugars. *C. bifermentans* and *C. sporogenes* are principally amino acid fermenters but will ferment some sugars. The other two species have a wide range of substrates, both sugars and amino acids.

TABLE 7.3 Clostridial Species in Silage

Clostridium bifermentans
Clostridium butyricum
Clostridium paraputrificum
Clostridium perfringens
Clostridium sphenoides
Clostridium sporogenes
Clostridium tyrobutyricum

SOURCE: Gibson 1965.

The optimum conditions for the growth of these bacteria are fortunately different from that of the lactic acid bacteria. Most of the silage clostridia have optimum temperatures around 37°C with the exception of *C. sphenoides* (45°C) (68, p. 80). Thus, heating of the crop is normally advantageous to clostridia.

The clostridia have a higher optimum pH (7.0 to 7.4) and lesser tolerance of low water activities than lactic acid bacteria (68, p. 78). These two factors interact such that the minimum pH for growth varies with moisture content. Based on *C. tyrobutyricum* data from Wieringa (127), Leibensperger and Pitt (61) derived a relationship for minimum water activity as a function of pH. Muck (83) used this relationship with water activity data by Greenhill (34) to develop an approximate minimum pH for clostridial growth as a function of crop dry matter content (Figure 7.1). The wetter the silage, the lower the pH required to stop clostridial activity.

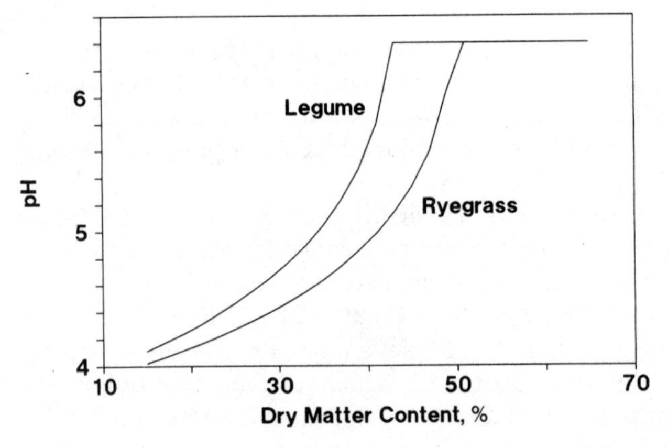

Figure 7.1 The pH at which the growth of *Clostridium tyrobutyricum* ceases as a function of the DM content of the crop. (*From Muck 1988.*)

The fermentation of both sugar and lactic acid are by a similar pathway. These substrates are converted to pyruvate and from there to butyric acid, CO_2, and hydrogen. The fermentation of either glucose or lactic acid results in substantial DM (51.1 percent) and energy (18.4 percent) losses (68, p. 174), reducing the percent of harvested crop available for animal consumption.

The fermentation of amino acids occurs by three general pathways: deamination, decarboxylation, and oxidation/reduction (Stickland reactions) (97). Deamination and many of the oxidation/reduction reactions will increase the ammonia content of the silage, raising silage pH. High ammonia content in silages usually indicates clostridial activity. Of greater concern is the decarboxylation of amino acids that produces various amines and the loss of CO_2. The presence of amines has been linked with reduced intake of silage by ruminants (93, 94).

Fungi

The fungi are usually not significant in silage fermentation. The majority of yeasts and molds are aerobic organisms and so are of importance in the aerobic degradation of silage when the silo is opened (135). However, anaerobic yeasts may be important in the fermentation of some silages.

A variety of genera of yeasts have been reported as dominant organisms in silages (Table 7.4). Many of these strains will ferment glucose and other sugars, but most with the exception of *Torulopsis*, prefer aerobic conditions (5).

The primary yeast fermentation observed in silages is that of sugars to ethanol. This fermentation (1 mole glucose to 2 moles ethanol and 2 moles CO_2) results in large DM losses (48.9 percent) but little energy loss (0.2 percent) (68, p. 174). Such a fermentation could be detrimental when sugar supplies in the crop are limited because sugar is converted to a product that does not lower pH. Furthermore, whereas ethanol production maintains the energy content of the silage, ethanol is not efficiently utilized by ruminants. Consequently, yeast growth during ensiling is not desirable.

Yeasts produce minor amounts of other alcohols (*n*-propanol, iso-

TABLE 7.4 Important Fungal Genera in Silage

Candida	*Saccharomyces*
Cryptococcus	*Torulopsis*
Hansenula	*Trichosporon*
Pichia	

SOURCE: Gibson 1965, McDonald 1981.

butanol, iso-pentanol) and volatile fatty acids (acetic, propionic, butyric, and iso-butyric) (68, p. 95). Presence of these alcohols, in particular, is evidence of yeast activity.

Enterobacteria

Another group of microorganisms that occasionally have significant effects on silage fermentation is the enterobacteria. They are gram-negative, non-spore-forming, facultative anaerobes. Those that proliferate in silage ferment sugars producing principally acetic acid (68, p. 93).

Species from four genera have been identified on forages and silages: *Erwinia herbicola, Escherichia coli, Hafnia alvei*, and *Klebsiella* species (28, 29, 116). *Erwinia herbicola* appears to be the dominant organism on grass prior to ensiling (30, 116), whereas *Klebsiella* has most often been cited as the dominant genus during ensiling (28, 29, 116).

The optimum pH for these bacteria is approximately 7.0 (68, p. 91). As a result, they are usually a factor at the beginning of ensiling or until the lactic acid bacteria start producing enough acids to rapidly drop pH (28; 29; 68, p. 91; 116). The enterobacteria appear to compete better with the lactic acid bacteria at moderate temperatures (22 to 30°C) than under warm conditions (40°C) (28).

As indicated earlier, the principal fermentation product of the enterobacteria is acetic acid. However, the fermentation of sugars may also result in formate, succinate, ethanol, 2,3-butanediol, lactate, hydrogen, and CO_2 (68, pp. 93–94). Beck (5) also indicated that enterobacteria may deaminate and decarboxylate amino acids. Recent work (116) indicates that some species (*E. coli, H. alvei*, and others) can reduce nitrate, producing nitrite and nitric oxide which may be important in inhibiting clostridial growth.

Other bacteria

Bacillus. *Bacillus* species are more likely to be important in aerobic deterioration of silages (135) than in fermentation. Nevertheless, three species (*B. coagulans, B. licheniformis*, and *B. polymyxa*) multiply during fermentation particularly under warmer conditions (30 to 40°C) (28). A later study of these organisms (134) indicated all three organisms could produce lactic and acetic acid. However, the principal fermentation products of *B. licheniformis* are glycerol and 2,3-butanediol, and *B. polymyxa* produces mainly ethanol and 2,3-butanediol. These products will not lower crop pH. These organisms are not suppressed by low pH (134) so that there appears to be little means to prevent their growth except for insuring conditions that will cause the lactic acid bacteria to proliferate.

Listeria. A microorganism occasionally found in silage is *Listeria monocytogenes*. Because it is both an animal and human pathogen, there is concern that it could survive in the silo or possibly multiply. Isolations of this organism have usually occurred in poorly fermented silages (133), but there is evidence of it in acceptable big-bale silages (23). In general, low pH (<4.2) is known to prevent the development of *Listeria* in silages (121).

Normal Ensiling Dynamics

Shifts in major microbial groups

Typical changes in the populations of microorganisms during a successful silage fermentation are shown in Fig. 7.2. Because the lactic acid bacteria are the primary fermenters in silage, their rapid rise and dominance would be expected. Frequently, however, the initially dominant microbial species are enterobacteria. In unwilted crops, enterobacteria reach maximum populations (10^7 to 10^9 cfu/g silage) after 1 to 4 days of ensiling (10, 28, 29, 56, 63, 69, 98, 112, 113, 116, 118). With moderate wilting (40 percent DM), their growth rate decreases, lengthening the time until maximum levels are reached (63). Moderate wilting has little effect on the magnitude of the population (63, 116). At 10^8 or more colony-forming units per gram of silage, enterobacteria would be expected to significantly affect the fermentation products, especially acetic acid, found in the silage (83). As silage

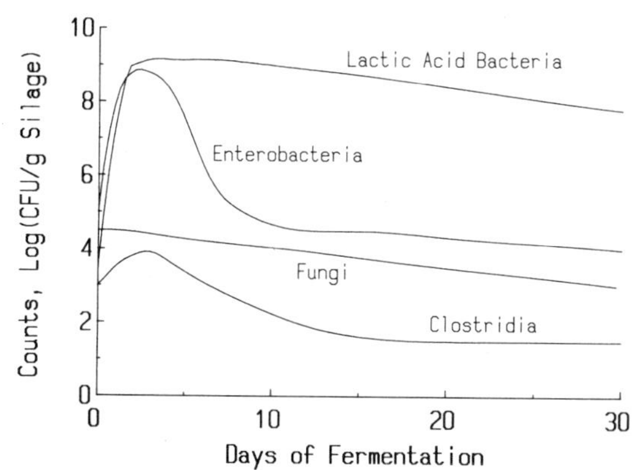

Figure 7.2 Typical changes in major microbial groups during the ensiling of unwilted forage crops.

pH drops, the numbers of enterobacteria decline rapidly to 10^5 cfu/g or less (10, 28, 29, 63, 98, 112, 113). As a result, they do not significantly affect the latter stages of fermentation.

Lactic acid bacteria develop rapidly as well, reaching higher numbers than the enterobacteria. In unwilted crops, peak populations (10^8 to 10^{10} cfu/g) are attained in 1 to 4 days (10, 18, 57, 67, 98, 112, 113). Wilting tends to reduce the level of the peak populations, but counts usually reach 10^8 cfu/g with the exception of very dry silages (>60 percent DM) (49, 85). In contrast to the enterobacteria, the decline in lactic acid bacteria numbers is gradual. Frequently, the levels after a month or more of storage are still 10^7 to 10^8 cfu/g (10, 18, 57, 67, 112, 113, 122).

The dominance of the lactic acid bacteria appears to be due to several factors. First, as indicated earlier, these bacteria are able to grow under a wide range of conditions. They are tolerant of low pH and low water activities. This gives them advantages over the enterobacteria and the clostridia, respectively. Second, some of the strains are among the fastest-growing natural anaerobic species. Finally, there is substantial evidence that some strains of lactic acid bacteria may produce antimicrobial compounds (17, 96), providing an additional competitive advantage.

The development of clostridia depends largely on the dry matter content of the crop. In unwilted forages as in Fig. 7.2, clostridial numbers may climb for the first 2 to 4 days of ensiling followed by a slow decline once silage pH has dropped sufficiently to inhibit growth (4, 28, 29, 69). In drier crops (35 percent DM or greater), clostridial numbers rarely multiply (122) because the water activity is too low for spore germination and growth.

With good ensiling technique, yeast and mold counts should be highest at the start of ensiling. Most often, both yeast and mold counts will decline slowly with time (18, 20, 95), although yeast counts may remain relatively constant throughout storage (10, 20, 95, 113). A more rapid decline, particularly in mold counts, during ensiling has been noted in legume silages (20, 102, 122) where mold counts after ensiling are below detectable level. The study of O'Kiely et al. (102) suggests that antifungal compounds are produced in legumes during fermentation.

Aerobic bacteria generally decline rapidly after the start of ensiling (28, 67). This may not always be true. With sufficient nitrate available, aerobes capable of nitrate reduction may maintain or increase their numbers for the first several days of fermentation (67).

Shifts in the dominant lactic acid bacteria

Substantial shifts in the dominant species of lactic acid bacteria normally occur during ensiling. The lactic acid bacteria on forage crops entering the silo are typically dominated by the cocci, particularly *Leuconostoc* and *Streptococcus* species (18, 56, 91, 103, 112, 120). These species multiply during the first few days in the silo but not as fast as the lactobacilli. By the time that silage pH is rapidly dropping, homofermentative lactobacilli are usually dominant, (4, 19, 56, 63, 103, 112) although the homofermentative pediococci may also be important (56, 103, 112). After the active fermentation period is over, the heterofermentative lactobacilli make up an increasing percentage of the lactic acid bacterial population (4, 19, 32, 56, 63, 112).

The causes of these shifts are thought to be principally related to environmental factors. Many of the species on the plant before ensiling apparently do not grow well in the silo (53). One possible reason for this may be a preference for aerobic conditions, as has been noted in some strains of *Leuconostoc mesenteroides* (125). A more likely cause for many of the species is pH tolerance. Lactobacilli are more tolerant of low cytoplasmic pH, and thus low external pH, than streptococci (44). Even streptococci selected for silage inoculation do not maintain their influence on fermentation once pH drops below 5.0 (63).

The shift from homofermenters to heterofermenters late in fermentation may not be due to pH directly but rather the effects of end products on growth. Beck (3) reported that *L. buchneri* and *L. brevis* were two to three times more tolerant of acetate than homofermentative species. Perhaps, as suggested by Kashket (44), acetate influx affects the lowering of cytoplasmic pH of homofermenters more than that of heterofermenters.

Another possible factor affecting the dominant species at the end of fermentation is substrate availability. Particularly in unwilted grasses and legumes, fermentation may be stopped by the lack of substrate. In these cases, the last remaining substrates, possibly plant organic acids and pentoses from hemicellulose hydrolysis, would favor heterofermenters such as *L. brevis, L. buchneri, L. fermentum*, and *L. cellobiosis* (115; 68, p. 71). *L. plantarum* and pediococci are the homofermenters that will typically use such substrates (115; 68, p. 71). All of these species have been cited as being present in significant numbers at the end of ensiling (4, 19, 20, 32, 56). Consequently, substrate availability may be important in determining the dominant species at the end of active fermentation.

Fermentation products

Because of shifts in general microbial populations and the dominant lactic acid bacterial species, fermentation products do not appear in uniform ratios throughout the ensiling process. Figure 7.3 provides a hypothetical time course of product accumulation in a good silage fer-

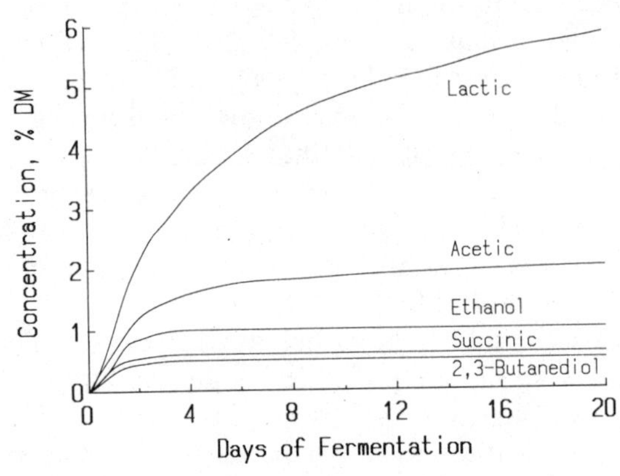

Figure 7.3 Typical accumulation of fermentation products during the ensiling of unwilted forage crops.

mentation. Actual successful fermentations may vary from this; however, this figure represents the most common observations regarding product appearances.

Of the two principal products of fermentation, lactate and acetate, acetate often appears before lactate or at least at a relatively high ratio with respect to lactate during the first day or two of fermentation (46, 57, 58, 103). There are several reasons for this. First, as indicated earlier, enterobacteria ferment sugars, producing acetate as their principal product. Second, one of the dominant lactic acid bacteria at the start of ensiling is the heterofermentative leuconostocs. Finally, acetate may also be the result of plant enzyme activity (68, p. 59). All three of these factors to a greater or lesser degree are probably operative in most forage silages.

The minor fermentation products (succinate, ethanol, formate, 2,3-butanediol) usually reach their final concentrations within the first day or two of fermentation (36; 46; 57; 85; Muck, unpublished data). All four products may be the result of lactic acid bacterial activity, but this seems unlikely because little or no further accumulation of these products normally occurs in untreated silages (36; 46; 57; 85; Muck, unpublished data). Enterobacteria may also produce these compounds (68, pp. 93–94). *Bacillus* species can produce ethanol and 2,3-butanediol (134), and various silage yeasts will ferment sugars to ethanol (5). Considering that these compounds are produced when enterobacteria, *Bacillus* species, and yeasts are normally at their highest levels, one or more of these microbial groups are most likely responsible for production of succinate, ethanol, formate, and 2,3-butanediol, with the possible exception of succinate, which occurs via plant activity.

After the lactic acid bacteria become the dominant microbial species, lactate and acetate are essentially the only fermentation products made. With homofermentative lactic acid bacteria typically dominating the active fermentation phase, lactate production would be expected to be several times that of acetate production. After active fermentation has ceased and heterofermenters are more prominent, acetate production relative to lactate would be expected to increase. This trend, however, has not been observed. Pitt et al. (108) in summarizing data from several studies found that the most homofermentative production of lactate was after the silage had reached a low pH. This suggests that the relative production of the two acids is not only a function of the dominant lactic acid bacterial species but also the growth conditions. This is supported by data on *Streptococcus bovis* (110). This organism is highly heterofermentative at low growth rates on glucose at near neutral pH's, but at lower pH (4.7) these bacteria shift to homofermentative production of lactate. If heterofermentative lactic acid bacterial species were similarly capable of homofermentation at low pH, it would explain the apparent contradiction between dominant species and ratio of fermentation products late in silage fermentation.

One common trend in good silages not shown in Fig. 7.3 occurs long after active fermentation has ceased. This is a slow decrease in lactate and an increase in acetate. Even in well-sealed silos, there is a slow infiltration of oxygen. Several species of lactic acid bacteria, *L. casei* and *L. plantarum*, can aerobically metabolize lactate to acetate and carbon dioxide (43).

Changes in the crop

The most obvious changes in the crop are the losses of sugars as they are fermented by lactic acid bacteria and other microorganisms. However, other carbohydrates and organic acids are lost. Especially with legumes that have relatively low sugar contents, fermentation products often exceed the original amount of sugar (85, 112).

As indicated earlier, some lactic acid bacteria can ferment citrate and malate. Playne and McDonald (109) found that both of these acids declined in Italian ryegrass and red clover silages. Levels of two other organic acids, fumaric and glyceric, decreased during fermentation, whereas levels of malonic, oxalic, and glycolic acids remained relatively constant. However, lactic acid bacteria were not shown to cause these changes.

Decreases in nonstructural complex carbohydrate content during ensiling have been noted. Fructans are the main storage carbohydrate in grasses. The fructan content drops to near trace levels (36) providing extra sugars for fermentation. The storage carbohydrate in legumes is starch. Starch is not as completely hydrolyzed as fructans. Starch hydrolysis in legume silages adds approximately 1 percent DM in sugars for fermentation (74, 85). The starch converted is proportional to the amount present at ensiling except at high DM contents (>60 percent) (85); however, the starch remaining after ensiling is usually between 0.5 and 1.0 percent DM (74, 85).

Breakdown of structural carbohydrate may also occur. Dewar et al. (22) extracted hemicellulases from perennial ryegrass and determined their activity by the increase in pentoses. This suggested that hemicellulose hydrolysis was possible. Several silage studies have found that the difference between neutral detergent fiber (NDF) and acid detergent fiber (ADF) often decreases during ensiling (2, 75). This loss has been attributed to the breakdown of hemicellulose, but may include pectin degradation. Several other studies have provided more direct evidence of hemicellulose loss (68, p. 53). In any case it appears that some degradation of structural carbohydrate occurs during ensiling.

Various changes in the nitrogen fraction of the crop occur in the silo. The largest change is the breakdown of protein to soluble nonprotein fractions (peptides, free amino acids, and ammonia). This is primarily from plant enzyme activity (97). This activity is greatest at the start of fermentation and declines such that there is little activity after 5 days (6, 72, 82). The amount of proteolysis depends on moisture content (82), pH (71) and temperature (25, 70, 86). However, soluble nonprotein nitrogen in good silages is most frequently between 50 and 75 percent of total nitrogen compared with 15 to 30 percent in the crop as it enters the silo (64, 82, 86, 97, 105). Lower amounts of proteolysis in untreated silages usually occur in dry (>60 percent DM) silages (82).

Breakdown of free amino acids to ammonia is mostly the result of microbial activity (97). In a successful fermentation, the increase in ammonia in nonclostridial silages is probably due to some species of lactic acid bacteria (9, 16, 60) and enterobacteria (5), as indicated earlier.

Factors Affecting Ensiling Dynamics

Oxygen

Complete exclusion of oxygen throughout storage is economically infeasible for silage systems. Oxygen will slowly diffuse through the walls and covers of a silo. Diurnal and seasonal heating and cooling of the silo will also cause air to be drawn into the structure. However, a well-maintained and sealed silo minimizes the amount of oxygen entering the silo.

Unfortunately, oxygen is the factor which most negatively affects the course of silage fermentation and preservation of the crop. Oxygen affects the growth rate and end products of lactic acid bacteria as well as the amount of sugars available for lactic acid bacterial growth. Excess oxygen directly or indirectly encourages the growth of the fungi, enterobacteria and clostrida. Finally, the heat generated from aerobic respiration may at sufficient levels damage the nutritive value of the crop through the formation of Maillard products or lead to spontaneous combustion in dry silages.

Oxygen affects the growth of lactic acid bacteria in several ways. Lactic acid bacteria, in general, are tolerant of aerobic conditions but differ in how they use oxygen. Oxygen is reduced to hydrogen peroxide or water (13, 43). In some of the streptococci and lactobacilli, hydrogen peroxide may build up to levels which inhibit growth (13, 45).

Oxygen also affects the end products of lactic acid bacterial growth. Many of the lactic acid bacteria will produce less lactate and more acetate under aerobic conditions (7, 13, 65). Enhanced concentrations of ethanol, acetoin, 2,3-butanediol, and diacetyl might also be expected (13, 65). Such shifts, decreased lactate and increased acetate concentrations, compared with controls have been measured in aerated experimental silages (57, 58, 103). This shift in fermentation products slows the rate of pH decline in the crop and may prevent a low final pH. After active fermentation is complete and sugar concentrations are low, oxygen will allow many lactic acid bacteria strains to oxidize lactate to acetate (13, 43, 57).

Early in ensiling, excess oxygen will contribute to unnecessary plant and microbial respiration. This respiration decreases the amount of sugar available for fermentation and heats the crop. The loss of sugar plus inefficient production of lactate by the lactic acid bacteria caused by the presence of oxygen may prohibit the crop from reaching a pH sufficiently low to prevent clostridial growth (57; 68, p. 105; 98), reducing silage feed value.

The presence of oxygen, particularly in the early stages of ensiling, encourages the growth of fungi (95, 103) and enterobacteria (56, 98). Of particular concern is the proliferation of lactate-assimilating yeasts (103) during storage. If these yeasts reach sufficient levels during storage, then the crop may be particularly vulnerable to aerobic deterioration (and the resultant heating and DM losses) when the crop is fed out. Furthermore, maintenance of an anaerobic environment is the principal means of inhibiting growth of fungi because low pH and normal fermentation products have little or no effect on preventing fungal growth (135). The effects of oxygen on enterobacteria are less critical. However, high populations of enterobacteria mean that more sugar is being converted to a weak acid, acetic acid, and more proteolytic activity may occur.

Initial microbial populations

Given good ensiling technique and management and sufficient sugars, lactic acid bacteria will dominate a silage fermentation almost without exception. This perhaps is surprising given the potential relative variation of the microbial groups, but understandable given their growth rates, the range of pH and water activities lactic acid bacteria can tolerate, and the antibiotic compounds some of the species produce.

In spite of this, a major silage additive, particularly in the United States, is the bacterial inoculant. Inoculants generally supply 10^4 to 10^6 cfu per gram of crop of homofermentative lactic acid bacteria. The most common species is *L. plantarum*. However, many products include *S. faecium* and/or *Pediococcus* species in addition (130). The purpose of these products is to guarantee a fast, efficient (i.e., homofermentative) fermentation. The efficacy of these products has been related in part to the natural or epiphytic level of lactic acid bacteria on the crop. Muck and O'Connor (1988) found that the lactic acid bacteria in a commercial inoculant did not affect the speed of fermentation or final silage characteristics in alfalfa if they were applied at less than 1 percent of the natural population. Epiphytic counts in that study were enumerated on Rogosa SL agar acidified to pH 4.5 to eliminate streptococci and other lactic acid bacteria which would be effective only in early fermentation. An inoculant applied at 10 percent or more of the epiphytic population consistently improved the rate of pH decline but did not always improve the final silage characteristics. More important to the farmer, Satter et al. (111) reported that improvements in milk production in dairy cattle being fed inoculated alfalfa silage did not occur unless the inoculant bacteria were at least 10 times the epiphytic population as measured on Rogosa medium. These results suggest that the use of silage inoculants can be maximized if the farmer knows when the natural lactic acid bacterial population is low.

Most of the typical crops ensiled by farmers have low lactic acid bacterial populations on the standing plant prior to harvest. Populations on growing grasses, cereals, and legumes have usually been reported as less than 100 cfu per gram of crop (24, 53, 84, 119, 120). The exception to this is corn (maize). At the normal maturity for ensiling, counts on corn are typically between 10^3 and 10^7 cfu per gram of crop (50, 79). There is ample evidence that the lactic acid bacterial population on forage crops increases as the plant matures (50, 79, 96).

When forage crops are mown and wilted prior to ensiling, lactic acid bacterial counts may increase or decrease (24, 36, 84, 104, 123). Factors related to lactic acid bacterial growth in wilting alfalfa have been temperature, length of wilting, and the rate of drying (84). Lactic acid bacterial numbers did not increase unless the average temperature was above 15°C, and numbers were usually less than 10^4 cfu per gram of crop unless wilting times were greater than 1 day except under very slow drying conditions.

A final factor affecting the number of lactic acid bacteria on the crop entering the silo is the forage harvester or chopper. There are numerous indications that the harvester may inoculate the crop if lactic acid bacteria populations are low (24, 84, 104, 120). The cause of this observation is unclear; however, the amount of inoculation in alfalfa appears related to harvest conditions (temperature, antecedent rainfall or humidity, crop moisture content) and ranges from 100 to 10^5 cfu per gram of crop (84, 87).

Currently, there are no simple methods for farmers to estimate when natural lactic acid bacteria counts are low, but some are under development for alfalfa under U.S. weather conditions (87). However, the cause of the apparent inoculation by the forage harvester must be solved before a reasonably accurate prediction can be guaranteed.

Moisture content of the crop

The moisture content of the crop significantly affects silage fermentation, plant enzyme activity, and effluent losses from the silo. Of these, effluent losses are the most noticeable.

Effluent production or the loss of plant juices from silage occurs because the crop is compressed such that the spaces between silage particles cannot hold all the juices expressed from the plant cells. With unwilted crops, effluent DM losses may reach 12 percent or more (68, p. 175; 78). This effluent contains soluble carbohydrates, minerals, nitrogen, and other nutrients, and thus represents a loss in feed value greater than the amount of DM lost.

The amount of effluent lost is related to the moisture content and density of the crop. Effluent production decreases as DM content increases (68, p. 120). For horizontal silos and small upright silos, little or no effluent occurs if the crop is wilted to at least 30 percent DM (68, p. 120; 78). With taller upright silos, pressures at the bottom of the silo are greater, creating increased densities. As a result, the DM content necessary for minimum effluent losses increases as silo height increases.

Moisture content affects microbial growth. As the DM content of the crop increases, water activity decreases (or osmotic pressure increases) (34), causing a subsequent reduction in microbial growth rates (44, 59, 127). Several studies have found that increasing DM content slows the rate of lactic acid bacterial growth and reduces peak populations in silages (42, 49, 85, 118). As a result, the rate of pH decline and acid production are reduced at higher DM contents (40, 42, 49, 85).

Based on the reduced lactic acid bacterial growth rates, wilting may not appear to be beneficial. However, wilting is an important method of preventing clostridial growth. As indicated in Fig. 7.1, wilting a crop to 40 to 50 percent DM prevents any clostridial fermentation from occurring. Increasing amounts of lactic acid bacterial fermentation are required to stop clostridial activity as DM content decreases. Several studies have demonstrated the benefit of wilting in reducing clostridial activity; the unwilted silages had high butyric acid concentrations and/or high final pH relative to the wilted silages (40, 63, 67).

Like the clostridia, the pH at which the growth of lactic acid bacteria ceases is affected by moisture level. Several studies have shown that final pH increases as DM content increases (40, 42, 49, 82). Similarly, the total amount of fermentation products is reduced at high DM contents (40, 42, 49, 85). Thus moisture level affects the substrate required for the cessation of fermentation by low pH, preventing bacterial growth. This will be considered in greater detail in the next section.

Another positive effect of wilting is reduced plant proteolytic activity. Proteolysis rate decreases linearly with increasing DM content (82); however, the total amount of proteolytic activity is not as strongly correlated with DM level. At less than 50 percent DM, total proteolysis may or may not be higher in wilted silages than in unwilted ones (35, 72, 82, 105, 117). Above 50 percent DM, the amount of proteolysis does decrease consistently as DM increases (82). The variability in the wetter silages is apparently due to the interaction of moisture level on both proteolysis and pH decline from fermentation since pH also affects proteolysis. At the higher DM contents, pH decline normally occurs after most proteolysis has ended, so that the effect of moisture level on proteolysis is more consistent.

Overall, a moderate amount of wilting (35 to 45 percent DM) is beneficial for ensiling. It prevents effluent losses and essentially inhibits the growth of clostridia. This amount of wilting may reduce proteolytic plant activity. Additional wilting might appear to be good for guaranteeing reduced proteolysis. However, the increased risk of weather damage during wilting, machinery losses (38) and heat damage (106) associated with higher DM contents make high-DM silages less likely to be profitable. It must also be noted that in some climates, such as those in northern Europe, the weather risks associated with even moderate wilting often outweigh its value. Consequently, wilting is an effective tool for ensuring a good silage only when it can be done without rain damage to the crop.

Sugar content

Sugar content of the crop is important, particularly at DM contents less than 30 percent. At high DM contents, clostridial activity is inhibited by the low water activity, and thus lactic acid bacterial fermentation is relatively less important to crop preservation in the silo. At moisture levels where clostridial growth can occur, sufficient lactic and acetic acid production is needed to lower pH below the point where clostridial activity ceases. As indicated in Fig. 7.1, the critical pH decreases with lower DM contents. Consequently, the amount of acid production and hence the sugars required for fermentation needed to prevent clostridial activity increase for lower DM contents.

An additional factor affecting the sugar requirement is the buffering capacity of the crop. Buffering capacity is typically defined as the equivalents of acid required to drop crop pH from 6 to 4 per unit of DM. Organic acids account for 70 to 80 percent of the buffering in ryegrass and clover; amino acids and inorganic ions are responsible for much of the remaining buffering capacity (109). As buffering capacity increases, the amount of sugar needed to lower pH to a given value increases proportionately. Similarly, final pH in alfalfa silages was negatively and linearly correlated with the sugar to buffering capacity ratio (74). The limit of such a relationship occurs when there is sufficient sugar such that pH is lowered to the point where lactic acid bacterial growth ceases.

Buffering capacity varies among crops, maturities, and growth conditions. Grasses have lower buffering capacities than legumes (250 to 450 versus 350 to 600 meq/kg DM) (68, p. 31; 74; 89; 109). Whole-plant corn, like grasses, also has a low buffering capacity (68, p. 33). With increasing maturity, buffering capacity declines (74, 89, 129). Buffering capacity is high when the crop is not water-stressed and under high fertility (89, 109, 114).

Currently, there are not adequate methods to enable the farmer to easily determine the sufficiency of the sugar content of the crop as it enters the silo. Nevertheless, some generalizations regarding conditions requiring additional sugars can be made. First, whole-plant corn is unlikely to have this problem because of its low buffering capacity relative to the amount of sugar present. Second, clostridial silages resulting from sugar limitations tend to occur in direct cut or lightly wilted grasses and legumes (e.g., 29, 63, 67, 69, 98, 112). Third, silage fermentation may be limited by sugar content in moderately wilted grasses and legumes (30 to 50 percent DM) (41, 85), but the likelihood of clostridial fermentation is small. In this range, sugar limitation would principally affect plant enzyme activity (41, 82) and the effectiveness of bacterial inoculants (41). Finally, in heavily wilted crops (>50 percent DM), sugar limitations to fermentation appear unlikely (41, 85).

Three different groups of additives are being marketed, principally in Europe, to guarantee that the crop has sufficient sugars to reach a given pH: sugars, acids, and enzymes. Sugars, including molasses, are a direct approach to the problem. However, the dominant additives are the acids (130), particularly formic and sulfuric. Application rates are such that crop pH is lowered to between 4.5 and 5.0. Then, natural fermentation further reduces pH. The acids not only reduce the sugar required to reach a given pH, but the immediate drop in pH may be beneficial in decreasing proteolytic activity (83). The third approach to solving sugar limitations is the use of enzymes, mostly cellulases and hemicellulases, to hydrolyze structural carbohydrates to hexoses and pentoses. At recommended levels, these products have shown some benefit in unwilted forages that would be marginally short of sugars, but higher enzyme concentrations would most likely be needed when more severe limitations occur in order to supply sufficient sugars during the active fermentation period (11, 37, 39, 47, 76).

Temperature

Although temperature affects microbial growth rates and enzyme activities, the overall effect of varying temperature between 10 and 40°C is relatively minor compared to the factors which have already been discussed. As expected, increasing temperature increases the speed of fermentation and as a result lowers pH more rapidly (28, 63, 86). However, the final silage characteristics are often not that different. In direct-cut to moderately wilted (40 percent DM) forages, final pH decreases as fermentation temperature increases. The amount of difference ranges from approximately 0.0 to 0.6 pH units in nonclostridial silages for a 20°C change in temperature (28, 63, 86). At 55 percent DM in alfalfa, either no change or a reverse trend with temperature has been reported (86). Effects on final acid content are somewhat less consistent than the trends with pH. Both lactic and acetic acid in wetter silages tend to be greater at high temperatures (63, 86). This suggests that more breakdown of complex carbohydrates, providing extra substrate for fermentation, may be occurring at higher temperatures. The opposite trend was noted at 55 percent DM (86).

An advantage of keeping the crop cool is the inhibition of clostridial activity. Gibson et al. (28) noted that little clostridial activity occurred in grass silage at 22°C as compared with silages at 30 and 40°C. In another study, a comparison of pH and lactic acid content in untreated silages incubated at 10 and 20°C suggested that there was a clostridial fermentation in the 20°C silage (63). Because many clostridia have higher optimum growth temperatures (37°C) than most lactic acid bacteria (25 to 35°C), as indicated earlier, such observations would agree with pure-culture data.

Another anticipated effect of temperature would be on plant enzyme activity, but few measurements have been made in silages. Wood and Parker (131) found that plant respiration approximately doubled for a 10°C rise in temperature between 5 and 25°C and remained constant between 25 and 45°C. Keeping the temperature low before sealing the silo would help in reducing losses of sugars by respiration.

The breakdown of hemicellulose from both chemical hydrolysis and plant enzyme hydrolysis is affected by temperature. Between 22 and 37°C, chemical hydrolysis increased linearly with temperature, whereas plant enzyme hydrolysis increases quadratically (22). The more significant activity is due to plant enzymes; however, their activity is limited to the first week of ensiling (22). Since this breakdown of hemicellulose provides additional pentoses for fermentation, higher temperatures early in fermentation would promote this activity.

Greenhill (33) found that temperature affected the time before plant cells ruptured in the silo, releasing sugars contained in cell contents for fermentation. At 40°C, this occurred at approximately 2 h in ryegrass, alfalfa, and clover. Decreasing the temperature to 25°C at least doubled the time for this process to occur.

Whereas higher temperatures may be beneficial in enhancing hemicellulose breakdown and the release of plant cell contents, elevated temperatures also increase proteolysis. Proteolytic activity between 10 and 40°C increases sharply with temperature; a 20°C rise causes a three to fourfold increase in rate (8, 25, 70). Because temperature also increases the fermentation rate and thus the rate of pH decline, the increase in the amount of proteolysis is much less than what might be expected from the effects on proteolytic activity. Muck and Dickerson (86) found that increasing temperature from 15 to 35°C increased the amount of soluble nonprotein nitrogen by 10 percentage units as a fraction of total nitrogen. Ammonia increased 10 to 20 percent between 15 and 25°C; however, inconsistent effects were noted between 25 and 35°C.

The final effect of elevated temperatures is formation of Maillard or browning reaction products. As temperature rises above 35 or 40°C, chemical reactions between amino acids and sugars occur, forming acid detergent-insoluble compounds. This process also liberates heat, potentially increasing the silage temperature even more. A moderate amount of this activity may be beneficial in improving the efficiency of nitrogen absorption by ruminants, but the process is not controllable with current technology. Excessive activity may lead to spontaneous combustion of the silage.

Weighing the positive and negative effects of elevated temperatures, the farmer is most often best served by maintaining crop temperature close to ambient. This can be accomplished by filling the silo quickly and sealing it well, preventing excessive aerobic respiration.

Chop length

The effect of forage particle size on fermentation and on animal intake and performance has been a concern. Today, most forages are chopped because of the ease of handling both in harvesting and in silo loading and unloading. Consequently, the degree of chopping rather than whether to chop is the matter of concern.

Laboratory studies have shown effects on fermentation due to chop length. The shorter the chop length and/or the greater the degree of laceration or mincing, the faster the lactic acid bacteria grow and produce fermentation products, the lower the final pH and the less susceptible the silage is to clostridial fermentation (66, 113, 118, 128). These studies suggest that decreasing chop length and increasing the physical damage to the crop makes sugars more readily available for fermentation.

Data on field-scale silos show no consistent effect due to particle size or degree of laceration. In some cases, the longer particles ensiled better (66), but in most cases there have been only slight or no improvement seen by decreasing particle length (31, 66, 100). Differences between laboratory and field-scale trials have been attributed to the compaction and pressures obtained in full-scale silos (14), which would help express plant juices and minimize the differences in chop length on fermentation. Consequently, chop length most likely has little practical significance in silage fermentation.

Models of Silage Fermentation

Silage fermentation simulation models have recently been developed. Despite their short existence, these models are proving useful in understanding the dynamics of the ensiling process and the conditions under which various silage additives will be beneficial.

Neal and Thornley (92) published the first simulation model. This model was not designed to accurately predict the processes occurring during fermentation. Rather, they sought to develop a simple model of anaerobic fermentation in the silo and determine if this model would qualitatively describe many of the changes observed in normal ensiling dynamics. Table 7.5 gives an overview of the model. Two groups of bacteria, lactic acid bacteria and clostridia, were modeled. The lactic acid bacteria metabolized water-soluble carbohydrate to solely lactic acid, whereas the clostridia metabolized both lactic acid and water-soluble carbohydrate to butyric acid. The rate of metabolism per unit mass of bacteria was affected by pH but not by substrate concentration unless substrates were exhausted. Moisture content only affected clostridial growth, and temperature effects were not directly considered. The pH of the silage was determined by the assumed buffering characteristics of the crop and the association-dissociation equilibria for lactic and butyric acid.

Considering the foregoing discussion of the ensiling process, such a simple model would hardly seem sufficient to describe the dynamics of the silage fermentation. However, the various simulations presented by Neal and Thornley suggested that the model described the changes in pH, fermentation acids, and bacterial groups surprisingly well. The development of clostridia lagged behind the lactic acid bacteria. Clostridial fermentation was reduced at higher DM contents. Changing the fractional output of lactic acid per unit of water-soluble carbohydrate to simulate the change from a homofermentative to a heterofermentative lactic acid bacterial population slowed the rate of pH decline and thus increased clostridial fermentation.

From a qualitative sense, the model failed in only two areas: the rate of pH decline did not vary with DM content and final pH in nonclostridial silages was unreasonably low. Clearly the effects of water activity on microbial growth needed to be included to provide more typical variation with DM content. The low final pH values were due to the assumed effect of pH on the growth of lactic acid bacteria, which could easily be adjusted.

The model of Pitt et al. (108) was more comprehensive than that of Neal and Thornley even though it did not consider clostridial fermen-

TABLE 7.5 Processes Simulated in Neal and Thornley Model

Process	Inputs	Outputs
Growth, lactic acid bacteria	WSC, LAB	LA, LAB
Growth, clostridia	WSC, LA, CB	BA, CB
Death, lactic acid bacteria	LAB	LAB
Death, clostridia	CB	CB

NOTE: WSC = water-soluble carbohydrate; LAB = lactic acid bacterial mass; LA = lactic acid concentration; CB = clostridial bacterial mass; BA = butyric acid concentration.

SOURCE: Neal and Thornley 1983.

tation. This model included plant activities as well as bacterial growth. Table 7.6 provides an overview of the phases and processes simulated in the model. Ensiling was separated into three phases: aerobic respiration, lag, and fermentation. In the first phase, oxygen was removed from the silo by plant respiration using sugars. Concurrently, the action of plant enzymes breaking down hemicellulose to pentoses and protein to soluble nonprotein nitrogen were modeled. Once an anaerobic environment was established, a lag period for the lysis of plant cells and subsequent lag in lactic acid bacterial growth was simulated. After the lag, the growth of the lactic acid bacteria on sugars already present and those produced through hemicellulose hydrolysis was modeled.

As indicated in Table 7.6, the model of Pitt et al. (108) included more environmental factors affecting the ensiling process. The growth of lactic acid bacteria was affected by temperature, pH, water activity, and substrate concentration. Similarly, the activities of other processes were modified by as many environmental factors as supported by literature. Moreover, a more thorough attempt was made to ensure that the effects of the various environmental factors were accurately described.

Like the model of Neal and Thornley, this model assumed only one

TABLE 7.6 Phases and Processes Simulated in the Pitt et al. Model

Phase	Process	Inputs to process	Outputs from process	Environmental factors affecting rate of process
Aerobic	Respiration	WSC,O_2	CO_2,H_2O,heat	T,pH,d,C_{CO_2},C_{O_2}
	Hemicellulose hydrolysis	HC	WSC	T,pH
	Proteolysis	PN	NPN	T,pH,d,C_{PN}
Lag	Plant cell lysis	—	—	T,A_w
	Lag in bacterial growth	—	—	A_w
	Hemicellulose hydrolysis	HC	WSC	T,pH
	Proteolysis	PN	NPN	T,pH,d,C_{PN}
Fermentation	Growth of LB	WSC	LB,L,A	T,pH,A_w,C_{WSC}
	Death of LB	LB	WSC	T,pH,A_w
	Change in pH	L,A	H^+	pH,B_h,C_L,C_A
	Hemicellulose hydrolysis	HC	WSC	T,pH
	Proteolysis	PN	NPN	T,pH,d,C_{PN}

NOTE: Symbols used: C_x = concentration of x; WSC = water-soluble carbohydrates; O_2 = oxygen; CO_2 = carbon dioxide; H_2O = water; T = temperature; d = dry matter content; HC = hemicellulose; PN = protein nitrogen; NPN = nonprotein nitrogen; A_w = water activity; LB = lactic acid bacteria; L = lactic acid; A = acetic acid; H^+ = hydrogen ions; B_h = buffer index of herbage.

SOURCE: Pitt et al. 1985.

group of lactic acid bacteria; however, the end products of fermentation were expanded to include acetic acid. Instead of assuming homo- or heterofermentative lactic acid bacteria dominated the fermentation, they regressed literature data to obtain relative lactic and acetic acid production as a function of pH and sugar content. They assumed that this approach would provide a typical natural mix of homo- and heterofermentative lactic acid bacteria.

The model was then validated against various reported studies where clostridial fermentation was not a factor. The time for and temperature rise produced by aerobic respiration were closely predicted by the model. Final pH, water-soluble carbohydrates, and lactic acid as a fraction of lactic plus acetic acid were predicted surprisingly well. The prediction of lactic acid was about 10 percent above the actual values, but the trend indicated that the simplified approach for predicting end-product ratios was reasonable. Proteolysis, though based on a sparse amount of data, was predicted well in two-thirds of the cases.

One area that was not predicted well were pH time courses. The model set maximum bacterial growth rate at pure-culture level. At this level, fermentation proceeded too quickly. In order to bring the model in line with actual studies, the rate had to be approximately 38 percent of the pure-culture rate.

The model of Pitt et al. (108) was enhanced by Leibensperger and Pitt (61). The primary addition was clostridial growth. Clostridia were assumed to use only lactic acid, producing butyric acid and carbon dioxide. Their growth was modified by temperature, pH, water activity, and lactic acid concentration. Ammonia production rate from proteolytic clostridia was calculated based on the production rate and concentration of butyric acid.

The growth of lactic acid bacteria was enhanced in several areas. End products of fermentation were expanded to include mannitol, ethanol, and carbon dioxide. Additional literature data were used to expand the range of temperatures and DM contents over which the model would reasonably predict lactic acid bacterial growth.

A sensitivity analysis of the model indicated that the occurrence of clostridial silages was affected by initial bacterial numbers, initial pH, temperature, DM content, and initial sugar to buffering capacity ratio. The degree of sensitivity to one particular factor was often mediated by other factors. For example, high sugar to buffering capacity ratios generally prevented clostridial silages. However, if lactic acid bacterial counts were sufficiently low with respect to clostridial numbers or if temperatures were above 30°C, no amount of sugar could prevent a clostridial silage from happening with wet direct cut silages (<15 percent DM). At 20 to 25°C, initial lactic acid bacterial levels had large effects on the range of conditions where silages went clostridial, whereas initial pH was relatively unimportant. At 29 to 32°C, the reverse was true. Increasing temperature between 27 and 35°C substantially increased the number of conditions under which a clostridial silage would occur, whereas temperature changes above or below these levels had relatively little effect on the range of conditions at which clostridial silages were predicted.

These results were very significant because they demonstrated the potential of a silage model to sort out the key variables and their interactions in particular silage problems. The model of Leibensperger and Pitt (61) has subsequently been used to analyze the effectiveness of several silage additives: inoculants, formic acid, and molasses (62, 107).

Meiering et al. (73) have developed the most recent silage model, based primarily on Pitt et al. (108), to predict the production of toxic nitrous oxide gases in silages. Like the earlier models, one group of lactobacilli was simulated. Clostridia, enterobacteria, and yeasts were added to this model. All three new microbial groups were assumed to ferment water-soluble carbohydrates. From this substrate the model simulates butyric acid and CO_2 production by clostridia; acetic acid, ethanol, and CO_2 production by enterobacteria; and ethanol and CO_2 production by yeasts. In addition, the model predicts clostridial fermentation of amino acids to ammonia, respiration of nitrate to nitrous oxides by enterobacteria, nitrate reductase activity by the crop, and chemical nitrite reduction by organic acids. Consequently, this model represents the most complete attempt at modeling the various microbial groups involved in silage fermentation.

The model parameters were adjusted to match values obtained from unwilted ryegrass silages made in minisilos. Then the model was used to simulate three other experimental ryegrass silage treatments. The model reasonably predicted pH, water-soluble carbohydrates, lactic acid, CO_2, and nitrous oxides with time. Simulations of microbial mass with time were not reported. Nevertheless, the model performed quite satisfactorily and should be of use in evaluating strategies for reducing toxic silage gases.

Research Needs

Computer models of silage fermentation have predicted pH and fermentation products well even though, for example, lactic acid bacteria have been modeled as a single group. This means that a relatively unsophisticated modeling approach to describing silage fermentation does in fact explain much of the observed variation in nature. This, however, does not mean that there are not research needs in this area.

The greatest need is in providing farmers with the necessary information to make day-to-day management decisions regarding the ensiling of crops. Computer models have demonstrated that much of the variation in silage quality can be explained by initial microbial numbers, crop buffering capacity, and sugar content. If the farmer knew these values as the crop entered the silo, then accurate decisions regarding the use of additives could be made. Currently, the refractometer has been demonstrated in grass to be an effective, quick method for estimating sugar content (101). However, similar methods are needed for lactic acid bacterial numbers, clostridial counts, and buffering capacity. Progress on prediction of lactic acid bacteria on alfalfa is being made under northern U.S. conditions (87), but such results may not apply to other climates and are highly unlikely to work with other crops.

This type of research unfortunately requires large numbers of samples in various climates and crops to develop predictive relationships. Because of this, such research should be restricted to those crops in which problems occur or in which there is economic opportunity for such knowledge. Whole-plant corn silage typically has low buffering capacity, high sugar content, and high numbers of lactic acid bacteria. Consequently, effort along these lines in this crop are of low priority. Legumes and grasses ensiled at high DM contents (>40 percent) are most likely to ensile well even if sugar contents are low because clostridial fermentation is unlikely. However, benefit from inoculation with lactic acid bacteria has been shown when the inoculant supplies lactic acid bacteria at sufficiently high levels above the natural population (Satter et al. 1987). For this reason, epiphytic lactic acid bacterial numbers would be important for these crops. Direct-cut or lightly wilted grasses and legumes may have a problem with clostridial fermentation. Thus, methods of predicting microbial numbers, sugar contents, and buffering capacity would be important for such conditions.

Another area of research is to determine the cause of animal performance responses to silages inoculated with lactic acid bacteria. Generally, these silages do not differ in nutrient and fiber content from well-preserved uninoculated silages, so that animal responses would not be anticipated. Determination of the factors leading to a response may improve silage inoculants.

There are several areas of basic research that may be fruitful. First, computer models consolidate the lactic acid bacteria into one group because insufficient data are available to differentiate growth of various species. Shifts in the dominant lactic acid bacterial species during fermentation may be better understood if the effects of pH, temperature, water activity, and fermentation end products on the growth kinetics and fermentation products of various lactic acid bacterial species were known. Second, little data on plant organic acids that cause much of the crop buffering capacity are available. Not only is variation within and between plant species needed, but information on which acids can be metabolized by lactic acid bacteria and other indigenous microorganisms would be helpful. Also, substrate preferences of lactic acid bacteria among the acids and sugars might provide insight into potentially beneficial additives not being considered currently. A final basic research area is that of structural carbohydrate breakdown in forage crops during ensiling. The potential for hemicellulose hydrolysis has been demonstrated (22), but more information on actual changes in hemicellulose, pectins, and cellulose would be useful in predicting the level of additional sugars made available by the plant for fermentation.

All of the above research needs relate principally to improving the understanding and management of ensiling. In the long term, the silo holds much greater possibilities than just preserving the crop. Forages stored dry are relatively inactive biologically. However, storing a wetter crop in a silo may provide opportunities to improve the value of the crop from that harvested through enzymes, microbial inoculation, etc. Certainly, silage researchers should be alert for potential means of enhancing crops in the silo.

References

1. Allen, L. A., and J. Harrison, "Anaerobic Spore-Formers in Grass and Grass Silage," *Annals Appl. Biol.* 24:148–153 (1937).
2. Balogu, D. O., "Influence of Moisture Levels and Reconstitution on Nutrient Content of Alfalfa and Corn Silages," Ph.D. Thesis, Mississippi State Univ., 1985.
3. Beck, Th., "Der gegenwärtige Stand der Mikrobiologie der Einsäuerung und Trockung von Futterpflanzen", Ber. des 3 Kongr. d. Europäischen Grünlandver. Braunwchweig, 1969, pp. 207–219.
4. Beck, Th., "Die quantitative und qualitative Zusammensetzung der Milchsäurebakterienpopulation im Gärfutter", *Landwirt. Forschung* 27:55–63 (1972).
5. Beck, Th., "The Micro-biology of Silage Fermentation," In McCullough, M. E., ed., *Fermentation of Silage—A Review*, National Feed Ingred. Assoc., W. Des Moines, Iowa, 1978, pp. 61–115.
6. Bergen, W. G., E. H. Cash, and H. E. Henderson, "Changes in Nitrogenous Compounds of the Whole Corn Plant during Ensiling and Subsequent Effects of Dry Matter Intake by Sheep," *J. Anim. Sci.* 39:629–637 (1974).
7. Blickstad, E., "Growth and End Product Formation of Two Psychrotrophic *Lactobacillus* spp. and *Brochothrix thermosphacta* ATCC 11509^T at Different pH Values and Temperatures," *Appl. Environ. Microbiol.* 46:1345–1350 (1983).
8. Brady, C. J., "The Leaf Protease of *Trifolium repens*," Biochem. J. 78:631–640 (1961).
9. Brady, C. J., "The Redistribution of Nitrogen in Silage by Lactic-Acid Producing Bacteria," *Aust. J. Biol. Sci.* 19:123–130 (1966).
10. Chamberlain, D. G., and J. Quig, "The Effects of the Rate of Addition of Formic Acid and Sulphuric Acid on the Ensilage of Perennial Ryegrass in Laboratory Silos," *J. Sci. Food Agric.* 38:217–228 (1987).

11. Chamberlain, D. G., P. C. Thomas, and S. Robertson, "The Effect of Formic Acid, Bacterial Inoculants, and Enzyme Additives on Feed Intake and Milk Production in Cows Given Silage of High or Moderate Digestibility with 2 Levels of Supplementary Concentrates," In *Eighth Silage Conference: Summary of Papers*, Agric. Food Res. Counc., Hurley, UK, 1987, pp. 31–32.
12. Christensen, M. D., M. N. Albury, and C. S. Pederson, "Variation in the Acetic Acid–Lactic Acid Ratio among the Lactic Acid Bacteria," *Appl. Microbiol.* 6:316–318 (1958).
13. Condon, S., "Responses of Lactic Acid Bacteria to Oxygen," *FEMS Microbiol. Rev.* 46:269–280 (1987).
14. Cowan, A. M., K. K. Barnes, and K. S. Allen, "Evaluation of Shredded Legume-Grass Silage," *Agric. Engin.* 38:588–605 (1957).
15. de Giori, G. S., G. F. de Valdez, A. P. de Ruiz Holgado, and G. Oliver, "Effect of Growth Temperature on Acid Production by Lactic Acid Bacteria," *Microbiol.-Aliments-Nutrition* 3:243–246 (1985).
16. de Giori, G. S., G. F. de Valdez, A. P. de Ruiz Holgado, and G. Oliver, "Effect of pH and Temperature on the Proteolytic Activity of Lactic Acid Bacteria," *J. Dairy Sci.* 68:2160–2164 (1985).
17. de Klerk, H. C., and J. N. Coetzee, "Antibiosis among Lactobacilli," *Nature* 192(4800):340–341 (1961).
18. Dellaglio, F., and S. Torriani, "Microbiological Variations in Silage of Lucerne and Italian Rye-Grass with Added Lactic Acid Bacteria or Formic Acid," *Microbiol.-Aliments-Nutrition* 3:273–282 (1985).
19. Dellaglio, F., and S. Torriani, "DNA-DNA Homology, Physiological Characteristics and Distribution of Lactic Acid Bacteria Isolated from Maize Silage," *J. Appl. Bact.* 60:83–92 (1986).
20. Dellaglio, F., S. Torriani, and E. Santi, "Ricerche Sulla Conservazione dei Foraggi Verdi per Insilamento. II. Presenza e Caratterizzazione Fisiologica di Batteri Lattici in Insilati di Prato e di Erba Medica," *Ann. Fac. Agr. Univ. Catt. Sacro Cuore* 23:171–182 (1983).
21. Dellaglio, F., M. Vescovo, L. Morelli, and S. Torriani, "Lactic Acid Bacteria in Ensiled High-Moisture Corn Grain: Physiological and Genetic Characterization," *System. Appl. Microbiol.* 5:534–544 (1984).
22. Dewar, W. A., P. McDonald, and R. Whittenbury, "The Hydrolysis of Grass Hemicelluloses during Ensilage," *J. Sci. Food Agric.* 14:411–417 (1963).
23. Fenlon, D. R., J. Wilson, and J. R. Weddell, "A Comparison of the Incidence of *Listeria monocytogenes* in Bagged and Wrapped Big Bale Silage," In *Eighth Silage Conference: Summary of Papers*, Agric. Food Res. Counc., Hurley, UK, 1987, pp. 113–114.
24. Fenton, M. P., "An Investigation into the Sources of Lactic Acid Bacteria in Grass Silage," *J. Appl. Bact.* 62:181–188 (1987).
25. Finley, J. W., C. Pallavicini, and G. O. Kohler, "Partial Isolation and Characterization of *Medicago sativa* Leaf Proteases," *J. Sci. Food Agric.* 31:156–161 (1980).
26. Gibson, T., "Clostridia in Silage," *J. Appl. Bact.* 28:56–62 (1965).
27. Gibson, T., and A. C. Stirling, "The Bacteriology of Silage," *N.A.A.S. Quarterly Rev.*, no. 44, 167–172 (1959).
28. Gibson, T., A. C. Stirling, R. M. Keddie, and R. F. Rosenberger, "Bacteriological Changes in Silage Made at Controlled Temperatures," *J. Gen. Microbiol.* 19:112–129 (1958).
29. Gibson, T., A. C. Stirling, R. M. Keddie, and R. F. Rosenberger, "Bacteriological Changes in Silage as Affected by Laceration of the Fresh Grass," *J. Appl. Bact.* 24:60–70 (1961).
30. Goodfellow, M., B. Austin, and D. Dawson, "Classification and Identification of Phylloplane Bacteria Using Numerical Taxonomy," In Dickson, C. H., and T. F. Preece, eds., *Microbiology of Aerial Plant Surfaces*, Academic Press, London, 1976, pp. 275–292.
31. Gordon, F. J., "The Effects of Degree of Chopping Grass for Silage and Method of Concentrate Allocation on the Performance of Dairy Cows," *Grass Forage Sci.* 37: 59–65 (1982).

32. Grazia, L., and G. Suzzi, "A Survey of Lactic Acid Bacteria in Italian Silage," *J. Appl. Bact.* 56:373–379 (1984).
33. Greenhill, W. L., "Plant Juices in Relation to Silage Fermentation. II. Factors Affecting the Release of Juices," *J. Brit. Grassl. Soc.* 19:231–236 (1964).
34. Greenhill, W. L., "Plant Juices in Relation to Silage Fermentation. III. Effect of Water Activity of Juice," *J. Brit. Grassl. Soc.* 19:336–339 (1964).
35. Henderson, A. R., P. McDonald, and D. H. Anderson, "The Effect of Silage Additives Containing Formaldehyde on the Fermentation of Ryegrass Ensiled at Different Dry Matter Levels and on the Nutritive Value of Direct-Cut Silage," *Anim. Feed Sci. Tech.* 7:303–314 (1982).
36. Henderson, A. R., P. McDonald, and M. K. Woolford, "Chemical Changes and Losses during the Ensilage of Wilted Grass Treated with Formic Acid," *J. Sci. Food Agric.* 23:1079–1087 (1972).
37. Henderson, A. R., R. McGinn, and W. D. Kerr, "The Effect of a Cellulase Preparation Applied With or Without an Inoculum of Lactic Acid Bacteria on the Chemical Composition of Lucerne Ensiled in Laboratory Silos," In *Eighth Silage Conference: Summary of Papers*, Agric. Food Res. Counc., Hurley, UK, 1987, pp. 29–30.
38. Hoglund, C. R., "Comparitive Storage Losses and Feeding Value of Alfalfa and Corn Silage Crops When Harvested at Different Moisture Levels and Stored in Gas-Tight and Conventional Tower Silos: an Appraisal of Research Results," Michigan State Univ. Agron. Econ. Publ. No. 947, East Lansing, Michigan, 1964.
39. Hopkins, J. R., and R. V. Bass, "Enzymes—Their Contribution as a Silage Additive," In *Eighth Silage Conference: Summary of Papers*, Agric. Food Res. Counc., Hurley, UK, 1987, pp. 23–24.
40. Ito, K., "Studies on the Maintaining Nutritive Value of Silage. III. Effect of Wilting and Treatment with Formaldehyde on the Inhibition of Ensiling Fermentation," *Bull. Akita Pref. Coll. Agr.* 9:129–135 (1983).
41. Jones, B. A., "Efficacy of Bacterial Inoculants and Substrate Additions on Alfalfa Silages," M. S. Thesis, Univ. Wisconsin—Madison, 1988.
42. Kamra, D. N., R. Singh, R. C. Jakhmola, and R. V. N. Srivastava, "Effect of Wilting and the Additives, Straw, Molasses, and Urea on the Fermentation Pattern of Maize Silage," *Anim. Feed Sci. Tech.* 9:185–196 (1983).
43. Kandler, O., "Carbohydrate Metabolism in Lactic Acid Bacteria," *Ant. van Leeuwenhoek* 49:209–224 (1983).
44. Kashket, E. R., "Bioenergetics of Lactic Acid Bacteria: Cytoplasmic pH and Osmotolerance," *FEMS Microbiol. Rev.* 46:233–244 (1987).
45. Keen, A. R., "Growth Studies on the Lactic Streptococci," *J. Dairy Res.* 39:151–159 (1972).
46. Kempton, A. G., and C. L. San Clemente, "Chemistry and Microbiology of Forage-Crop Silage," *Appl. Microbiol.* 7:362–367 (1959).
47. Kennedy, S. J., "The Effect of Enzyme Treatments on the Preservation of First and Third Harvest Grass and on Subsequent In-Silo Loss, Silage Intake and Performance of Beef Cattle," In *Eighth Silage Conference: Summary of Papers*, Agric. Food Res. Counc., Hurley, UK, 1987, pp. 25–26.
48. Kibe, K., J. M. Ewart, and P. McDonald, "Chemical Studies with Silage Microorganisms in Artificial Media and Sterile Herbages," *J. Sci. Food Agric.* 28:355–364 (1977).
49. Kibe, K., E. Noda, and Y. Karasawa, "Effect of Moisture Level of Material Grass on Silage Fermentation," *J. Fac. Agr., Shinshu Univ.* 18:145–154 (1981).
50. Koch, G., A. Morwarid, and M. Kirchgessner, "Zum Einfluβ der Mikroorganismen der Maispflanzen auf die Stabilität der Silagen," *Wirtschaft.* Futter 19:15–20 (1973).
51. Koegel, R. G., R. J. Straub, and R. P. Walgenbach, "Quantification of Mechanical Losses in Forage Harvesting," *Trans. ASAE* 28:1047–1051 (1985).
52. Kozeluhova, K., and J. Prikryl, "(The Effect of the Application of Preservatives on the Dynamics of Microorganisms during the Fermentation of Clover-Grass Silage)," *Zivoc. Vyr.* 30:463–474 (1985).

53. Kroulik, J. T., L. A. Burkey, and H. G. Wiseman, "The Microbial Population on the Green Plant and of the Cut Prior to Ensiling," *J. Dairy Sci.* 38:256–262 (1955).
54. Langston, C. W. and C. Bouma, "A Study of the Microorganisms from Grass Silage. I. The Cocci," *Appl. Microbiol.* 8:212–222 (1960).
55. Langston, C. W., and C. Bouma, "A Study of the Microorganisms from Grass Silage. II. The Lactobacilli," *Appl. Microbiol.* 8:223–234 (1960).
56. Langston, C. W., C. Bouma, and R. M. Conner, "Chemical and Bacteriological Changes in Grass Silage during the Early Stages of Fermentation. II. Bacteriological Changes," *J. Dairy Sci.* 45:618–624 (1962).
57. Langston, C. W., H. Irvin, C. H. Gordon, C. Bouma, H. G. Wiseman, C. G. Melin and L. A. Moore, "Microbiology and Chemistry of Grass Silage," U.S. Dept. Agric. Tech. Bull. No. 1187, Washington, D.C., 1958.
58. Langston, C. W., H. G. Wiseman, C. H. Gordon, W. C. Jacobson, C. G. Melin, and L. A. Moore, "Chemical and Bacteriological Changes in Grass Silage during the Early Stages of Fermentation. I. Chemical Changes," *J. Dairy Sci.* 45:396–402 (1962).
59. Lanigan, G. W., "Silage Bacteriology. I. Water Activity and Temperature Relationships of Silage Strains of *Lactobacillus plantarum, Lactobacillus brevis*, and *Pedicoccus cerevisiae*," *Aust. J. Biol. Sci.* 16:606–615 (1963).
60. Law, B. A., and J. Kolstad, "Proteolytic Systems in Lactic Acid Bacteria," *Ant. van Leeuwenhoek* 49:225–245 (1983).
61. Leibensperger, R. Y., and R. E. Pitt, "A Model of Clostridial Dominance in Ensilage," *Grass Forage Sci.* 42:297–317 (1987).
62. Leibensperger, R. Y., and R. E. Pitt, "Modeling the Effects of Formic Acid and Molasses on Ensilage," *J. Dairy Sci.* 71:1220–1231 (1988).
63. Lindgren, S., K. Pettersson, A. Jonsson, P. Lingvall, and A. Kaspersson, "Silage Inoculation: Selected Strains, Temperature, Wilting and Practical Application," *Swed. J. Agric. Res.* 15:9–18 (1985).
64. Livingston, A. L., R. E. Knowles, A. Amella, and G. O. Kohler, "Nutrient Changes during Alfalfa Wilting and Dehydration," *J. Agric. Food Chem.* 25:779–783 (1977).
65. Manderson, G. J., and H. W. Doelle, "The Effect of Oxygen and pH on the Glucose Metabolism of *Lactobacillus casei* var. *rhamnosus* ATCC 7469," *Ant. van Leeuwenhoek* 38:223–240 (1972).
66. Marsh, R. "A Review of the Effects of Mechanical Treatment of Forages on Fermentation in the Silo and on the Feeding Value of the Silages," *N.Z. J. Exp. Agric.* 6:271–278 (1978).
67. Masuko, T., T. Uchimura, T. Otani, and K. Awaya, "Studies on the Disappearance of Nitrate in Forage Crops during Ensilage," *J. Japan. Grassl. Sci.* 29:73–81 (1983).
68. McDonald, P., *The Biochemistry of Silage*, John Wiley and Sons, Chichester, 1981.
69. McDonald, P., A. C. Stirling, A. R. Henderson, and R. Whittenbury, "Fermentation Studies on Wet Herbage," *J. Sci. Food Agric.* 13:581–590 (1962).
70. McKersie, B. D., "Proteinases and Peptidases of Alfalfa Herbage," *Can. J. Plant Sci.* 61:53–59 (1981).
71. McKersie, B. D., "Effect of pH on Proteolysis in Ensiled Legume Forage," *Agron. J.* 77:81–86 (1985).
72. McKersie, B. D., and J. Buchanan-Smith, "Changes in the Levels of Proteolytic Enzymes in Ensiled Alfalfa Forage," *Can. J. Plant Sci.* 62:111–116 (1982).
73. Meiering, A. G., M. G. Courtin, S. F. Spoelstra, G. Pahlow, H. Honig, R. E. Subden, and E. Zimmer, "Fermentation Kinetics and Toxic Gas Production of Silages," *Trans. ASAE* 31:613–621 (1988).
74. Melvin, J. F., "Variations in the Carbohydrate Content of Lucerne and the Effect on Ensilage," *Aust. J. Agric. Res.* 16:951–959 (1965).
75. Merchen, N. R., and L. D. Satter, "Changes in Nitrogenous Compounds and Sites of Digestion of Alfalfa Harvested at Different Moisture Levels," *J. Dairy Sci.* 66:789–801 (1983).
76. Merry, R. J., and G. D. Braithwaite, "The Effect of Enzymes and Inoculants on the Chemical and Microbiological Composition of Grass and Legume," In *Eighth Si-*

lage Conference: Summary of Papers, Agric. Food Res. Counc., Hurley, UK, 1987, pp. 27–28.
77. Middelhoven, W. J., and M. M. Franzen, "The Yeast Flora of Ensiled Whole-Crop Maize," *J. Sci. Food Agric.* 37:855–861 (1986).
78. Miller, W. J., and C. M. Clifton, "Relation of Dry Matter Content in Ensiled Material and Other Factors to Nutrient Losses by Seepage," *J. Dairy Sci.* 48:917–923 (1965).
79. Miskovic, K., and B. Rasovic, "Quantitative Participation of Lactic Acid Bacteria in the Epiphytic Microflora of Sorghum, Maize, Lucerne, and Silo Maize," *Savremena Poljoprivreda* 20:45–54 (1972).
80. Morrison, I. M., "Influence of Some Chemical and Biological Additives on the Fibre Fraction of Lucerne on Ensilage in Laboratory Silos," *J. Agric. Sci. Camb.* 111:35–39 (1988).
81. Moser, L. E., "Quality of Forage as Affected by Post-Harvest Storage and Processing," In Hoveland, C. S., ed., *Crop Quality, Storage, and Utilization*, American Society of Agronomy, Madison, Wisconsin, 1980, pp. 227–260.
82. Muck, R. E., "Dry Matter Level Effects on Alfalfa Silage Quality I. Nitrogen Transformations," *Trans. ASAE* 30:7–14 (1987).
83. Muck, R. E., "Factors Influencing Silage Quality and Their Implications for Management", *J. Dairy Sci.* 71:2992–3002 (1988).
84. Muck, R. E., "Initial Bacterial Numbers on Lucerne Prior to Ensiling," *Grass Forage Sci.* 44:19–25 (1989).
85. Muck, R. E., "Dry Matter Level Effects on Alfalfa Silage Quality II. Fermentation Products and Starch Analysis," *Trans. ASAE* 33:373–381 (1990).
86. Muck, R. E., and J. T. Dickerson, "Storage Temperature Effects on Proteolysis in Alfalfa Silage," *Trans. ASAE* 31:1005–1009 (1988).
87. Muck, R. E., and P. L. O'Connor, "Prediction of Lactic Acid Bacterial Numbers on Alfalfa," ASAE Paper No. 87-1550, Amer. Soc. Agric. Engrs., St. Joseph, Michigan, 1987.
88. Muck, R. E., and P. L. O'Connor, "Effect of Inoculation Level on Alfalfa Silage Quality," ASAE Paper No. 88-1070, Amer. Soc. Agric. Engrs., St. Joseph, Michigan, 1988.
89. Muck, R. E., and R. P. Walgenbach, "Variations in Alfalfa Buffering Capacity," ASAE Paper No. 85-1535, Amer. Soc. Agric. Engrs., St. Joseph, Michigan, 1985.
90. Mundt, J. O., "Lactic Acid Bacteria Associated with Raw Plant Food Material," *J. Milk Food Tech.* 33:550–553 (1970).
91. Mundt, J. O., and J. L. Hammer, "Lactobacilli on Plants," *Appl. Microbiol.* 16: 1326–1330 (1968).
92. Neal, H. D. St C., and J. H. M. Thornley, "A Model of the Anaerobic Phase of Ensiling," *Grass Forage Sci.* 38:121–134 (1983).
93. Neumark, H., "On the Areas of the Stomach of Sheep That Are Sensitive to Formic Acid and Histamine," *J. Agric. Sci., Camb.* 69:297–305 (1967).
94. Neumark, H., and A. Tadmor, "The Effect of Histamine Combined with Formic or Acetic Acid on Food Intake and Rumen Motility, When Infused into the Omasum of a Ram," *J. Agric. Sci., Camb.* 71:267–270 (1968).
95. Nilsson, P. E., "Some Characteristics of the Silage Microflora," *Archiv Mikrobiol.* 24:396–411 (1956).
96. Nilsson, G., and P. E. Nilsson, "The Microflora on the Surface of Some Fodder Plants at Different Stages of Maturity," *Archiv Mikrobiol.* 24:412–422 (1956).
97. Ohshima, M., and P. McDonald, "A Review of the Changes in Nitrogenous Compounds of Herbage During Ensilage," *J. Sci. Food Agric.* 29:497–505 (1978).
98. Ohyama, Y., T. Morichi, and S. Masaki, "The Effect of Inoculation with *Lactobacillus plantarum* and Addition of Glucose at Ensiling on the Quality of Aerated Silages," *J. Sci. Food Agric.* 26:1001–1008 (1975).
99. Orla-Jensen, S., *The Lactic Acid Bacteria*, Andr. Fred. Hort U. Sohn-Verl, Kopenhagen, 1919.
100. O'Kiely, P., A. V. Flynn and D. B. R. Poole, "Some New Silage Developments from Ireland," ASAE Paper No. 86-1528, Amer. Soc. Agric. Engrs., St. Joseph, Michigan, 1986.

101. O'Kiely, P., A. V. Flynn, and R. K. Wilson, "Ensilability of Grass", In: Animal Production, Research Report 1985, An Foras Taluntais, Dublin, 1986, pp. 62–63.
102. O'Kiely, P., R. E. Muck, and P. L. O'Connor, "Aerobic Deterioration of Alfalfa and Maize Silage," ASAE Paper No. 86-1526, Amer. Soc. Agric. Engrs., St. Joseph, Michigan, 1986.
103. Pahlow, G., "O_2-abhängig Veränderungen der Mikroflora in Silagen mit Lactobacterienzusatz," *Landwirt. Forschung* 37, (Kongressband for 1984):630–639 (1985).
104. Pahlow, G., and B. Dinter, "Epiphytic Lactic Acid Bacteria of Forages—Methods of Evaluation and First Results," In *Eighth Silage Conference: Summary of Papers*, Agric. Food Res. Counc., Hurley, UK, 1987, pp. 1–2.
105. Papadopolous, Y. A., and B. D. McKersie, "A Comparison of Protein Degradation During Wilting and Ensiling of Six Forage Species," *Can. J. Plant Sci.* 63:903–912 (1983).
106. Pitt, R. E., "Mathematical Prediction of Density and Temperature of Ensiled Forage," *Trans. ASAE* 26:1522–1527, 1532 (1983).
107. Pitt, R. E., and R. Y. Leibensperger, "The Effectiveness of Silage Inoculants: A Systems Approach," *Agric. Syst.* 25:27–49 (1987).
108. Pitt, R. E., R. E. Muck, and R. Y. Leibensperger, "A Quantitative Model of the Ensilage Process in Lactate Silages," *Grass Forage Sci.* 40:279–303 (1985).
109. Playne, M. J., and P. McDonald, "The Buffering Constituents of Herbage and of Silage," *J. Sci. Food Agric.* 17:264–268 (1966).
110. Russell, J. B., and T. Hino, "Regulation of Lactate Production in *Streptococcus bovis*: a Spiraling Effect that Contributes to Rumen Acidosis," *J. Dairy Sci.* 68: 1712–1721 (1985).
111. Satter, L. D., J. A. Woodford, B. A. Jones, and R. E. Muck, "Effect of Bacterial Inoculants on Silage Quality and Animal Performance," In *Eighth Silage Conference: Summary of Papers*, Agric. Food Res. Counc., Hurley, UK, 1987, pp. 21–22.
112. Seale, D. R., A. R. Henderson, K. O. Pettersson, and J. F. Lowe, "The Effect of Addition of Sugar and Inoculation with Two Commercial Inoculants on the Fermentation of Lucerne Silage in Laboratory Silos," *Grass Forage Sci.* 41:61–70 (1986).
113. Seale, D. R., C. M. Quinn, and P. A. Whittaker, "Microbiological and Chemical Changes During the Ensilage of Long, Chopped and Minced Grass," *Irish J. Agric. Res.* 21:147–158 (1982).
114. Shockey, W. L., and A. L. Barta, "Relationship Between Alkaline Mineral Concentration and Rate of pH Decline in Ensiled Greenhouse-Grown Alfalfa," *Agron. J.* 79:307–310 (1987).
115. Sneath, P. H. A., N. S. Mair, and M. E. Sharpe, eds., *Bergey's Manual of Systematic Bacteriology*, Vol. 2, Williams and Wilkins, Baltimore, 1986.
116. Spoelstra, S. F., "Degradation of Nitrate by Enterobacteria During Silage Fermentation of Grass," *Neth. J. Agric. Sci.* 35:43–54 (1987).
117. Stallings, C. C., R. Townes, B. W. Jesse, and J. W. Thomas, "Changes in Alfalfa Haylage During Wilting and Ensiling with and Without Additives," *J. Anim. Sci.* 53:765–773 (1981).
118. Stirling, A. C., "Bacterial Changes in Experimental Laboratory Silage," *Proc. Soc. Appl. Bact.* 14:151–156 (1951).
119. Stirling, A. C., "Lactobacilli and Silage-Making," *Proc. Soc. Appl. Bact.* 16:27–29 (1953).
120. Stirling, A. C., and R. Whittenbury, "Sources of the Lactic Acid Bacteria Occurring in Silage," *J. Appl. Bact.* 26:86–90 (1963).
121. Vetter, R. L., and K. N. Von Glan, "Abnormal Silages and Silage Related Disease Problems," In McCullough, M. E., ed., *Fermentation of Silage—A Review*, National Feed Ingred. Assoc., W. Des Moines, Iowa, 1978, pp. 281–332.
122. Weinberg, Z. G., G. Ashbell, and A. Azrieli, "The Effect of Applying Lactic Acid Bacteria at Ensilage on the Chemical and Microbiological Composition of Vetch, Wheat, and Alfalfa Silages," *J. Appl. Bact.* 64:1–7 (1988).
123. Wermke, M., U. Küntzel, and F. Weise, "Untersuchungen zur Konservierungseignung von Futterpflanzen. I. Beziehungen zwischen Kohlenhydrat-

gehalt, epiphytischem Keimbesatz und Konservierungsablauf beim Wiesenschwingel (*Festuca pratensis*)," *Z. Acker-und Pflanzen.* 137:174–190 (1973).
124. Whittenbury, R. "An Investigation of the Lactic Acid Bacteria," Ph.D. Thesis, Edinburgh University, 1961.
125. Whittenbury, R., "A Study of the Genus *Leuconostoc*," *Archiv. Mikrobiologie* 53: 317–327 (1966).
126. Whittenbury, R., P. McDonald, and D. G. Bryan-Jones, "A Short Review of Some Biochemical and Microbiological Aspects of Ensilage," *J. Sci. Food Agric.* 18:441–444 (1967).
127. Wieringa, G. W., "The Effect of Wilting on Butyric Acid Fermentation in Silage," *Neth. J. Agric. Sci.* 6:204–210 (1957).
128. Wieringa, G. W., "Some Factors Affecting Silage Fermentation. II. Influence of Degree of Laceration and of the Bacterial Flora from the Grass," *Neth. J. Agric. Sci.* 7:237–241 (1959).
129. Wilkinson, J. M., and R. H. Phipps, "The Development of Plant Components and Their Effects on the Composition of Fresh and Ensiled Forage Maize 2. The Effect of Genotype, Plant Density, and Date of Harvest on the Composition of Maize Silage," *J. Agric. Sci., Camb.* 92:485–491 (1979).
130. Wilkinson, M., *Silage UK*, 3d ed., Chalcombe Publications, Marlow, Bucks, UK, 1986.
131. Wood, J. G. M., and J. Parker, "Respiration During the Drying of Hay," *J. Agric. Engin. Res.* 16:179–191 (1971).
132. Woodford, M. M., "The Silage Fermentation," In Wood, B. J. B., ed., *Microbiology of Fermented Foods*, vol. 2, Elsevier, New York, 1985, pp. 85–112.
133. Woolford, M. K., "Some Aspects of the Microbiology and Biochemistry of Silage Making," *Herb. Abstr.* 42:105–111 (1972).
134. Woolford, M. K., "Studies on the Significance of Three *Bacillus* Species to the Ensiling Process," *J. Appl. Bact.* 43:447–452 (1977).
135. Woolford, M. K., "Managing Aerobic Deterioration in Silage," In: McCullough, M. E., and K. K. Bolsen, eds., *Silage Management*, National Feed Ingredients Assoc., W. Des Moines, Iowa, 1984, pp. 42–77.

Chapter

8

Use of Mixed Cultures for the Production of Commercial Chemicals

Paul J. Weimer

U.S. Department of Agriculture
Agricultural Research Service
U.S. Dairy Forage Research Center

Department of Bacteriology
University of Wisconsin
Madison, Wisconsin

Introduction

Mixed cultures of microorganisms have been used by humans for thousands of years in the production of food and drink. However, only recently have they been exploited for the production of commercial chemicals. This is certainly due in part to the relatively recent development of the chemical industry. However, it is also due largely to the fact that process development for the biological production of chemicals has been historically built around monocultures; mixed cultures were only investigated in hopes of overcoming some of the practical limitations of established (or at least well-studied) monoculture systems.

This chapter will focus on the use of mixed cultures not only in current commercial processes for the manufacture of chemicals but also on reported experimental data which show the promise of mixed cultures in improving monoculture-based processes.[1] Efforts will be made

[1]This work was supported by U.S. Department of Agriculture grant CWU 3655-33000-003-D.

to place the information on individual products and processes into two larger perspectives: (1) the commercial viability of the mixed-culture process relative to other biological and nonbiological processes, and (2) the general principles of microbial physiology which govern interactions among microorganisms.

David Harrison (1978) in beginning his review on the industrial use of mixed cultures, wrote:

> The use of mixed microbial cultures in industrial processes is a subject which is, at the same time, too large to be covered sensibly in one chapter and somewhat slight in scientific content. The number of occurrences of mixed cultures in industrial processes is legion if both accidental and deliberate mixtures are considered. However, the number of mixed cultures in commercial use which are well-studied and understood, and deliberately constituted, is extremely small.

In certain respects these words still hold true a dozen years later. However, more recent work in applied microbiology has improved the outlook for mixed-culture processes, and basic studies on the physiology of microbial mixed cultures have deepened our understanding of the metabolic relationships among different microbial species. Thus, while the mixed cultures discussed in this chapter will include both undefined and defined (with respect to species composition) mixtures, the defined mixed cultures will be emphasized, owing to their greater flexibility in design and manipulation, and our ability to dissect the relevant microbial interactions at the physiological and ecological levels.

The particular products under consideration here will include both commodity (bulk) chemicals used in the manufacture of various end-use products, and certain specialty (fine) chemicals which are ordinarily employed as pharmaceutical agents. However, the extensive and highly successful use of mixed cultures directly in food and beverage production (e.g., cheesemaking, single-cell protein, and brewing) will not be addressed here.

For the purposes of this review, a "mixed culture" will be regarded as one which contains more than one microbial species (or more than one strain of the same species), regardless of whether the organisms are inoculated simultaneously or at intervals; these cultures may be either defined or undefined with respect to species or strain composition. The term "coculture" will be reserved for mixed cultures of defined species or strain composition whose metabolic interactions have been elucidated, and which contribute to the formation of the desired fermentation in a synergistic fashion, regardless of whether or not the culture can be maintained by sequential transfer. The term "fermentation" will be used in its broadest sense, i.e., a chemical transformation carried out by microorganisms.

Production of Commercial Chemicals by Mixed Cultures

Theoretical aspects

The criteria for successful commercial production of chemicals using mixed cultures are identical to those based on monocultures or on purely chemical (abiotic) processes: reproducibly high yield of product of the required purity, at a total cost (for substrates, downstream processing, labor, energy, capital equipment, marketing, sales, etc.) sufficient to guarantee a reasonable profit.

In terms of the direct and indirect cost items beyond the initial fermentation step, there is generally little difference between microbial systems based on mixed cultures versus those based on monocultures. Any advantage of using a mixed culture is instead reflected in the improvement of the initial (production) step. These improvements may include: (1) elimination of the need for a multistep fermentation process employing two or more monocultures; and (2) metabolic interactions which result in a more favorable product yield, a reduced demand for cofactors, or the elimination of an unwanted side product.

The general lack of use of mixed cultures in commercial processes has resulted primarily from a lack of bench-scale investigation into the useful properties thereof. Historical development of processes used in industrial microbiology has usually followed a common scenario: (1) identification of individual microbial strains producing a desired compound, followed by (2) laboratory-scale environmental and/or genetic manipulation to improve yield, productivity, final product concentration, etc., followed by (3) development of industrial-scale fermentation technology based on laboratory results. Attempts to improve these processes by random combination of different species were not likely to be successful, and rational design of mixed cultures had to await a more thorough knowledge of the principles of microbial physiology and ecology to guide the selection of potentially useful coculture organisms for a given process.

Of the seven common types of interactions among species—competition, predation, parasitism, amensalism, neutralism, commensalism, and mutualism—only the last two are likely to yield a mixed culture having properties superior to those of monocultures (Harrison, 1978). However, since our view of superiority is particularly anthropocentric, we must realize that in at least some of the mixed cultures we will describe, the individual species within the mixtures may not interact strongly enough to demonstrably affect such characteristic benchmarks of microbial activity as growth rate, growth yield, or overproduction of a particular metabolite, but may interact in a subtler fashion to enhance (or inhibit) the survival of the species in its particular environment.

Production of commodity chemicals

Commodity chemicals are defined here as those produced in excess of one million pounds (4.5×10^5 kg) per year. They are generally produced by chemical syntheses, using fossil fuels or (occasionally) biomass as starting materials in elaborate, energy-intensive processes. When manufactured by fermentation processes, commodity chemicals are almost exclusively produced either as primary catabolic products (e.g., ethanol, acetone, acetic acid) or important intermediates in major amphibolic pathways (e.g., citric acid, malic acid).

Despite considerable research efforts, production of commodity chemicals by mixed cultures has yet to be realized at the commercial level. The discussion which follows will thus summarize current information on laboratory-scale mixed cultures which have been investigated for chemical production.

Organic acids. Organic acids that are produced by mixed cultures, or that show promise for being so produced, include tricarboxylic acid-cycle intermediates (malic and citric), as well as acetic and lactic acids. All are used primarily as acidulents in food processing, or as additives to other industrial chemical products. The major problem for fermentation-based (monoculture or mixed culture) processes for production of organic acids is economical recovery of product, at the required purity, from the dilute aqueous fermentation broth.

Malic acid. Mixed-culture fermentations for L-malic acid generally employ a fungus or yeast which converts a primary substrate (glucose or paraffins, respectively) to fumaric acid, along with a second microbe having a high level of the enzyme fumarase (L-malate hydrolyase, E.C. 4.2.1.2). Inoculation of the primary and secondary organisms may be either simultaneous or staggered. Examples of such fermentations from the literature are shown in Table 8.1. Note that these processes are outstanding in terms of yields and final product concentrations. Nevertheless, both parameters could probably be enhanced by applying recent improvements in the technology for production of fumarate by *Rhizopus arrhizus* (Ng et al. 1986). The latter process has reached yields of 1.5 g dicarboxylic acids (mostly fumarate but also small amounts of succinate and malate) per gram of glucose, and final

TABLE 8.1 Examples of Literature Reports on the Production of L-Malic Acid by Mixed Cultures of Microorganisms

Organisms: Primary	Organisms: Secondary	Substrate	Inoculation	Yield, %	Final conc., g/L	Reference
Candida hydrocarbofumarica	*Candida utilis*	*n*-paraffins	Staggered	72		Furukawa et al. 1978
"	*Pichia membranefaciens*	"	"	70		Furukawa et al. 1978
"	*Mycotorula japonica*	"	"		15.8	Yamada and Furukawa 1975
Rhizopus arrhizus	*Proteus vulgaris*	Glucose	"	70–75		Takao and Hotta 1970
"	"	"	"	84	126	Takao and Hotta 1977
R. arrhizus NRRL 1526	*Paecilomyces varioti*	"	Simultaneous	>60	>48	Takao et al. 1983

product concentrations of ~ 150 g/L. Integration of these process improvements into a mixed-culture system containing *R. arrhizus* and any of the secondary organisms listed in Table 8.1 may prove rewarding.

Citric acid. Mixed cultures of the yeast *Candida lipolytica* IFO0717 and the fungus *Schizophyllum commune* IFO4918 have been reported to produce up to 98.3 g citrate per liter in a mineral salts medium containing 62 g paraffins and 1 g peptone per liter, following incubation for 10 days at 30°C (Tachibana 1972). In addition, small amounts of *d*-isocitric acid and 4 mg of riboflavin per liter were reportedly produced by the mixed culture. However, no further information is available on the scale-up of these pilot studies, which were performed in shake flasks at 60-mL scale.

Acetic acid and its salts. The commercial production of food-grade acetic acid (vinegar) using aerobic, ethanol-oxidizing *Acetobacter* or *Gluconobacter* species, represents one of fermentation technology's oldest success stories. This process is traditionally operated without elaborate attempts to maintain biological purity, and instead relies upon the very low pH and general austerity of the growth medium to control invasion by unwanted organisms. As such, the process may actually represent a complex, undesigned mixed-culture system wherein contaminating microorganisms may subtly affect product quality. Because the true microbial status of the vinegar fermentation is uncertain, and because as a food it extends beyond the scope of this chapter, our consideration of acetic acid production will instead focus upon its potential production as a commodity chemical by other microorganisms.

Khan et al. (1981) have demonstrated that a cellulolytic enrichment culture from sewage sludge, which normally produces CH_4 and CO_2, could be converted to an acetic acid–producing culture by heating it to 80°C for 15 min. This heat treatment apparently results in selective removal of the methanogenic population. The amounts of acetate produced were rather low (1.0 to 2.2 g/L) and several coproducts were also formed, including propionate, CO_2, and traces of ethanol and H_2.

One of the most widely studied mixed-culture systems for acetic acid production is based on *Clostridium thermocellum*, a thermophilic, anaerobic spore-forming bacterium having an optimum growth temperature near 60°C. As a result of its ability to degrade rather recalcitrant forms of cellulose, *C. thermocellum* has been proposed as a central component of several mixed-culture systems for production of other bulk chemicals, including methane (Weimer and Zeikus 1977), lactic acid (see below), and ethanol (see below). Its potential role in an acetic acid fermentation is in fact an outgrowth of its tendency to divert most of its substrate carbon to acetate, rather than ethanol, when cultivated on crude biomass materials such as wood or straw (Wang et al. 1981). The general catabolic scheme for carbohydrate fermentation by this bacterium is shown in Fig. 8.1. *C. thermocellum* utilizes a mixed-acid fermentative pathway, which results in the unproductive diversion of substrate carbon away from the desired end product (in this case acetic acid). Organisms selected for coculture with *C. thermocellum* have been chosen on the basis of their ability to direct the latter's fermentation pathway toward the product of interest as well as to consume the considerable amount of reducing sugars left by the *Clostridium* during cellulose hydrolysis (Zeikus 1980).

In an effort to develop an acetic acid process, *C. thermocellum* has been cocultured with the homoacetate-fermenting bacterium *Acetogenium kivui* (LeRuyet et al. 1984). When grown on crystalline cellulose powder at 60°C and pH 6.8, the coculture produces 2.7 mol acetate per mole of anhydroglucose, which translates to a weight yield of 100 percent. However, the final concentrations of acetate anion and free acetic acid were only 10 and 0.1 g/L, respectively. This inability of the strict anaerobes in general, and thermophilic ones in particular, to produce high acetate concentrations in solution at low pH has bedeviled engineers concerned with product recovery. That the acetate an-

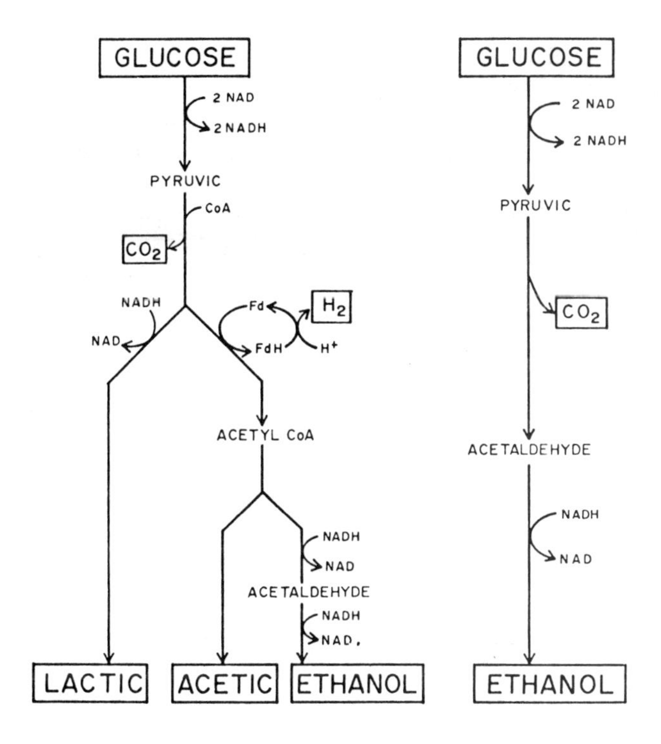

Figure 8.1 Comparison of the basic features of the catabolic pathways of the mixed-acid-producing cellulolytic anaerobic bacterium *Clostridium thermocellum*, and the ethanol-producing yeast *Saccharomyces cerevisiae*. The former performs a phosphoroclastic cleavage of the pyruvate intermediate to ultimately yield a mixture of products, including some ethanol. The yeast performs a direct decarboxylation of pyruvate to ultimately yield two moles of ethanol per mole of glucose. *(From Zeikus 1980. Reproduced, with permission, from the Annual Reviews of Microbiology, Vol. 34, © Annual Reviews, Inc.)*

ion can be recovered only with difficulty and at considerable cost has forced these anaerobic processes for acetic acid production to await quantum improvements in recovery technology. To put this fact in perspective, consider that modern vinegar fermentations produce acetic acid at 13 to 14 percent (w/v) at pH 2 (i.e., where the acetic acid is completely protonated), yet costs of recovering pure acetic acid from this solution are such that even the vinegar process cannot compete economically with chemical synthetic routes to chemical-grade (glacial) acetic acid.

One potential niche for fermentation acetate is in the form of the salt calcium magnesium acetate (CMA), a material which may displace the more corrosive and environmentally undesirable rock salt

(NaCl) as a road deicer (Jain et al. 1989). Examination of the above fermentations from this perspective suggests that pH control with Ca and Mg-containing alkalis would probably increase final acetate concentration and productivity and would ease product recovery, all of which should brighten process economics. Indeed, a two-stage fermentation process to CMA has been proposed (Brumm 1987) based on the mesophile *Lactobacillus acidophilus* and the thermophile *Clostridium thermoaceticum*. In this process, the sugar is first fermented by the *Lactobacillus* at pH 5.0 to 5.5, resulting in lactate concentrations of 60 to 90 g/L and volumetric productivities of >3 g/L · h. The resulting lactate-containing broth is fermented by the *Clostridium* at 55 to 60°C and pH 6.5. These environmental conditions, combined with the simplicity of the fermentation medium, virtually remove the threat of contamination with unwanted organisms. Dolomite is used to adjust the pH of both product streams, resulting in a CMA solution amenable to settling and evaporative drying.

It might be useful to explore development of a CMA process based on related lactic-clostridial mixed cultures. For example, *Streptococcus lactis* and *Clostridium formicoaceticum* have been shown (Tang et al. 1988) to convert lactose to acetic acid via a lactic acid intermediate at impressive yields (90 percent by weight at pH 7.0, and 95 percent at pH 7.6). Batch fermentations of whey permeate [an inexpensive and abundant (Moulin and Galzy 1984) waste product of cheese manufacture containing 5 percent lactose (w/v)] yielded 20 g acetate per liter after only 20 h, and 30 g acetate plus 20 g lactate per liter after 80 h. Twofold dilution of the substrate resulted in complete fermentation of the 2.5% lactose to acetic acid within 30 h. Again, acetate anion rather than free acetic acid is the primary fermentation product, which would favor conversion of the process for CMA production.

Lactic acid. Lactic acid is commercially produced by homofermentative lactic acid bacteria (e.g., *Lactobacillus delbreuckii*) in yields that approach 100 percent (weight basis) and product concentrations of 120 g/L. As the fermentation is carried out at relatively low pH, the product is concentrated directly to a 60 percent (w/w) syrup for shipment and use (Miall 1987). Mixed cultures have not yet deeply penetrated the lactic acid market, although the potential for using mixed cultures of different lactobacilli may be considerable.

Mixed cultures of some lactic acid bacteria have been reported to produce higher lactate yields than do the corresponding monocultures. For example, a mixed culture of *L. bulgaricus* AU and *L. casei* 9649 produced lactic acid yields of 74 percent, compared to yields by the same strains in monoculture of 44 and 62 percent, respectively (Tiwari et al. 1977). However, there is no evidence that lactic acid yields may be improved by combinations of any of the high-yielding commercial strains currently in use.

Efforts to improve the economics of the lactic acid fermentation have included attempts to replace the hexose substrates with inexpensive polysaccharides. Cellulose is an economically attractive substrate for chemical production because of its wide availability as woody biomass, agricultural residues, or municipal waste. Cocultures of the cellulolytic fungus *Trichoderma reesei* and the lactic acid bacterium *Lactobacillus delbreuckii* IFO3534 have been reported (Sabe et al. 1978) to convert cellulose (165 g/L) to lactic acid (41 g/L) in a complex medium during incubation at 45°C for 72 h. It is clear that considerable improvements in both yield and final product concentration are required for development of a commercially useful process.

As mentioned above, the thermophilic anaerobic cellulolytic bacterium *Clostridium thermocellum* has been proposed as the basis of a cellulose-based lactic acid fermentation. One potential coculture organism is *Thermoanaerobium brockii*, a noncellulolytic saccharolytic anaerobe which actively scavenges reducing sugars released by the *Clostridium*, and which contains a very active lactate dehydrogenase (E.C. 1.1.1.27) which is activated by fructose-1,6-bisphosphate. When grown on glucose, *T. brockii* has been shown (Lamed and Zeikus 1980) to produce lactic acid at yields of 47 percent (w/w), although the sensitivity of the organism to low pH may ultimately prevent its commercial use. The same might be true for the coculture of *C. thermocellum* with the facultative anaerobe *Bacillus coagulans*. The latter is a moderate thermophile which performs a homolactic fermentation on hexoses, but a heterolactic fermentation of pentoses. Inoculation of *B. coagulans* into an active *C. thermocellum* fermentation of Solka Floc, has reportedly (Wang et al. 1981) resulted in lactate yields of 35 percent (w/w), although the process required pH control (~6.8) and produced final lactate concentrations of only 7 g/L (equivalent to a free lactic acid concentration of <0.1 g/L).

The relatively slow hydrolysis of cellulose has led others to explore other polysaccharides as substrates for lactic acid production. Kurosawa et al. (1988) have described an interesting starch-based immobilized cell system involving the aerobic, amylolytic fungus *Aspergillus awamori* and the aerotolerant anaerobic lactic acid bacterium *Streptococcus lactis*. The two species were differentially localized within the Ca alginate beads, with the fungus at or near the bead surface and the bacterium buried within the bead. In a continuous culture operated at 30°C, 1 ppm dissolved O_2, and an initial starch concentration of 50 g/L, the culture produced lactic acid at a volumetric productivity of 0.34 to 0.43 g/L · h, although the effluent lactic acid concentration was relatively low (8 to 10 g/L).

Ethanol. Commercial production of ethyl alcohol has gone through considerable changes in the last few decades. Although traditionally produced by fermentation, chemical synthesis routes based on the hydration of then-inexpensive ethylene nearly buried fermentation routes to ethanol during the period from the 1950s to the mid-1970s. Subsequent escalation in fossil fuel prices turned the economic tide back in favor of the fermentation route. Most industrial (nonbeverage) ethanol produced both in the United States and the developing world is now derived from fermentation of sugars. These fermentation processes are all based on the fermentation of glucose materials by any of a number of yeast strains, coupled to various product recovery schemes based on distillation, vapor recompression, etc. An admirably detailed review of fermentation ethanol technology has been presented by Kosaric et al. (1983).

All of the fermentation ethanol produced in the United States is currently via processes based on sugar-fermenting yeast, such as *Saccharomyces cerevisiae*. These organisms will probably continue to dominate the ethanol industry, although some competition from the mesophilic bacterium *Zymomonas mobilis* might well be expected. This latter organism possesses a yeastlike pyruvate decarboxylase system (Fig. 8.1), and thus unlike "mixed-acid" fermentative bacteria, converts glucose to ethanol and CO_2 without the formation of unwanted side products such as acetic or lactic acids. The purported advantages of *Zymomonas* over the yeasts include its rapid growth rates and high specific and volumetric productivities, all of which are 2 to 2.5 times those reported for yeast (Rogers et al. 1980); the purported disadvantages of *Zymomonas* include greater cooling costs required to accommodate the organism's lower temperature optimum, and difficulty in maintaining culture viability on an industrial scale.

Several mixed cultures of different yeast species have shown promise in bench-scale experiments. A mixed culture of *Saccharomyces uvarum* and a brewer's yeast, grown on beet sugar (sucrose) produced ethanol to a final concentration of 7.9% after 48 h and 8.5% after 60 h (Jones et al. 1982). Mixed cultures of *Kluyveromyces fragilis* and *S. cerevisiae*, grown on an inexpensive mixture of corn mash and whey permeate pretreated with β-amylase, produced ethanol at a concentration of 9.5% by volume and a yield of 0.5 L ethanol per kilogram of corn, over a run period of 60 to 72 h (Whalen and Shahani 1985). In this mixed culture, *Kluyveromyces* fermented the lactose from the whey, while the *Saccharomyces* fermented the maltose released from the corn mash by enzymatic treatment; it was advantageous to delay the addition of the *Saccharomyces* until the *Kluyveromyces* fermentation was well underway. This process was eventually piloted at the 6500-L scale, but to this reviewer's knowledge has not been commercialized.

Bench-scale continuous culture experiments with immobilized mixed cultures have been shown to produce ethanol at very impressive volumetric productivities, presumably as a result of the high cell densities achieved on the bead surfaces. A mixture of alginate-immobilized bottom yeasts and vermiculite-immobilized *Z. mobilis* have produced ethanol at 13 g/L · h (Hitachi 1983a), and an immobilized mixed culture of brewer's yeast and an unidentified ethanol-producing bacterium displayed productivities of 16 g/L · h (Hitachi 1983b). However, like all continuous processes, effluent product concentrations were relatively low. Because the costs of ethanol recovery increase in a dramatic and nonlinear fashion with decreasing ethanol concentration in the fermentation broth, these processes will probably not be commercially viable in their present configuration.

A number of mixed-culture studies of ethanol production have involved attempts to utilize polysaccharides such as starch or cellulose as fermentation substrates. The conversion of a partially purified wood cellulose (Solka Floc BW200) to ethanol has been demonstrated in a multistep process using cultures of *Trichoderma reesei* C30 and *S. cerevisiae* (Hahn-Haegerdahl and Haeggstroem 1985). The system produced ethanol in relatively high yield (83 percent of theoretical) and at a moderate final product concentration (40 g/L), but at poor volumetric productivity (0.2 g/L · h).

The group of Kurosawa et al. has coimmobilized either *Z. mobilis* (1986) or *S. cerevisiae* (1989) with *Aspergillus awamori* in Ca alginate beads, in a method similar to their lactic acid coculture described above. As in the above case, the *Aspergillus* mycelia predominated at the bead surface, and the ethanol producer was localized primarily within the bead where the oxygen tension was low. The *Z. mobilis* coculture converted 100 g starch per liter to 25 g ethanol per liter, while the *Saccharomyces* coculture, run in the presence of a biocide to control the surface *Saccharomyces* population, converted 40 g starch per liter to 12.3 g ethanol per liter. These yields and product concentrations are well below those of the current commercial yeast-based monocultures using sugars as substrate.

The potential use of thermophilic anaerobic mixed cultures for ethanol production has been reviewed from both biological (Weimer 1986) and engineering (Lynd 1989) perspectives. Most of this work has involved pure cultures, although mixed-culture fermentations of both cellulose and starch have been described.

In cellulose-containing media, *C. thermocellum* and *C. thermohydrosulfuricum* form a stable coculture which can be repeatedly transferred. Fermentation of 1 percent (w/v) Solka Floc has yielded ethanol concentrations as high as 4.5 g/L [a weight yield of 45 percent (Zeikus et al. 1983, Ng et al. 1981)]. However, considerable improvements in fermentation rate and final product concentration (a reflection of the poor ethanol tolerance of wild-type strains) are required for the development of a commercially viable process. Cocultures of *C. thermocellum* and *Thermoanaerobacter ethanolicus* also show enhanced cellulose fermentation relative to the clostridial monoculture. Wild-type strains converted steam-exploded wood (1%, w/v) to 44 mM (2 g/L) ethanol over a relatively long (14 days) incubation (Ljungdahl 1981). Ethanol yields and product concentrations from other woods, forages, and agricultural residues were even lower (Carreira and Ljungdahl 1983).

Cocultures of *Clostridium thermocellum* S-7 and *C. thermosaccharolyticum* HG-6-62 have been reported to ferment ground mesquite beans to ethanol (Wang et al. 1981). In a fed-batch mode with a staggered inoculation (HG-6-62 at time-zero and S-7 at 42 h of an 80-h fermentation), 100 g of mesquite beans (which contained 11 g of cellulose and 54 g of other carbohydrate, mostly fructose) per liter yielded 30 g of ethanol per liter at a volumetric productivity of 0.44 g/L · h.

Cocultures of *Clostridium thermohydrosulfuricum* 39E and *C. thermsulfurogenes* 4B have been shown to ferment starch at a yield of 48 percent (weight basis) and a final product concentration of 13 g/L (Hyun and Zeikus 1985). This coculture represents an interesting example of a mutualistic interaction, due to the production of different polysaccharide-degrading enzymes by the two species. While both strains produced glucoamylase (E.C. 3.2.1.3), strain 39E produced pullulanase ("debranching enzyme," E.C. 3.2.1.41) and strain 4B produced β-amylase (E.C.3.2.1.2). As a result of this cooperative interaction, the concentration of low-molecular-weight glucose oligomers remained below the levels which cause repression of enzyme synthesis. This resulted in a more rapid and complete utilization of the starch, and thus a higher product yield and concentration.

From a commercial standpoint, there is little doubt that the strictly anaerobic thermophilic bacterial processes exemplified above are severely deficient in several important cost-related factors such as: (1) low (generally $<2\%$, w/v) ethanol concentration in the fermentation broth; (2) low volumetric productivities due to the low cell densities obtained during the anaerobic fermentation; and (3) moderate diversion of substrate to the production of uneconomical side products such as acetic and lactic acids, particularly when using crude cellulosic substrates. These problems currently outweigh any advantages gained by the use of inexpensive cellulosic substrates, or the savings gained in cooling and stirring costs.

The problems alluded to above—low product concentration, low productivity, and the undesired formation of coproducts—appears to be a nearly universal phenomenon among clostridia and other saccharolytic bacteria lacking a pyruvate decarboxylase system. Any gains achieved in improvement of one of these areas is almost always balanced by losses in one of the others. This notion is reinforced by experiments performed with "wild" fermentations (i.e., fermentations in which the complex and undefined microflora are derived from the unsterilized substrate). Segers et al. (1981) have reported that a wild mesophilic mixed culture, grown in a continuous mode on sucrose, shifted its products away from butyric acid and toward ethanol as the pH is lowered in stages from 7 to 4. At pH 4 to 4.5, ethanol production accounted for ~ 40 percent of the influent chemical oxygen demand of 25.5 g/L, although the total amount of ethanol produced was similar in both cultures due to differences in substrate consumption. Volfova et al. (1985) have reported that cellulose fermentation by a wild thermophilic mixed culture showed ethanol-acetate ratios of 1.1, 4.1, and 6.1 at pH values of 7.0, 6.5, and 6.0, respectively. However, this gain in ethanol yield with decreasing pH was tempered by a decrease in the total amount of ethanol produced: final ethanol concentrations from the 2% cellulose feed declined from 4.6 to 5.1 g/L (pH 7) to <1 g/L (pH 6).

As a result of the complex regulation of product formation in mixed-acid anaerobic bacteria, their use for the commercial production of ethanol will probably not be forthcoming until their metabolic characteristics have been modified by genetic manipulation.

Acetone-butanol

The acetone-butanol fermentation of sugars by *Clostridium acetobutylicum* has been resurrected in several countries as a viable manufacturing process for these solvents (see Gibbs 1983 and Hastings 1978 for reviews). In an attempt to reduce substrate costs, Fond et al. (1983) have examined the fermentation of cellulose by a mesophilic coculture of *C. acetobutylicum* and *C. cellulyticum*. Although the cellulolytic bacterium degraded the polymer (Solka Floc SW40) four times more rapidly in the coculture than in the monoculture, the rate of solvent production was nevertheless limited by the slow rate of cellulose hydrolysis. As a result, the formation of acids predominated over solvent production. A 13-day batch culture run resulted in the conversion of 30 g Solka Floc per liter to 14 g butyric acid, 4 g acetic acid, 3 g ethanol, and 1 g butanol per liter.

More rapid and efficient solvent production from cellulosic materials thus must await a more active cellulolytic coculture organism.

Conversion of cellulose to acetone and butanol has also been demonstrated using *C. acetobutylicum* and *C. thermocellum* (Yu et al. 1985). Because of the different temperature requirements of the two species, the process was operated in a sequential mode, with the solvent-producer inoculated later in the fermentation, concurrent with a reduction in the operating temperature. Nevertheless, the mixed culture ended up producing 1.7- to 2.6-fold more total fermentation products than did the monoculture, due to the efficient utilization of the reducing sugars left behind from cellulose hydrolysis. Although the fermentation products were mostly acetic and butyric acids, solvent production was improved by early addition of butyric acid to the culture.

Production of acetone, butanol, and ethanol has also been demonstrated in a sequential mixed culture of *C. acetobutylicum* and *Kluyveromyces fragilis* (Ballerini et al. 1985). Sugar beet juice containing 110 g sucrose per liter and supplemented with small amounts of corn steep liquor, acetic acid, and NH_4OH was inoculated with the *Clostridium* and incubated for 10 h at 35°C. The culture was then inoculated with the yeast and incubated for 30 h at 30°C; the final product concentrations were (per liter) 12 g butanol, 4 g acetone, and 7.5 g ethanol.

Production of specialty chemicals

Specialty chemicals are those having a high value in use and—in terms of absolute quantity—a relatively small market. Unlike commodity chemical production, in which biological processes must compete against well-established and relatively inexpensive chemical processes, the production of specialty chemicals by purely chemical routes is often difficult and expensive. Consequently, biological routes are the preferred methods of manufacturing many specialty chemicals. The production phase of fine chemical manufacture is often less problematic than are the subsequent downstream processing steps required for recovery of the product at the very high purities usually required for their demanding end use (e.g., as pharmaceuticals).

Specialty chemicals produced by microbial action are generally either the terminal products of long primary or secondary biosynthetic pathways (e.g., vitamins, antibiotics), or are the products of biotransformation of an exogenously added, structurally related precursor (e.g., in steroid biotransformations). Manufacture of many specialty chemicals is dominated by microbial monocultures, owing to the fact that strains producing high yields were developed gradually from promising individual strains by painstaking selection procedures.

There are few theoretical reasons why a mixed culture should be inherently superior to a properly selected monoculture for the production of any end product of either a primary or secondary biosynthetic pathway, and consequently their potential has not been systematically examined by industrial microbiologists. However, for biotransformations involving one or a few specific reactions, mixed cultures have certain advantages which have led to their exploitation.

Propionic acid. Food-grade propionic acid (produced by fermentation from food products) is widely used as a mold inhibitor in foods, but also has a variety of other end uses (Jain et al. 1989). The high cost of the food-grade propionic acid has stimulated a search for efficient fermentation routes for its production. Mixed cultures of *Lactobacillus* and *Propionibacterium* species have been shown to convert glucose to propionic acid at good yield. While each organism can ferment glucose (to produce lactic and propionic acids, respectively), the *Propionibacterium* will preferentially utilize lactate if it is available. Consequently, mixed cultures of the two organisms exhibit a mutualistic interaction in which the *Lactobacillus* ferments glucose to lactic acid, and the *Propionibacterium* converts this inhibitory product to propionic acid (Lee et al. 1976, Liu and Moon 1982, Parker and Moon 1982). Bodie et al. (1987) have extended this work by using whey, an inexpensive and abundant liquid waste product of cheese manufacture, as the fermentation substrate. Whey contains approximately 5% lactose, and strains of *L. casei* and *P. shermanii* exhibit the same type of interaction on this substrate as they do on glucose. As is commonly observed in other acidogenic fermentations, pH control greatly increases final product concentration. In this case, propionate concentrations of 22 and 30 g/L have been obtained in 3-L fermenters at pH 6.9 from whey solids concentrations of 70 and 120 g/L, respectively. These product concentrations are approximately five times higher than those in similar fermentations not subjected to pH control.

As already mentioned, the inherent difficulty in adapting fermentative organisms to efficient production of organic acids at low pH has stimulated research in economic recovery of organic acids directly from dilute aqueous solutions containing the homologous anionic species (e.g., by electrodialysis; see Jain et al. 1989). The ultimate commercial utility of such fermentations will rest on the success of these recovery processes, except in cases where the value of the final product (e.g., food-grade acids) are sufficiently high to offset even large recovery costs.

Vitamins. With the exception of vitamin B_{12} (the most structurally complex known vitamin), all vitamins are commercially produced either by extraction from natural sources or by chemical synthetic routes.

Vitamin B_{12}. Commercial production of vitamin B_{12} by microbial fermentation occurs exclusively via one of several monoculture-based processes (Wuest and Perlman 1967). However, there have been hundreds of literature or patent descriptions of fermentation processes which have not proceeded to commercialization. Among these are several employing mixed cultures. These are summarized in Table 8.2.

Carotenoids. Naturally occurring carotenoids include several hundred distinct compounds which share a basic tetraterpenoid structure having eight isoprene units. Two of the more important carotenoids, β-carotene and lycopene (Fig. 8.2), can be produced in moderate amounts by fermentation.

The β-carotene process is interesting in that optimal production is attained during the sexual reproductive process of zygospore formation by a fungus, *Blakeslea trispora*. In its preferred configuration (Ninet and Renaut 1979) the process involves the coculture of relatively large amounts of the producing strain NRRL 2457 (−) and smaller amounts of the nonproducing sexual partner strain NRRL 2456 (+). The fermentation is run in a submerged, batch mode in a complex medium (corn starch, distillers solubles, soybean meal, and cottonseed oil) amended with various "activators" whose exact mode of action remains unclear. These latter compounds include trisporic acid C, isoniazid, and β-ionone (Fig. 8.2). Other additions include kerosene (whose function is unknown but which doubles the yield of product)

TABLE 8.2 Reported Processes for the Production of Vitamin B_{12} by Mixed Cultures

Organisms	Growth substrate	B_{12} concentration, mg/L	Reference
Eubacterium limosum + *Pseudomonas stutzeri*, + unidentified bacterium	Methanol + butyrate + yeast extract	39	Ebina et al. 1987
Mixed methanogenic culture	Methanol + DMBI[a]	35[b]	Szemler and Szekely 1969
Corynebacterium sp. IFO12320 + *Rhodopseudomonas sphaeroides*	Paraffins + corn steep liquor	2.3[c]	Nakao et al. 1979
Propionibacterium shermanii + *Mycobacterium filiforme*	Lactic acid	2[c]	Baranova et al. 1968
Propionic acid bacteria + *Acetobacter aceti*	Serum + hydrolyzed milk + DMBI[a]		Bogdanov and Grudzinskaya 1969
Propionibacterim freundrichii + *Lactobacillus casei*	Whey powder	4.3[c]	Leviton and Hargrove 1952
Proteus sp. + *Pseudomonas* sp.	Cottonseed meal	0.42	Hodge et al. 1952

[a]DMBI = 5,6-dimethylbenzylimidazole
[b]Also produced 10 mg etiocobalamin per liter
[c]Mixed cultures which produced more B_{12} than did the homologous monocultures.

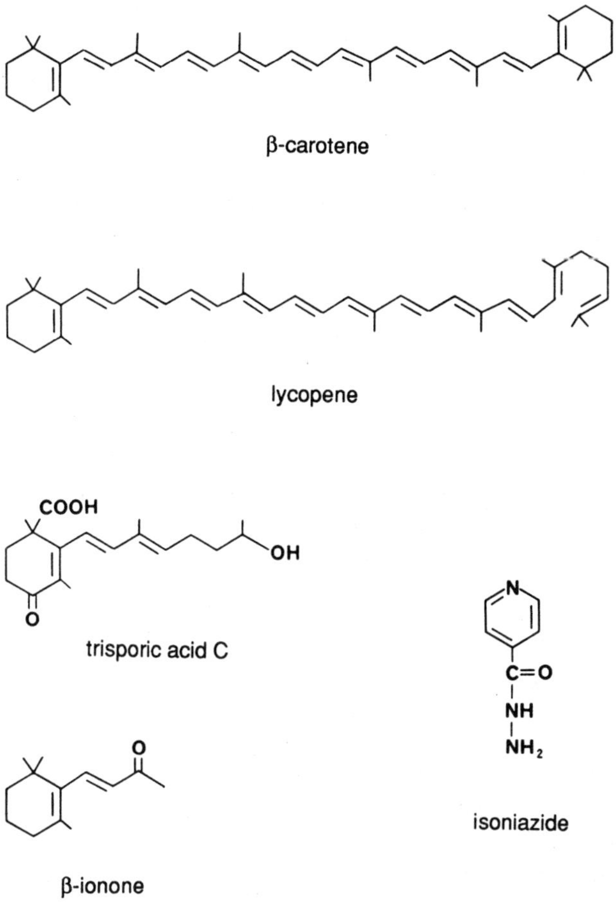

β-carotene

lycopene

COOH
OH
O
trisporic acid C

N
C=O
NH
NH_2
isoniazide

O
β-ionone

Figure 8.2 Structures of two carotenoid vitamins and some activators used in their fermentative production. (See text for discussion.)

and an antioxidant to stabilize the β-carotene during the prolonged (8-day) run time. β-Carotene concentrations of up to 3.0 g/L have been obtained.

The fermentative production of lycopene also involves a mixture of two strains of *B. trispora*, NRRL 2896 (−) and NRRL 2895 (+). In this case, triethylamine is added to inhibit closure of the terminal portions of the molecule into β-ionone rings. Up to 0.40 g lycopene per liter have been reported for 5-day fermentation runs (Crueger and Crueger 1984).

Other vitamins. Riboflavin has been produced in a two-step process employing a primary (fermentation) stage with the fungi *Ashbya gossypii* or *Eremothecium ashbyii* in a grain stillage–glucose–peptone medium, followed by a recovery step in which the oxidized form of the vitamin is reduced by certain streptococci (Hickey 1946). The reduced product is relatively insoluble in water, thus concentrating the vitamin into the easily recovered cell pellet.

Saccharomyces cerevisiae has been reported to produce 3.5 to 5.06 g of riboflavin per liter in monocultures, and slightly higher amounts when cocultured with propionic acid bacteria (Kim 1974). Traces of riboflavin have also been detected in a citric acid producing coculture.

Sewage sludge (which may be regarded as a sort of ultimate mixed culture) has been reported to contain considerable amounts of biotin, and the supernatant from the digestion process contains ~10 μg biotin per gram dry suspended solids (Nuejahr and Hartwig 1961).

Antibiotics. Mixed cultures of *Aspergillus nidulans* GH79 and *A. flavus* CMI 91019B have been shown to produce hydroaspergillic acid, although neither species produces this compound in monoculture (Perry et al. 1984). Mixed cultures of *Trichoderma* sp. and *Fusarium oxysporum* or *F. solani* produced trichodermaol (Adachi et al. 1983).

Production of mycoheptin by *Streptomyces mycoheptinicum* was enhanced by coculture with any of five different yeasts: *Schizosaccharomyces romys, Rhodotorula mucilaginosum, Candida utilis, C. tropicalis*, and *Monilia nivea* (Kuznetsova 1979). However, 30 other microorganisms were shown to supress mycoheptin production, and another 15 reduced the activity of the antibiotic. These findings serve as useful reminders that production of a particular compound in mixed culture is neither simple nor straightforward.

Biosynthesis of mixed levorin antibiotics by *Actinomyces levoris* has been shown to be modulated by the presence of various coculture organisms (Yakovleva and Sokolova 1973). For example, *Candida tropicalis, C. parapsilosis*, and *Torula utilis* increased the proportion of levorin A and decreased that of levorin B. By contrast, *S. cerevisiae, S. carlsbergensis*, and *Candida metalondinsis* decreased antibiotic production and the relative proportion of levorin A. At least some of these changes may have been due to simple changes in medium pH resulting from the metabolic activity of the coculture organism (Yakovleva 1976).

In addition to the effects of coculture organisms on specific antibiotic fermentations noted above, several screening studies have demonstrated production of unidentified antibiotics by microbial mixed cultures. Likhacheva et al. (1987) observed that 14 of 53 *Streptomyces* strains studied established a distinctive interaction with the green alga *Chlorella*, as evidenced by the formation of a lichenlike thallome. In these cases, it was noted that the mixed cultures demonstrated enhanced antibiotic activity relative to the *Streptomyces* monocultures. Along a similar line, a mixed culture of *Lactobacillus bulgaricus* and *Bifidobacterium bifidum* produced more activity against enteric bacteria than did either species in monoculture (Khamnaeva et al. 1985).

Steroid transformations. Modification of steroids by selective, stereospecific microbial transformation has been performed for several decades as an alternative to elaborate and relatively less specific chemical syntheses. The organisms used in these mixed-culture transformations are exclusively mesophilic aerobes and are representatives of the same groups of microbes which are famous for their secondary metabolic capabilities in general, and their steroid-transforming ability in particular: fungi (*Aspergillus, Curvularia*), filamentous bacteria (*Arthrobacter, Corynebacterium, Nocardia, Streptomyces*) and *Bacillus*.

Two fundamentally different types of mixed-culture steroid transformations have been distinguished by Ryu et al. (1969). They term the first of these "sequential steroid transformation," in which a substrate S_r is converted to an intermediate S_i, which is in turn converted to a product S_p:

$$S_r \xrightarrow[E_1]{M_1} S_i \xrightarrow[E_2]{M_2} S_p$$

Each conversion is mediated by enzymes (E_1 and E_2) produced by two different organisms (M_1 and M_2, respectively). In this type of transformation, there is not a significant metabolic interaction between the two organisms; M_1's function is merely to provide substrate for M_2. Consequently, addition of M_2 to the reactor need not occur until the conversion of S_r to S_i by organism M_1 is essentially complete. Such a system might improve the commercial viability of the process by reducing capital outlay for additional fermentors, or by eliminating costs associated with purifying the intermediate S_i. The second type of steroid transformation is a true "mixed-culture fermentation," in which both organisms are present throughout the fermentation period. In this case, M_1 and M_2 establish a distinct metabolic interaction in which the expression of particular enzymes in one organism is controlled by the metabolic behavior of the other organism. Although most of the mixed-culture transformations of steroids that have been described have not been examined from a mechanistic standpoint, it would appear that most are of the first (sequential) type. The earliest work on steroid transformations by mixed cultures was performed with the corticosteroid class. McAleer et al. (1958, 1959) demonstrated

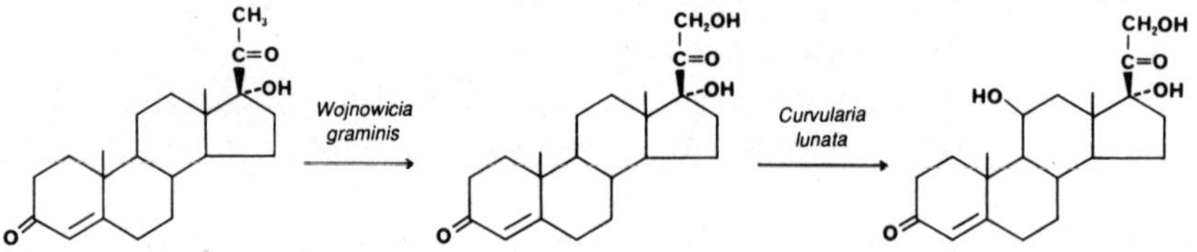

Figure 8.3 Conversion of progesterone to hydrocortisone via a 21-hydroxyprogesterone intermediate, using a mixed culture of fungi. (*From McAleer et al. 1959 and Shull 1959.*)

the conversion of progesterone to hydrocortisone via a 21-hydroxyprogesterone intermediate (Fig. 8.3). A variety of sequential corticosteroid transformations have since been demonstrated. The pathways involved in some of these conversions are shown in Fig. 8.4, and the particular organisms responsible for the various sequential conversions summarized in Table 8.3. Inspection of the table reveals that a wide variety of microbial taxa are represented.

Shull (1959) was the first to demonstrate a useful nonsequential mixed-culture steroid transformation, using *Curvularia lunata* and *Mycobacterium phlei* to convert cortexolone to prednisolone. Kimura (1962) reported a similar transformation by a different pathway: *Bacillus sphaericus* converted cortexolone to Δ^1-hydrocortisone, and *C. lunata* dehydrogenated the latter compound to prednisolone. Subsequent research has verified and extended the utility of the nonsequential mixed-culture conversion mode.

Cocultures of *Streptomyces roseochromogenes* and *Arthrobacter simplex* exemplify an interesting interaction (Fig. 8.5; Lee et al. 1969). In pure culture, the former organism dehydrogenates 9α-fluorohydrocortisone at the 1 position. The latter organism in pure culture can hydroxylate the resulting 1-dehydro-9α-fluorohydrocortisone to triamcinolone, but yields are low due to significant reduction of the keto group at position 20. However, in coculture the enzyme responsible for this reduction is almost completely repressed, resulting in triamcinolone molar yields of nearly 100 percent at 9α-fluorohydrocortisone conversions of 70 to 89 percent. Although the mechanism of the repression was not elucidated, the authors proposed that the *Streptomyces* removed materials from the complex soy-based medium which ordinarily derepress the 20-ketoreductase of the *Arthrobacter*. While this interaction technically fits our definition of a coculture (see the introduction), it should be noted that the streptomycete strongly inhibited the net growth of the *Arthrobacter*. The fermentation technology of the process has been described in detail (Lee et al. 1969). A kinetic model for this conversion has been developed based on the increased reaction rate which results from pH changes in the medium during growth (Yoshida et al. 1981a, b).

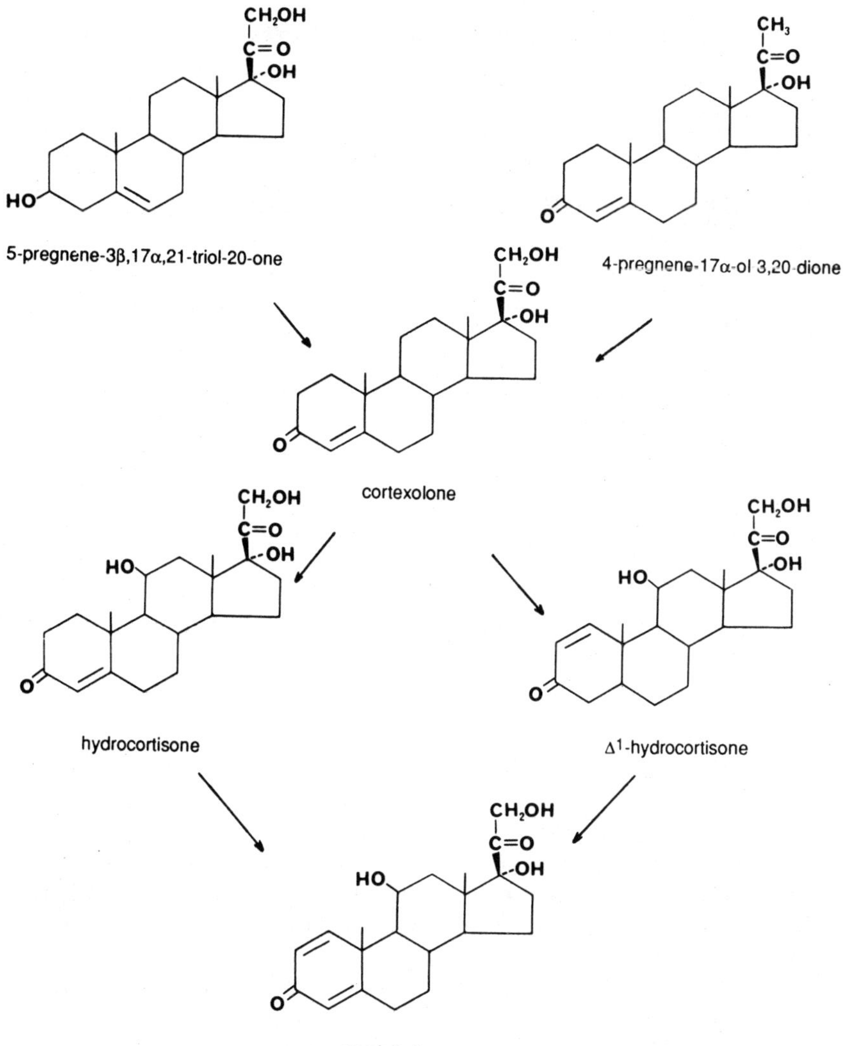

5-pregnene-3β,17α,21-triol-20-one

4-pregnene-17α-ol 3,20-dione

cortexolone

hydrocortisone

Δ1-hydrocortisone

prednisolone

Figure 8.4 Conversion of corticosteroids by sequential mixed cultures. (See Table 8.3 for organisms mediating these transformations.)

Lee et al. (1970) have also shown that *S. roseochromogenes* represses the formation of both the 20-ketoreductase and the 1-dehydrogenase of *Nocardia restrictus*. In both of the cocultures. The pattern of derepression—and thus product formation—was shown to be highly dependent upon the steroid substrate, medium composition, and the size and solubility of the steroid particles (Lee et al. 1969, 1970). Thus the use of mixed cultures holds promise for a large num-

TABLE 8.3 Examples of Transformations of Corticosteroids by Sequential Mixed Cultures[a]

S_r	M_1	S_i	M_2	S_p	Reference
Progesterone	*Wojnowicia graminis*	21-Hydroxy-progesterone	*Curvularia lunata*	Hydrocortisone	McAleer et al. 1958, 1959
Cortexolone	*Curvularia lunata*	Hydrocortisone	*Mycobacterium phlei*	Prednisolone	Shull 1959
4-Pregnene-17α-ol-3,20-dione	*Ophiolobus herpotricus*	Cortexolone	*Alternaria passiflorae*	Hydrocortisone	CIBA 1960
5-Pregnene-3β,17α-21-triol-20-dione	*Corynebacterium mediolanum*	Cortexolone	*Cunninghamella blakesleana*	Hydrocortisone	Spalla et al. 1962

[a]Adapted from Ryu et al. 1969. (See text for discussion of sequential transformations, and Fig. 8.4 for chemical structures.)

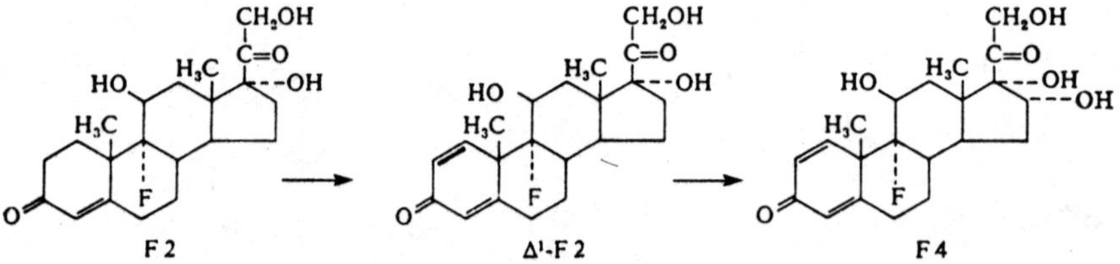

Figure 8.5 Conversion of 9α-fluorohydrocortisone (F2) to 1-dehydro-16α-hydroxy-9α-fluorohydrocortisone (F4) by a mixed culture of *Arthrobacter simplex* and *Streptomyces roseochromogenes*. *(From B. K. Lee et al. 1969. Used with permission.)*

ber of steroid transformations, as long as the regulatory patterns are understood and the fermentation conditions adequately controllable.

Several other described mixed cultures incorporate *Arthrobacter simplex* to carry out hydrogenations or dehydrogenations at positions 1 and/or 4 of the A ring. The other member of the mixed culture performs any of a number of other reactions at other parts of the molecule. Lee et al. (1970) have demonstrated the conversion of 16α-hydrocortexolone-16,17-acetonide to Δ^1-11β,16α-dihydroxycortexolone-16,17-acetonide (an intermediate in triamcinolone synthesis) using cocultures of *A. simplex* and *Curvularia lunata* (Fig. 8.6).

A mixture of *A. simplex* strain A-1 and *Nocardia* strain 76816 have been shown to dehydrogenate 17α-methyl-17β-hydroxyandrosta-1,4-dien-3-one (Chen et al. 1983). The conversion was run for 96 h following an initial preculture of the mixed culture prior to the addition of the steroid substrate. In 50 mL of complex medium containing mineral salts, glucose, corn steep liquor, and yeast extract (pH = 6.7), 250 mg of substrate was converted at a yield of 71 percent.

Mixed cultures of *A. simplex* and a strain of *Nocardia* have been

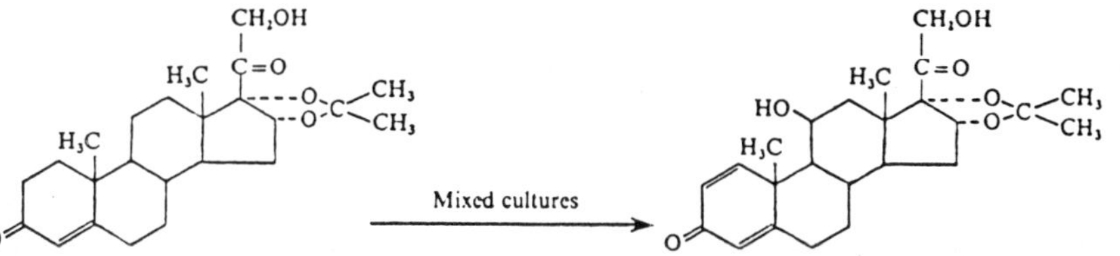

Figure 8.6 Conversion of 16α-hydroxycortexolone-16,17-acetonide to its 1-dehydro-11β-hydroxy product by a mixed culture of *Arthrobacter simplex* and *Curvularia lunata*. The latter is an intermediate in the chemical synthesis of triamcinolone. (*From B. K. Lee et al. 1970. Used with permission.*)

shown to perform different types of reactions on certain very similar steroid substrates (Zhang et al. 1981). For example, the dehydrogenation occurs at positions 1 and 4 of both 16β-methyl-5α-$\Delta^{9(11)}$-pregnene-3β,17α,21-triol-20-one-3β,21-diacetate and 17α-methyl-17β-hydroxy-5α-androstan-3-one. However, the former compound is also deacetylated at position 21 (and, to a lesser extent, epoxidated at position 9), while the latter is subjected to an opening of the B ring.

A mixed culture of *A. simplex* and *Corynebacterium equi* has been reported to convert a mixture of sterols (β-sitosterol, stigmasterol, and campestrol) to $\Delta^{1,4}$-androstadien-3,17-dione (Yamamoto et al. 1979).

An intriguing report suggests that both steroids and glycoalkaloids can be produced by mixed cultures of nightshade plant cells and certain cyanobacteria (Gorelov 1986). This opens up the possibility of combining the tremendous biosynthetic capabilities of plants with the quite different secondary metabolic capabilities of microorganisms to yield useful specialty chemicals.

Summary and Conclusion

A summary look at the fermentation parameters of described mixed-culture systems for commodity chemical production (Table 8.4) reveals that at present all are inferior to proven monoculture systems. Thus the use of mixed cultures for the production of most commodity chemicals does not represent a viable commercial opportunity at this time, and probably will not in the near future. Inherent limitations in the physiological properties (e.g., low product tolerance) of the organisms in these culture mixtures, coupled with engineering constraints (e.g., recovery of product from dilute aqueous solution) will probably require that successful mixed cultures be built around individual strains which have proven their utility in monoculture systems.

The outlook for the use of mixed cultures for the production of specialty chemicals appears much brighter. In several cases, parallel

TABLE 8.4 Comparison of Best-Case Data for Commercial Monoculture-Based Fermentation Processes and for Laboratory-Scale Mixed Culture-Based Fermentation Processes for Commodity Chemical Production[a]

	Product conc., g/L		Yield, %		Productivity, g/L · h	
Chemical	Mono	Mixed	Mono	Mixed	Mono	Mixed
Acetic acid	130–140	41				
Acetone	6	4			0.12	0.05
Butanol	15	12			0.3	
Citric acid	100	98				
Ethanol	100–160	85	47		15	16
Lactic acid	120					
Malic acid	130	126		84		

[a]Data for different parameters for the same product may not be from the same process (e.g., laboratory-scale mixed-culture-based ethanol processes, depending on their configuration, may produce 85 g ethanol per liter, or at a productivity of 16 g/L · h, but a single process will not do both).

comparisons reveal that mixed cultures can outperform their homologous monocultures. Thus, efforts to improve monoculture-based specialty chemical products by incorporating a coculture organism may prove rewarding. However, it should be noted that genetic engineering technology, though currently restricted to relatively few microbial taxa, may ultimately replace the need for mixed cultures if the desirable properties of individual species may be combined into a single organism.

References

Abe, S., M. Takagi, and S. Suzuki. Jpn. Patent 78/113085 (1978).

Adachi, T., H. Aoki, T. Osawa, M. Namiki, T. Yamane, and T. Ashida. Structure of trichodermaol, antibacterial substance produced in combined culture with *Fusarium oxysporum* or *F. solani*. Chem. Letts. 6:923–926 (1983).

Ballerini, D., D. R. Marchal, M. Hermann, D. Blanchette, and J. P. Vandecasteele. Fr. Patent. 2,550,222 A1 (1985).

Baranova, N. A., L. I. Vorob'eva, Z. A. Arkad'eva, and D. Dadabaeva. Biosynthesis of vitamin B_{12} in mixed propionic acid bacteria and *Mycobacterium* culture. *Biol. Nauki* 7:101–104 (1968) (cited through *Chemical Abstracts* 69:65311w).

Bodie, E. A., N. Goodman, and R. D. Schwartz. Production of propionic acid by mixed cultures of *Propionibacterium shermanii* and *Lactobacillus casei* in autoclave-sterilized whey. *J. Industr. Microbiol.* 1:349–353 (1987).

Bogdanov, V. M., and E. E. Grudzinskaya. USSR Patent 195597 (1969). (cited through *Chemical Abstracts* 72:131104w).

Brumm, P. J. A dual fermentation process for the production of aliphatic acids. *Biotechnol. Bioeng.* 30:784–787 (1987).

Carreira, L. H., and L. G. Ljungdahl. Production of ethanol from biomass using thermophilic anaerobic bacteria. In D. L. Wise, ed. *Liquid Fuel Developments*, CRC Press, Boca Raton, Fla., 1983, pp. 1–29.

Chen, M., Y. Zhou, and L. Zhang. Preparation of 17α-methyl-17β-hydroxy-androsta-1,4-dien-3-one by microbial transformation. *Yiyao Gonge* 4:7–8 (1983) (cited through *Chemical Abstracts* 99:156756f).

CIBA Ltd. Br. Patent 827,182 (1960).

Crueger, W., and A. Crueger. *Biotechnology: A Textbook of Industrial Microbiology*. Eng. Ed. transl. by T. D. Brock., Sinauer Associates, Sunderland, Mass., 1984.

Ebina, S., I. Terao, and I. Nagai. Jpn. Patent 87/44172 A2 (1987).

Fond, O., E. Petitdemange, H. Petitdemange, and J.-M. Engasser. Cellulose fermentation by a coculture of a mesophilic cellulolytic *Clostridium* and *Clostridium acetobutylicum*. *Biotechnol. Bioeng. Symp. Ser.* 13:217–224 (1983).

Furukawa, T., T. Nakahara, and K. Yamada. Utilization of hydrocarbons by microorganisms. XX. Conversion of fumaric acid to malic acid by the association of two kinds of yeast. *Agr. Biol. Chem.* 34:1833–1838 (1970).

Gibbs, D.F. The rise and fall (...and rise?) of acetone/butanol fermentations *Trends Biotechnol.* 1:12–15 (1983).

Gorelov, O. A. Biosynthesis and accumulation of steroid compounds by nightshade cells under mixed cultivation conditions with cyanobacteria *Kultura Kletok. Rast. i. Bioteknol.*, M. 261–263. In *Ref. Zh. Fiz.-Khim. Biol. Bioteknol.* 1986, Abstract 12E344.

Hahn-Haegerdahl, B., and M. Haeggstroem. Production of ethanol from cellulose, Solk Floc BW200, in a fedbatch mixed culture of *Trichoderma reesei* C30 and *Saccharomyces cerevesiae*. *Appl. Microbiol. Biotechnol.* 22:187–189 (1985).

Harrison, D. E. F. Mixed cultures in industrial fermentation processes. *Adv. Appl. Microbiol.* 24:129–164 (1978).

Hastings, J. J. H. Acetone-butyl alcohol fermentation. In A. H. Rose, ed. *Economic Microbiology*, vol. 2, *Primary Products of Metabolism*, Academic Press, London, 1978, pp. 31–45.

Hickey, R. J. Precipitation of riboflavin from aqueous solution by bacteriological reduction. *Arch. Biochem* 11:259–267 (1946).

Hitachi Shipbuilding and Engineering Co. Ltd. Jpn. Patent 83/129984 A2 (1983a).

Hitachi Shipbuilding and Engineering Co. Ltd. Jpn. Patent 83/129983 A2 (1983b).

Hodge, H. M., C. T. Hincon, and R. J. Allgeier. Animal protein factor supplement produced by direct bacterial fermentation: production and evaluation. *Ind. Eng. Chem.* 44:132–135 (1952).

Hyun, H. H., and J. G. Zeikus. Simultaneous and enhanced production of thermostable amylases and ethanol from starch by cocultures of *Clostridium thermosulfurogenes* and *Clostridium thermohydrosulfuricum*. *Appl. Environ. Microbiol.* 49:1174–1181 (1985).

Jain, M. K., R. Datta, and J. G. Zeikus. High-value organic acids fermentation—emerging processes and products. In T. K. Ghose, ed. *Bioprocess Engineering*, Ellis-Horwood Ltd., 1989 pp. 366–389.

Jones, J. P., D. Alexander, and J. E. Zajic. Ethanol production from sucrose using a mixed culture of *Saccharomyces* spp. *Dev. Ind. Microbiol.* 23:366–377 (1982).

Khamnaeva, M. I., I. S. Khamagaeva, V. F. Tovarov, V. I. Sharobaiko. Antibiotic properties of a mixed inoculum. *Molochn. Prom-St.* 3:31–32 (1985). (cited through *Chemical Abstracts* 103:51089p)

Khan, A. W., D. Wall, and L. Van den Berg. Fermentative conversion of cellulose to acetic acid and cellulolytic enzyme production by a bacterial mixed culture obtained from sewage sludge. *Appl. Environ. Microbiol.* 41:1214–1218 (1981).

Kim, V. I. Biosynthesis of B vitamins by microorganisms on culture media. *Izv. Akad. Nauk Kaz. SSR, Ser. Biol.* 12:66–69 (1974) (cited through *Chemical Abstracts* 81: 134835g).

Kimura, T. On the transformation of Reichstein's substance S to prednisolone with the cooperative action of fungi and bacteria. *Shionogi Res. Lab. Report* 12:180 (1962).

Kosaric, N., A. Wieczorek, G. P. Cosentino, R. J. Magee, and J. E. Prenosil. Ethanol fermentation. In H. Dellweg, ed. *Biotechnology*, vol. 3a. Verlag Chemie, Weinheim, 1983, pp. 257–385.

Kurosawa, H., H. Ishikawa, and H. Tanaka. L-lactic acid production from starch by co-immobilized mixed culture system of *Aspergillus awamori* and *Streptococcus lactis*. *Biotechnol. Bioeng.* 31:183–187 (1988).

Kurosawa, H., N. Nomura, and H. Tanaka. Ethanol production from starch by a co-immobilized mixed culture system of *Aspergillus awamori* and *Saccharomyces cerevisiae*. *Biotechnol. Bioeng.* 33:716–723 (1989).

Kuznetsova, N. A. Cocultivation of a mycoheptin producer with different microorganisms. *Antibiotiki (Moscow)* 19:122–123 (1974) (cited through *Chemical Abstracts* 81: 36403v).

Lamed, R., and J. G. Zeikus. Glucose fermentation pathway of *Thermoanaerobium brockii. J. Bacteriol.* 141:1251–1257 (1980).

Lee, B. K., D. D. Y. Ryu, R. W. Thoma, and P. A. Diassi. *Fr. Demande* 2,000,716 (1969).

Lee, B. K., D. Y. Ryu, R. W. Thoma, and W. E. Brown. Induction of repression by steroid hydroxylases and dehydrogenases in mixed culture fermentation. *J. Gen. Microbiol.* 55:145–153 (1969).

Lee, B. K., W. E. Brown, D. Y. Ryu, H. Jacobson, and R. W. Thoma. Influence of mode of steroid addition on the conversion of steroid and growth characteristics in mixed-culture fermentation. *J. Gen. Microbiol.* 61:97–105 (1970).

Lee, I. H., A. G. Fredrickson, and H. M. Tsuchiya. Dynamics of mixed cultures of *Lactobacillus plantarum* and *Propionibacterium shermanii. Biotechnol. Bioeng.* 18: 513–526 (1976).

LeRuyet, P., H. C. Dubourgier, and G. A. Albagnac. Homoacetogenic fermentation of cellulose by a coculture of *Clostridium thermocellum* and *Acetogenium kivui. Appl. Environ. Microbiol.* 48:893–894 (1984).

Leviton, A., and R. E. Hargrove. Microbiological synthesis of vitamin B_{12} by propionic acid bacteria. *Ind. Eng. Chem.* 44:2651–2655 (1952).

Likhacheva, A. A., G. M. Zenova, and L. V. Kalakutskii. The interaction of streptomycetes and algae in mixed cultures. *Mikrobiologiya* 56:309–313 (1987).

Liu, J. P., and N. J. Moon. Commensalistic interaction between *Lactobacillus acidophilus* and *Propionibacterium shermanii. Appl. Environ. Microbiol.* 44:715–722 (1982).

Lynd, L. R. Production of ethanol from lignocellulosic materials using thermophilic bacteria: critical evaluation of potential and review. In: A. Fiechter, ed., *Adv. Biotechnol./Biochem. Eng.*, vol. 38, Springer-Verlag, Berlin, 1989, pp. 1–52.

Ljungdahl, L. G., L. Carreira, and J. Wiegel. Production of ethanol from carbohydrates using thermophilic bacteria. *The Eckman Days Int. Symp. Wood and Pulp Chem.* Stockholm 4:23 (1981).

McAleer, W., E. Dulaney, and E. L. Dulaney. U.S. Patent 2,831,789 (1958).

McAleer, W., E. Dulaney, and E. L. Dulaney. U.S. Patent 2,875,132 (1959).

Miall, L. M. Organic acids. In A. H. Rose, ed. *Economic Microbiology*, vol. 2. *Primary Products of Metabolism*. Academic Press, London, pp. 47–119 (1978).

Moulin, G., and P. Galzy. Whey, a potential substrate for biotechnology. *Biotechnol. Gen. Eng. Revs.* 1:347–374 (1984).

Nakao, Y., K. Hisano, T. Kanemura, and T. Yamoto. Jpn. Patent 74/15796 (1974).

Ng, T. K., R. J. Hesser, B. Stieglitz, B. S. Griffiths, and L. B. Ling. Production of tetrahydrofuran/1,4-butanediol by a combined biological and chemical process. *Biotechnol. Bioeng. Symp. Ser.* 17:355–364 (1986).

Ng, T. K., A. Ben-Bassat, and J. G. Zeikus. Ethanol production by thermophilic bacteria: fermentation of cellulosic substrates by a coculture of *Clostridium thermocellum* and *Clostridium thermohydrosulfuricum. Appl. Environ. Microbiol.* 41: 1337–1343 (1981).

Ninet, L., and J. Renaut. Carotenoids, In H. J. Peppler and D. Perlman, eds. *Microbial Technology*, vol. 1, 2d ed. Academic Press, New York, 1979 pp. 529–544.

Nuejahr, H. Y., and J. Hartwig. On the occurrence of biotin in different fractions of municipal sewage. *Acta. Chem. Scand.* 15:954–955 (1961).

Parker, J. A., and N. J. Moon. Interactions of *Lactobacillus* and *Propionibacterium* in mixed culture. *J. Food. Protection* 45:326–330 (1982).

Perry, M. J., J. F. Makins, M. W. Adlard, and G. Holt. Aspergillic acids produced by mixed cultures of *Aspergillus flavus* and *Aspergillus nidulans. J. Gen. Microbiol.* 130: 319–323 (1984).

Rogers, P. L., K. J. Lee, and D. E. Tribe. Ethanol production in *Zymomonas mobilis. Adv. Biochem. Eng./Biotechnol.* 23:27 (1980).

Ryu, D. Y., B. K. Lee, R. W. Thoma, and W. E. Brown. Transformation of steroids by mixed cultures. *Biotechnol. Bioeng.* 11:1255–1270 (1969).

Segers, L., L. Verstrynge, and W. Verstreate. Product patterns of non-axenic sucrose fermentation as a function of pH. *Biotechnol. Letts.* 3:635–640 (1981).

Shull, G. M. Ger. Pat. 1,050,335 (1959).
Shull, G. M. U.S. Patent 2,908,616 (1959).
Spalla, C., R. Modelli, and A. M. Amici. U.S. Patent 3,030,278 (1962).
Szemler, L. L., and A. D. Szekely. Vitamin B_{12} from sewage sludge. *Process Biochem.* 4(12):25–27 (1969).
Tachibana, S. Jpn. Patent 72/16879 (1972).
Takao, S., A. Yakota, and M. Tanida. L-malic acid fermentation by a mixed culture of *Rhizopus arrhizus* and *Paecilomyces varioti. J. Ferm. Technol.* 61:643–645 (1983).
Takao, S., and K. Hotta. Conversion of fumaric acid fermentation to L-malic acid fermentation by the association of *Rhizopus arrhizus* and *Proteus vulgaris. Hakko Kogaku Zasshi* 54:197–204 (1970) (cited through *Chemical Abstracts* 85:19029t)
Takao, S., and K. Hotta. L-malic acid fermentation by mixed culture of *Rhizopus arrhizus* and *Proteus vulgaris. Agr. Biol. Chem.* 41:945–950 (1977).
Tanaka, H., H. Kurosawa, and H. Murakami. Ethanol production from starch by a coimmobilized mixed culture system of *Aspergillus awamori* and *Zymomonas mobilis. Biotechnol. Bioeng.* 28:1761–1768 (1986).
Tang, I. C., S. T. Yang, and M. R. Okos. Acetic acid production from whey lactose by cocultures of *Streptococcus lactis* and *Clostridium formicoaceticum. Appl. Microbiol. Biotechnol.* 28:138–143 (1988).
Tiwari, K. P., A. Pandey, and N. Mishra. Lactic acid production from molasses by mixed population of *Lactobacillus bulgaricus* and *L. casei. Proc. Natl. Acad. Sci., India, Sect. A.* 47:130–132 (1977).
Volfova, O., O. Suchardova, J. Panos, and V. Krummphanzl. Ethanol formation from cellulose by thermophilic bacteria. *Appl. Microbiol. Biotechnol.* 22:246–248 (1985).
Wang, D. I. C., C. L. Cooney, A. L. Demain, R. F. Gomez, and A. J. Sinskey. Degradation of cellulosic biomass and its subsequent utilization for the production of chemical feedstocks. U.S. Dept. of Energy Report COO-4198-11 (1981).
Wang, D. I. C., C. L. Cooney, A. L. Demain, R. F. Gomez, and A. J. Sinskey. Degradation of cellulosic biomass and its subsequent utilization for the production of chemical feedstocks. U.S. Dept. of Energy Report COO-4198-13 (1981).
Weimer, P. J. Use of thermophiles for the production of fuels and chemicals. In T. D. Brock, ed. *Thermophiles: General, Molecular, and Applied Microbiology.* John Wiley and Sons, New York, 1986, pp. 217–255.
Weimer, P. J., and J. G. Zeikus. Fermentation of cellulose and cellobiose by *Clostridium thermocellum* in the absence and presence of *Methanobacterium thermoautotrophicum. Appl. Environ. Microbiol.* 33:289–297 (1977).
Whalen, P. J., and K. M. Shahani. Optimization of a process for ethanol production via cofermentation of cheese whey and corn. In *Agricultural Waste Utilization Management*, Am. Soc. Agr. Eng. Publication 13-85, pp. 29–36 (1985).
Wuest, H. M., and D. Perlman. Industrial production and preparation. In W. H. Sebrell and R. S. Harris, eds. *The Vitamins*, vol. 2, 2d ed., Academic Press, New York, 1967, pp. 139–144.
Yakovleva, E. P. Effect of different pH values of the medium on antibiotic syntheses by *Actinomyces levoris* and yeasts grown together. *Antibiotiki (Moscow)* 21:494–499 (1976) (cited through *Chemical Abstracts* 85:107422w).
Yakovleva, E. P., and E. N. Sokolova. Comparative characteristics of levorin preparations produced during mixed cultivation of *Actinomyces levoris* with yeasts. *Antibiotiki (Moscow)* 18:485–489 (1973) (cited through *Chemical Abstracts* 79:76918g).
Yamada, K., and T. Furukawa. L-malic acid by microorganism. Jpn. Patent 75/19631 (1975).
Yamamoto, M., H. Hashiba, N. Watanabe, M. Nagasawa, N. Iguchi, H. Ariba, and M. Tamura. Jpn. Patent 79/3954 (1979).
Yoshida, T., H. Taguchi, S. Kulprecha, and N. Nilubol. Kinetics and optimization of steroid transformation in a mixed culture. In C. Vezina and K. Singh, eds., *Adv. Biotechnol.: Proc. 6th Int. Ferm. Symp.* vol 3, Pergamon Press, Toronto, 1981a, pp. 501–506.
Yoshida, T., M. Sueki, H. Taguchi, S. Kulprecha, and N. Nilubol. Modeling and optimization of steroid transformation in a mixed culture. *Eur. J. Appl. Microbiol. Biotechnol.* 11:81–88 (1981b).

Yu, E. K. C., M. K. H. Chan, and J. N. Saddler. Butanol production from cellulosic substrates by sequential coculture of *Clostridium thermocellum* and *Clostridium acetobutylicum*. *Biotechnol. Lett.* 7:509–514 (1985).

Zeikus, J.G. Chemical and fuel production by anaerobic bacteria. *Ann. Rev. Microbiol.* 34:423–464.

Zeikus, J. G., T. K. Ng, A. Ben-Bassat, and R. J. Lamed. U.S. Patent 4,400,472. (1983).

Zhang, L., E. Zhang, and Z. Wu. Microbial transformation of 16β-methyl-5α-$\Delta^{9(11)}$-pregnene-3β,17α,21-triol-20-one-3β,21-diacetate and 17α-methyl-17β-hydroxy-5α-androstan-3-one. *Yaoxue Xuebao* 16:356–360. (1981) (cited through *Chemical Abstracts* 97:70723q).

Chapter

9

Mixed Cultures in Enzymatic Degradation of Polysaccharides

Badal C. Saha

Michigan Biotechnology Institute
Lansing, Michigan

Department of Food Science and Human Nutrition
Michigan State University
East Lansing, Michigan

Introduction

The utilization of polysaccharides by mixed cultures has been studied for numerous applications such as production of enzymes, fermentable sugars, single-cell protein (SCP), ethanol, liquid fuels and chemicals, and biogas. Various enzymes are involved in the enzymatic degradation of polysaccharides, such as cellulose, hemicellulose, starch, pectin, and chitin. Synergistic action of various enzymes is needed for efficient degradation of polysaccharides. In natural environments, microbial degradation of various polysaccharides involves the combined activity of many different microorganisms. These mixed cultures are often stable owing to their synergistic metabolic interactions. Different mechanisms of microbial interaction such as competition, commensalism, and mutualism are involved in the degradation of polysaccharides by mixed cultures. A large variety of microorganisms produce extracellular enzymes that depolymerize polysaccharides into fermentable sugars. However, the enzymatic hydrolysis of some polysaccharides is strongly regulated by the end-product inhibition. Thus, mixed cultures are very useful for overcoming feedback regulation and catabolic repression, as one organism consumes the material produced by another organism. Mixed-culture utilization of polysaccharides has been used to produce a variety of products of biotechnological interest such as animal food and fermentation enzymes, and in waste utilization. In this chapter, the current state and potential of mixed-culture processes for enzymatic degradation of polysaccharides is reviewed.

Enzymatic Degradation Mechanisms and Application in Biomass Conversion

Cellulosic materials

Cellulosic materials in the form of cellulose, hemicellulose, and lignin are the most abundant biopolymers in the world. Cellulose is a linear polymer of 8000 to 12,000 glucose units connected by β-1,4 linkages. The complete degradation of cellulose requires the synergistic action of at least three enzymes, which are collectively known as cellulase. These enzymes are endo-β-glucanase, exo-β-glucanase, and β-glucosidase. Endoglucanase randomly cleaves internal β-1,4 linkages in cellulose, resulting in a slow release of reducing sugars. Exoglucanase cleaves cellobiose and glucose units from the non-reducing end of the glucan, and β-glucosidase hydrolyzes the resulting cellobiose and cello-oligosaccharides to glucose. There are many different kinds of hemicellulose, which is a heteropolymer and may have up to five or six different sugar components including D-xylose, L-arabinose, D-galactose, D-mannose and D-glucose and their uronic acids. Xylan is the main component of hemicellulose and is composed of linear D-xylose units linked by β-1,4 glycosidic bonds. At least two types of xylanolytic enzymes are involved in the degradation of xylan: endo-β-xylanase and β-xylosidase. Endoxylanase randomly cleaves internal bonds in xylan and β-xylosidase is able to split the disaccharide or higher oligosaccharides of xylose. Lignin is a polymer of p-hydroxyphenylpropane units connected by C—C and C—O linkages.

Cellulolytic enzymes are produced by a large variety of microorganisms. *Trichoderma reesei* is one of the most potent producers of cellulase enzyme systems. However, the enzyme activity is deficient in β-glucosidase activity (Ryu and Mandels 1980). The hydrolysis product of cellulose by this enzyme system is cellobiose, a strong competitive inhibitor of exo-β-glucanase activity. As a result, there is a decrease in the rate of sugar production and in the final concentration of sugar produced. However, the performance of *Trichoderma* cellulase is greatly enhanced by the addition of supplemental β-glucosidase (Bisset and Sternberg 1978). In fact, β-glucosidase and hemicellulase are needed at optimum levels for practical separation of pretreated agricultural residues (Ghose and Bisaria 1979). Several species of *Aspergillus* produce high levels of β-glucosidase (Sternberg et al. 1977, Srivastava et al. 1981). Mixed cultivation of *T. reesei* and *Aspergillus* sp. produces a cellulase complex which shows enhanced β-glucosidase activity and greatly improves cellulose hydrolysis (Panda et al. 1983, Ghose et al. 1985, Duff et al. 1985). Ghose et al. (1985) studied a mixed-culture process for enhancing the synthesis of cellulase and hemicellulase using *T. reesei* D1-6 (produces high amounts of cellulase and xylanase) and *A. wentii* Pt 2804 (high β-glucosidase producer). Significant increases in extracellular cellulase and hemicellulase activities were observed in the biosynthesis of cellulase enzymes in mixed culture when the *A. wentii* inoculum was phased by 15 h (Table 9.1). *A. wentii* produced mannanase, which affected the release as well as activity of cellulase enzyme in mixed culture. Mannanase inactivated filter paper (FP) and endoglucanase (EG) activities but had no effect on xylanase and β-glucosidase activities. The level of cellulase and xylanase produced by the mixed culture was found to be a function of the individual biomass concentration of the two fungi in the mixed culture. Panda et al. (1989) developed a simple differential method based on measurement of an intracellular pigment produced by *A. wentii* for estimation of growth of the individual two fungi, *T. reesei* and *A. wentii*, in mixed culture fermentation of an insoluble cellulosic substrate.

Duff et al. (1986) evaluated the hydrolytic potential of crude cellulase from mixed cultivation of *T. reesei* and *A. phoenicis* and demonstrated its enhanced activity against cellulose. The mixed-culture enzyme was more resistant to end-product inhibition by glucose than the *Trichoderma* enzyme and hydrolysis of steam-exploded aspen wood (SEAW) by the mixed-culture cellulase was markedly improved over that achieved by *Trichoderma* cellulase. The mixed cultivation of the two organisms eliminates the need for costly fermentation and recovery steps traditionally used to produce β-glucosidase. Duff et al.

TABLE 9.1 Production of Cellulase and Xylanase by Single and Mixed Cultures of *T. reesei* and *A. wentii* under Optimum Conditions

	Enzyme activities, U/ml			
Culture(s)	FP	EG	Xylanase	β-glucosidase
T. reesei D1-6	2.56	12.10	9.70	3.47
A. wentii Pt 2804	0.41	3.25	2.89	5.34
Mixed culture of:				
Tr D1-6 and *Aw* Pt 2804				
(a) No phasing of *Aw*	2.05	9.00	10.10	6.90
(b) 15-h phasing of *Aw*	4.11	40.38	22.58	6.46

SOURCE: From Ghose et al. 1985. Reproduced with permission.

(1987) also studied the effect of media composition and growth conditions on production of cellulase and β-glucosidase by the mixed fungal fermentation. The optimum temperature and pH were 27°C and 4.6. The authors were able to reduce the salt concentration (calcium, magnesium, and potassium) by 50 percent in shake flask experiments with no adverse effect on enzyme activity in the broth. The type and concentration of carbon source was found to have a strong effect on the relative concentration of various components of the cellulase system produced (Duff et al. 1985). As shown in Table 9.2, the cellulase produced by the mixed culture of *T. reesei* RUT C30 and *A. phoenicis* had a higher β-glucosidase activity than cellulase from *T. reesei* alone cultured in the same medium. Duff et al. (1987) studied this effect further by varying the starch concentration from 1 to 10 g/L at a fixed concentration of cellulose (10 g/L). The β-glucosidase activity of the resultant cellulase preparation increased with increasing starch concentration, but the overall cellulase activity, as measured by FP activity, decreased. The authors mentioned that the increase in β-glucosidase activity is probably due to increased growth of the *Aspergillus* component of the mixed culture. It is known that starch is an effective substrate for growth and enzyme production by *A. phoenicis* (Sternberg et al. 1977). *Aspergillus* sp. produce a variety of metabolites including proteases (Cohen 1977) and other unknown proteins (Wase et al. 1986) that can reversibly inhibit the activity of cellulase (Wase et al. 1986). The response of *Aspergillus* to competitive pressure from the *Trichoderma* component of the mixed culture may result in the production of one or more of the known compounds. This has been observed in mixed culture of *Trichoderma* with *Actinomyces, Fusidium, Fusarium*, and *Cellvibrio* (Lamot and Vadamme 1982). As mentioned earlier, Ghose et al. (1985) showed that *A. wentii* produces mannanase in mixed fermentation with *T. reesei* D1-6. They proposed that this enzyme is responsible for the reduction in FP and EG activity in

TABLE 9.2 Enzyme Production by *Trichoderma reesei* and *Aspergillus phoenicis* Grown Separately and in Mixed Culture on Varied Concentrations of Cellulose and Starch

Cellulose conc., g/L	Starch conc., g/L	Inoculum	Final FPA, IU/ml	Final β-glucosidase activity, IU/ml
—	10	Mixed	1.45	2.78
10	—	Mixed	2.83	1.07
10	10	Mixed	3.0	2.86
10	10	*Trichoderma*	3.5	0.7
10	10	*Aspergillus*	0.0	4.5

SOURCE: From Duff et al. 1985. Reproduced with permission.

Trichoderma culture. Cellulase has been reported to be a glycoprotein having mannose-containing saccharide as the main carbohydrate moiety (Gum and Brown 1976). On the other hand, it was observed that the mannose residues of cellulase from *Trichoderma* were not only very resistant to glycosidase activity (Gum and Brown 1976) but also that removal of the glycosidic side chains did not adversely affect enzyme activity (U.S. Department of Energy Alcohol Fuels Program Review, 1984). Other factors besides mannanase appear to be involved in the regulation of the synthesis and release of cellulase and hemicellulase in the mixed-culture fermentation. Trivedi and Desai (1984) investigated cellulase and β-glucosidase production by mixed-shake cultivation of *Scytalidium lignicola* and *T. longibrachiatum* and also by monocultures of each strain. Activity against cotton (exoglucanase) and carboxymethyl cellulose (CMC) (endoglucanase) was highest in the mixed culture. However, the β-glucosidase activity was not increased in comparison with that from a monoculture of *S. lignicola*. Lamot and Vadamme (1982) investigated the cellulolytic activity of actinomycetes in axenic mixed fermentations. The degradation of insoluble cellulose with cocultures of cellulolytic actinomycetes and fungi occurred much faster and more completely than with the corresponding monocultures. The fungi affected the initial cellulose degradation, while the actinomycetes subsequently metabolized the formed soluble oligosaccharides. Both the fungi and actinomycetes produced active cellulases in cocultures, resulting in increased cellulase activity of the culture filtrates as compared with monocultures. Friedrich et al. (1987) used *T. reesei* and *A. awamori*, separately and in mixed culture, to convert apple distillery waste into microbial biomass. *Trichoderma* contributed to good fiber degradation and protein enrichment of the biomass as well as cellulolytic, xylanolytic, and pectolytic activities of the filtrate. *Aspergillus* contributed to improved filtration, chemical oxygen demand (COD) reduction, and β-glucosidase synthesis. Peitersen (1975) examined the production of cellulases and cell protein from mixed-culture fermentations of NaOH-treated straw containing *T. viride* and a yeast (*Saccharomyces cerevisiae* or *Candida utilis*). The yeast was inoculated 24 to 32 h after the fungus. The presence of yeast accelerated cellulase and protein production and reduced the production time for maximum yield of cellulases and cell protein by several days. Takagi et al. (1977) developed a method for producing alcohol directly from cellulose using mixed cultures of *T. reesei* and yeast. The cellulolytic enzyme system of *T. reesei* produces glucose from cellulose and the yeast consumes the glucose and produces ethanol. Hagerdal and Haggstrom (1985) demonstrated that ethanol can be produced directly from cellulose with a mixed culture of *T. reesei* and *S. cerevisiae*. Up to

40 g/L ethanol was obtained with the mixed cultures, achieving a productivity of 0.2 g/L · h and a maximum yield of 83 percent. The limiting factor in the process was the activity of cellulolytic enzymes produced by *T. reesei* C30. It was suggested that oxygen could be used in this type of a mixed culture as an external regulator to maintain an optimal cellulolytic enzyme activity in the system, thus optimizing an overall conversion of cellulose to ethanol.

Wayman et al. (1987) studied the simultaneous saccharification and fermentation of SO_2-prehydrolyzed wood by mixed cultures of *Brettanomyces clausenii* and *Pichia stipitis* R. They used pilot-scale Wenger and Stake II reactors for prehydrolyzing aspen and coniferous wood chips in the presence of SO_2 catalyst to obtain highly digestible lignocellulosic substrates from which about 90 percent yields of hemicellulose (mostly in monomeric form) could be recovered. *B. clausenii* was identified as a potent cellobiose fermenter, and its use with a pentose fermenter *P. stipitis* R in simultaneous saccharification and fermentation without β-glucosidase resulted in rapid and efficient fermentation of these SO_2-prehydrolyzed woods.

Freer and Wing (1985) investigated the potential for enhancing ethanol production from cellodextrins by employing a mixed culture of *Candida wickerhamii* and *S. cerevisiae*. The cellodextrins were produced by trifluoroacetic acid hydrolysis of cellulose and contained compounds that were slightly inhibitory to *C. wickerhamii*. In this case, the mixed-culture fermentations produced 12 to 45 percent more ethanol than a pure-culture *C. wickerhamii* fermentation. However, if the substrate was treated with Darco G-60 charcoal, the toxic materials were apparently removed and the pure-culture *C. wickerhamii* fermentations performed as well as the mixed-culture fermentations. Thus, depending on the nature of the inhibitory compounds and the extent and cost of the treatments required to remove them from substrates produced from biopolymers, the use of mixed-culture fermentations may be of real advantage.

Halsall and Gibson (1985) studied cellulose decomposition and associated nitrogen fixation by mixed cultures of *Cellulomonas gelida* and *Azospirillum* sp. or *Bacillus macerans*. They followed this cooperative process over 30 days in sand-based cultures in which the breakdown of 20 percent cellulose and 28 to 30 percent straw resulted in the fixation of 12 to 14.6 mg N per gram of cellulose and 17 to 19 mg N per gram of straw consumed. *Cellulomonas* sp. could degrade cellulose at oxygen concentrations as low as 1% O_2 (v/v), which allowed a close association between cellulose-degrading and microaerobic diazotrophic microorganisms. The cellulolytic activity of *Cellulomonas* sp. could greatly stimulate the nitrogen-fixing activities of both *Azospirillum* sp. and *B. macerans* (Halsall and Gibson 1985, 1986). Scanning electron micrographs of straw incubated with a mixed culture for 8 days showed cells of both species in close proximity to each other (Halsall and Gibson 1986). Thus the organisms cooperated to promote straw breakdown under field conditions and enhance the nitrogen status of the soil (Halsall and Goodchild 1986). Similar systems involving aerobic *T. harzinum* and anaerobic *Clostridium butyricum* (Veal and Lynch 1984), and *Penicillium corylophilum* and *C. butyricum* (Lynch and Harper 1983) have also been studied.

Sounder and Chandra (1987) isolated an integrated mixed bacterial culture consisting of four strains by a batch enrichment technique on cellulose. The cellulolytic strain was a *Cellulomonas* sp. and the others were noncellulolytic (one of which was identified as *Arthobacter* sp). The authors found that the interaction between the two strains was pronounced and appeared to involve an exchange of reducing sugars and growth factors. They concluded that the symbiotic relationship of this naturally occurring mixed culture is, therefore, one of mutualism. The FP cellulase and CMCase activities in extracellular fluid were high but β-glucosidase activity was low (Table 9.3). However, the mixed culture digested a variety of lignocellulosic substrates efficiently. Han (31) investigated the interactions of different cellulolytic mixed systems of *Cellulomonas* sp. with *Alcaligenes faecalis* and cellobiose-utilizing yeasts. Interaction was found to be

TABLE 9.3 Enzyme Production by Strain D in Pure and Mixed Culture

Culture medium and inoculum	Percent cellulose utilization	Reducing sugar in culture filtrate, mg/ml	Extracellular protein, mg/ml	FPase, mg sugar/ml/h	CMCase, mg sugar/ml/h	β-glucosidase, mg PNP/ml/h
1% w/v FP cellulose + YE (0.05%, w/v), sterile	NT	0	0.001	NT	NT	NT
1% w/v FP cellulose + strain D	0.05	0.002	0.001	ND	ND	ND
1% w/v FP cellulose + YE (0.05%, w/v) + strain D	49.00	0.048	0.029	27.10	0.19	0.003
1% w/v FP cellulose + strain D + strain C	62.02	0.031	0.044	32.42	0.23	0.005

SOURCE: From Soundar and Chandra 1987. Reproduced with permission.

more pronounced in the growth of mixed cultures containing *A. faecalis* than with yeasts as interacting species. It was proposed that the removal of cellobiose from the culture medium by *A. faecalis* increased the production of cellulase, and consequently increased the cell mass of both bacteria. De la Torre and Campillo (1984) isolated a mixed culture of two bacterial species identified as *Cellulomonas flavigena* and *Xanthomonas* sp. by using a batch culture enrichment technique, and tested the ability of both bacteria to grow on alkaline-pretreated sugar cane bagasse or cellobiose as pure cultures in a mineral medium. It was found that *C. flavigena* as pure culture was able to grow on both substrates only when yeast extract or biotin and thiamine were added to the culture medium. *Xanthomonas* sp. could not grow on sugarcane bagasse, but assimilated cellobiose if yeast extract was supplied. The two organisms in mixed culture grew very well on both substrates and did not require any growth factor, but the interaction mechanism was not established. Callihan and Dunlap (1971) investigated SCP production on sugarcane bagasse by *Cellulomonas* sp. and a mixed culture of this strain and *Alcaligenes faecalis*. Kriestensen (1978) studied a continuous mixed culture of *Cellulomonas* sp. and *Candida utilis* using barley straw pretreated with caustic soda. Pietersen (1975), as mentioned earlier, also investigated the mixed culture of *T. viride* Q119123 and *S. cerevisiae* or *C. utilis* using barley straw pretreated with caustic soda. In both studies, inhibition of enzymatic hydrolysis of cellulose to glucose was depressed by assimilation of glucose by the yeast, and the protein content was increased. Nigam et al. (1987) investigated the mixed-culture fermentation for bioconversion of whole bagasse into microbial protein. They used two basidiomycete cultures, *Polyporus* BH1 and BW1 in conjunction with *Trichoderma* and *Cellulomonas* strains. It was found that the biomass yields in mixed cultures were much higher than in monocultures, especially in *Cellulomonas-Polyporus* mixtures.

Clostridium thermocellum produces an extracellular cellulolytic system which hydrolyzes cellulose to cellodextrins and cellobiose, and these saccharides are converted intracellularly by phosphorylases to mostly glucose-1-phosphate and some glucose (Ljungdahl et al. 1981). The glucose may accumulate and inhibit cellobiose phosphorylase, causing cellobiose to accumulate, which in turn inhibits the utilization of cellulose by *C. thermocellum*. The fermentation of cellulose proceeded rapidly by using cocultures of *C. thermocellum* and other cellobiose, glucose, and xylose fermenting ethanologenic bacteria such as *C. thermohydrosulfuricum*, *C. thermosaccharolyticum*, and *Thermoanaerobacter ethanolicus*. Three stages are involved in the fermentation of cellulose by a defined mixed culture (Carreira and Ljungdahl

1984). In the first stage, the induction of the cellulase system of *C. thermocellum* occurs and hydrolysis of cellulose starts. In the second stage, the hydrolysis products of cellulose are utilized by the ethanologenic bacteria. Both cellulolytic and ethanologenic bacteria are active in this stage. In the third stage, the cellulolytic and ethanologenic bacteria are inhibited by final fermentation products and the fermentation comes to a stop.

Ng et al. (1981) examined the fermentation of various saccharides derived from cellulosic biomass to ethanol in mono- and cocultures of *C. thermocellum* strain LQR1 and *C. thermohydrosulfuricum* strain 39E. They obtained a stable coculture that contained nearly equal numbers of both organisms. The ethanol yield was about twofold higher than in monoculture fermentations. The coculture activity fermented MN 300 cellulose, Avicel, Solka-Floc, SO_2-treated wood, and steam-exploded wood (Table 9.4). The metabolic basis for the enhanced fermentation effectiveness of the coculture on Solka-Floc cellulose included the ability of *C. thermocellum* cellulase to hydrolyze α-cellulose and hemicellulose; the enhanced utilization of mono- and disaccharides by *C. thermohydrosulfuricum*; increased cellulose consumption; threefold increase in the ethanol production rate; and twofold decrease in acetate production rate (Fig. 9.1). The mixed-culture fermentation of cellulose by *C. thermocellum* with *C. thermosaccharolyticum* or *T. ethanolicus* were also studied (Ljungdahl et al. 1981, Avgerinos and Wang 1980). These cocultures ferment a wide variety of cellulosic materials with ethanol as a main product. Weimer and Zeikus (1977) studied the kinetics of cellulose and cellobiose fermentation to ethanol and acetate or methane and acetate by defined cocultures of *C. thermocellum* and *Methanobacterium thermoautotropicum*. When grown on cellulose, the coculture exhibited a shorter lag before initiation of growth and cellulolysis than did the monoculture. Cellulase activity appeared earlier in the coculture than in the monoculture. However, after growth had ceased, cellulase activity was greater in the monoculture. Monocultures produced primarily ethanol, acetic acid, H_2, and CO_2. Cocultures produced more H_2 and acetic acid and less ethanol than did the monoculture. In the coculture, conversion of H_2 to methane was usually complete, and most of the methane produced was derived from CO_2 reduction rather than from acetate. The methanogen acts as a sink for hydrogen, enabling increased growth rates, altered end-product ratios, and an improved range of substrate utilization (Saddler and Chan 1984). Saddler and Chan (1984) investigated the conversion of pretreated lignocellulose substrates to ethanol by *C. thermocellum* in mono- and coculture with *C. thermosaccharolyticum* or *C. thermohydrosulfuricum*. Coculturing *C. thermocellum* with *C. thermohydrosulfuricum*

TABLE 9.4 Comparison of the Mono- and Coculture Fermentations of Cellulosics by *C. thermocellum* and *C. thermohydrosulfuricum*[a]

Substrate	Condition	Product formed, m*M*		CMCase, U/ml	Reducing sugar, mg/ml	Substrate consumed, %	Ethanol/acetate ratio, mol/mol
		Ethanol	Acetate				
MN 300	Monoculture	31.2	22.5	2.66	1.28	60	1.39
	Coculture	88.0	4.2	0.79	0.08	100	21.10
Solka-Floc	Monoculture	30.8	27.3	4.61	2.89	50	1.13
	Coculture	98.7	11.8	1.54	0.15	80	8.73
SO_2-treated wood	Monoculture	16.0	13.6	2.14	1.55	—	1.24
	Coculture	54.3	11.2	1.0	0.64	—	4.85
Steam-exploded wood	Monoculture	14.9	15.2	2.97	1.58	—	0.98
	Coculture	63.9	6.6	0.72	0.12	—	9.69
Untreated wood (control)	Monoculture	22.2	6.4	—	—	—	0.34
	Coculture	8.4	9.3	—	—	—	0.90

[a]Experiments were performed in anaerobic culture tubes that contained 10 ml of GS medium and 1% substrate except for MN300 cellulose (0.8%). Tubes were incubated without shaking at 62°C for 120 h.

SOURCE: From Ng et al. 1981. Reproduced with permission.

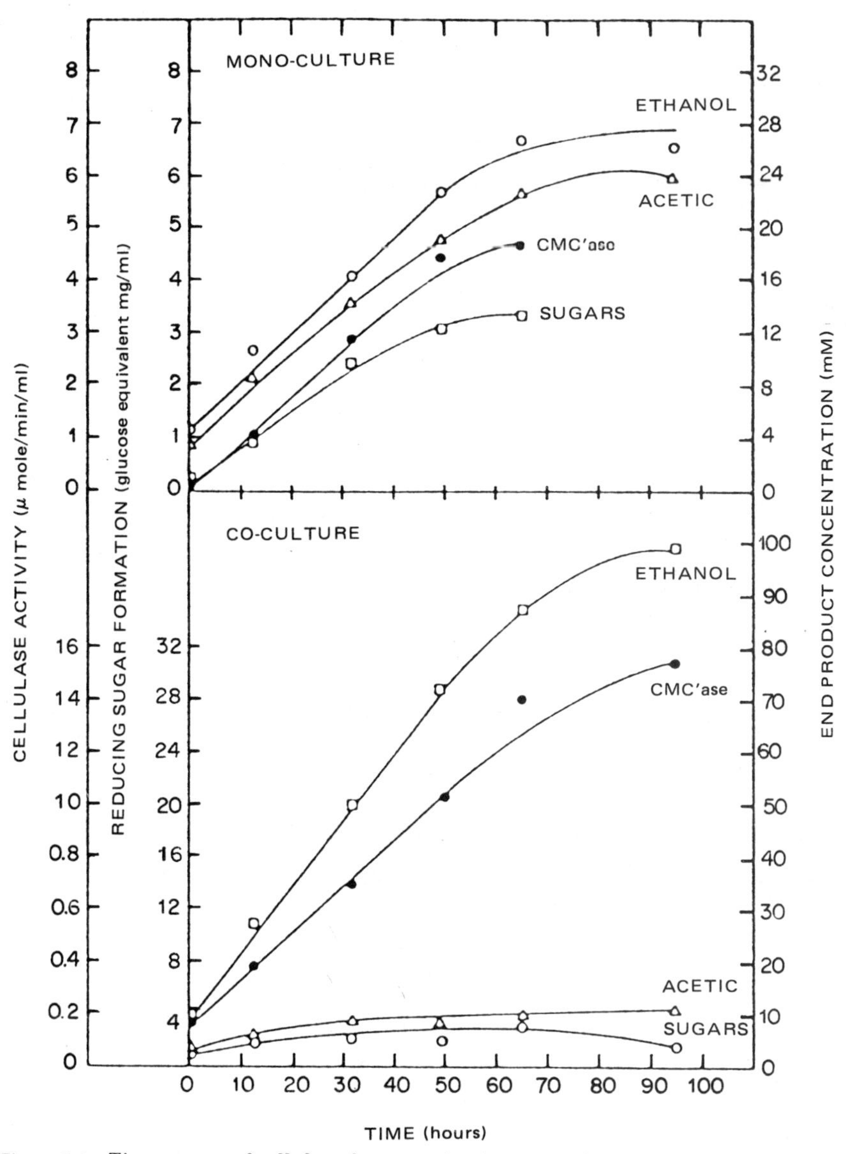

Figure 9.1 Time course of cellulose fermentation in monoculture of *C. thermocellum* LQR1 and in coculture with *C. thermohydrosulfuricum* strain 39E. Experiments were performed in anaerobic serum vials that contained 80 ml of GS medium with 1% Solka Floc. Experiments were incubated at 60°C without shaking and pH control. (*From Ng et al. 1981. Reproduced with permission.*)

resulted in enhanced substrate hydrolysis and almost complete utilization of all the soluble reducing sugars (Table 9.5). A wide range of pretreated substrates could be utilized by cocultures of the above organisms; however, their practicality was limited by the problems of relatively low ethanol production, restricted growth at high substrate concentrations, and the presence of soluble inhibitors in pretreated lignocellulosic substrates. Volfova et al. (1985) studied ethanol production from cellulose by cocultures of thermophilic bacteria using a mixture of five obligately thermophilic bacteria of the genus *Bacillus* and a single cellulolytic species of *Clostridium*. The coculture fermented crystalline cellulose and produced ethanol. The ethanol yield decreased with increasing substrate concentration. High ethanol concentration in the medium suppressed cellulose degradation by 50 percent and inhibited the actual production of ethanol by the coculture. LeRuyet (1984) studied the homoacetogenic fermentation of cellulose by a coculture of *C. thermocellum* and *Acetogenium kivui*. *A. kivui* is a newly described thermophilic hydrogen-oxidizing acetogenic bacterium. At 65°C, *C. thermocellum* TC11 and *A. kivui* LKT-1 formed a stable association at pH 6.8 that was successfully maintained by subculturing in buffered cellulose medium. Hydrogen consumption by *A. kivui* resulted in a quasicomplete shift of the catabolic pathway of *C. thermocellum* to acetate synthesis. Yu et al. (1985) investigated the butanediol production from lignocellulosic substrates by *Klebsiella pneumoniae* grown in sequential coculture with *C. thermocellum*. Both pentose and hexose sugars accumulated in the culture medium when *C. thermocellum* was grown on various lignocellulosic substrates. *K. pneumoniae* can readily utilize all the pentose and hexose sugars known to be present in biomass hydrolysis. The butanediol and

TABLE 9.5 Fermentation Products of Mono- and Cocultures of Two Strains of *C. thermocellum* Grown with *C. thermosaccharolyticum* for 6 Days on Different Pretreated Wheat Straw Fractions[a]

		Ethanol yield, mg/ml		Acetic acid, mg/ml	
Substrate, %	*C. thermocellum* strains	Monoculture	Coculture	Monoculture	Coculture
SF	NRCC 2688	0.9	1.8	1.0	1.4
	ATCC 31924	3.8	3.7	0.5	1.2
SES	NRCC2688	0.6	1.4	1.0	1.5
	ATCC 31924	0.9	1.2	0.7	1.1
SEAW	NRCC 2688	0.6	1.4	1.0	1.5
	ATCC 31924	1.2	1.5	0.6	1.0
SESWA	NRCC 2688	1.1	2.1	1.6	1.9
	ATCC 31924	2.7	2.7	0.9	1.1

[a]Based on percent composition of cellulose and pentosan in 1% substrate, theoretical.
SOURCE: From Saddler and Chan 1984. Reproduced with permission.

ethanol levels obtained from the sequential coculture system using Solka-Floc and SEAW surpassed the monoculture system by sixfold (Fig. 9.2). Murray (1986) studied the symbiotic relationship of a naturally occurring mesophilic coculture of *Bacteroides cellulosolvens* and *C. saccharolyticum* in cellulose fermentation. The coculture was isolated from a cellulose enrichment culture started from sewage sludge. The noncellulolytic spore former *C. saccharolyticum* depended upon *B. cellulosolvens* for the production of sugars from cellulose and the production of a growth factor replaceable by yeast extract (Murray et al. 1982). In coculture, *B. cellulosolvens* and *C. saccharolyticum* fermented 33 percent more cellulose than did the monoculture of *B. cellulosolvens* (Fig. 9.3). The symbiotic relationship of this naturally occurring coculture is one of mutualism, in which the cellulolytic microbe supplies the saccharolytic microbe with nutrients, and in return the saccharolytic microbe removes a secondary metabolite toxic to the primary microbe. Latham and Wolin (1977) investigated the fermentation of cellulose by *Ruminococcus flavefaciens* in the presence and absence of *Methanobacterium ruminatium*. The anaerobic cellulolytic rumen bacterium *Ruminococcus flavefaciens* normally produced succinic acid as a major fermentation product together with acetic and formic acids, H_2, and CO_2. When grown on cellulose and in the presence of *M. ruminatium*, acetate was the major fermentation product, succinate was formed in small amounts, little formate was detected, H_2 did not accumulate, and large amounts of CH_4 were formed. In mixed culture, the methanobacterium utilized the H_2 and possibly the formate produced by the *Ruminococcus* and in so doing stimulated the flow of electrons generated during glycolysis by the *Ruminococcus* toward H_2 formation and away from formation of succinate. Khan et al. (1981) developed a procedure for the production of cellulolytic enzymes and acetic acid in a cellulose-enriched bacterial mixed culture obtained from sewage sludge under anaerobic condition. In this procedure, the culture which converts cellulose to CH_4 and CO_2 was mixed with a synthetic medium and cellulose, and heated to 80°C for 15 min before incubation. The end products formed were acetic acid, propionic acid, CO_2, and traces of ethanol and H_2. Supernatants from 6 to 10-day cultures contained 16 to 36 m*M* acetic acid. The cellulolytic enzymes in the supernatant had the ability to hydrolyze CMC, a microcrystalline cellulose, cellobiose, xylan, and FP to reducing sugars. Maki (1954) studied the microbiology of cellulose decomposition in a municipal sewage plant and found that the maximum rate of cellulose hydrolysis by one isolate from the sewage digester in pure culture was 260 mg cellulose per liter per day. This rate of cellulose degradation was increased to 660 mg cellulose per liter per day when a noncellulolytic *Clostridium* sp. from the digester was grown with it.

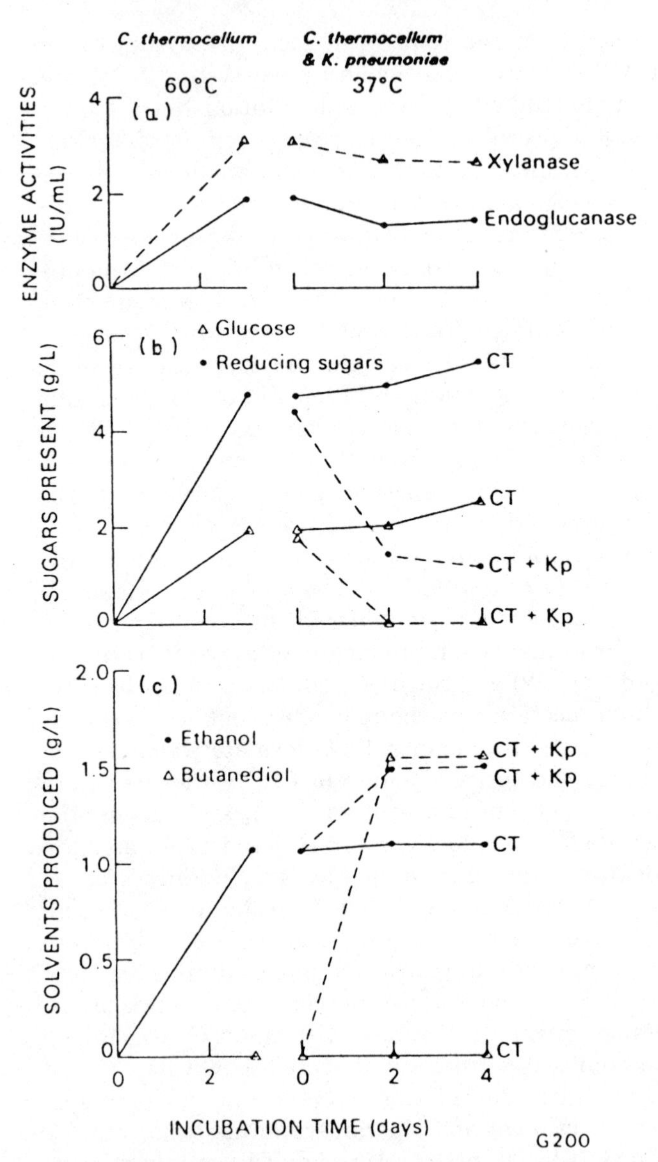

Figure 9.2 Enzyme activities, sugar utilization, and solvent production during sequential coculture of *C. thermocellum* and *K. pneumoniae* grown on Solka-Floc (1%, w/v). CT, monoculture of *C. thermocellum*; CT + Kp, coculture of *C. thermocellum* and *K. pneumoniae*. (*From Yu et al. 1985 Reproduced with permission.*)

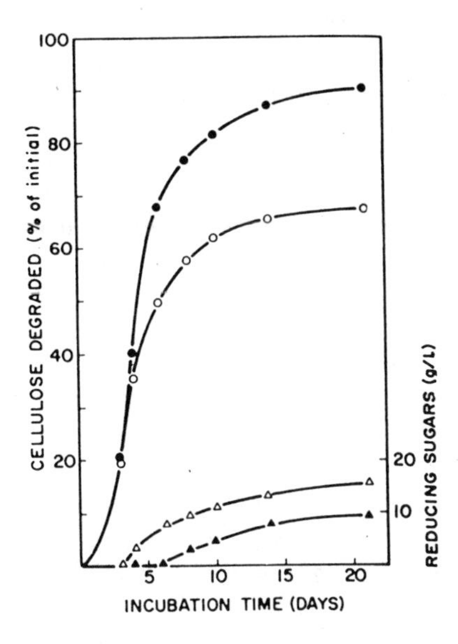

Figure 9.3 Cellulose degradation (○ and ●) and sugar production (△ and ▲), by the *B. cellulosolvens–C. saccharolyticum* coculture (● and ▲) and by *B. cellulosolvens* alone (○, △). (*From Murray, 1986. Reproduced with permission.*)

Stutzenberger et al. (1970) investigated the cellulolytic activity of microorganisms in municipal solid waste composting. The cellulase activity of the compost increased tenfold at a logarithmic rate, while the cellulose content decreased to 50 percent. Three organisms, *A. fumigatus*, a *Bacillus* sp., and a *Thermoactinomyces* sp. of the Actinobifida were identified from the compost. Glansen et al. (1986) developed a two-step biodegradation of lignin by means of a mixed culture of yeasts. They refluxed corn stover in dilute sulfuric acid which resulted in hemicellulose and small amounts of cellulose. This hydrolyzate containing lignin and xylose was used for the growth of a mixed culture of the yeasts *Trichosporon fermentans* and *Pachysolen tannophilus*. *P. tannophilus* fermented xylose to ethanol, which in turn was the primary source of energy for the growth of *T. fermentans*. *T. fermentans* also depolymerized lignin during coculture.

Starchy materials

Starch contains about 15 to 30 percent amylose and 70 to 85 percent amylopectin. Amylose, a linear polymer, consists of long chains of α-D-1,4-linked glucose residues. Amylopectin, a branched polymer, is an α-D-1,4-linked glucan with α-D-1,6-linked branching points. There are three types of amylolytic enzymes produced by microorganisms: endoamylase (α-amylase), exoamylase (β-amylase, glucoamylase), and debranching enzymes (pullulanase, isoamylase). All of these types of amylolytic enzyme activities are essential for the complete breakdown of starch molecules into glucose.

Mixed cultures have been used for the production of SCP, ethanol, 2,3-butanediol, and lactic acid from starchy materials. Yeast biomass was produced in a fermentation process in which *Saccharomycopsis fibuligera* and *Candida utilis* were grown in a mixed batch culture (Jarl 1969, Skogman 1978). *S. fibuligera*, an amylolytic yeast, hydrolyzed starch to simple sugars by means of an enzyme complex consisting of glucoamylase, α-amylase, and maltase (Jarl 1969), and *C. utilis*, a nonamylolytic yeast, used these sugars for its growth. The main interactions between *S. fibuligera* and *C. utilis* were commensalism and competition for glucose. Moreton (1978) examined the growth of *C. utilis* and *S. fibuligera* in mixed batch cultures and observed that amylase production might be limited. Lemmel et al. (1980) studied mixed-culture growth of *C. utilis* and *S. fibuligera* in single-stage continuous culture on blancher water generated during potato processing. Food processing wastewater containing starch can be an excellent feedstock for yeast production. By converting waste starch sugar and other nutrients to yeast cells, valuable microbial biomass is produced as the wastewater strength (biological oxygen demands, BOD) is reduced. It was observed that a single-stage continuous mixed-culture fermentation is not efficient as amylase production appeared to be limiting. Adamassu et al. (Adamassu and Korus 1983, Adamassu et al. 1984) then used a two-stage, continuous, associative fermentation to remove this limitation with a simulated potato processing waste feed. In the first stage, *S. fibuligera* was cultivated in pure culture to produce amylase. The effluent from this stage, which contained cells, spent medium, and amylase, then flowed into the second stage which contained the mixed culture. A supplemental flow of fresh medium was also introduced into this second stage. Maximum biomass production occurred at a second-stage dilution rate, D_2 of 0.27 h^{-1} at a volumetric ratio (V_1/V_2) of 0.57. The principal advantages of the two-stage process are a greater reduction in feedstock carbohydrates and a product which is primarily *C. utilis* with an established food or feed usage. Later, Pasari et al. (1989) developed a mathematical model to describe mixed-culture interactions in continuous, single-stage, and two-stage yeast fermentations. The model incorporates the essential features of enzyme production, starch hydrolysis, substrate consumption, and yeast growth.

Starchy materials are easily converted to ethanol by at least two prefermentation steps—gelatinization of starch by heating under acid conditions (cooking) and subsequent saccharification by amylolytic enzymes (Faust et al. 1983). Attempts were also made to reduce or eliminate the cooking process by using amylase preparations (Ueda and Koba 1980, Saha and Ueda 1983). Direct fermentation of potato starch to ethanol by cocultures of *Aspergillus niger* and *Saccharomyces cerevisiae* was investigated (Abouzeid and Reddy 1986). *A. niger*, an amylolytic fungus, converts starch to glucose and *S. cerevisiae*, a nonamylolytic yeast, efficiently ferments glucose to ethanol. The amylolytic activity, rate, amount of starch utilization, and ethanol yields increased severalfold in coculture compared to monoculture due to synergistic metabolic interactions between the species. It was shown that the amylase activity in *Aspergillus* species tested was not subject to catabolic repression, but was subject to feedback inhibition of glucose (Reddy and Abouzied 1986). This feedback inhibition of amylase activity could be relieved when *Aspergillus* sp. were cocultured in starch-containing medium with a rapid sugar utilizer such as *S. cerevisiae*. Del Rosario and Wong (1984) investigated microbial conversion of dextrinized cassava starch into ethanol using cultures of *A. awamori* and *S. cerevisiae* in a one-stage combined culture and a two-stage sequential process. The batch culture of the combined microorganisms produced 4.3 percent alcohol by weight from 15 percent cassava flour slurry in 39 h. Two-stage continuous fermentation was performed using *A. awamori* in an airlift fermenter and *S. cerevisiae* in a tower fermenter. A residence time of 12.5 h for the first stage resulted in 12.5% sugar concentration and a saccharification efficiency of 88 percent. A residence time of 5 to 6 h for the second stage gave an alcohol concentration of 5.3 percent and a starch-into-ethanol conversion efficiency of 72.5 percent. Amin et al. (1985) studied direct alcoholic fermentation of starchy biomass using amylolytic yeast strains. *S. diastaticus* secretes multiple forms of glucoamylase, but no α-amylase and debranching enzyme activity (Tamaki 1980, Sills et al. 1984). *Schwanniomyces castellii* produces α-amylase and glucoamylase (Sills and Stewart 1982). The conversion efficiency of the batch fermentation with *S. diastaticus* was improved by using a mixed culture with *S. castellii*. From 400 g/L of dextrin, *S. cerevisiae* produced 120 g/L ethanol. With a mixed culture of *S. diastaticus* and *S. castellii*, 136 g/L of ethanol was produced. It is assumed that the *S. diastaticus* amylolytic system was limiting fermentation and this could be alleviated by using α-amylase-producing *S. castellii*. Wilson et al. (1982) described the associative fermentation of starch, using *Schw. alluvius* for its amylolytic activity and *S. uvarum* for ethanol production. Dostalek and Haggstrom (1983) studied the conversion of starch to ethanol in a mixed culture of *S. fibuligera* and an anaerobic bacterium, *Zymomonas mobilis*. It was found that the interactions between the component cultures were commensalism and competition for glucose. No accumulation of reducing sugars was observed in the mixed culture when compared to monoculture of *S. fibuligera* grown on starch. The final concentration of ethanol, 9.7 g/L produced from 30 g/L of starch, shows that all the glucose available from starch hydrolysis was converted to ethanol. The glucose formed was instantly used by *Z. mobilis* for ethanol production and the glucose inhibition hydrolysis of non-glucose-reducing sugars was relieved. Glucose production from starch was the rate-limiting reaction in the system, causing a lower ethanol production rate in mixed culture than in monoculture of *Z. mobilis* grown on glucose. Tanaka et al. (1986) have investigated the production of ethanol from starch by a coimmobilized mixed-culture system of an aerobic fungus *A. awamori* and *Z. mobilis*, in calcium-alginate gel beads under aerobic culture conditions. By controlling the mixing ratio of the microorganism in the inoculum size, a desirable coimmobilized mixed-culture system was achieved in which the aerobic mycelia grew on and near the oxygen-rich surface of the gel beads, while the anaerobic bacteria grew mainly in the oxygen-deficient central part of the gel beads. A schematic diagram of the coimmobilized mixed-culture system of *A. awamori* and *Z. mobilis* in calcium-alginate gel beads is shown in Fig. 9.4. Under optimum conditions in the flask culture, the ethanol produced from 100 g/L starch was 25 g/L and the yield coefficient for ethanol, $Y_{1/s}$, was 0.38. Hyun and Zeikus (1985) reported simultaneous and enhanced production of thermostable amylases and ethanol from starch by cocultures of *C. thermosulfurogenes* and *C. thermohydrosulfuricum*. Both species produced ethanol and amylases with different components as primary metabolites of starch fermentation. Under given monoculture fermentation conditions, neither species completely degraded starch during the time course of study, whereas in coculture starch was completely degraded. In monoculture starch fermentation, *C. thermohydrosulfuricum* produced lower levels of pullulanase and glucogenic amylase, whereas *C. thermosulfurogenes* produced lower levels of β-amylase and glucogenic amylase. In coculture fermentation, improve-

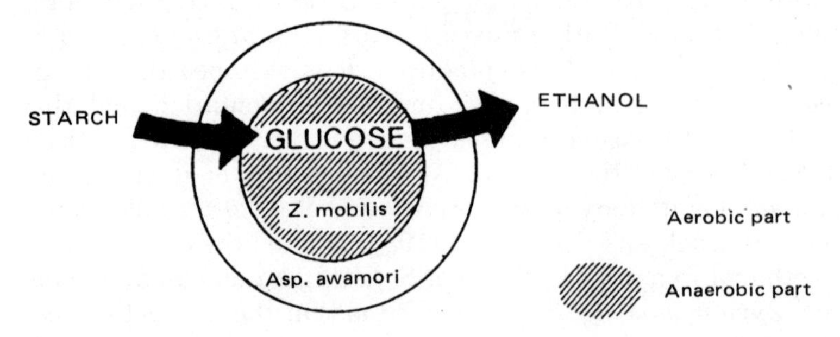

Figure 9.4 Schematic diagram of the coimmobilized mixed-culture system of *Aspergillus awamori* and *Zymomonas mobilis* in Ca-alginate gel beads. (*From Tanaka et al. 1986. Reproduced with permission.*)

ment of starch metabolism by each species was noted in terms of increased amounts and rates of starch consumption, amylase production, and ethanol formation (Table 9.6). The single-step coculture fermentation completely degraded 2.5% starch in 30 h at 60°C and produced 9 U of β-amylase per milliliter, 1.3 U of pullulanase per milliliter, 0.3 U of glucogenic amylase per milliliter, and >120 mM ethanol with a yield of 1.7 mol/mol of glucose in starch. It was concluded that the improvement in coculture fermentation was due to coordinate action of amylolytic enzymes and synergistic metabolic interactions between the species.

Kurosawa et al. (1988) investigated the production of lactic acid from starch by the coimmobilized mixed-culture system using *A. awamori* and *Streptomyces lactis* as anaerobic bacterium. The scheme for correlation of metabolism between *A. awamori* and *S. lactis* in a coimmobilized mixed-culture system is shown in Fig. 9.5. The final concentration of lactic acid produced from 50 g/L starch was 25 g/L and the yield coefficient for lactic acid, $Y_{P/S}$, was 0.66. The authors concluded that this aerobic-anaerobic coimmobilized mixed-culture system has great potential and can be widely used.

Suprinto et al. (1989) investigated the effect of mixed culture of *Rhizopus* sp., *Saccharomycopsis* sp. and *Streptococcus* sp. on the liquefaction of glutinous rice and aroma formation in *tape* preparation. It was found that steamed glutinous rice was liquefied and saccharified by the amylase produced by *Rhizopus* sp. and aroma was formed by the mixed

TABLE 9.6 Comparison of Starch Metabolism Parameters in Monoculture and Coculture Fermentations of *C. thermohydrosulfuricum* 39E and *C. thermosulfurogenes* 4B[a]

Condition[b]	Culture strain	Growth (OD_{660})[c]	pH	Starch remaining, g/L	Reducing sugar remaining, g/L
Starch (0.65%)	39E	1.9	6.1	0.5	0.3
	4B	1.3	4.3	2.3	0.2
	39E and 4B	1.5	4.7	0.0	0.0
Starch (3.0%)	39E	2.7	6.0	22	8.7
	4B	2.7	4.3	22	8.6
	39E and 4B	3.8	4.7	19	14.0

[a]Errors in measurement of reported values were less than 5%.

[b]Conditions are as follows: 26-ml pressure tubes contained a complex medium with trypticase (1% BB, Microbiology Systems) and yeast extract (0.3%) and a $N_2 - CO_2$ head space and were incubated for 36 h at 60°C without shaking.

[c]OD_{660}, optical density at 660 nm

SOURCE: From Hyun and Zeikus, 1985. Reproduced with permission.

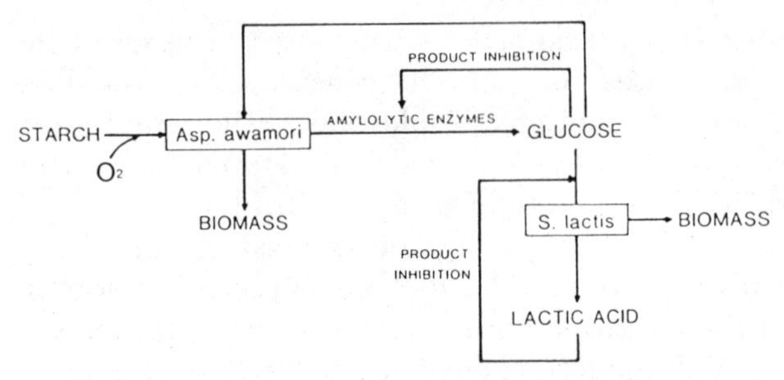

Figure 9.5 Scheme for correlation of metabolism between *A. awamori* and *S. lactis* in coimmobilized mixed-culture system. *(From Kurosawa et al. 1988. Reproduced with permission.)*

culture of *Rhizopus* and *Saccharomycopsis* sp. It is interesting that liquefaction was not caused by the amylases of *Saccharomycopsis*, even though it produced high activity of α-amylase. A high level of aroma was formed by inoculation of *Streptococcus* in a mixed culture of *Rhizopus* and *Saccharomycopsis* sp.

Other polysaccharides

Chitin, a major structural polymer in invertebrates, is a polysaccharide composed of β-1,4-linked monomers of N-acetyl-D-glucosamine (NAG). Billions of tons of chitin are produced annually in the oceans (Goodrich and Morita, 1977). Chitin is an abundant byproduct of the fishing (crab, shrimp, and prawn) and fermentation industries.

Boyer (1986) examined the anaerobic pathway of chitin decomposition by chitinoclastic bacteria with an emphasis on end-product coupling to other salt marsh bacteria. Mixed-culture experiments with various combinations of chitinoclasts, sulfate reducers, and methanogens were performed. The chitinoclastic bacteria could degrade chitin and produce CO_2, H_2, and acetate. Some chitinoclastic bacteria could also produce a series of volatile fatty acids including propionate and butyrate. Neither sulfate reducers nor methanogens could grow on chitin as sole carbon source or produce any measurable degradation products. Coculture experiments of chitinoclasts and sulfate reducers showed two distinct pathways of sulfate reduction. One was the acetoclastic route, by which acetate was oxidized to CO_2 and sulfide was produced. The other one was a H_2-utilizing route, by which sulfate was reduced directly. Mixed cultures of chitinoclasts and methanogens produced CH_4, with decreases in H_2 and CO_2, although, these methanogens could not ferment acetate to CH_4. The combination of chitinoclasts, methanogens, and sulfate reducers produced methane and sulfide simultaneously. Methane production was increased in mixed cultures containing CO_2-reducing methanogens and acetoclastic sulfate reducers because of less interspecific H_2 consumption (Tanaka et al. 1986). Pel et al. (1989) also studied the degradation of chitin by pure and cocultures of the chitinolytic *Clostridium* strain 9.1 and various nonchitinolytic sugar-fermenting bacteria, facultative anaerobe strain HA 8.1, GH 8.2; *Klebsiella aerogenes* ATCC 15380; *C. acetobutylicum* ATCC 824; a wild type *Escherichia coli*; and sulfate-reducing bacteria (*Desulfovibrio* sp. strain HL21, *Desulfovibrio desulfuricans* ATCC 27774, *Desulfovibrio* sp. strain 20028). Chitin degradation by the mixed cultures was enhanced five to eight fold (Fig. 9.6). This enhancement was not due to alleviation of inhibition of chitinolytic enzyme system by the chitin hydrolysis products such as NAG and NAG-oligomers, but due to the release of growth factors essential to chitinolytic *Clostridium* strain 9.1 by these secondary bacterial populations. Pel and Gottschal (1989) demonstrated that chitinolysis in pure cultures of *Clostridium* strain 9.1 was accelerated by the addition of thioredoxin. Thioredoxin is a thermostable dithiol enzyme capable of reducing disulfide proteins (Holmgren 1981). The extracellular fluid of cocultures of chitinolytic *Clostridium* strain 9.1 and nonchitinolytic saccharolytic bacteria contained a thermostable factor of high molecular weight also capable of stimulating chitinolysis. It is suggested that the secondary species released the compound(s) involved which is identical or closely related to

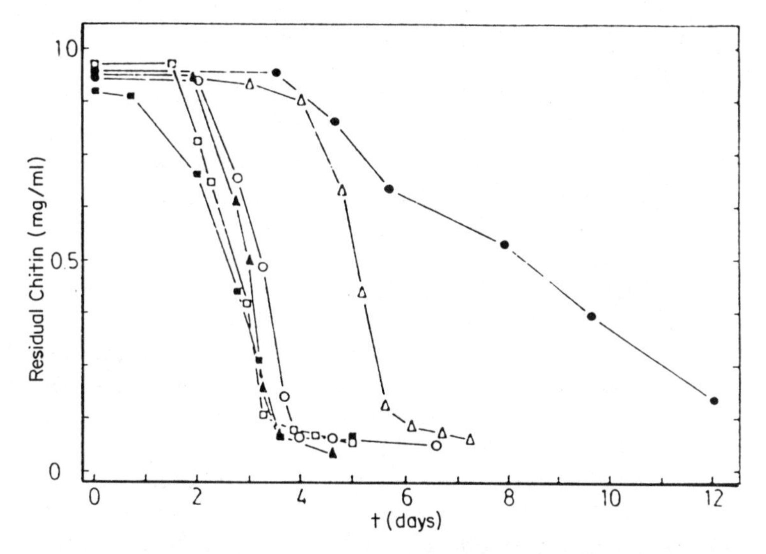

Figure 9.6 Time course of chitinolysis in pure culture of *Clostridium* sp. strain 9.1 (●–●) and in coculture with HA 8.1 (○–○), strain GH 8.2 (■–■), *Klebsiella aerogenes* (□–□), *Clostridium acetobutyricum* (▲–▲) and *Escherichia coli* (△–△) *(From Pel et al. 1989. Reproduced with permission.)*

thioredoxin (Pel and Gottschall 1989). In addition to the probable involvement of thioredoxin in the enhancement of chitinolysis, Pel et al. (1989) found that there was an additional requirement of the organism for a low-molecular-weight fraction of yeast extracts. Apparently these growth factors were also released by the saccharolytic bacteria because the chitinolysis in cocultures without yeast extract could proceed at an appreciable rate.

In mixed-culture fermentation of chitin by *Clostridium* sp. strain 9.1 and sulfate-reducing bacteria such as *Desulfovibrio* sp. strain HL 21 or *Desulfovibrio desulfuricans*, the fermentation of chitin was about sixfold more rapid than in pure cultures of strain 9.1 (Pel et al. 1989). Sulfite was also formed in these cocultures. This indicates that the sulfate-reducing species were metabolically active; however, both of the sulfate reducers were unable to use NAG or NAG oligomers. Pel and Gottschal (1989) concluded that the stimulatory interaction is of a very general nature because it was observed with so many different, unrelated bacteria in anaerobic mixed cultures. Pel and Gottschal (1989) hypothesized that essential sulfhydryl groups in the chitinolytic system of strain 9.1 are reduced by thioredoxin and similar thiol disulfide transhydrogenases present in the cell-free extracts and spent media, resulting in an acceleration of chitin hydrolysis and fermentation.

Pectin is a complex carbohydrate containing pectic acid (polygalacturonic acid) esterified with methyl alcohol. Pectin-degrading enzymes produced by a variety of microorganisms include both esterases and depolymerases.

Fellows and Worgan (1984a,b) demonstrated that *Saccharomycopsis fibuligera* could partially degrade pectin and *Candida utilis* could grow on D-galacturonic acid (Fellows and Worgan 1987). They studied the growth of *S. fibuligera* and *C. utilis* in mixed culture on pectic materials (Fellows and Worgan 1987). It was found that pectin hydrolysis by *S. fibuligera* and growth of *C. utilis* on pectin hydrolyzate could proceed well in a three-stage sequential culture of the two yeasts. *S. fibuligera* was first grown under aerobic conditions to generate cell mass. The concentration of dissolved oxygen was then reduced to promote pectolytic activity and reduce the number of viable cells in the culture. Finally, culture conditions were adjusted to promote the growth of *C. utilis* in mixed culture with *S. fibuligera*. The rate of pectolytic activity by *S. fibuligera* was increased by the presence of *C. utilis* (Table 9.7). It was assumed that *C. utilis* assimilated the products of pectin hydrolysis by *S. fibuligera* which would otherwise have inhibited the pectolytic activity. In anaerobic ecosystems, pectin biogradation enables synergistic metabolic communication between hydrolytic and methylotropic bacterial species. The mixed culture of

TABLE 9.7 Effect of Aeration and the Presence of *Candida utilis* on the pectolytic activity of *Saccharomycopsis fibuligera*.

Aeration	Reduction in apparent viscosity, %/h	
	S. fibuligera monoculture	*S. fibuligera* + *C. utilis*
Continuous	0.42	0.56
Intermittent	0.69	0.83

SOURCE: From Fellows and Worgan, 1987a. Reproduced with permission.

C. butyricum and *Methanosarcina barkeri* degraded pectin completely with only H_2, methanol, and acetate being formed as intermediary metabolites, whereas the pectin was partially degraded by *C. butyricum* in pure culture (Schink and Zeikus 1982).

Conclusion

Although our understanding of the enzymatic degradation of polysaccharides by mixed-culture systems has advanced, much remains to be done to make mixed-culture degradation of polysaccharides attractive and economically feasible in biotechnology. Multiple control mechanisms of enzyme production are involved in polysaccharide degradation by mixed cultures. There are definitely certain merits of mixed cultures, however, there are certain difficulties to establish suitable conditions for mixed cultures of two strains, because the strains do not always have the same optimum culture conditions for pH, temperature, nutrients, oxygen demand, stability, etc. Research on design, control, and optimization of the mixed-culture system will be necessary for effective degradation of polysaccharides. Mixed cultures in enzymatic degradation of polysaccharides will find widespread applications in sewage and industrial waste utilization.

References

Abouzeid, M. M., and C. A. Reddy. 1986. Direct fermentation of potato starch to ethanol by cocultures of *Aspergillus niger* and *Saccharomyces cerevisiae*. *Appl. Environ. Microbiol.* 52:1055–1059.

Adamassu, W., and R. A. Korus. 1983. Two-stage continuous fermentation of *Saccharomycopsis fibuligera* and *Candida utilis*. *Biotechnol. Bioeng.* 25:2641–2651.

Adamassu, W., R. A. Korus, and R. C. Heimsch. 1984. Continuous fermentation of *Saccharomycopsis fibuligera* and *Candida utilis*. *Biotechnol. Bioeng.* 26:1511–1513.

Amin, G., R. De Mot, K. Van Dijck, and H. Verachtert. 1985. Direct alcoholic fermentation of starchy biomass using amylolytic yeast strains in batch and immobilized cell systems. *Appl. Microbiol. Biotechnol.* 22:237–245.

Avgerinos, G. C., and Wang, D. I. C. 1980. Direct microbiological conversion of cellulosic to ethanol. *Ann. Rep. Ferm. Proc.* 4:165–191.

Bisset, F., and D. Sternberg. 1978. Immobilization of *Aspergillus* β-glucosidase on chitosan. *Appl. Environ. Microbiol.* 35:750–755.

Boyer, J. N. Boyer. 1986. End products of anaerobic chitin degradation by salt march bacteria as substrates for dissimilatory sulfate reduction and methanogenesis. *Appl. Environ. Microbiol.* 52:1415–1418.

Callihan, C. D., and C. E. Dunlap. 1971. Construction of a chemical-microbial pilot plant for production of single cell protein from cellulosic wastes. Report PB 203620, National Technical Information Service, U.S. Department of Commerce.

Carreira, L. H., and L. G. Ljungdahl. 1984. Production of ethanol from biomass using anaerobic thermophilic bacteria. pp. 1–29 in D. L. Wise (ed.), *Liquid Fuel Developments*, CRC Press, Boca Raton, Florida.

Cohen, B. L. 1977. Genetics and physiology of *Aspergillus*. pp. 282–292. in J. E. Smith and J. A. Pateman (eds.), Academic Press, London.

del Rosario, E. J., and R. L. Wong. 1984. Conversion of dextrinized cassava starch into ethanol using cultures of *Aspergillus awamori* and *Saccharomyces cerevisiae*. *Enzyme Microb. Technol.* 6:60–64.

de La Torre, M., and C. C. Campillo. 1984. Isolation and characterization of a symbiotic cellulolytic mixed bacterial culture. *Appl. Microbiol. Biotechnol.* 19:430–434.

Dostalek, M., and M. H. Haggstrom. 1983. Mixed culture of *Saccharomycopsis fibuligera* and *Zymomonas mobilis* on starch—use of oxygen as a regulator. *Eur. J. Appl. Microbiol. Biotechnol.* 17:269–274.

Duff, S. J. B., D. G. Cooper, and O. M. Fuller. 1985. Cellulase and β-glucosidase production by mixed culture of *Trichoderma reesei* RUT C30 and *Aspergillus phoenicis*. *Biotech. Lett.* 7:185–190.

Duff, S. J. B., D. G. Cooper, and O. M. Fuller. 1986. Evaluation of the hydrolytic potential of a crude cellulase from mixed cultivation of *Trichoderma reesei* and *Aspergillus phoenicis*. *Enzyme Microb. Technol.* 8:305–308.

Duff, S. J. B., D. G. Cooper, and O. M. Fuller. 1987. Effect of media composition and growth conditions on production of cellulase and β-glucosidase by a mixed fungal fermentation. *Enzyme Microb. Technol.* 9:47–52.

Faust, U., P. Prave, and M. Schlingmann. 1983. An integral approach to power alcohol. *Process Biochem.* 18:31–37.

Fellows, P. J., and J. T. Worgan. 1984a. Studies on the growth of *Saccharomycopsis fibuligera* on pectic materials. *Enzyme Microb. Technol.* 6:350–354.

Fellows, P. J., and J. T. Worgan. 1984b. An investigation on the pectolytic activity of the yeast *Saccharomycopsis fibuligera*. *Enzyme Microb. Technol.* 6:405–410.

Fellows, P. J., and J. T. Worgan. 1987a. Growth of *Saccharomycopsis fibuligera* and *Candida utilis* in mixed culture on pectic materials. *Enzyme Microb. Technol.* 9:430–433.

Fellows, P. J., and J. T. Worgan. 1987b. Growth of *Saccharomycopsis fibuligera* and *Candida utilis* in mixed culture on apple processing wastes. *Enzyme Microb. Technol.* 9:434–437.

Freer, S. N., and R. E. Wing. 1985. Fermentation of cellodextrins to ethanol using mixed culture fermentations. *Biotechnol. Bioeng.* 27:1085–1088.

Friedrich, J., A. Cimerman, and A. Perdih. 1987. Mixed culture of *Aspergillus awamori* and *Trichoderma reesei* for bioconversion of apple distillery waste. *Appl. Microbiol. Biotechnol.* 26:299–303.

Ghose, T. K., and V. S. Bisaria. 1979. Studies on the mechanism of enzymatic hydrolysis of cellulosic substances. *Biotechnol. Bioeng.* 21:131–146.

Ghose, T. K., T. Panda, and V. S. Visaria. 1985. Effect of culture phasing and mannanase on production of cellulase and hemicellulase by mixed culture of *Trichoderma reesei* D1-6 and *Aspergillus wentii* Pt 2804. *Biotechnol. Bioeng.* 27:1353–1361.

Glanser, M., L. Dvoracek, and S. Ban. 1986. Two-step biodegradation of lignin by means of a mixed culture of yeasts. *Proc. Biochem.* 21:169–173.

Goodrich, T. D., and R. Y. Morita. 1977. Incidence and estimation of chitinase activity associated with marine fish and other estuarine samples. *Mar. Biol.* 41:349–353.

Gum, E. K., and R. D. Brown. 1976. Structural characterization of a glycoprotein cellulase, 1,4-β-D-glucan cellobiohydrolase from *Trichoderma viride*. *Biochem. Biophys. Acta* 446:371–386.

Hagerdal, B. H., and M. Haggstrom. 1985. Production of ethanol from cellulose, Solka

Folc BW 200, in a fedbatch mixed culture of *Trichoderma reesei*, C30 and *Saccharomyces cerevisiae*. *Appl. Microbiol. Biotechnol.* 22:187–189.
Halsall, D. M., and A. H. Gibson. 1985. Cellulose decomposition and associated nitrogen fixation by mixed cultures of *Cellulomonas gelida* and *Azospirillum* species or *Bacillus macerans*. *Appl. Environ. Microbiol.* 50:1021–1026.
Halsall, D. M., and A. H. Gibson. 1986. Comparison of two strains of *Cellulomonas* spp. and their interaction with *Azospirillum brasilense* in degradation of wheat straw and associated nitrogen fixation. *Appl. Environ. Microbiol.* 51:855–861.
Halsall, D. M., and D. J. Goodchild. 1986. Nitrogen fixation associated with development and localization of mixed populations of *Cellulomonas* sp. and *Azospirillum brasilence* grown on cellulose or wheatstraw. *Appl. Environ. Microbiol.* 51:849–854.
Han, Y. W. 1982. Nutritional requirements and growth of *Cellulomonas* species on cellulosic substrates. *J. Ferment. Technol.* 60:99–104.
Holmgren, A. 1985. Thioredoxin. *Ann. Rev. Biochem.* 54:237–271.
Hyun, H. H., and J. G. Zeikus. 1985. Simultaneous and enhanced production of thermostable amylases and ethanol from starch by cocultures of *Clostridium thermosulfurogenes* and *Clostridium thermohydrosulfuricum*. *Appl. Environ. Microbiol.* 49:1174–1181.
Jarl, K. 1969. Production of microbial food from low cost starch materials and purification of industry's waste starch effluents through the Symba yeast process. *Food Technol.* 23:1009–1012.
John Wase, D. A., S. Raymahasay, and S. Green. 1986. Inhibition of β-D-glucosidase and endo-1,4-β-D-glucanase produced by *Aspergillus fumigatus* IMI 255091. *Enzyme Microb. Technol.* 8:48–51.
Khan, A. W., D. Wall, and L. van den Berg. 1981. Fermentative conversion of cellulose to acetic acid by a mixed culture obtained from sewage sludge. *Appl. Environ. Microbiol.* 41:1214–1218.
Kristensen, T. D. 1978. Continuous single cell production from *Cellulomonas* sp. and *Candida utilis* grown in mixture on barley straw. *Eur. J. Appl. Microbiol. Biotechnol.* 5:155–163.
Kurosawa, H., H. Ishikawa, and H. Tanaka. 1988. L-Lactic acid production from starch by coimmobilized mixed-culture system of *Aspergillus awamori* and *Streptococcus lactis*. Biotechnol. Bioeng. 31:183–187.
Lamot, E., and E. J. Vadamme. 1982. Improved cellulolytic activity of actinomycetes in axenic mixed fermentations. *J. Chem. Technol. Biotechnol.* 32:735–743.
Latham, M. J., and M. J. Wolin. 1977. Fermentation of cellulose by *Ruminococcus flavefaciens* in the presence and absence of *Methanobacterium ruminantium*. *Appl. Environ. Microbiol.* 34:297–301.
Lemmel, S. A., R. C. Heimsch, and L. L. Edwards. 1979. Optimizing the continuous production of *Candida utilis* and *Saccharomycopsis fibuligera* on potato processing wastewater. *Appl. Environ. Microbiol.* 37:227–232.
Lemmel, S. A., R. C. Heimsch, and R. A. Korus. 1980. Kinetics of growth and amylase production of *Saccharomycopsis fibuligera* on potato processing wastewater. *Appl. Environ. Microbiol.* 39:387–393.
Le Ruyet, P., H. C. Dubourguier, and G. Albagnac. 1984. Homoacetogenic fermentation of cellulose by a coculture of *Clostridium thermocellum* and *Acetogenium kivui*. *Appl. Environ. Microbiol.* 48:893–894.
Ljungdahl, L. G., L. Carreira, and J. Wiegel. 1981. Production of ethanol from carbohydrates using anaerobic thermophilic bacteria. The Ekman-Days 1981, Int. Symp. Wood and Pulping Chem. Stockholm, 4:23–28.
Lovitt, R. W., B. H. Kim, G.-J. Shen, and J. G. Zeikus. 1988. Solvent production by microorganisms. *CRC Crit. Rev. Biotechnol.* 7:107–186.
Lynch, J. M., and S. H. T. Harper. 1983. Straw as a substrate for cooperative nitrogen fixation. *J. Gen. Microbiol.* 129:251–253.
Maki, L. R. 1954. Experiments on the microbiology of cellulose decomposition in a municipal sewage plant. *Antonio van Leeuwenhock* 20:185–200.
Moreton, R. S. 1978. Growth of *Candida utilis* on enzymatically hydrolyzed potato waste. *J. Appl. Bacteriol.* 44:373–382.
Murray, W. D. 1986. Symbiotic relationship of *Bacteroides cellulosolvens* and

Clostridium saccharolyticum in cellulose fermentation. *Appl. Environ. Microbiol.* 51: 710–714.

Murray, W. D., A. W. Khan, and L. van den Berg. 1982. *Clostridium saccharolyticum* sp. nov., a saccharolytic species from sewage sludge. *Int. J. Syst. Bacteriol.* 32:132–135.

Ng, T. K., A. Ben-Bassat, and J. G. Zeikus. 1981. Ethanol production by thermophilic bacteria: fermentation of cellulosic substrates by cocultures of *Clostridium thermocellum* and *Clostridium thermohydrosulfuricum*. *Appl. Environ. Microbiol.* 41: 1337–1343.

Nigam, P., A. Pandey, and K. A. Prabhu. 1987. Mixed-culture fermentation for bioconversion of whole bagasse into microbial protein. *J. Basic Microbiol.* 27:323–327.

Panda, T., V. S. Visaria, and T. K. Ghose. 1983. Studies on mixed fungal culture for cellulase and hemicellulase production. Part 1: Optimization of medium for the mixed culture of *Trichioderma reesei* D1-6 and *Aspergillus wentii* Pt 2804. *Biotech. Lett.* 5:767–772.

Panda, T., V. S. Visaria, and T. K. Ghose. 1989. Method to estimate growth of *Trichoderma reesei* and *Aspergillus wentii* in mixed culture on cellulosic substrates. *Appl. Environ. Microbiol.* 55:1044–1046.

Pasari, A. B., R. A. Korus, and R. C. Heimsch. 1989. A model for continuous fermentations with amylolytic yeasts. *Biotechnol. Bioeng.* 33:338–343.

Peitersen, N. 1975. Cellulase and protein production from mixed cultures of *Trichoderma viride* and a yeast. *Biotechnol. Bioeng.* 17:1291–1299.

Pel, R., and J. C. Gottschal. 1989. Interspecies interaction based on transfer of a thioredoxin-like compound in anaerobic chitin-degrading mixed cultures. *FEMS Microbiol. Ecol.* 62:349–358.

Pel, R., G. Hessels, H. Aalfs, and J. C. Gottschal. 1989. Chitin degradation by *Clostridium* sp. strain 9.1 in mixed cultures with saccharolytic and sulfur-reducing bacteria. *FEMS Microbiol. Ecol.* 62:191–200.

Reddy, C. A., and M. M. Abouzied. 1986. Glucose feedback inhibition of amylase activity in *Aspergillus* sp. and release of this inhibition when cocultured with *Saccharomyces cerevisiae*. *Enzyme Microb. Technol.* 8:659–664.

Ryu, D. D. Y., and M. Mandels. 1980. Cellulases: biosynthesis and applications. *Enzyme Microb. Technol.* 2:91–102.

Saddler, J. N., and M. K.-H. Chan. 1984. Conversion of pretreated lignocellulosic substrates to ethanol by *Clostridium thermocellum* in mono- and co-culture with *Clostridium thermosaccharolyticum* and *Clostridium thermohydrosulfuricum*. *Can. J. Microbiol.* 30:212–220.

Saha, B. C., and S. Ueda. 1983. Alcoholic fermentation of raw sweet potato by a nonconventional method using *Endomycopsis fibuligera* glucoamylase preparation. *Biotechnol. Bioeng.* 25:1181–1186.

Schink, B., and J. G. Zeikus. 1982. Microbial ecology of pectin decomposition in anoxic lake sediments. *J. Gen. Microbiol.* 128:393–404.

Sills, A. M., and G. G. Stewart. 1982. Production of amylolytic enzymes by several yeast species. *J. Inst. Brew.* 88:313–316.

Sills, A. M., M. E. Sauder, and G. G. Stewart. 1984. Isolation and characterization of the amylolytic system of *Schwanniomyces castellii*. J. Inst. Brew. 90:311–314.

Skogman, H. 1978. Production of Symba yeast from potato wastes. pp. 167–179 in G. C. Birch, K. J. Parker, and J. T. Worgen (eds.), *Food from Waste*, Applied Science Publishers, London.

Soundar, S., and T. S. Chandra. 1987. Cellulose degradation by a mixed bacterial culture. *J. Ind. Microbiol.* 2:257–265.

Srivastava, S. K., K. B. Ramachandran, and K. S. Gopalkrishnan. 1981. Beta-glucosidase production by *Aspergillus wentii* in stirred tank bioreactors. *Biotech. Lett.* 3:477–480.

Sternberg, D., P. Vijay Kumar, and E. T. Reese. 1977. Beta-glucosidase: microbial production and effect on enzymatic hydrolysis of cellulose. *Can. J. Microbiol.* 23:139–147.

Stutzenberger, F. J., A. J. Kaufman, and R. D. Lossin. 1970. Cellulolytic activity in municipal solid waste composting. *Can. J. Microbiol.* 16:553–560.

Suprinto, R. Ohba, T. Koga, and S. Ueda. 1989. Liquefaction of glutinous rice and aroma formation in Tape preparation by Ragi. *J. Ferment. Technol.* 67:249–252.

Takagi, M., S. Abe, S. Suzuki, G. H. Emert, and N. Yata. 1977. A method for production of alcohol directly from cellulose using cellulase and yeast. pp. 551–571 in *Proc. Bioconversion. Symp., IIT*, Delhi.

Tamaki, H. 1980. Purification of glucoamylase isoenzymes produced by *Saccharomyces diastaticus. Doshisha Joshi Daigaku Gakujitsu Kenkyo Nenpo* 31:270–286.

Tanaka, H., H. Kurosawa, and H. Murakami. 1986. Ethanol production from starch by a coimmobilized mixed culture system of *Aspergillus awamori* and *Zymomonas mobilis. Biotechnol. Bioeng.* 28:1761–1768.

Trivedi, S. M., and J. D. Desai. 1984. Cellulases and β-glucosidase production by mix-shake cultivation of *Scytalidium lignicola* and *Trichoderma longibrachiatum. J. Ferment. Technol.* 62:211–215.

Ueda, S., and Y. Koba. 1980. Alcoholic fermentation of raw starch without cooking by using black-koji amylase. *J. Ferment. Technol.* 58:237–242.

US Department of Energy Alcohol Fuels Program Review. 1984. The Solar Energy Research Institute, pp. 29–30. Golden, Colorado.

Veal, D. A., and J. M. Lynch. 1984. Associative cellulolysis and dinitrogen fixation by cocultures of *Trichoderma harzianum* and *Clostridium butyricum. Nature* (London) 310:694–697.

Volfova, O., O. Suchardova, J. Panos, and V. Krumphanzl. 1985. Ethanol formation from cellulose by thermophilic bacteria. *Appl. Microbiol. Biotechnol.* 22:246–248.

Wayman, M., R. S. Parekh, and S. R. Parekh. 1987. Simultaneous saccharification and fermentation by mixed cultures of *Brettanomyces clausenii* and *Pichia stipitis* R of SO_2-prehydrolyzed wood. *Biotech. Lett.* 9:435–440.

Weimer, P. J., and J. G. Zeikus. 1977. Fermentation of cellulose and cellobiose by *Clostridium thermocellum* in the absence and presence of *Methanobacterium thermoautotrophicum. Appl. Environ. Microbiol.* 33:289–297.

Wilson, J. J., G. G. Khachatourians, and W. M. Ingledew. 1982. *Schwanniomyces*: SCP and ethanol from starch. *Biotech. Lett.* 4:333–338.

Yu, E. K. C., M. K.-H. Chan, and J. N. Saddler. 1985. Butanediol production from lignocellulosic substrates by *Klebsiella pneumoniae* grown in sequential coculture with *Clostridium thermocellum. Appl. Microbiol. Biotechnol.* 22:399–404.

Chapter

10

Mixed-Culture Interactions in Methanogenesis

Jürgen H. Thiele

Gesellschaft für Biotechnologische Forschung mbH, D3300 Braunschweig Federal Republic of Germany

Introduction

Methanogenesis is one of the most important anaerobic biodegradation processes on earth. Large quantities of methane are formed annually by complex microbial consortia in anoxic aquatic environments such as marsh areas, paddy fields, and peat lands (Conrad et al. 1987a). Methane is also produced to a smaller extent in the intestinal tracts of animals such as ruminants or termites (Breznak 1982, Hungate 1966). Methanogenic bacteria can live as endo- and episymbiontic bacteria in sediment protozoa and rumen ciliates (Vogels et al. 1980, van Bruggen et al. 1983). Biomethanation, the technical use of methanogenic ecosystems, has gained increasing importance as a cost-effective and environmentally preferred waste-treatment technology.[1] As all natural and industrial methanogenic processes are effected by joint metabolic reactions of similar mixed consortia of anaerobic bacteria, it is important to study their mutual microbial interactions in order to gain better control and stability in anaerobic digestors and in environmental natural processes. The study of anaerobic and especially methanogenic bacteria has focused

---

[1]Gosh and Klass 1978, McCarty 1982, Sahm 1984, Lettinga et al. 1984, Lettinga et al. 1985, Pohland and Harper 1985, Harper and Pohland 1986.

on the ultrastructure, physiology, and genetics of purified strains. Many excellent reviews are available.[2] Much less is known about the microbial ecology–mixed-culture biology of methanogenesis. Many review articles exist about selected aspects of the ecophysiology of methanogenic reactions and interactions in nature,[3] but a comprehensive understanding of the microbial interactions during methanogenesis in nature and digestors is still lacking. This is understandable, because the variety of microbial species involved in each methanogenic habitat and the variety of known environmental methanogenic habitats are large. Cryophilic, mesophilic, thermophilic, freshwater, marine, and hypersaline methanogenic ecosystems are known with a large variety of important microbial partnerships and interactions. It is beyond the scope of this chapter to provide an encyclopedic review of the existing literature about the different interactions during methanogenesis in the various natural and technical habitats. Instead, we will stress selected important aspects of the complex microbial mixed-culture interactions that occur during biomethanation in those habitats that have been extensively studied, i.e., freshwater sediments and mesophilic anaerobic digestors. Emphasis will be put on microbial interactions that enhance or lessen biomethanation rates and stabilities. These topics are most important for practical industrial applications of biomethanation ecosystems.

Definition of Microbial Interactions

Microbial interaction prototypes in anaerobic ecosystems are conveniently classified into mutualism, commensalism, amensalism, competition, and prey-predator relationships. All five interactions are part of the complex interplay that regulates the carbon and electron flow and the microbial population dynamics in different anaerobic ecosystems. Each interaction is, in a strict sense, only defined for two-member mixtures, and interactions in multispecies microbial mixtures may be described by combinations of the different prototypes.

Mutualism defines interactions, where both members of the mixture derive some advantage from each other's presence in terms of increased growth rates or increased population sizes.

Commensalism is the situation, where only one member of a community benefits from the presence of the second species, which itself

[2]Zeikus 1977, Balch et al. 1979, Daniels et al. 1984, Thauer and Morris 1984, Zeikus et al. 1985, Jones et al., 1987, Dolfing 1988, Widdel 1988.

[3]Hungate 1966, Bryant and Wolin 1975, Mah 1982, Wolin 1982, Zeikus 1983, Ward and Winfrey 1985, Grotenhuis et al. 1986, Widdel 1988, Dolfing 1988, Thiele and Zeikus 1988d, Oremland 1989.

does not derive any advantage or disadvantage from the presence of the first species.

Amensalism is an interaction, where growth of one population is restricted by the presence of a second, which itself is unaffected by the metabolism of the inhibited population.

Competition defines the situation, where growth rate and final population size of both populations are limited by a common dependence on an external growth factor.

Prey-predator relationships describe situations where one member of the mixed culture, the predator, gains directly on the expense of living biomass of the second member in the culture, the prey.

Geochemical Parameters

Anoxic conditions in microbial habitats are predominantly caused by rapid oxygen consumption through aerobic and facultative anaerobic microorganisms at the habitat surface. This results in a diffusion-limited oxygen transport from the atmosphere to the interior of the respective microbial habitat, and the formation of anaerobic ecological niches. The niches are found wherever microbial carbon and energy sources are available in excess and efficient environmental mixing is hindered by physical barriers. Examples for such barriers are

- Stratified limnic ecosystems
- Stratified freshwater and marine sediments
- Capillary solute transport in soils or plant tissues
- Formation of microbial biofilms, aggregates, and dental plaques
- Partial compartmentalization due to the formation of intestinal organs as part of the metazoan gut ecosystem.

Anaerobic microorganisms can colonize these anoxic habitats owing to their possession of (1) various fermentation pathways which allow substrate-level phosphorylation during the disproportionation of organic matter into more oxidized and more reduced carbon compounds (see, for example, Table 10.2) and (2) anaerobic oxidation pathways of the reduced fermentation products with inorganic terminal electron acceptors other than oxygen (Zehnder, 1988). These reactions cover a range of redox potentials from approximately −300 to +400 mV and are presented in ascending sequence of standard redox potentials:

- Reduction of CO_2 to acetate
- Reduction of CO_2 to methane

- Reduction of sulfate and oxidized sulfur compounds to sulfide
- Microbial Fe(III) reduction
- Fumarate reduction to succinate
- Reduction of nitrate/nitrite to ammonia and dinitrogen

The interplay of these different anaerobic oxidation reactions is currently a subject of intensive studies. Depending on its geochemical availability, an electron acceptor with a more positive redox potential can exclude the simultaneous reaction of anaerobic oxidations with lower redox potentials. This is partly due to bioenergetic reasons, as more free energy is gained with increasing standard redox potentials. Thus, based on a common substrate-like acetate, nitrate-reducing bacteria could outcompete sulfate-reducing bacteria and sulfate-reducing bacteria could outcompete methanogenic bacteria since microbial growth under anaerobic conditions is mostly energy-limited (Schönheit et al. 1982).

Another explanation for the mutual exclusion of the different inorganic electron acceptors in natural environments is based on the concept that H_2 is a common electron donor for all anaerobic oxidation reactions. Because H_2 concentrations are extremely low in anoxic environments,[4] microorganisms compete for this easily metabolized energy source. It has been demonstrated in well-controlled laboratory studies that sulfate-reducing bacteria outcompete H_2-consuming methanogens because the sulfate-reducing bacteria exhibit a higher substrate affinity and a higher specific growth yield.[5]

Presently, the most convincing explanation for mutual exclusion of the various oxidation reactions in natural anaerobic environments was suggested by the observation that in the presence of excess electron acceptors, anaerobic oxidation reactions lowered the H_2 partial pressure in the ecosystem only down to a reaction-specific minimum threshold value, which could not be further lowered (Lovley et al. 1982, Lovley 1985, Cord-Ruwisch et al. 1988). Reactions with higher standard redox potentials and lower free energies for H_2 consumption displayed consistently lower minimum threshold values (Table 10.1). The threshold concept is therefore independent of the population density or microbial substrate affinities and is solely based on the differences in the redox chemistry and thermodynamics of the various reactions. Detailed studies with a large variety of pure and mixed

[4]Strayer and Tiedje, 1978; Robinson et al., 1981; Lovley et al., 1982; Conrad et al., 1985, 1986.

[5]Winfrey and Zeikus 1977, Abram and Nedwell 1978a, b; Kristiansson et al. 1982; Lovley et al. 1982; Lovley and Klug 1983; Lupton and Zeikus 1984; Robinson and Tiedje 1984; Widdel 1988.

TABLE 10.1 Threshold H_2 Concentrations of Different Hydrolytic, Acetogenic and Methanogenic Microorganisms Growing on Organic Substrates or H_2

Electron acceptor, ox./red	Microbial genera	Substrates	H_2, ppm
CO_2/acetate	*Sporomusa*, *Acetobacterium*	Methanol, H_2	430–950
CO_2/CH_4	*Methanospirillum*, *Methanobrevibacter*, *Methanobacterium*, *Methanococcus*	Formate, H_2	25–100
Sulfur/sulfide	*Desulfovibrio*, *Wolinella*	Lactate, H_2	5–24
Sulfate/sulfide	*Desulfovibrio*, *Desulfobulbus*	Lactate, H_2	8–19
Fumarate/ succinate	*Wolinella*, *Desulfovibrio*	Fumarate, H_2	0.02–0.9
Nitrate/ammonia	*Desulfovibrio*, *Wolinella*	Lactate, H_2	0.02–0.03

SOURCE: Adapted from Cord-Ruwisch et al., 1988.

anaerobic cultures supported this threshold concept (Table 10.1). It was shown in defined cocultures that microbial species with a "lower H_2 threshold" metabolism could pull electrons away from substrate oxidation reactions that occurred in cocultured species with higher H_2 threshold values. *Desulfovibrio vulgaris* grown on lactate + sulfate transferred 24 to 30 percent of the reducing equivalents to *Wolinella succinogenes* under NO_3^--reducing conditions, but not to *Methanospirillum hungatei*, which had higher H_2 threshold values (Cord-Ruwisch et al. 1988). Similar effects were observed during acetate and methanol catabolism by *Methanosarcina barkeri* and *D. vulgaris* in the presence of sulfate, where the sulfate reducer scavenged reducing equivalents from the methanogen metabolism (Phelps et al. 1985). These results further implied that dissolved H_2 functioned as an extracellular reaction intermediate in these coculture studies. Microbial competition governed by metabolic thresholds is thus an important interaction between anaerobic bacteria with different terminal electron acceptors. Support for the hydrogen threshold concept came also from direct environmental studies (Lovley and Phillips 1987). Sediments from the Potomac River estuary had H_2 partial pressures of 0.82, 0.17, and 0.03 Pa under methanogenic, sulfate-reducing, and Fe(III)-reducing conditions, respectively. Addition of Fe(III) to methanogenic or sulfate-reducing sediments inhibited both reactions by more than 90 percent, but did not alter the total redox equivalents that were simultaneously consumed by all three ongoing anaerobic oxidation reactions. Fe(III) reduction could thus pull electrons away from methanogenesis and sulfate reduction and did not merely inhibit the other anaerobic oxidation reactions, supporting the threshold concept. It is possible that the threshold concept may be extended to reduced metabolites other than H_2. Distinct threshold values for acetate were recently found in pure cultures of methanogenic bacteria (Westermann et al. 1989) and the values recorded (69 to 1180 μM) were significantly higher than those reported for *Desulfobacter postgatei* (Ingvorsen et al. 1984). Also, formate is an important reduced anaerobic fermentation product which is redox equivalent to H_2 (Thiele and Zeikus 1988d). Threshold values for formate are not yet established, but these should be checked in order to further validate the H_2 threshold concept.

The actual occurrence of a particular redox reaction in anoxic environments, however, depends more on the availability of the respective inorganic electron acceptor rather than on its position on the microbial H_2 threshold scale (Table 10.1). Thus, geochemical and ecophysiological parameters are as important for the expression of particular microbial interactions as bioenergetic considerations of individual microbial metabolism.

All inorganic electron acceptors except bicarbonate are transported to the anoxic habitat by diffusion from the environment, whereas bicarbonate is produced in the environment through fermentative and methanogenic reactions (Fig. 10.1). In theory, the reaction rate of anaerobic oxidation reactions with terminal electron acceptors other than bicarbonate is thus (according to Fick's laws of diffusion) a direct function of the respective electron acceptor diffusivity and of the concentration gradient in the habitat (see below). As a consequence, stratified anoxic habitats such as sediments, microbial mats, biofilms, and large microbial flocs develop layers of metabolically similar microbial cells perpendicular to the diffusional flow direction of inorganic electron acceptors and display negative concentration gradients of the respective electron acceptors along the diffusional flow direction. The different microbial cells within each metabolic layer will compete for common electron donors with the consequence that electron acceptors allowing higher microbial growth yields or lower threshold values are preferentially consumed in the top zones. Electron acceptor supply is diffusion-limited and all inorganic electron acceptors of relevance have similar diffusivities. Microorganisms with higher threshold values are restricted to the anoxic habitat zones below, because their respective electron acceptors are not consumed in

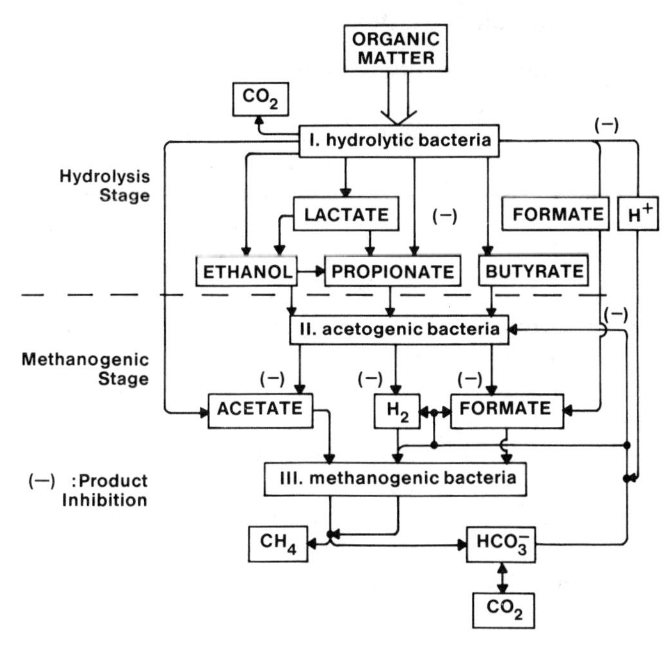

Figure 10.1 Carbon and electron flow in relation to important trophic groups in the anaerobic digestion of organic matter (Thiele and Zeikus, 1988a). Note the different product inhibited steps indicated by (-). (- - - -): Preferred separation of hydrolytic and syntrophic biomethanation stages in two stage anaerobic digestion systems.

the top zone. Metabolically layered anaerobic habitats are the result. Spatial considerations (Wimpenny 1989) should thus be included in studies of microbial interactions in anaerobic ecosystems. A good example of environmental metabolic stratification occurs in anaerobic freshwater and marine sediments in the presence of sulfate, where sulfate reduction occurs only in the top few centimeters and sulfate exhaustion allows for methanogenesis in the layers below (Cappenberg 1974, Winfrey and Zeikus 1977, Mountfort et al. 1980). Spatial separation is thus an important factor influencing diffusion-controlled electron acceptor flow in anoxic habitats. Bicarbonate should not be overlooked as a potential electron acceptor since it is generated by methanogenic metabolism within the anoxic habitat. This formation bypasses the diffusion-limited transport of available inorganic electron acceptors from the environment and enables bicarbonate-consuming bacteria to outcompete sulfate-reducing bacteria. This may lead to formation of metabolic subcompartments within the anoxic habitat, independent of bioenergetic factors or H_2 threshold values.

Metabolic and Ecophysiological Context

A very good qualitative and quantitative picture of the general carbon flow pattern of anaerobic digestion has been obtained through ^{14}C tracer studies (Jeris and McCarty 1965, Chartrain and Zeikus 1986). The anaerobic digestion of complex organic matter in the absence of sulfate or nitrate occurs generally in three sequential biodegradation steps represented by separate trophic groups of hydrolytic fermentative, syntrophic acetogenic, and methanogenic microorganisms. Trophic group I effects the breakdown of complex organic matter into simple low-molecular-weight fermentation products such as lactate, ethanol, acetate, formate, H_2, propionate, and butyrate (Fig. 10.1). Trophic group II converts these fermentation products into acetate, formate, and hydrogen. Microorganisms of the terminal trophic group III, the methanogenic bacteria, continuously remove resulting acetate, formate, and hydrogen equivalents from the environment (Fig. 10.1). Each trophic group contains a large variety of different microbial species, which represent a broad spectrum of organism-specific reactions (Fig. 10.1, Table 10.2). A functional interspecies metabolite transfer between the different species and the functioning of various mutualistic microbial interactions among the different trophic groups are thus the basis for stable methanogenesis. Although methanol plays a role as methanogenic precursor during the methanogenic degradation of pectinic biomass (Schink and Zeikus 1982, Phelps and Zeikus 1985) and methylamines are significant methanogenic precursors in marine ecosystems, metabolism of these compounds has been omitted from this chapter for the purpose of clarity.

Hydrolytic fermentative bacteria

Organisms within trophic group I may include eubacteria, fungi or protozoa depending on the individual ecosystem and its trophic status.[6] Anaerobic bacteria are the prevalent hydrolytic fermentative microorganisms in digestors and sediments and these bacteria compete for common nutrients and carbon sources. In natural and technical ecosystems, group I metabolism is always affected by several species. This is partly due to the complex nature of the decaying organic matter, which creates several simultaneous ecophysiological niches for fermentative microorganisms. Stably co-existing multiple fermentative species were also found in continuous mixed-culture experiments with defined carbon sources like glucose, cellulose, or whey.[7]

---

[6]Zeikus 1979, Mountfort 1986, Bornemann et al. 1989, van Bruggen et al. 1983.

[7]Zoetemeyer et al. 1982, Cohen et al. 1985, Murray 1986, Chartrain and Zeikus 1986b, Zellner et al. 1987, Kisaalita et al. 1987.

TABLE 10.2 Important Microbial Reactions in Methanogenic Ecosystems

	Reaction	G, kJ/reaction
A. Fermentative reactions:		
(1)	$C_6H_{12}O_6 + 3\ H_2O \rightarrow 3\ CH_4 + 3\ HCO_3^- + 3\ H^+$	- 403.6
(2)	$C_6H_{12}O_6 + 2\ H_2O \rightarrow 2$ ethanol $+ 2\ HCO_3^- + 2\ H^+$	- 225.4
(3)	$C_6H_{12}O_6 \rightarrow 2$ lactate $+ 2\ H^+$	- 198.1
(4)	$C_6H_{12}O_6 + 2\ H_2O \rightarrow$ butyrate $+ 2\ HCO_3^- + 3\ H^+ + 2\ H_2$	- 254.4
(5)	$C_6H_{12}O_6 \rightarrow 3$ acetate $+ 3\ H^+$	- 310.6
(6)	3 lactate $\rightarrow$ 2 propionate + acetate $+ HCO_3^- + H^+$	- 164.8
B. Syntrophic acetogenic reactions:		
(7)	lactate $+ 2\ H_2O \rightarrow$ acetate $+ 2\ H_2 + HCO_3^- + H^+$	- 3.96
(8)	ethanol $+ 2\ HCO_3^- \rightarrow$ acetate + 2 formate $+ H_2O + H^+$	+ 7.0
(9)	ethanol $+ H_2O \rightarrow$ acetate $+ 2\ H_2 + H^+$	+ 9.6
(10)	butyrate $+ 2\ H_2O \rightarrow 2$ acetate $+ 2\ H_2 + H^+$	+ 48.1
(11)	benzoate $+ 7\ H_2O \rightarrow 2$ acetate $+ 3\ H_2 + HCO_3^- + 2\ H^+$	+ 53.0
(12)	propionate $+ 3\ H_2O \rightarrow$ acetate $+ 3\ H_2 + HCO_3^- + H^+$	+ 76.1
C. Methanogenic reactions:		
(13)	acetate $+ H_2O \rightarrow$ methane $+ HCO_3^-$	- 31.0
(14)	$4\ H_2 + HCO_3^- + H^+ \rightarrow$ methane $+ 3\ H_2O$	- 135.6
(15)	4 formate $+ H^+ + H_2O \rightarrow$ methane $+ 3\ HCO_3^-$	- 130.4
D. Sulfidogenic reactions:		
(16)	2 lactate $+ SO_4^{2-} \rightarrow 2$ acetate $+ 2\ HCO_3^- + HS^- + H^-$	- 161.1
(17)	2 lactate $+ 3\ SO_4^{2-} \rightarrow 6\ HCO_3^- + 3\ HS^- + H^+$	- 255.3
(18)	$4\ H_2 + SO_4^{2-} + H^+ \rightarrow 4\ H_2O + HS^-$	- 152.2
(19)	acetate $+ SO_4^{2-} \rightarrow 2\ HCO_3^- + HS^-$	- 47.6
(20)	4 propionate $+ 3\ SO_4^{2-} \rightarrow 4$ acetate $+ 4\ HCO_3^- + 3\ HS^-$	- 150.6
(21)	2 butyrate $+ SO_4^{2-} \rightarrow 4$ acetate $+ HS^- + H^+$	- 28.0

This indicated the existence of mutualistic metabolic interactions among trophic group I members which are required to balance the population size of a mixed hydrolytic flora in a given environment. This possibility was clearly illustrated by the symbiotic relationship between *Bacteroides cellulosolvens* and *Clostridium saccharolyticum* during cellulose degradation (Murray 1986). In coculture, *B. cellulosolvens* fermented 33 percent more cellulose because *C. saccharolyticum* removed an unknown toxic secondary fermentation metabolite from the medium. *C. saccharolyticum* also grew during the cellulose fermentation because *B. cellulosolvens* provided it with monosaccharides and a growth factor. Growth of hydrolytic fermentative and acetogenic bacteria is usually inhibited by high end-product concentrations, especially organic acid anions, protons, and hydrogen.[8] Thus, efficient mutualistic end-product removal by trophic groups downstream in the metabolic chain is an important prerequisite for proliferation of trophic group I. Energy metabolism of group I bacteria involves substrate-level phosphorylations during glycolysis and the disposal of the generated reducing equivalents, with the organic fermentation end products produced including ethanol and lactate [Table 10.1, reactions (2) and (3)]. Ethanol and lactate were the most important fermentation products found in methanogenic ecosystems in nature and digestors.[9] Additional ATP equivalents for fermenting organisms can be generated through further substrate oxidation to acetyl-CoA and a phosphoroclastic cleavage of acetyl-CoA with generation of ATP and acetate (Thauer et al. 1977, Thauer and Morris 1984). Excess reducing equivalents are discharged from fermenting cells as H_2 or formate. In defined cocultures metabolic coupling of fermentative H_2 production and H_2-consuming reactions was found to increase acetate production and energy gain for H_2-producing partners.[10] Interspecies H_2 transfer to methanogens is thus also advantageous for fermentative bacteria in biomethanation ecosystems.

Group I is less sensitive to acidified environments than trophic groups II and III. Hydrolytic fermentative bacteria produce large amounts of hydrogen, formate, and acetate, which may inhibit syntrophic acetogenic bacteria.[11] Rapid growth and metabolism of trophic group I members is thus responsible for the souring of anaerobic digestors under overload conditions. Uncontrolled organic acid production continues at low pH values and inhibits acid-removal reactions of the methanogenic bacteria. This is an example for amensalistic interactions in anaerobic ecosystems.

Syntrophic acetogenic bacteria

Trophic group II, the syntrophic acetogenic bacteria (SAB), are eubacterial species that oxidize hydrolytic fermentation products like ethanol, propionate, butyrate, and benzoate to acetate. The reducing equivalents generated by the oxidation are used to reduce protons to

---

[8]Chung 1976, Kaspar and Wuhrmann 1978, Schink and Zeikus 1982, Dolfing and Tiedje 1988, Ahring and Westermann 1988, Thiele and Zeikus 1988c, Beaty and McInerney 1989.

[9]Phelps and Zeikus 1985, Chartrain and Zeikus 1986b, Zellner et al. 1987, Goodwin and Zeikus 1987, Chartrain et al. 1987.

[10]Ianotti et al., 1973; Weimer and Zeikus 1977, Ben-Bassat et al. 1981, Chen and Wolin 1977.

[11]Bryant et al. 1967, Reddy et al. 1972a, McInerney et al. 1979, Boone and Bryant 1980, Mountfort and Bryant 1982, Ahring and Westermann 1988, Thiele and Zeikus 1988c, Thiele 1988.

H_2 or CO_2 to formate [Table 10.2, reactions (8) and (9)]. SAB-catalyzed fermentation reactions are characterized by a positive chemical free energy change under standard conditions (Table 10.2). SAB grow and metabolize only in the presence of metabolically active H_2/formate (XH_2) consuming bacteria (syntrophy), providing H_2/formate as an energy source. Some XH_2-consuming methanogens also require acetate as an assimilatory carbon source and their growth thus depends on a functioning SAB metabolism. Methanogens or sulfate-reducing bacteria function as metabolic partners of SAB. This essential mutualistic interaction between syntrophic acetogenic and XH_2-consuming bacteria has been termed interspecies hydrogen transfer (Wolin and Miller 1982), although "interspecies electron transfer" or "XH_2 transfer" is more appropriate because the transferred molecule may be H_2 or formate (Thiele and Zeikus 1988d). The large negative free energy changes of the XH_2-consuming reactions [Table 10.2, reactions (14), (15), and (18)] result in very low steady-state XH_2 levels during the anaerobic digestion process. Excess acetogenic XH_2 production may lead to product accumulation and inhibition of syntrophic acetogenic metabolism. The growth rate of a mutualistic coculture of an acetogenic and methanogenic partner may also be controlled by the maximum specific growth rate of the methanogenic partner (Archer and Powell 1985). A complete inhibition of the SAB metabolism was observed when the free energy change for syntrophic acetogenesis became positive (Thiele and Zeikus 1988c, Thiele 1988, Dwyer et al. 1988). It should not be overlooked that acetate is also an inhibitory end product of the SAB metabolism and that interspecies acetate transfer to methanogenic bacteria contributes to the overall reaction bioenergetics of SAB (Kaspar and Wuhrmann 1978, Dolfing and Tiedje 1988, Ahring and Westermann 1988). Acetogenic hydrogen production via proton reduction by a ferredoxin:NAD or methylviologen-dependent hydrogenase has been demonstrated in cell-free extracts and whole cells of SAB (Reddy et al. 1972b, Beaty et al. 1989) and in defined cocultures of SAB with methanogens and/or sulfate-reducing bacteria.[12] Formate production from acetogenic CO_2 reduction may also be significant in methanogenesis since the acetogenic production of formate and/or hydrogen is redox-equivalent [Table 10.2, reactions (8) and (9)]. Syntrophic acetogenic CO_2 reduction to formate was demonstrated during the ethanol degradation in a defined coculture and in complex digestor flocs (Thiele and Zeikus 1988c, Thiele 1988). The possibility for CO_2 reduction to formate during syntrophic acetogenic butyrate degradation and fermentative metabolism in the rumen has

[12]Bryant et al. 1967, 1977; Boone and Bryant 1980; McInerney et al. 1979; Mountfort et al. 1982; Lee and Zinder 1988.

been also proposed by others (Bryant and Wolin 1975, Beaty et al. 1989). Current reviews of the syntrophic acetogenic metabolism in defined cocultures and complex ecosystems are given elsewhere.[13] A very interesting interaction between an acetogenic bacterium and an H_2/formate-consuming methanogen was recently described (Zinder and Koch 1984). Strain AOR from a thermophilic digestor was able to oxidize acetate to CO_2 and H_2 in coculture with an H_2-consuming methanogen. The acetate oxidation reaction was in this case the reverse reaction of homoacetogenic acetate formation from H_2 and CO_2. This unusual coculture achieved acetate cleavage to methane and CO_2 [Table 10.2, reaction (13)], extending the range of syntrophic acetogenic substrates from acetate to stearate (Table 10.3). The importance of the AOR organism in biomethanation ecosystems needs to be determined by further studies.

SAB species, initially isolated from anaerobic environments,[14] appear to be specialized for the oxidation of either ethanol, propionate, butyrate, or benzoate. New SAB species with a broader substrate spectrum have been isolated from biomethanation ecosystems. New carbon substrates and electron acceptors have also been discovered (Table 10.3) that allow some SAB to grow in pure culture without a XH_2-consuming partner.[15] This clearly identifies SAB as a group of functionally similar microorganisms in the ecosystem, that were obtained from the environment using a common isolation strategy, but may not necessarily grow as SAB in nature and digestors. The existence of nonsyntrophic substrates for some SAB is only an indication that they can exist in nature without a XH_2-consuming partner, and the environmental significance of some of these "nonsyntrophic" substrates is doubtful, because ethanol, propionate, isobutyrate, and butyrate constitute the majority of the physiological substrates for SAB in the environment.

Methanogenic bacteria

So far more than 30 different species of methanogens have been isolated, which have been classified into 14 genera and 5 families (Jones et al. 1987). In anaerobic digestors, more than 60 percent of the biological methane originates from acetoclastic methanogenesis and the remainder from H_2, formate, and methanol (Jeris and McCarty 1965,

[13]Mah 1982, Widdel 1988, Dolfing 1988, Thiele and Zeikus 1988d.

[14]Reddy et al. 1972a, Boone and Bryant 1980, McInerney et al. 1981, Mountfort and Bryant 1982.

[15]Reddy et al. 1972a, Bryant et al. 1977, Schink and Stieb 1983, Stieb and Schink 1985, Roy et al. 1986, Dubourguier et al. 1986, Chartrain and Zeikus 1986b, Beaty and McInerney 1987, Lee and Zinder 1988.

TABLE 10.3 **Carbon Substrate Range of Syntrophic Acetogenic Bacteria**

	Acetogenic carbon substrates		
	Syntrophic growth		
Organisms	Alcohols, aldehydes	Organic acids	Pure culture
S-organism (Reddy et al., 1972a)	Ethanol-pentanol, acetaldehyde	Oxaloacetate	Pyruvate
Pellobacter carbinolicus (Schink, 1984; Dubourguier, et al., 1986)	Ethanol-isobutanol, propanediol, butanediol	—	Acetoin, ethylene glycol
Desulfovibrio vulgaris (Bryant et al., 1977, Chartrain and Zeikus, 1986a)	Ethanol, lactate	—	Ethanol + sulfate, lactate + sulfate
Syntrophobacter wolinii (Boone and Bryant, 1980)	—	Propionate	—
Syntrophomonas wolfei (McInerney et al., 1979, 1981; Beaty and McInerney, 1987)	—	Butyrate-caprylate isoheptanoate	Crotonoate
Clostridium bryantii (Stieb and Schink, 1985)	—	Butyrate-heptanoate	—
Syntrophomonas sapovorans (Roy et al., 1986)	—	Butyrate-stearate	—
Strain NSF-2 (Shelton and Tiedje, 1984)	—	Butyrate-caproate	—
Strain SF-1 (Shelton and Tiedje, 1984)	—	Butyrate-caproate	—
Syntrophus buswellii (Mountfort et al., 1984)	—	Benzoate	—
Strain BZ-2 (Shelton and Tiedje, 1984)	—	Benzoate	—
AOR-organisms (Zinder and Koch, 1984; Lee and Zinder, 1988)	Ethanol	Acetate	Ethylene glycol

Smith and Mah 1966, Chartrain and Zeikus 1986b). Acetoclastic methanogens perform an important function for the pH homeostasis of the ecosystem by acetic acid removal and also produce CO_2/HCO_3^- [Table 10.2, reaction (13)] as an important electron acceptor for CO_2-reducing, SAB- and XH_2-removing methanogens (Thiele and Zeikus

1988b). A mutualistic interaction may exist between SAB and acetoclastic methanogens forming a syntrophic microniche together with XH_2-consuming methanogens (Thiele and Zeikus 1988c). SAB produce most of the acetate equivalents for acetoclastic methanogenesis and receive CO_2 as an electron acceptor from the methanogens in return. Important acetoclastic species belonging to the family Methanosarcinaceae are *Methanosarcina barkeri, Methanosarcina mazei, Methanosarcina acetivorans*, and *Methanothrix söhngenii*. Major differences between these species exist in their maximum growth rate, K_s, for acetate utilization and acetate threshold concentration (Table 10.4). It is evident that all species of the genus *Methanosarcina* have a higher K_s and higher acetate threshold value than *Methanothrix* (Table 10.4). On the other hand, *Methanosarcina* species have a significantly faster growth rate than *Methanothrix* species. Thus, the possibility for substrate competition exists among the different acetoclastic methanogenic species. *Methanosarcina* will out compete *Methanothrix* at high acetate concentrations, but not at acetate concentrations below 200 to 400 μM. These patterns have been frequently observed in different digestor ecosystems (Zinder et al. 1984, Wiegant and de Man 1986). More than 20 different H_2/formate oxidizing species of methanogens have been described in the literature and most of them have very similar substrate affinities (Jones et al. 1987, Thiele and Zeikus 1988c). Substrate competition between different XH_2-consuming methanogenic species is thus improbable. On the other hand, several methanogens isolated from digestors or the rumen ecosystem required coenzyme M as an essential growth factor, whereas other XH_2-consuming methanogens did not (Lovley et al. 1984, Miller et al. 1986, Jones et al. 1987). Free coenzyme M from lysed bacteria is frequently found in rumen contents or in the liquid content of high-rate anaerobic digestors. A commensal relationship

TABLE 10.4 Kinetic Parameters for Important Acetate Consuming Methanogens

	Acetate Transformation Parameters	
Organism	K_s, μM	Threshold, μM
Methanosarcina sp., strain CALS-1	ND[a]	200–700
Methanosarcina barkeri 227	5000	1180
Methanosarcina mazei S-6	ND	397
Methanosarcina barkeri, strain Fusaro	3000	ND
Methanothrix sp., strain CALS-1	ND	12–21
Methanothrix sp.	ND	69
Methanothrix söhngenii	700	ND

[a]ND = not determined.

SOURCE: Huser et al., 1982; Westermann et al., 1989; Min and Zinder, 1989; Schönheit et al., 1982.

exists between CoM-producing and CoM-dependent methanogenic bacteria.

Careful quantitative immunological analysis of various mesophilic and thermophilic anaerobic digestors revealed a surprising diversity of species and at the same time a striking similarity of the methanogenic population patterns in the various digestors (Macario and Conway de Macario 1988, Macario et al. 1989). *Methanobacterium formicicum* and *Methanobrevibacterium arboriphilum* strain AZ were consistently found in more than 90 percent of the digestors. This indicated a special importance of these two species for biomethanation ecology. *M. formicicum* but not *M. arboriphilum* AZ consumes formate for methanogenesis (Zeikus and Henning 1975). The coexistence of formate- and H_2-consuming methanogenic species supports the importance of both reduced chemical species as methanogenic precursors.

Sulfidogenic (sulfate-reducing) bacteria

A detailed review of the physiology, microbiology, and ecology of sulfate-reducing bacteria (SRB) has been given by Widdel (1988). Nearly every methanogenic substrate can be oxidized by SRB, which occurs by two different patterns of metabolism—incomplete substrate oxidation to acetate, bicarbonate, and HS^-, and complete oxidation to bicarbonate and HS^- [Table 10.2, reactions (16) to (21)]. For reasons outlined above, SRB can usually outcompete methanogens for carbon sources if sulfate is present in excess. Incomplete SRB substrate oxidation to acetate resembles the metabolism of SAB with the difference that the generated reducing equivalents are used for sulfate reduction to sulfide instead of CO_2 reduction to methane. SRB are thus often referred to as sulfidogens. Lactate- and ethanol-oxidizing SRB like *Desulfovibrio vulgaris* could also function as SAB in coculture with methanogenic bacteria in which the SRB transferred the corresponding reducing equivalents to the methanogens either by interspecies H_2 transfer or formate transfer (Bryant et al. 1977, Chartrain and Zeikus 1986, Thiele and Zeikus 1986b). The "dual lifestyle" of SRB—syntrophic and sulfidogenic—makes it very difficult to assign them a defined role in the freshwater ecosystem. In fact, the fluctuating low sulfate levels in freshwater and wastewater environments would force SRB to adapt to syntrophic growth when sulfate is limiting, and to sulfate reduction when sulfate is available. Lactate and ethanol are major methane precursors in Lake Mendota sediments (Phelps and Zeikus 1985) and more than 95 percent of the syntrophic redox metabolism in the sediments is independent of the dissolved H_2 gas pool (Conrad et al. 1985). Sulfate is continuously limiting in the sediments during most of the year and rises sharply only once per year during fall turnover (Phelps and Zeikus 1985). Thus it was speculated that SRB in Lake Mendota sediments exist as SAB during most of the season. When sulfate was added to Lake Mendota sediments, a strong inhibition of methane formation from H_2 was observed only after 45 h, but not immediately after sulfate addition (Conrad et al. 1987b). This indicated that the SRB in the sediment did not actively reduce sulfate immediately after sulfate addition. During the transition, H_2 gas turnover increased and dissolved H_2 gas concentration dropped severalfold, and methanogenesis became sulfate-inhibited. This indicated that sulfate reducers and methanogens did not immediately compete for H_2. Likewise, H_2 turnover was $CHCl_3$-sensitive immediately after sulfate addition, indicating the involvement of methanogenic bacteria in H_2 consumption in the presence of sulfate. These data indicate that SRB in the sediments existed as SAB in the absence of sulfate (Conrad et al. 1985), and that these syntrophic SRB shifted their ethanol and lactate metabolism from syntrophic growth with methanogens to sulfate reduction when sulfate became available. The data do not indicate whether formate or H_2 transfer was the prevalent mechanism of syntrophic interaction of the SRB with the methanogens in the absence of sulfate. However, the relative insignificance of the H_2 pool turnover in the sediments suggests that formate may have been the syntrophic intermediate. This is analogous to syntrophic ethanol oxidation by *D. vulgaris* and *M. formicicum* in microbial flocs from a whey digestor (Thiele and Zeikus 1986c).

Similar observations were made during a study of the anaerobic digestion of sulfate containing wastewaters in upflow reactors with polyurethane sponges as carriers (Isa et al. 1986a,b). In these reactions SRB could not outcompete the methanogens despite the presence of excess sulfate in the waste and in the effluent. Only 10 to 20 percent of the total electron flow was scavenged by the SRB. Competitiveness of SRB was higher at low substrate concentrations due to their lower K_s values. The methanogenic bacteria could regain dominance of the reactors despite the presence of excess sulfate, even when sulfate reducers were the initial colonizers and methanogens were added later. The reasons for the competitiveness of methanogens in these experiments were not completely clear, but the authors speculated that better adherence of methanogens to the polyurethane foam carriers in the reactors constituted the major selective advantage over SRB. The proportion of SRB carrying out syntrophic growth in the reactors was not indicated, but syntrophic substrates like ethanol were clearly the prevalent carbon substrates of the SRB in the reactors. Formate and acetate supported SRB growth only very poorly. The SRB may only grow as SAB in the center of the polyurethane foam particles where the sulfate supply from the medium would be diffusion-limited. Likewise, availability of other nutrients including Fe may have been diffusion-limited in the center portion of the sponge particles. Thus, reaction-diffusion dynamics and not basic competition kinetics were important competitive factors in this anaerobic ecosystem. These last two examples should illustrate how difficult it is to predict practical microbial interactions based on conclusions from laboratory experiments. Reaction-diffusion dynamics and microbial adherence mechanisms were more important microbial parameters than were, for example, K_s or V_{max} values determined under artificial chemostat conditions.

Lastly, undissociated H_2S has been shown to be a strong inhibitor of methanogenesis from acetate and of syntrophic propionate degradation.[16] Fifty percent inhibition was observed at approximately 100 to 150 mg/L undissociated H_2S. Thus, sulfide production by SRB in wastewaters with high sulfate and sulfide contents should also be considered as an amensalistic microbial interaction between methanogens and SRB.

Methanogenic Carbon Flow Coordination

The prevalent carbon flow routes of methanogenesis and some important branch points are shown in Fig. 10.1. This pattern has been consistently observed in different methanogenic environments. The organization of carbon flow into three separate steps demands a coordinated metabolism of the three trophic groups in order to avoid fermentation end-product accumulation and inhibition of syntrophic acetogenic and methanogenic bacteria (see above). The rate-limiting step in biomethanation depends on the major carbon substrates and physical parameters in the environment. Fermentative hydrolysis of plant cell wall components and plant biopolymers was a rate-limiting step in freshwater sediments (Winfrey and Zeikus 1979, Schink and Zeikus 1982, Zeikus 1983). Acetoclastic methanogenesis was biomethanation rate limiting in a laboratory digestor degrading whey waste (Chartrain and Zeikus 1986) and during sewage digestion (Jeris and McCarty 1965). In industrial digestors and in farm digestors syntrophic acetogenic propionate degradation may be rate-limiting for biomethanation as indicated by the accumulation of propionate in the digestor contents. But in ecoengineered high-rate biomethanation reactors treating volatile fatty acid (VFA) containing wastewaters, methanogenic acetate degradation showed the highest specific COD-

[16]Mountfort and Asher 1979, Koster et al. 1986, Karhadkar et al. 1987, Rinzema and Lettinga 1988.

removal rate and syntrophic propionate degradation was rate-limiting (Thiele and Zeikus 1988a). Thus, the nature of the biomethanation rate-limiting step depends strongly on ecosystem-specific parameters.

Each species in methanogenic ecosystems appears to operate at the highest possible metabolic rate and is usually restricted by the accumulation of self-generated metabolic end products (Fig. 10.1). Rapid fermentative metabolism can continue under substrate overload conditions and inhibit the terminal carbon degradation reactions in methanogenic ecosystems. Methods that allow control of the hydrolytic fermentative metabolism in anaerobic mixed cultures would be important to improve anaerobic digestor stability against organic shock loads.

A second strategy to enhance the stability of biomethanation carbon flow is enhancement of the mutualistic interactions of the terminal organic acid consuming SAB and methanogens. The stimulation of mutualistic interactions increases the metabolic throughput of the terminal acid-removing carbon degradation reactions independent of the initial hydrolytic fermentation step, thereby reducing the steady-state concentrations of inhibitory acetate, hydrogen, and formate. The following example illustrates this interaction. Imagine an anaerobic digestor where hydrolytic fermentation reactions and acid-removing syntrophic biomethanation are balanced. The COD conversion rate R at each step of the balanced carbon flow follows the well-known general equation

$$R = R_{max} * \frac{[S]}{K_s + [S]}$$

where R_{max} = maximum volumetric COD conversion rate
K_s = substrate constant
$[S]$ = substrate concentration

It is evident that an increased R_{max} of the syntrophic metabolism will decrease the syntrophic substrate concentration $[S]$ that must accumulate for a certain COD conversion rate (R). Thus, if R_{max} is very large, $[S]$ will become very small for acetogens. If methanogenic R_{max} is large, then acetogenic product inhibition will be insignificant. If R_{max} becomes so large that $[S] << K_s$, the maximum dynamic response toward organic shock loads will be obtained and R can simply be increased by a corresponding small accumulation of the syntrophic carbon substrates S. Such anaerobic ecosystems are relatively stable toward fluctuations in organic loading rates. If, on the other hand, R_{max} is very small, the reactor operates close to R_{max} and has no degradation rate reserve to respond to fluctuating organic loading rates. Such anaerobic ecosystems are very unstable. The same kinetic logic applies to the hydrolytic fermentation reactions.

Different technical solutions exist to increase R_{max} of anaerobic carbon flow. These are

- Metabolic compartmentalization on the bioreactor level
- "Species juxtapositioning" of metabolic partners within microbial aggregates
- Metabolic compartmentalization on the biocatalyst level
- Anaerobic pathway optimization.

Metabolic compartmentalization on the bioreactor level

Two-stage anaerobic digestion systems (Gosh and Klass 1978) separate hydrolytic fermentative and syntrophic biomethanation reactions into two continuously stirred sequential digestor tanks. Hydrolytic fermentation products from the first tank are dosed to the biomethanation reactor according to the existing R_{max} of the established syntrophic biomethanation flora in the second tank. This avoids toxification of the syntrophic methanogenic flora, minimizes the inhibition by hydrolytic fermentation products, and maximizes R_{max} because these products are carefully dosed to the syntrophic biomethanation flora independent of hydrolytic fermentation rates (Cohen et al. 1980, 1982). In a different approach, hydrolysis and preacidification of certain organic wastes (for example, brewery wastewater) was obtained in the waste holding tank and the preacidified wastewater was pumped in upflow mode through a single-stage anaerobic reactor (Lettinga et al. 1984, 1985). The upflow mode (UASB, upflow anaerobic sludge blanket reactor) resulted in selection and retention of large microbial aggregates in the reactor bottom and washout of finely dispersed bacteria. R_{max} in this case was maximized through the high biomass concentration in the sludge blanket in the first 50 percent of the reactor volume. Bacteria in well-functioning UASB reactors grow as granular sludge with extremely high biomass concentrations (>40 g/L biomass, Hulshoff Pol et al. 1983).

As a further example, an anion-exchange membrane bioreactor combination was described where syntrophic acetogenic and methanogenic bacteria were retained in a closed reactor environment and syntrophic carbon sources were supplied by membrane filtration and anion-exchange of volatile fatty acids and bicarbonate (Thiele and

Zeikus 1988a). This maximized the methane productivity of the syntrophic methanogenic flora because organisms from the syntrophic biomethanation reactor could not be washed out by the waste flow. The reaction conditions were optimized for syntrophic biomethanation reactions, because volatile fatty acid substrates were stoichiometrically supplied to syntrophic acetogenic bacteria by anion exchange with bicarbonate. Syntrophic acetogenic and methanogenic bacteria in anion-exchange substrate shuttle reactors grew also as very dense granular aggregates with extremely high biomethanation rates (Thiele and Zeikus 1988a).

Species juxtapositioning within microbial aggregates

Further microbial strategies to increase R_{max} were indicated by the spatial organization of the syntrophic methanogenic metabolism via "species juxtapositioning" inside anaerobic digestor flocs from a whey-degrading methane digestor (Thiele et al. 1988b). The syntrophic ethanol degradation inside the flocs operated at much higher specific activities and at simultaneously significantly reduced H_2, ethanol, and propionate levels when compared to syntrophic species that were not juxtaposed inside the digestor flocs. The short interspecies distance of the juxtaposed species allowed high metabolic fluxes between acetogens and methanogens with very flat interspecies concentration gradients (Thiele et al. 1988b). This enhanced the carbon flux and minimized acetogenic product inhibition. Species juxtapositioning by coordinated microbial growth in microniches inside microbial aggregates is a means to increase maximum syntrophic methanogenic reaction rates. Species juxtapositioning of syntrophic acetogenic bacteria and methanogens was also observed in anaerobic granular sludge from UASB reactors (Dubourguier et al. 1987). Distinct microcolonies of juxtaposed acetogens and methanogens were found by electron microscopy of different sludges. A characteristic ratio of methanogen/*Syntrophobacter* of 2.33 to 2.46 and methanogen/*Syntrophomonas* of 0.48 to 0.71 was reported inside these microcolonies (Dubourguier et al. 1987).

The kinetic consequences of species juxtapositioning for syntrophic methanogenesis have been studied by several authors. Conrad and coworkers (Conrad et al. 1985, 1986) hypothesized that juxtaposed acetogens and methanogens inside microbial aggregates would form distinct internal H_2 gas pools with high internal turnover in digestors and sediments. These pools would be kinetically uncoupled from the rest of the ecosystem. Data collected from sewage digestors and lake sediments supported the hypothesis because only 5 to 6 percent of the measured H_2 gas turnover in these environments could account for the simultaneously measured methanogenic H_2 consumption indicating separated H_2 gas pools (Conrad et al. 1985). In a separate study the authors analyzed the bioenergetics of syntrophic acetogenesis in these environments and speculated that the H_2 concentration in these internal H_2 pools with the high turnover would be much lower than in the environment (Conrad et al. 1986). The calculated free energy changes based on measured parameters from the external H_2 pool were too positive to explain active syntrophic acetogenesis. They hypothesized that one function of syntrophic aggregates with low internal H_2 concentrations was to provide favorable growth conditions for SAB. Detailed reaction-diffusion studies with computer models, however, demonstrated that syntrophic microbial aggregates develop internal H_2 concentrations that are higher or equal to but not lower than the H_2 pool in the environment (Ozturk et al. 1989). Figure 10.2a shows a typical time course simulation of dissolved H_2 gas levels during the syntrophic ethanol degradation in such aggregates. The internal turnover of H_2 gas accounted for more than 95 percent of the methanogenic CO_2 reduction in these digestor flocs due to the short interspecies distance (Thiele et al. 1988b, Ozturk et al. 1989), but the simultaneous H_2 concentration levels in the aggregate did not improve syntrophic acetogenesis. Lower H_2 gas levels in the aggregates could only be obtained if the ethanol diffusion from the environment into the aggregate was restricted by very large floc diameters (Fig. 10.2b). A diffusion-limited carbon substrate supply from the environment indicated unfavorable growth conditions for SAB in these large aggregates. Species juxtapositioning of syntrophic acetogens and methanogens in microbial aggregates enhances the microbial substrate turnover, but does not always improve growth conditions for SAB.

Metabolic compartmentalization on the biocatalyst level

Other microbial interactions to increase R_{max} and thus enhance the stability and carbon flow coordination of anaerobic digestion were discovered when the ecophysiology of syntrophic ethanol biomethanation in the digestor flocs was biochemically analyzed (Thiele and Zeikus 1988c). A new reaction mechanism was found that metabolically compartmentalized the syntrophic electron transfer by unique reaction intermediates (Fig. 10.3). Ethanol was oxidized to acetate by *D. vulgaris* in the flocs and the resulting reducing equivalents were used to reduce bicarbonate to formate by CO_2 reductase instead of protons to H_2 in the *D. vulgaris* cells [Table 10.2, reaction (8)]. Formate was excreted by *D. vulgaris* and was consumed for methanogenesis by *M.*

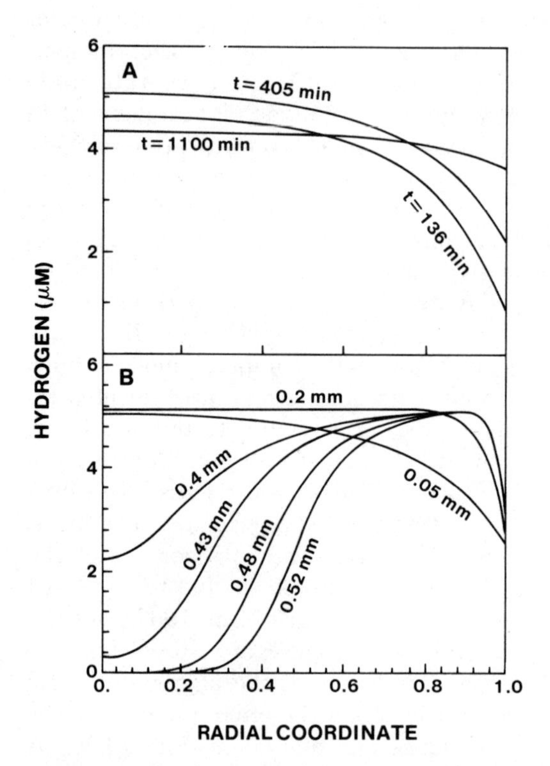

Figure 10.2 Steady-state hydrogen gas profiles during syntrophic ethanol metabolism in anaerobic digestor flocs (Özturk et al., 1989). (*a*) Time-dependent H_2 profiles inside the digester flocs during ethanol digestion experiments. The profiles calculated were for flocs of 100 μm diameter. The ethanol concentration in the bulk decreased during this experiment from 13.5 m*M* to 3.5 m*M*. Note elevated H_2 concentration inside the flocs at any time. (*b*): Sensitivity analysis of H_2 profiles with increasing floc diameters. The floc radii were varied between 0.05 and 0.52 mm. Ethanol and H_2 concentrations in the bulk were 13.5 m*M* and 2.5 μ*M*, respectively. Note the sharp transition in the H_2 profiles at floc diameters above 0.4 mm. The ethanol metabolism inside the flocs changed from reaction controlled to diffusion controlled indicated by the "bathtub" shaped H_2 concentration profiles.

formicicum. The result was regeneration of 3 bicarbonate from 4 formate (Table 10.2, reaction (15)]. The formate:bicarbonate electron cycle model (Fig. 10.3) provided the basis for a previously unknown syntrophic acetogenic reaction, the syntrophic acetogenic CO_2 reduction to formic acid. H_2 gas was not a direct syntrophic acetogenic end product, but was produced as a side reaction by formate:H_2 lyase. The acetogenic energy metabolism according to this model depends on the

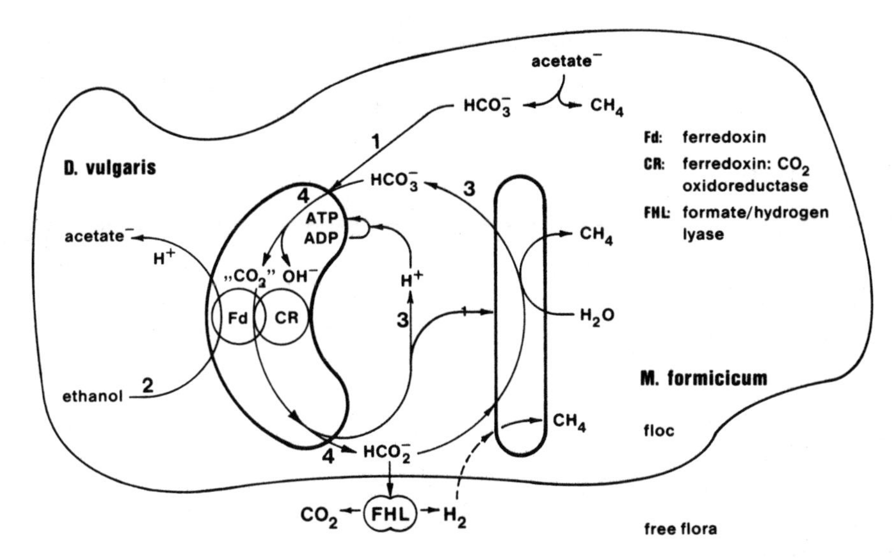

Figure 10.3 Model of the formate:bicarbonate electron shuttle mechanism during syntrophic methanogenesis from ethanol. The juxtaposition of *Desulfovibrio vulgaris* and *Methanobacterium formicicum* in the flocs enabled the methanogenic CO_2 regeneration to be coupled with acetogenic CO_2 reduction to formic acid, providing interspecies electron flow. Energy conservation in the acetogen is postulated to occur as a consequence of a putative cytosolic carbonic anhydrase and a membrane linked CO_2 reduction, which leads to the generation of an alkaline cell interior, an acidic exterior, and a proton-motive force to drive membrane coupled ATP synthesis.

bicarbonate:formate ratio in the ecosystem and not directly on the H_2 partial pressure. The independence of syntrophic acetogenesis from the H_2 partial pressure in the flocs was experimentally demonstrated in the study (Thiele and Zeikus 1988c) and gave further credibility to the model. The high ratio of bicarbonate to formate provided additional free energy for syntrophic acetogenic alcohol oxidation. Thus, high bicarbonate (CO_2) concentrations and low formate levels pulled the syntrophic acetogenic metabolism independent of the H_2 partial pressure (Thiele and Zeikus 1988c). Formate transfer provided a much faster interspecies electron flow to the methanogens due to its much higher solubility compared with H_2 and due to the high substrate affinity of formate-consuming methanogens (Thiele and Zeikus 1988c). The apparent H_2 independence of syntrophic methanogenesis in sediments and sewage digestors (Conrad et al. 1985) supported a metabolic electron flow compartmentalization by interspecies formate transfer. Unfortunately, formate transfer was never measured in these studies. The available mixed-culture data of other experiments in the literature that have claimed interspecies H_2 transfer as syntrophic reaction mechanisms were also consistent with the formate transfer hypothesis, but formate was also not measured in these studies (Thiele and Zeikus 1988d). Further support for formate: bicarbonate electron flow control as an important syntrophic mechanism during anaerobic digestion came from other studies and diffusion calculations in a defined coculture of *S. wolfei* and *M. formicicum* (Boone et al. 1989). The authors calculated (but did not measure) that H_2 transfer in the defined coculture of *S. wolfei* and *M. formicicum* accounted only for a fraction of the measured syntrophic methanogenesis, whereas interspecies formate transfer could theoretically account for a 98-fold faster interspecies electron transfer. Measurements of at least fivefold higher formate transfer versus H_2 transfer were reported from a defined syntrophic coculture of *D. vulgaris* and *M. formicicum* during ethanol degradation (Thiele 1988).

The significance of these findings for the understanding of microbial interactions in biomethanation ecosystems is supported by the bicarbonate dependence of syntrophic fatty acid degradation (unpublished results). Thus, bicarbonate is an important electron acceptor for syntrophic acetogenic bacteria and anaerobic digestor process control should also include the alkalinity and bicarbonate concentration in the reactor independent of H_2 control (Harper and Pohland 1986). The importance of acetoclastic methanogenesis for stable biomethanation is not only for organic acid removal and pH homeostasis in the environment, but also for bicarbonate renewal. Mutualistic syntrophic interactions exist not only between syntrophic acetogens and formate-consuming methanogens but also between bicarbonate-consuming SAB and bicarbonate-producing acetoclastic methanogens [Table 10.2, reaction (13)]. This led to the proposal of a formate:bicarbonate–controlled syntrophic microniche (Thiele and Zeikus 1988d). Acetoclastic methanogens outside the microniche provide bicarbonate molecules, which diffuse into the niche as electron acceptor for formate-coupled syntrophic methanogenesis. Consistent with this view, high syntrophic reaction rates were observed in granular sludge samples with high rates of acetoclastic methanogenesis (Dubourguier et al. 1987, Thiele and Zeikus 1988a).

Ecoengineering anaerobic pathway optimization

The three steps of anaerobic digestion suggest another possibility to improve R_{max} for anaerobic digestor stability—ecoengineering of anaerobic pathway design (Thiele and Zeikus 1988a). This approach tries to avoid unproductive carbon flow pathways in the anaerobic digestion process by selective stimulation and fast turnover of biomethanation intermediate production. It would be especially desirable to suppress the formation of propionic acid, as propionate is an anaerobic metabolite with a very low maximum turnover rate (Kaspar and Wuhrmann 1978, Dubourguier et al. 1987, Thiele and Zeikus 1988a). One possibility to suppress propionate formation is to construct digestor ecosystems based on purified starter cultures of known reaction specificity that do not produce propionate (Winter and Wolfe 1980, Jones et al. 1984, Chartrain et al. 1987). Practical applications of this approach, however, are difficult because most organic waste materials are nonsterile and continuous inoculation with waste microorganisms leads quickly to a waste-acclimated flora independent of the composition of the respective starter cultures. Cohen et al. (1985) tried to influence the hydrolytic fermentative flora in two-stage digestors by repeated starvation periods, but this treatment increased propionate formation rate instead of reducing it. During the ecoengineering of a whey waste-treatment process (Thiele et al. 1990), addition of a clostridial butyrogenic starter culture (*C. butyricum*) to deproteinized whey waste suppressed lactate and propionate formation in the whey by approximately 60 percent. Treatment of this biologically preconditioned waste increased the COD removal rate of different biomethanation reactors by 44 to 88 percent during nearly three months of continuous operation. The process in this case was unique, because the whey deproteinization method allowed starter culture inoculation of a pasteurized waste. This example showed that it is possible to engineer an anaerobic ecosystem in order to enhance R_{max} of the rate-limiting biomethanation steps.

Summary and Outlook

This chapter described several examples of microbial interactions during the anaerobic digestion process. However, it is evident that our present understanding of methanogenic ecosystems is insufficient to understand all the microbial mixed-culture interactions. The formation of microniches and diffusion-controlled microbiology in gradient ecosystems contributes to this complexity. Other kinetic concepts, like species juxtapositioning and the formation of kinetically uncoupled diffusion-controlled microenvironments in sediments and digestor sludges will be of prime importance to understand the microbial ecology of anaerobic digestion. These research areas are still in their infancies, and although we know a lot about the microbial adhesion in dental plaque formation (Mergenhagen and Rosan 1985) and microbial adhesion to metazoan tissue surfaces, we know little about the factors that control adhesion of anaerobes during the biofilm formation in fixed-film reactors or pelletization in anaerobic sludge bed reactors.[17] More fundamental research needs to be done in these areas in order to provide new insights into the microbial interactions during anaerobic digestion.

References

Abram, J. W., and D. B. Nedwell. 1978a. Inhibition of methanogenesis by sulfate reducing bacteria competing for transferred hydrogen. *Arch. Microbiol.* 117:89–92.

Abram, J. W., and D. B. Nedwell. 1978b. Hydrogen as a substrate for methanogenesis and sulfate reduction in anaerobic salt marsh sediments. *Arch. Microbiol.* 117:93–97.

Ahring, B. K., and P. Westermann. 1988. Product inhibition of butyrate metabolism by acetate and hydrogen in a thermophilic coculture. *Appl. Environ. Microbiol.* 54:2393–2397.

Archer, D. B., and G. E. Powell. 1985. Dependence of the specific growth rate of methanogenic mutualistic cocultures on the methanogen. *Arch. Microbiol.* 141:133–137.

Balch, W. E., G. E. Fox, L. J. Magrum, C. R. Woese, and R. S. Wolfe. 1979. Methanogens: Re-evaluation of a unique biological group. *Microbiol. Rev.* 43:260–296.

Beaty, P. S., and M. J. McInerney. 1987. Growth of *Syntrophomonas wolfei* in pure culture on crotonate. *Arch. Microbiol.* 147:389–393.

Beaty, P. S., and M. J. McInerney. 1989. Effects of organic acid anions on the growth and metabolism of *Syntrophomonas wolfei* in pure culture and in defined consortia. *Appl. Environ. Microbiol* 55:977–983.

Beaty, P. S., N. Q. Wofford, and M. J. McInerney. 1989. The metabolism of formate by the syntrophic bacterium *Syntrophomonas wolfei*. Abstr. I31 of the 89th Annual meeting of the American Society for Microbiology, New Orleans, Louisiana.

Ben-Bassat, A., R. Lamed, and J. G. Zeikus. 1981. Ethanol production by thermophilic bacteria: metabolic control of end product formation in *Thermoanaerobium brockii*. *J. Bacteriol.* 146:192–199.

Boone, D. R., and M. P. Bryant. 1980. Propionate-degrading bacterium, *Syntrophobacter wolinii* sp. nov. gen. nov. from methanogenic ecosystems. *Appl. Environ. Microbiol.* 40:626–632.

Boone, D. R., R. L. Johnson, and Y. Liu. 1989. Diffusion of the interspecies electron carriers H_2 and formate in methanogenic ecosystems and its implication in the measurement of K_M for H_2 or formate uptake. *Appl. Environ. Microbiol.* 55:1735–1741.

Bornemann, W. S., D. E. Akin, and L. G. Ljundahl. 1989. Fermentation products and plant cell wall degrading enzymes produced by monocentric and polycentric anaerobic rumen fungi. *Appl. Environ. Microbiol.* 55:1066–1073.

Breznak, J. A. 1982. Intestinal microbiota of termites and other xylophagous insects. *Ann. Rev. Microbiol.* 36:323–343.

van Bruggen, J. J. A., C. K. Stumm, and G. D. Vogels. 1983. Symbiosis of methanogenic bacteria and sapropelic protozoa. *Arch. Microbiol.* 136:89–95.

Bryant, M. P., E. A. Wolin, M. J. Wolin, and R. S. Wolfe. 1967. *"Methanobacillus omelianskii,"* a synbiotic association of two species of bacteria. *Arch. Microbiol.* 59:20.

Bryant, M. P., and M. J. Wolin. 1975. Rumen bacteria and their metabolic interactions. Pages 297–306 in T. Hasegawa (ed.), *Proceedings of the First International Congress of the International Association Microbiological Society*, vol. 2, *Developmental Microbiology, Ecology*. Science Council for Japan, Tokyo.

Bryant, M. P., L. L. Campbell, C. A. Reddy, and M. R. Crabill. 1977. Growth of

[17]Dolfing et al. 1985; Wiegant 1988; Dubourguier et al. 1988; Grotenhuis et al 1986, 1988; Guiot et al. 1988; San-Soon et al. 1988; Verrier et al. 1988.

Desulfovibrio in lactate and ethanol media low in sulfate in association with H_2 utilizing methanogenic bacteria. *Appl. Environ. Microbiol.* 33:1162–1169.
Cappenberg, T. A. 1974. Interrelations between sulfate reducing and methane producing bacteria in bottom deposits of a fresh water lake. I. Field observations. *Ant. v. Leeuwenhoek* 40:285–295.
Chartrain, M., and J. G. Zeikus. 1986a. Microbial ecophysiology of whey biomethanation: Characterization of trophic populations and prevalent species composition in continuous culture. *Appl. Environ. Microbiol.* 51:188–196.
Chartrain, M., and J. G. Zeikus. 1986b. Microbial ecophysiology of whey biomethanation: Intermediary metabolism of lactose degradation in continuous culture. *Appl. Environ.* Microbiol. 51:180–187.
Chartrain, M., L. Bhatnagar, and J. G. Zeikus. 1987. Microbial ecophysiology of whey biomethanation. Comparison of carbon transformation parameters, species composition, and starter culture performance in continuous culture. *Appl. Environ. Microbiol.* 53:1147–1156.
Chen, M., and M. J. Wolin. 1977. Influence of CH_4 production by *Methanobacterium ruminantium* on the fermentation of glucose and lactate by *Selenomonas ruminantium. Appl. Environ. Microbiol.* 34:756–759.
Chung, K. T. 1976. Inhibitory effects of H_2 on the growth of *Clostridium cellobioparum. Appl. Environ. Microbiol.* 31:342–348.
Cohen, A., A. M. Breure, J. G. van Andel, and A. van Deursen. 1980. Influence of phase separation on the anaerobic digestion of glucose. I. Maximum COD-turnover rate during continuous operation. *Water Res.* 14:1239–1248.
Cohen, A., A. M. Breure, J. G. van Andel, and A. van Deursen. 1982. Influence of phase separation on the anaerobic digestion of glucose. II. Stability and kinetic responses to shock loadings. *Water Res.* 16:449–455.
Cohen, A., B. Distel, A. van Deursen, A. M. Breure, and J. G. van Andel. 1985. Role of anaerobic spore forming bacteria in the acidogenesis of glucose: changes induced by discontinuous or low rate feed supply. *Ant. v. Leeuwenhoek* 51:179–192.
Conrad, R., T. J. Phelps, and J. G. Zeikus. 1985. Gas metabolism evidence in support of juxtaposition of hydrogen-producing and methanogenic bacteria in sewage sludge and lake sediments. *Appl. Environ. Microbiol.* 50:595–601.
Conrad, R., B. Schink, and T. J. Phelps. 1986. Thermodynamics of H_2-consuming and H_2-producing reactions in diverse methanogenic environments under in situ conditions. *FEMS Microbiol. Ecol.* 38:353–360.
Conrad, R., H. Schütz, and M. Babbel. 1987a. Temperature limitation of hydrogen turnover and methanogenesis in anoxic paddy soil. *FEMS Microbiol. Ecol.* 45:281–289.
Conrad, R., F. S. Lupton, and J. G. Zeikus. 1987b. Hydrogen metabolism and sulfate-dependent inhibition of methanogenesis in a eutrophic lake sediment (Lake Mendota). *FEMS Microbiol. Ecol.* 45:107–115.
Cord-Ruwisch, R., H. J. Seitz, and R. Conrad. 1988. The capacity of hydrogenotrophic anaerobic bacteria to compete for traces of hydrogen depends on the redox potential of the terminal acceptor. *Arch. Microbiol.* 149:350–357.
Daniels, L., R. Sparking, and G. D. Sprott. 1984. The bioenergetics of methanogenesis. *Biochem. Biophys. Acta* 768:113–163.
Dolfing, J., A. Griffioen, R. W. van Nerven, and L. P. T. M. Zeevenhuisen. 1985. Chemical and bacteriological composition of granular methanogenic sludge. *Can. J. Microbiol.* 31:744–750.
Dolfing, J. 1988. Acetogenesis. Pages 417–468 in A. J. B. Zehnder (ed.), *Biology of Anaerobic Microorganisms*, J. Wiley Interscience, New York.
Dolfing, J., and J. M. Tiedje. 1988. Acetate inhibition of methanogenic syntrophic benzoate degradation. *Appl. Environ. Microbiol.* 54:1871–1873.
Dubourguier, H. C., E. Samain, G. Prensier, and G. Albagnac. 1986. Characterization of two strains of *Pelobacter carbinolicus* isolated from anaerobic digestors. *Arch. Microbiol.* 145:248–253.
Dubourguier, H. C., G. Prensier, and G. Albagnac. 1988. Structure and microbial activities of granular anaerobic sludge. Pages 18–34 in G. Lettinga, A. J. B. Zehnder,

J. T. C. Grotenhuis, and L. W. Hulshoff Pol (eds.), *Granular Anaerobic Sludge: Microbiology and Technology*. Centre for Agricultural Publishing and Documentation, Wageningen, The Netherlands.

Dwyer, D. F., E. Weeg-Aerssens, D. R. Shelton, and J. M. Tiedje. 1988. Bioenergetic conditions of butyrate metabolism by a syntrophic anaerobic bacterium in coculture with hydrogen-oxidizing methanogenic and sulfidogenic bacteria. *Appl. Environ. Microbiol.* 54:1354–1359.

Gosh, S., and D. L. Klass. 1978. Two-phase anaerobic digestion. *Proc. Biochem.* 13:15–24.

Grotenhuis, J. T. C., F. P. Houwen, C. M. Plugge, and A. J. B. Zehnder. 1986. Microbial interactions in granular sludge. In *Perspectives in Microbial Ecology*. Proceedings of the Fourth International Symposium on Microbial Ecology, Slovene Society for Microbiology, Ljubljana.

Grotenhuis, J. T. C., J. van Lier, C. M. Plugge, and A. J. M. Stams. 1988. Effect of calcium removal on size and strength of methanogenic granules. Pages 117–120 in E. R. Hall and P. N. Hobson (eds.), *Anaerobic Digestion 1988*, Pergamon Press, New York.

Guiot, S. R., S. S. Gorur, and K. J. Kennedy. 1988. Nutritional and environmental factors contributing to microbial aggregation during upflow anaerobic sludge bed filter (UBF) reactor start-up. Pages 47–53 in E. R. Hall and P. N. Hobson (eds.), *Anaerobic Digestion* 1988, Pergamon Press, New York.

Harper, S. R., and F. G. Pohland. 1986. Recent developments in hydrogen management during anaerobic biological wastewater treatment. *Biotechnol. Bioeng.* 28:585–602.

Hulshoff Pol, L. W., W. J. de Zeeuw, C. T. M. Velzeboer, and G. Lettinga. 1983. Granulation in UASB reactors. *Water Sci. Technol.* 15:291–304.

Hungate, R. E. 1966. *The Rumen and Its Microbes*. Academic Press, New York.

Huser, B. A., K. Wuhrmann, and A. J. B. Zehnder. 1982. *Methanothrix soehngenii* gen. nov. sp. nov., a new acetotrophic non-hydrogen-oxidizing methane bacterium. *Arch. Microbiol.* 132:1–9.

Ianotti, E. K., P. Kafkewitz, M. J. Wolin, and M. P. Bryant. 1973. Glucose fermentation products of *Ruminococcus albus* grown in continuous culture with *Vibrio succinogenes*: Changes caused by interspecies transfer of H_2. *J. Bacteriol.* 114:1231–1240.

Ingvorsen, K., A. J. B. Zehnder, and B. B. Jorgensen. 1984. Kinetics of sulfate and acetate uptake by *Desulfobacter postgatei*. *Appl. Environ. Microbiol.* 47:403–408.

Isa, Z., Grusenmeyer, S., and Verstraete, W. 1986a. Sulfate reduction relative to methane production in high rate anaerobic digestion: Technical aspects. *Appl. Environ. Microbiol.* 51:572–579.

Isa, Z., S. Grusenmeyer, and W. Verstraete. 1986b. Sulfate reduction relative to methane production in high rate anaerobic digestion. Microbiological aspects. *Appl. Environ. Microbiol.* 51:580–587.

Jeris, J. S., and P. L. McCarty. 1965. The biochemistry of methane fermentation using ^{14}C-tracers. *J. Water Poll. Contr. Fed.* 37:178–192.

Jones, W. J., J. P. Guyot, and R. S. Wolfe. 1984. Methanogenesis from sucrose by defined immobilized consortia. *Appl. Environ. Microbiol.* 47:1–6.

Jones, W. J., D. P. Nagle, and W. P. Whitman. 1987. Methanogens and the diversity of Archaebacteria. *Microbiol. Rev.* 51:135–177.

Karhadkar, P. P., J. M. Audix, G. M. Faup, and P. Khanna. 1987. Sulfur and sulfate inhibition of methanogenesis. *Water Res.* 21:1061–1066.

Kaspar, H. F., and K. Wuhrmann. 1978. Product inhibition in sludge digestion. *Microbiol. Ecol.* 4:241–248.

Kisaalita, W. S., K. L. Pinder, and K. V. Lo. 1987. Acidogenic fermentation of lactose. *Biotechnol. Bioeng.* 30:88–95.

Koster, I. W., A. Rinzema, A. L. de Vegt, and G. Lettinga. 1986. Sulfide inhibition of the methanogenic activity of granular sludge at various pH levels. *Water Res.* 12:1561–1567.

Kristiansson, J. K., P. Schönheit, and R. K. Thauer. 1982. Different K_s values for hydrogen of methanogenic bacteria and sulfate reducing bacteria: an explanation for the apparent inhibition of methanogenesis by sulfate. *Arch. Microbiol.* 131:278–282.

Lee, M. J., and S. H. Zinder. 1988. Isolation and characterization of a thermophilic bacterium which oxidizes acetate in syntrophic association with a methanogen and which grows acetogenically on H_2-CO_2. *Appl. Environ. Microbiol.* 54:124–129.

Lettinga, G., L. W. Hulshoff Pol, I. W. Koster, W. M. Wiegant, W. J. de Zeeuw, A. Rinzema, P. C. Grin, R. E. Roersma, and S. W. Hobma. 1984. High rate anaerobic wastewater treatment using the UASB reactor under a wide range of temperature conditions. *Biotechnol. Gen. Eng. Rev.* 2:253–284.

Lettinga, G., W. de Zeeuw, P. L. Hulshoff, W. Wiegant, and A. Rinzema. 1985. Anaerobic digestion 1985. Pages 279–301 in *Proceedings of the Fourth International Symposium on Anaerobic Digestion*, Guangzhou, China.

Lovley, D. R., D. F. Dwyer, and M. J. Klug. 1982. Kinetic analysis of competition between sulfate reducers and methanogens for hydrogen in sediments. *Appl. Environ. Microbiol.* 43:1373–1379.

Lovley, D. R., and M. J. Klug. 1983. Sulfate reducers can out compete methanogens at freshwater sulfate concentrations. *Appl. Environ. Microbiol.* 43:1373–1379.

Lovley, D. R., R. Greening, and J. G. Ferry. 1984. Rapidly growing rumen methanogenic organism that synthesizes coenzyme M and has high affinity for formate. *Appl. Environ. Microbiol.* 48:81–87.

Lovley, D. R., and E. J. P. Phillips. 1987. Competitive mechanisms for inhibition of sulfate reduction and methane production in the zone of ferric iron reduction in sediments. *Appl. Environ. Microbiol.* 53:2636–2641.

Lupton, F. S., and J. G. Zeikus. 1984. Physiological basis for sulfate dependent hydrogen competition between sulfidogens and methanogens. *Curr. Microbiol.* 11:7–12.

Macario, A. J. L., and E. Conway de Macario. 1988. Quantitative immunological analysis of the methanogenic flora of digestors reveals a considerable diversity. *Appl. Environ. Microbiol.* 54:79–86.

Macario, A. J. L., E. Conway de Macario, U. Ney, S. M. Schoberth, and H. Sahm. 1989. Shifts in methanogenic subpopulations measured with antibody probes in a fixed-bed loop anaerobic bioreactor treating sulfite evaporator condensate. *Appl. Environ. Microbiol.* 55:1996–2001.

Mah, R. A. 1982. Methanogenesis and methanogenic partnerships. *Phil. Trans. R. Soc. Lond.* B297:599–616.

McCarty, P. L. 1982. One hundred years of anaerobic treatment. Pages 3–22 in D. E. Hughes et al. (eds.), *Anaerobic Digestion 1981*, Elsevier, New York.

McInerney, M. J., M. P. Bryant, and N. Pfennig. 1979. Anaerobic bacterium that degrades fatty acids in syntrophic association with methanogens. *Arch. Microbiol.* 132: 129–135.

McInerney, M. J., M. P. Bryant, R. B. Hespell, and J. W. Costerton. 1981. *Syntrophomonas wolfei* gen. nov. spec. nov., an anaerobic syntrophic fatty acid oxidizing bacterium. *Appl. Environ. Microbiol.* 41:1029–1039.

Mergenhagen, S. E., and B. Rosan (eds.). 1985. *Molecular Basis of Oral Microbial Adhesion*. American Society for Microbiology, Washington, D.C.

Miller, T. L., M. J. Wolin, Z. Hongxue, and M. P. Bryant. 1986. Characteristics of methanogens isolated from bovine rumen. *Appl. Environ. Microbiol.* 51:201–202.

Min, H., and S. H. Zinder. 1989. Kinetics of acetate utilization by two thermophilic acetotrophic methanogens: *Methanosarcina* sp. strain CALS-1 and *Methanothrix* sp. strain CALS-1. *Appl. Environ. Microbiol.* 55:488–491.

Mountfort, D. O., and R. A. Asher. 1979. Effect of inorganic sulfide on the growth and metabolism of *Methanosarcina barkeri* strain DM. *Appl. Environ. Microbiol.* 37:670–675.

Mountfort, D. O., R. A. Asher, E. L. Mays, and J. M. Tiedje. 1980. Carbon and electron flow in mud and sandflat intertidal sediments in Delaware Inlet, Nelson, New Zealand. *Appl. Environ. Microbiol.* 39:686–694.

Mountfort, D. O., and M. P. Bryant. 1982. Isolation and characterization of an anaerobic syntrophic benzoate-degrading bacterium from sewage sludge. *Arch. Microbiol.* 133: 249–256.

Mountfort, D. O., W. J. Brulla, L. R. Krumholz, and M. P. Bryant. 1984. *Syntrophus bruswellii* gen. nov. spec. nov.: a benzoate catabolizer from methanogenic ecosystems. *Int. J. Syst. Bacteriol.* 34:216–217.

Mountfort, D. O. 1986. Role of anaerobic fungi in the rumen ecosystem. Pages 176–179 in F. Megusar and M. Gantar (eds.), *Perspectives in Microbial Ecology*, Proceedings of the Fourth International Symposium on Microbial Ecology, Sloven Society for Microbiology, Ljubljana.

Murray, W. D. 1986. Symbiotic relationship of *Bacteroides cellulosolvens* and *Clostridium saccharolyticum* in cellulose fermentation. *Appl. Environ. Microbiol.* 51:710–714.

Oremland, R. S., and G. M. King. 1989. Methanogenesis in hypersaline environments. In Y. Cohen and E. Rosenberg (eds.), *Microbial Mats—Physiological Ecology of Benthic Microbial Communities*, American Society for Microbiology, Washington, D.C.

Ozturk, S. S., B. O. Palsson, and J. H. Thiele. 1989. Control of interspecies electron transfer flow during anaerobic digestion: Dynamic diffusion reaction models for hydrogen gas transfer in microbial flocs. *Biotechnol. Bioeng.* 33:745–757.

Phelps, T. J., and J. G. Zeikus. 1985. Effects of fall turnover on terminal carbon metabolism in Lake Mendota sediments. *Appl. Environ. Microbiol.* 50:1285–1291.

Phelps, T. J., R. Conrad, and J. G. Zeikus. 1985. Sulfate dependent interspecies H_2 transfer between *Methanosarcina barkeri* and *Desulfovibrio vulgaris* during coculture metabolism of acetate or methanol. *Appl. Environ. Microbiol.* 50:589–594.

Pohland, F. G., and S. H. Harper. 1985. *Bi*ogas developments in North America. Pages 41–81 in *Proceedings of the Fourth International Symposium on Anaerobic Digestion*, Guangzhou, China.

Reddy, C. A., M. P. Bryant, and M. J. Wolin. 1972a. Characterization of S-organism isolated from *Methanobacillus omelianskii. J. Bacteriol.* 109:539–545.

Reddy, C. A., M. P. Bryant, and M. J. Wolin. 1972b. Ferredoxin and nicotinamide adenin dinucleotide dependent H_2 production from ethanol and formate in extracts of S-organism isolated from *Methanobacillus omelianskii. J. Bacteriol.* 110:126–132.

Rinzema, A., and G. Lettinga. 1988. The effect of sulfide on the anaerobic degradation of propionate. *Environ. Tech. Lett.* 9:83–88.

Robinson, J. A., R. F. Strayer, and J. M. Tiedje. 1981. Method for measuring dissolved hydrogen in anaerobic ecosystems: applications to the rumen. *Appl. Environ. Microbiol.* 41:545–548.

Robinson, J. A., and J. M. Tiedje. 1984. Competition between sulfate reducing and methanogenic bacteria for H_2 under resting and growing conditions. *Arch. Microbiol.* 137:26–32.

Roy, F., E. Samain, H. C. Dubourguier, and G. Albagnac. 1986. *Syntrophomonas sapovorans* sp. nov., a new obligately proton reducing anaerobe oxidizing saturated and unsaturated long chain fatty acids. *Arch. Microbiol.* 145:142–147.

Sahm, H. 1984. Anaerobic wastewater treatment. *Adv. Biochem. Eng. Biotechnol.* 29: 84–115.

Sam-Soon, P. A. L. N. S., R. E. Lowenthal, P. L. Dold, and G. V. R. Marais. 1988. Pelletization in upflow anaerobic sludge bed reactors. Pages 55–60 in E. R. Hall and P. N. Hobson (eds.), *Anaerobic Digestion 1988*. Pergamon Press, New York.

Schink, B., and J. G. Zeikus. 1982. Microbial ecology of pectin decomposition and anoxic lake sediments. *J. Gen. Microbiol.* 128:393–404.

Schink, B., and M. Stieb. 1983. Fermentative degradation of polyethylene glycol by a strict anaerobic, Gram-negative, non-sporeforming bacterium. *Appl. Environ. Microbiol.* 45:1905–1913.

Schink, B. 1984. Fermentation of 2,3 butanediol by *Pelobacter carbinolicus* sp. nov. and *Pelobacter propionicus* sp. nov. and evidence for propionate formation from C_2 compounds. *Arch. Microbiol.* 137:33–41.

Schönheit, P., J. K. Kristijansson, and R. K. Thauer. 1982. Kinetic mechanism for the ability of sulfate reducers to outcompete methanogens for acetate. *Arch. Microbiol.* 132:285–288.

Shelton, D. R., and J. M. Tiedje. 1984. Isolation and partial characterization of bacteria in an anaerobic consortium that mineralizes 3-chlorobenzoic acid. *Appl. Environ. Microbiol.* 48:840–848.

Stieb, M., and B. Schink. 1985. Anaerobic oxidation of fatty acids by *Clostridium bryantii* spec. nov., a sporeforming, obligately syntrophic bacterium. *Arch. Microbiol.* 140:387–390.

Strayer, R. F., and J. M. Tiedje. 1978. Kinetic parameters of the conversion of methane precursors to methane in hypereutrophic lake sediment. *Appl. Environ. Microbiol.* 36:330–340.

Thauer, R. K., K. Jungermann, and K. Decker. 1977. Energy conservation in chemotrophic anaerobic bacteria. *Bacteriol. Rev.* 41:100–180.

Thauer, R. K., and J. G. Morris. 1984. Metabolism of chemotrophic anaerobes. Old views and new aspects. Pages 123–168 in D. P. Kelly and N. G. Carr (eds.), *The Microbe 1984*, Part 2 (Symp. 100th Meeting Soc. Gen. Microbiol., Warwick), Cambridge University Press, Cambridge.

Tiedje, J. M., A. J. Sexstone, T. B. Parkin, N. P. Revsbeck, and D. R. Shelton. 1984. Anaerobic processes in soil. *Plant Soil* 76:197–212.

Thiele, J. H. 1988. Formate is an intermediate of syntrophic ethanol conversion by *Desulfovibrio vulgaris* and *Methanobacterium formicicum*. Abstr. I8 on the 88th Annual Meeting of the American Society for Microbiology, Miami Beach, Florida.

Thiele, J. H., and J. G. Zeikus. 1988a. The anion-exchange substrate shuttle process: A new approach to two-stage biomethanation of organic and toxic wastes. *Biotechnol. Bioeng.* 31:521–535.

Thiele, J. H., M. Chartrain, and J. G. Zeikus. 1988b. Control of interspecies electron flow during anaerobic digestion: Role of floc formation in syntrophic methanogenesis. *Appl. Environ. Microbiol.* 54:10–19.

Thiele, J. H., and J. G. Zeikus. 1988c. Control of interspecies electron flow during anaerobic digestion: Significance of formate transfer versus hydrogen transfer during syntrophic methanogenesis in flocs. *Appl. Environ. Microbiol.* 54:20–29.

Thiele, J. H., and J. G. Zeikus. 1988d. Interactions between hydrogen and formate producing bacteria and methanogens during anaerobic digestion. Pages 537–595 in L. E. Erickson and D. Y. C. Fung (eds.), *Handbook on Anaerobic Fermentations*, Marcel Dekker, New York.

Thiele, J. H., W.-M. Wu, H. Grethlein, and J. G. Zeikus. 1990. Whey treatment with ecoengineered biomethanation catalysts: Biological process design with waste conditioning and byproduct recovery. *Biol. Wastes* (submitted).

Verrier, D., B. Mortier, H. C. Dubourguier, and G. Albagnac. 1988. Adhesion of anaerobic bacteria to inert supports and development of methanogenic biofilms. Pages 61–69 in E. R. Hall and P. N. Hobson (eds.), *Anaerobic Digestion 1988*, Pergamon Press, New York.

Vogels, G. D., W. Hoppe, and C. K. Stumm. 1980. Association of methanogenic bacteria with rumen ciliates. *Appl. Environ. Microbiol.* 40:608–612.

Ward, D. M., and M. R. Winfrey. 1985. Interactions between methanogenic and sulfate reducing bacteria in sediments. Pages 141–179 in H. W. Jannasch and P. J. B. Williams (eds.), *Advances in Aquatic Microbiology*, vol. 3, Academic Press, Orlando, Florida.

Weimer, P. J., and J. G. Zeikus. 1977. Fermentation of cellulose and cellobiose by *Clostridium thermocellum* and *Methanobacterium thermoautotrophicum*. *Appl. Environ. Microbiol.* 33:289–297.

Westermann, P., B. K. Ahring, and R. A. Mah. 1989. Threshold acetate concentrations for acetate catabolism by aceticlastic methanogenic bacteria. *Appl. Environ. Microbiol.* 55:514–515.

Widdel, F. 1988. Microbiology and ecology of sulfate and sulfur-reducing bacterium. Pages 469–585 in A. J. B. Zehnder (ed.), *Biology of Anaerobic Microorganisms*, Wiley Interscience, New York.

Wiegant, W. M., and A. W. A. de Man. 1986. Granulation of biomass in thermophilic anaerobic sludge blanket reactors treating acidified wastewaters. *Biotechnol. Bioeng.* 28:718–727.

Wiegant, W. M. 1988. The "spaghetti theory" on anaerobic sludge formation, or the inevitability of granulation. Pages 146–153 in G. Lettinga, A. J. B. Zehnder, J. T. C.

Grotenhuis, and L. W. Hulshoff Pol (eds.), *Granular Anaerobic Sludge: Microbiology and Technology*, Center for Agricultural Publishing and Documentation, Wageningen.

Winfrey, M. R., and J. G. Zeikus. 1977. Effect of sulfate on carbon and electron flow during microbial methanogenesis in freshwater sediments. *Appl. Environ. Microbiol.* 33:275–281.

Winfrey, M. R., and J. G. Zeikus. 1979. Anaerobic metabolism of immediate methane precursors in Lake Mendota. *Appl. Environ. Microbiol.* 37:244–253.

Winter, J. U., and R. S. Wolfe. 1980. Methane formation from fructose by syntrophic association of *Acetobacterium woodii* and different strains of methanogens. *Arch. Microbiol.* 124:73–79.

Wimpenny, J. 1989. Laboratory model systems for the experimental investigation of gradient communities. Pages 366–383 in Y. Cohen and E. Rosenberg (eds.), *Microbial Mats—Physiological Ecology of Benthic Microbial Communities*. American Society for Microbiology, Washington, D.C.

Wolin, M. J. 1982. Hydrogen transfer in microbial communities. Pages 323–356 in A. T. Bull and J. H. Slater (eds.), *Microbial Interactions and Communities*, vol. 1.

Wolin, M. J., and T. L. Miller. 1982. Interspecies H_2 transfer: 15 years later. *ASM News* 48:561–565.

Zehnder, A. J. B. (ed.) 1988. Geochemistry and biogeochemistry of anaerobic habitats. In *Biology of Anaerobic Microorganisms*, Wiley Interscience, New York.

Zeikus, J. G., and D. L. Henning. 1975. *Methanobacterium arboriphilum* sp. nov. an obligate anaerobe isolated from wetwood of living trees. *Ant. v. Leeuwenhoek* 41:543–552.

Zeikus, J. G. 1977. The biology of methanogenic bacteria. *Bacteriol. Rev.* 41:514–541.

Zeikus, J. G. 1979. Microbial populations in digestors. Pages 75–103 in D. A. Stafford et al. (eds.), *First International Symposium on Anaerobic Digestion*, A. D. Cardiff Scientific Press, Cardiff, United Kingdom.

Zeikus, J. G. 1980. Chemical and fuel production by anaerobic bacteria. *Ann. Rev. Microbiol.* 34:423–464.

Zeikus, J. G. 1983. Metabolism of one carbon compound by chemotrophic anaerobes. *Adv. Microb. Physiol.* 24:215–299.

Zeikus, J. G. 1983. Metabolic communications between biodegradative populations in nature. Pages 423–462 in J. H. Slater, R. Whittenbury, and J. W. T. Wimpenny (eds.), *Microbes in Their Natural Environment*, Symp. 34, Society General Microbiology, Ltd., Cambridge University Press, Cambridge, England.

Zeikus, J. G., R. Kerby, and J. A. Krzycki. 1985. Single-carbon chemistry of acetogenic and methanogenic bacteria. *Science* 227:1167–1173.

Zellner, G., P. Vogel, H. Kneifel, and J. Winter. 1987. Anaerobic digestion of whey and whey permeate with suspended and immobilized complex and defined consortia. *Appl. Microbiol. Biotechnol.* 27:306–314.

Zinder, S. H., and M. Koch. 1984. Non-aceticlastic methanogenesis from acetate: acetate oxidation by a thermophilic syntrophic coculture. *Arch. Microbiol.* 138:263–272.

Zinder, S. H., S. C. Cardwell, T. Anguish, M. Lee, and M. Koch. 1984. Methanogenesis in a thermophilic anaerobic digestor. *Methanothrix* as an important aceticlastic methanogen. Appl. Environ. Microbiol. 47:796–807.

Zoetemeyer, F. J., J. C. van den Heuvel, and A. Cohen. 1982. pH influence on acidogenic dissimilation of glucose in an anaerobic digestor. *Wat. Res.* 16:303–311.

Chapter

11

Mixed Cultures in Detoxification of Hazardous Waste

Lakshmi Bhatnagar

Michigan Biotechnology Institute
Lansing, Michigan

Department of Microbiology and Public Health
Michigan State University
East Lansing, Michigan

Babu Z. Fathepure

Department of Civil & Environmental Engineering
Michigan State University
East Lansing, Michigan

Introduction

Microorganisms possess catabolic pathways for degradation of naturally occurring organic compounds (lignocellulosics, pectin, etc.) that enable recycling of global organic carbon. Synthetic chemicals (xenobiotics) with novel structures are often recalcitrant to microbial attack in nature, perhaps because these chemicals are recent introductions to nature and the microbial world has not evolved catabolic pathways to degrade these recalcitrant compounds. Large quantities of such xenobiotics have been produced and released into the environment over the past few decades. Many toxic chemicals used as solvents, refrigerants, plasticizers, gasoline and other petrochemicals, biocides, pesticides, wood preservatives, etc., have ended up in industrial effluents, spills, waste dumps, sediments, groundwater, and air. The potential advantage for use of mixed cultures over pure cultures in the degradation of toxic compounds in hazardous wastes is becoming apparent, especially when we know that traditionally most waste-treatment systems use mixed cultures (e.g., activated sludge systems, anaerobic digestors). Mixed cultures are particularly important when the emphasis is placed on complete mineralization of toxic organics to CO_2 (under aerobic conditions) or to CO_2 and either CH_4, H_2S, or N_2 (under anaerobic conditions). Many pure-culture studies have shown that toxic intermediates accumulate during biodegradation, because a single organism may not have the ability to completely mineralize the xenobiotic. Mineralization of xenobiotics has been documented by aerobic microorganisms in pure culture (*Pseudomonas* sp., *Flavobacterium* sp., etc.) (Rochkind-Dubinsky et al. 1987, Leisinger and Brunner, 1986). Pure-culture degradation appears to be rare with anaerobic organisms since complete mineralization usually requires one or more trophic groups (such as fermentative-hydrolytic; sulfidogenic, acetogenic, and/or methanogenic groups) to convert toxic chemicals by a multistep reaction to mineralized end products (Young 1984, Schink 1988). In this chapter, we define microbial detoxification as a process of biodegradation in which toxic xenobiotics are either converted to nontoxic stable intermediates or degraded completely to mineralized end products by microorganisms.

Biodegradation studies essentially start with developing mixed microbial cultures capable of detoxifying a toxic chemical by employing enrichment and selection techniques. Microbial biochemists and geneticists work with pure cultures in order to understand the basic biochemical mechanisms and pathways of microbial detoxification, which are not well known. As a follow-up of such studies, several purified enzymes (e.g., oxygenases or dehalogenases) and their corresponding genes have been described. In this chapter, we will restrict ourselves to mixed-culture studies and emphasize the importance of microbial communities in biodetoxification, though pure-culture studies will be mentioned when necessary. There are several recent reviews on biodegradation of toxic chemicals by pure and some mixed cultures (Gibson and Subramanian 1984, Rochkind-Dubinsky et al. 1987, Reineke and Knackmuss 1988). Most of these studies have been on aerobes and very few reports have appeared on anaerobes.

One of the key aspects of mixed cultures and microbial communities is the interrelationships between different species in respect to their degradative capabilities toward toxic chemicals. This information is important in establishing the catabolic reaction sequence of a toxicant and the synergistic role of individual microbial species or groups in facilitating the rate of xenobiotic biodegradation. Notably, Slater and coworkers (1978, 1984) and a few other groups (Pfennig 1978, Whittenbury, 1978) have developed ecological concepts on microbial communities and their potential applications in biotechnological industries.

Ecological Role of Microbial Communities

Suitable techniques are now available to isolate and characterize stable microbial communities. It still remains an open question whether such isolated microbial communities perform differently in their natural habitats. It has been demonstrated that mixed cultures more effectively detoxify a chemical when compared to individual members of the mixed population. Many questions remain unanswered, including the following: What is the effect of environmental conditions versus idealized laboratory conditions on the composition and structure of microbial communities? What is the role of microbes not directly involved in detoxification in the microbial community and are they essential for the stability of the microbial population? What other microbial or physicochemical factors are required for the stability of a microbial community? These questions have led to distinguishing microbial communities based on different primary physiological interactions observed between the component species.

Types of communities

In spite of their distinctive properties, spatial locations, and metabolic activities, the growth abilities of different microorganisms overlap. The basic biological mechanisms of mixed-culture growth and activity are still not understood. Important concepts and properties of mixed microbial cultures may have been overlooked (Slater 1978, Bull 1980, Slater and Lovatt 1984). These factors make it difficult to simulate in the laboratory conditions that occur in the natural environment. Moreover, there is confusion over the terminology used to describe different interrelationships and communities, e.g., "loosely" interacting communities and "tightly" interacting communities. We will not go into details of these classifications, as excellent reviews have appeared during the last few years (Slater 1978, Slater and Lovatt 1984). In this chapter we will briefly describe the different types of microbial communities with special emphasis on the "tightly" interacting communities where biodetoxification can be achieved through complete mineralization of xenobiotic by sequential catabolic activities of two or more microorganisms. Studies with these kinds of communities can allow elucidation of basic mechanisms for biodegradation and the role of microbial components in the overall process of biodetoxification.

Efforts have been made to classify various types of microbial communities (Slater and Lovatt 1984). At least seven different types of microbial communities have been described based on various metabolic interactions: (1) provision of specific nutrients, (2) removal of growth-inhibitory products, (3) modification of individual organisms' basic growth parameters, (4) combined (concerted) metabolic attack, (5) cometabolism, (6) hydrogen (electron) transfer, and (7) presence of more than one primary substrate utilizer. Others have compared growth modes such as free-living versus attached populations in biofilms or aggregates (Thiele et al. 1988).

Biodegradative function of microbial communities

The first three types of microbial communities based on metabolic interactions are of importance for degradation of simple organic compounds, whereas the last four categories include microbial association involved in the catabolism of more complex organic compounds and xenobiotics. As an example of the first category, Stirling et al. (1976) isolated a stable community containing a *Nocardia* sp. and a pseudomonad from enrichments on cyclohexane. From this two-member microbial community, the *Nocardia* sp. could oxidize cyclohexane alone but it did not grow unless the pseudomonad was also present indicating that *Nocardia* sp. required growth factors, particularly biotin. Thermophilic microbial populations growing on hexadecane were isolated from soil at 55 to 65°C and grew in defined mineral medium containing hexadecane but failed to grow when individual colonies were transferred in the same medium. It has also been shown in other studies that thermophiles generally require exogenous sources of organic growth factors (Slater and Lovatt 1984).

A four-member community growing on methane that was used for single-cell protein production was isolated by Wilkinson et al. (1974), representing the second type of interaction. Methanol inhibited the methane-oxidizing pseudomonad unless it was consumed by *Hyphomicrobium* sp. present in the community. The other two members of the community were *Flavobacterium* sp. and *Acinetobacter* sp. Removal of inhibitory products is also an important characteristic of communities involved in sulfate reduction and sulfide oxidation (Widdel 1988).

A three-member community was developed from enrichments on orcinol (3,5-dihydroxy toluene) (Bull and Brown, 1979) which included a *Pseudomonas* sp. that degraded orcinol. The other two members of the community *Brevibacterium linens* and *Curtobacterium* sp. did not grow on orcinol unless the primary degrader *Pseudomonas* sp. was present. This represents the third category of microbial community based on the modification of an individual organism's basic growth parameters.

The fourth category of microbial community based on a combined or concerted metabolic attack is extremely important in the degradation of xenobiotics. The component organisms may not separately have the capacity to transform or detoxify the toxic chemical, whereas collectively the microbial community can completely mineralize the compound. This is probably because none of the individual species has a complete set of enzyme systems or genetic information for degradation of a xenobiotic. A multimember community was isolated by growth on the herbicide Dalapon (Senior et al. 1976). It was shown that the rate of Dalapon degradation was over 20 percent greater than the combined rates of the individual organisms. Gunner and Zuckerman (1968) were among the first investigators to report synergistic metabolic activity between an *Arthrobacter* sp. and *Streptomyces* sp., which degraded the insecticide diazinon (O,O-diethyl O-2-isopropyl-4-methyl-6-pyrimydyl thio-phosphate) in soil. None of the organisms could use diazinon as sole carbon source when grown separately. A four-member community comprising *Pseudomonas putida, Pseudomonas alcaligenes, Arthrobacter globiformis*, and *Serratia marcescens* was enriched from activated sludge on the surfactant LAS (linear alkyl benzene sulfonates) (Slater and Lovatt 1984). The rate of LAS degradation and ring cleavage was much higher when two or all the members of the community were present compared to the individual species. Interestingly, this community also formed flocs only when all the four members were present. Flocs may concentrate the chemical and facilitate contact with the degrading organisms (Swisher 1970).

A mixed microbial community isolated from soil was shown to degrade styrene (Sielicki et al. 1978). Intermediates such as phenylethanol and phenylacetic acid were detected in mixed cultures. Phenylacetic acid was further metabolized by well-known mechanisms (Dagley 1975) by an organism of the community which was unable to grow directly on styrene. In these experiments, 4-tert-butyl catechol was used to prevent free radical polymerization of styrene to polystyrene. Interestingly, an insoluble white material appeared and then disappeared from the culture during growth of the mixed culture. This mixed culture was shown to metabolize 4-tert-butyl catechol, thus allowing the polymerization of styrene, suggesting that this consortia had the capacity to degrade the polymer.

The fifth type of microbial community is based on cometabolism. In general, many microorganisms growing on one substrate may be able to transform or degrade a different cosubstrate in a reaction or sequence of reactions that is not directly associated with the organism's energy generation, carbon assimilation, and biosynthesis or growth

(Horvath 1972, Alexander 1979). Cometabolism is not clearly understood and its importance in nature has probably been underestimated because of the difficulties in designing appropriate enrichment and selection procedures. Cometabolism was formerly termed cooxidation, but the concept also includes reactions not necessarily involved in catabolism of the growth substrate. It is important to note that in a microbial community one organism generates a compound which it cannot use itself but which can be used as a metabolite for other organisms in the community. If these organisms cannot use the initial cometabolic substrate, then potential exists for an interacting microbial community.

Cycloalkanes are metabolized readily by mixed cultures in natural habitats, but pure cultures grow poorly on these compounds, suggesting the need for a microbial community for degradation (Beam and Perry 1974). Mixed cultures were enriched on parathion (O,O-diethyl-O-P-nitrophenyl phosphorothioate), a widely used insecticide using glucose as carbon source (Daughton and Hsieh 1977a,b). The community mainly contained organisms of the genera *Pseudomonas, Xanthomonas*, and *Brevibacterium*, none of which could individually use parathion as carbon and energy source for growth.

The sixth type of community is based on interspecies electron (H_2 or formate) or other nutrient transfer (Wolin 1982, Zeikus 1983a,b, Thiele and Zeikus 1988a,b). These communities function on the principle that under anaerobic condition organisms require a sink to dispose of excess reducing equivalents. One of the classical examples of this type of community came with the discovery of the closely associated two-member methanogenic community *Methanobacillus omelianski* (Bryant et al. 1967). This community comprised the "S organism" which oxidized ethanol to acetate and hydrogen and a methanogen *Methanobacterium* strain MOH which used hydrogen to reduce CO_2 to methane. This association permitted continuous ethanol metabolism, as the methanogen present prevented accumulation of inhibitory levels of hydrogen. Similar communities have been isolated (Bryant et al. 1977) from anaerobic environments. In other anaerobic mixed cultures, syntrophs degrade alcohols and fatty acids into acetate and formate which are then degraded by methanogens (Thiele and Zeikus 1988a,b).

Anaerobic communities have been described which degrade a variety of aromatic hydrocarbons (Evans 1977). A methanogenic community was isolated which degraded benzoate to methane and CO_2 in the absence of nitrate and sulfate (Ferry and Wolfe 1976, Tarvin and Buswell 1934). Addition of o-chlorobenzoate inhibited benzoate degradation but had no effect on the acetate metabolism or methanogenesis. Healy and Young (1978, 1979) isolated microbial communities growing on lignoaromatic compounds and demonstrated ring cleavage of aromatics such as catechol, cinnamic acid, ferulic acid, phenol, protocatechuic acid, syringaldehyde, and vanillic acid. The community structure was similar to that isolated from rumen degrading benzoic acid (Balba and Evans 1977), suggesting that communities have evolved the most appropriate species composition irrespective of their habitat. An anaerobic microbial community was isolated in which the degradation of phenol was not linked to methanogenesis but to nitrate reduction (Bakker 1977). Degradation of benzene, toluene, ethyl benzene, and xylenes has been recently reported under denitrifying conditions by mixed cultures (Olsen et al. 1989). Recently, a chlorobenzoate-degrading consortium comprising three organisms was described (Tiedje and Stevens 1988). Strain DCB1 dechlorinated chlorobenzoate to benzoate, which was then degraded by strain BZ-2 to acetate, hydrogen, and CO_2. The third partner of this consortium was a methanogen, *Methanosprillum* (strain PM-1).

The potential for the degradation of naturally occurring complex organic molecules such as lignin (Kirk et al. 1975, Zeikus et al. 1982) and pectin (Schink and Zeikus 1982) has been documented. Anaerobic communities based on hydrogen and formate transfer have been described and their degradative potential demonstrated for organic acids and alcohols (Zeikus 1983a, Chartrain et al. 1987, Thiele and Zeikus 1988a,b). The relative importance of interspecies hydrogen versus formate transfer has recently been documented in mixed anaerobic systems (Thiele et al. 1988; Thiele and Zeikus 1988a,b). Recently, a stable granular anaerobic community that mineralized pentachlorophenol (PCP) to methane and CO_2 was described (Bhatnagar et al. 1989a,b). This community contained mainly syntrophic acetogens and methanogens.

The last class of microbial communities is based on the presence of more than one primary utilizer. Many continuous culture enrichments have resulted in stable communities that contain more than one species capable of growing on the sole carbon and energy source provided. These communities are different from combined metabolism communities since the substrate can be completely metabolized by each of the primary utilizers. Often secondary organisms which are unable to metabolize the primary substrates are members of this stable community. The presence of several primary organisms and some secondary organisms indicates that interactions must exist between these two in order to stabilize free competition and prevent culture wash out. Such communities have been isolated on the herbicide Dalapon (Senior et al. 1976) and benzoic acid (Cossar et al. 1981). It was shown that Dalapon was degraded at much higher rates by the community than by pure cultures of the primary organisms, which may account for the success of the whole community (Slater and Lovatt 1984). A microbial community degrading the herbicide Fenuron (N,*N*-dimethyl-*N*-phenyl-urea) was isolated from estuarine sediments using continuous enrichment techniques (Minney et al. 1980). This community contained three coryneforms, one pseudomonad, and an *Alcaligenes* sp.

Little evidence is available suggesting that an isolated microbial community exists and functions in nature as they do in the laboratory, with respect to the degradation of xenobiotics. It seems likely that specific trophic groups rapidly associate to form a particular community to meet the specific environmental and metabolic challenge designed by the in vitro conditions.

Detoxification by Aerobic Mixed Cultures

Over the last few decades enormous quantities of industrial chemicals have been released into the environment. Microorganisms are essentially capable of catabolizing any organic molecule that structurally resembles a natural product. However, in recent years, microbes have encountered numerous novel compounds rarely found in nature but which are released into the biosphere by human activities. Fortunately, microorganisms collectively exhibit a remarkable ability to degrade a wide range of synthetic chemicals employing enzymatic systems that have evolved over 4 billion years for the catabolism of naturally occurring compounds. The evidence that these organisms will degrade many anthropogenic compounds can be attributed to their relaxed enzyme specificity (Dagley 1971, 1975) However, there are many synthetic chemicals of recent origin that are recalcitrant to microbial attack. This is particularly true of halogenated compounds. In terms of geological or evolutionary time scale, microorganisms have not had enough time to evolve the necessary enzymes to degrade such chemicals. However, considering the microbial ability to evolve catabolic systems for degradation of organic compounds, it is likely that many environmental toxic pollutants can be destroyed.

Many catabolic systems involved in waste degradation use molecular oxygen. Microorganisms require oxygen for two different functions. In one case it is used as terminal electron acceptor where electrons are released during oxidation of organic substrates and energy is generated. In the second case, oxygen is a substrate for a biochemical reaction and is incorporated into organic product which is further catabolized. The metabolic importance of oxygen is basically due to its high redox potential (E_0^1 = +0.86 V). Under normal conditions, oxygen is kinetically quite inert. The low reactivity of O_2 is due to its electronic structure (Leisinger and Brunner 1986). Oxygen can be activated by the introduction of external electrons to its π-orbit to form a radical of high reactivity (Malmstrom 1982). Oxygen can only react readily with organic substrates after this activation process. There are more than 100 oxygenases that are known to catalyze O_2 incorporation into organic substrates. Oxygenases are categorized into two groups based on the number of O_2 atoms incorporated into an organic molecule. Dioxygenases catalyze the insertion of both atoms of the dioxygen molecule and monooxygenases catalyze the insertion of a single atom of dioxygen, while the other oxygen is reduced to water.

Degradative mechanisms

Aromatic and halogenated aromatic hydrocarbons.

Aromatic hydrocarbons. Eukaryotes such as fungi and mammals employ monooxygenases to introduce a single atom of molecular O_2 into aromatic or aliphatic hydrocarbons. The resulting epoxide intermediate undergoes hydration with water to form a trans-1,2-dihydroxy-1,2-dihydro intermediate, which is converted to a trans-dihydroxy compound (Gibson 1976) (Fig. 11.1). Similarly, it is now generally

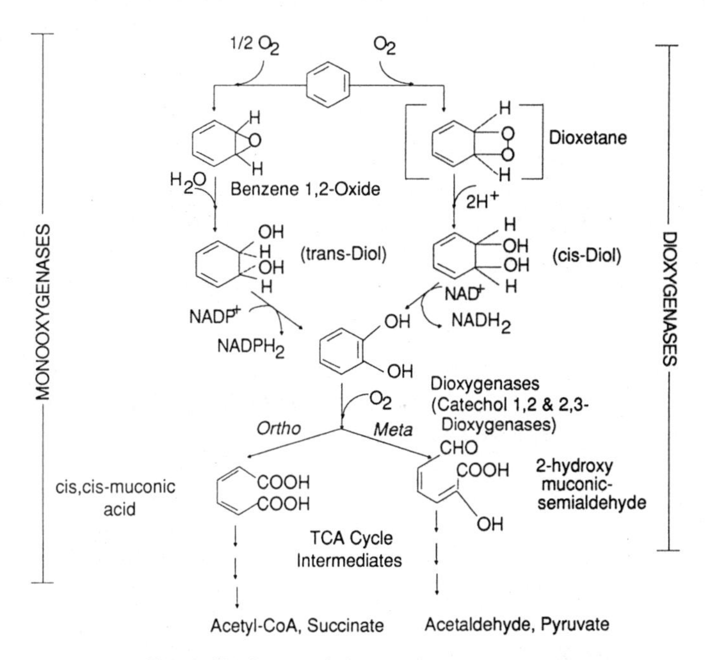

Figure 11.1 Generalized metabolic pathways of aromatics by eukaryotic and prokaryotic organisms. Formation of catechol as a central metabolite of aromatic compound oxidation. Degradation of catechol via *ortho* and *meta* pathways to tricarboxylic acid cycle intermediates. (*Adapted from Leisinger and Brunner 1986, Rochkind-Dubinsky et al. 1987, Gibson and Subramanian 1984, Ribbons and Eaton 1982, Fewson 1981.*)

accepted that degradation of aliphatic hydrocarbons proceeds via initial oxidation by monooxygenases, yielding primary or secondary alcohols (Britton 1984, Leisinger and Brunner 1986). In contrast, many bacteria catalyze oxidation of aromatic hydrocarbons by incorporating both atoms of molecular oxygen simultaneously to form cis-dihydrodiols as the first metabolite, which then loses two hydrogens to form catechol (Fig. 11.1). The formation of catechol via cis intermediate was identified first in 1968 (Gibson et al. 1968) and since then it has been shown to occur in most bacterial species studied (Fewson 1981, Leisinger and Brunner 1986, Ribbons and Eaton 1982, Gibson and Subramanian 1984, Rochkind-Dubinsky et al. 1987).

Substituted benzenes such as toluene, xylene, and other alkylbenzenes are also oxidized to cis-dihydrodiol by bacteria. In the case of substituted benzenes, oxidation is initiated through either the methyl group or the aromatic ring, depending upon the species. Kitagawa (1956) first reported the oxidation of toluene at the methyl moiety which proceeded via benzyl alcohol, benzaldehyde, benzoic acid, and catechol intermediates. Alternately, most bacteria oxidize substituted benzenes to yield alkyl catechol.[1] The oxidation of higher substituted benzenes, such as ethyl benzene, isopropyl benzene, isobutylbenzene and *n*-butylbenzene by various species of *Pseudomonas* was similar to the type of reactions observed for benzene or toluene metabolism in which the aromatic nucleus is oxidized.[2] However, long-chain alkylbenzenes are oxidized at the terminal methyl moiety. As the alkyl-chain length grows and the substituent portion gets larger, bacteria attack these chemicals like substituted alkanes rather than substituted benzenes. For example, *n*-alkylbenzenes are converted to phenyl alkanoic acids which are then metabolized by oxidation (Hou 1982).

Aromatic ring cleavage. Catechols or substituted catechols are the products of initial aromatic hydrocarbon oxidation. The ring cleavage is catalyzed by dioxygenases in which 1 mol of O_2 is incorporated into the same substrate (catechol or substituted catechol). In most microorganisms, ring opening proceeds via two pathways, namely, *ortho* and *meta*. In the case of *ortho* (intradiol), ring cleavage occurs between the two hydroxyl groups of catechol, while in the *meta* (extradiol) pathway, cleavage can occur next to one of the hydroxyl groups as shown in Fig. 11.1. In the latter case, the bond split can be distal or proximal depending on the primary substituent. The mode of cleavage is dependent on many factors, such as the type of aromatic hydrocarbon, bacterial species, and the mode of induction. Many times the

[1] Gibson and Subramanian 1984, Leisinger and Brunner 1986, Rochkind-Dubinsky et al. 1987, Hou 1982.

[2] Hou 1982, Omori et al. 1975, Jigami et al. 1975, Baggi et al. 1972.

same bacterium can use both pathways independently (Rochkind-Dubinsky et al. 1987). This kind of relaxed specificity is important and advantageous particularly when xenobiotics are to be degraded. For example, catechol 1,2-oxygenase from different organisms can use 3-methyl, 4-methyl, and 3-isopropylcatechol as substrates (Fewson 1981).

The ring fission products from the *ortho* or *meta* pathways are further metabolized to intermediates of the tricarboxylic acid cycle. The *ortho* cleavage leads to the formation of succinate and acetyl-CoA via 3-ketoadipic acid (Fig. 11.1) (Gibson and Subramanian 1984, Rochkind-Dubinsky et al. 1987). The *meta* cleavage pathway of catechol leads to the formation of pyruvic acid and acetaldehyde (Leisinger and Brunner 1986, Rochkind-Dubinsky 1987). The substituted catechols such as 3-methyl catechol and 4-methyl catechol yield acetate and formate, respectively (Bayly and Dagley 1969). Compounds that contain two hydroxyl groups located opposite to each other (para position) such as gentisic acid, are cleaved between the hydroxyl and adjacent side chain such as a carboxyl group. The products formed from this pathway include fumarate and pyruvate (Leisinger and Brunner 1986, Rochkind-Dubinsky et al. 1987).

Halogenated aromatics. As a result of rapid industrial developments over the past few decades, large quantities of organochlorine compounds have been introduced into the environment. These compounds are used in large quantities in a variety of industrial and agricultural applications such as intermediates in chemical synthesis, lubricants, insulators, heat-transfer media, plasticizers, pesticides, and others. The annual world production of chlorinated benzenes is estimated to be 8×10^5 metric tons (Merian and Zander 1982). Many of these chemicals end up in the environment either deliberately or accidentally, and due to their xenobiotic structure, most of these compounds remain undegraded for longer time periods in the nature. Under certain conditions, some of these compounds can be biotransformed by microorganisms employing existing enzymes that possess broad specificity.

The biodegradation of halogenated compounds can be considered complete only when the ring is broken down to intermediary metabolites and the organic halogen is mineralized. The single most important and rate-limiting step in biodegradation is the removal of the halogen substituent from the organic compound. This has been shown to occur mainly in two ways: (1) elimination of halogen during the initial stages of degradation via reductive, hydrolytic, or oxygenolytic removal mechanism, and (2) generation of nonaromatic structures which spontaneously lose halide by hydrolysis or hydrogen halide by β-elimination (Reineke and Knackmuss 1988, Reineke, 1984).

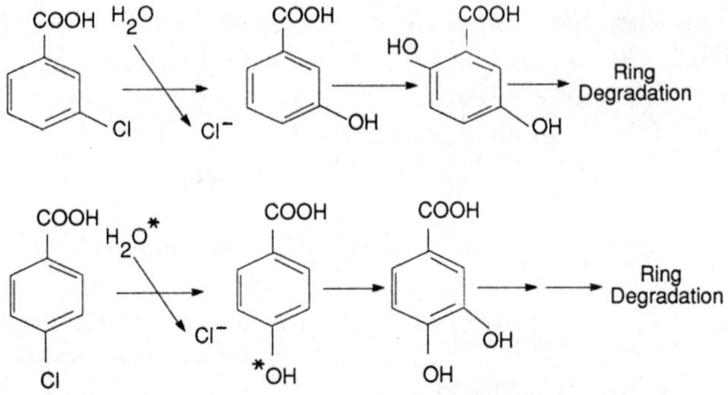

Figure 11.2 Hydrolytic dechlorination of 3- and 4-chlorobenzoate. (*Adapted from Reineke 1984, Reineke and Knackmuss 1988.*)

The arene-halogen bond is considered to be chemically inert and extremely resistant to nucleophilic displacement reactions. Hydrolysis usually requires extremely basic conditions. In the presence of a strong base, removal of a proton to promote the elimination of a halide ion forms an arene, which is then converted to an aromatic compound by nucleophilic addition (Reineke and Knackmuss 1988, Reineke 1984, Knackmuss, 1981). In situations where a haloaromatic compound contains an electron-withdrawing substituent such as NO_2 or SO_3H, the nucleophilic substitution proceeds by a two-step addition elimination mechanism. Because of the requirement of high activation energy ($E_A{}^{\#}$ for chlorobenzene ≥ 40 Kcal/mol) and a strong nucleophilic catalyst, it is unlikely that microorganisms contain enzymatic systems for direct hydrolysis of the carbon-halogen bond. Therefore, removal of halogen usually occurs only after the halogenated compound has undergone several transformations resulting in a labile carbon-halogen bond (Leisinger and Brunner, 1986).

Dehalogenation prior to ring-cleavage. Johnston et al. (1972) reported that a *Pseudomonas* sp. catalyzed dehalogenation of 3-chlorobenzoate to 3-hydroxybenzoate and 2,5-dihydroxybenzoate. In 1975 Chapman (1975) obtained several *Micrococcus* spp. capable of converting 4-chlorobenzoate to 4-hydroxybenzoate. Later, several studies made similar observations.[3] Recent work by Marks et al. (1984) and Mueller et al. (1984) employing $^{18}O_2$ and $H_2{}^{18}O$ have shown that the dechlorination reaction involves water and not molecular O_2 for cleavage of the carbon-halogen bond. Figure 11.2 shows the hydrolytic dechlorination of 3- and 4-chlorobenzoate. Similar observations of chlorine displacement by a hydroxyl group has been found for pentachlorophenol degradation by *Flavobacterium* sp. (Steiert and

[3]Klages and Lingens 1979; Zaitsev and Karasevich 1980a,b; Reineke and Knackmuss 1988; Reineke 1984.

Crawford 1986) and *Rhodococcus chlorophenolicus* (Apajalahati and Salkinoja-Salonen 1987).

Fortuitous elimination of chlorine substituent by dioxygenases is an alternate route of removing halogen from haloaromatic compounds. Oxygen is regioselectively introduced into an aromatic ring resulting in a spontaneous elimination of halogen. Goldman et al. (1967) reported such a mechanism in a *Pseudomonas* sp. that metabolized fluorobenzoate. This organism used two different pathways of 2-fluorobenzoate degradation catalyzed by nonselective benzoate-1,2-dioxygenase. Oxygen was introduced into the aromatic ring in such a way that a mixture of 2- and 6-fluoro-1,2-dihydro-1,2-dihydroxy benzoate were formed. In this reaction, the major portion (>85 percent) of 2-fluorobenzoate underwent spontaneous defluorination resulting in catechol, which was further degraded via the 3-oxoadipate pathway (Reineke and Knackmuss 1988, Reineke 1984). This nonenzymatic means of fluorine elimination from 2-fluoro-1,2-dihydro-1,2-dihydroxy benzoate probably was dehalogenation-catalyzed by the dioxygenase, which originally functioned for introducing dioxygen into benzoate. This type of reaction normally requires enzymatic systems which hydroxylate exclusively in the proximate position relative to the carboxylic group (Fewson 1981).

Halide elimination after ring-cleavage. Since haloaromatics are resistant to nucleophilic reactions, dehalogenation can occur after the haloarene has undergone initial transformation to a nonaromatic intermediate resulting in a weak halogen carbon bond. A common feature of these pathways is the removal of halide after the haloaromatic compound has been converted to 3-chlorocatechol and has undergone ring cleavage using degradative enzymes that are meant for nonhalogenated analogues. At this point the halide may be removed spontaneously because of the labilized halogen carbon bond.

The majority of haloaromatics are degraded via halocatechols, and the 3-isomers appear to be the major metabolite. Chlorobenzenes, chloroanilines, chlorobenzoates, and chlorophenols including pentachlorophenol, chlorophenoxy acetates (including 2,4-D, 2,4,5-T, and 2-methyl-4-chlorophenoxy acetates [MCPA]), and chlorinated biphenyls are ultimately converted to chlorocatechols (Reineke and Knackmuss 1988, Reineke 1984). In the case of phenoxyacetates, various soil bacteria have been shown to cleave the ether linkage to produce 2,4-dichlorophenol from 2,4-D, 2-methyl-4-chlorophenol from MCPA, and 4-chlorophenol from 4-chlorophenoxyacetate. All bacterial systems which utilize haloaromatics employ the enzymes of the *ortho* cleavage pathway. The key enzyme of the *meta* pathway, 2,3-dioxygenases, are irreversibly inactivated by acyl-halides generated from 3-chlorocatechol (Fig. 11.3).

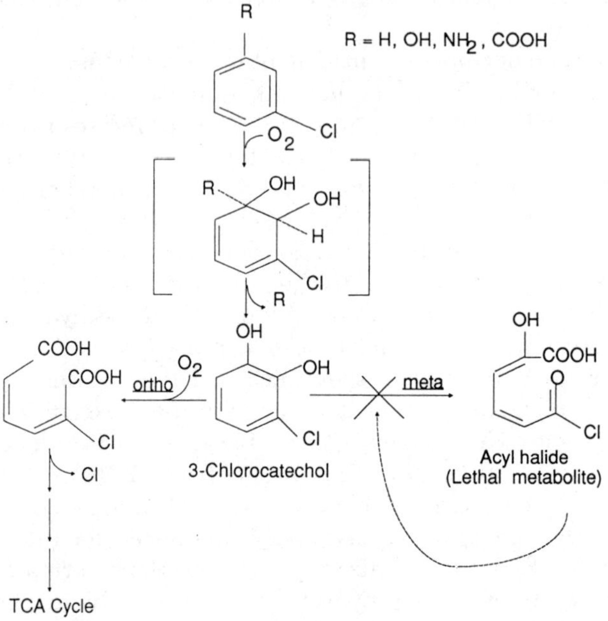

Figure 11.3 Halide elimination after ring cleavage. (*Adapted from Leisinger and Brunner 1986, Reineke 1984, Reineke and Knackmuss 1988.*)

Aliphatic and halogenated aliphatic hydrocarbons.

Aliphatic hydrocarbons. Aliphatic long-chain hydrocarbons are the major components of crude oils and petroleum products that enter the environment through unintentional spillage or disposal from refineries and other petrochemical industries (Atlas 1981). In recent times, interest in this class of chemicals has grown because of their potential threats to the environment (Singer and Finnerty 1984). The reader may refer to key publications and recent reviews[4] for the background, ecological aspects (Bartha and Atlas 1977, Colwell and Walker 1977), and biodegradation by pure cultures not covered in this chapter.

The first hydrocarbon-metabolizing organism, *Botrytis cinerea*, was reported in 1895 (Miyoshi 1895). This was followed by identification of a methane-utilizing organism, *Pseudomonas methanica*, by Sohngen in 1906 (Hou 1982). Since then several reports have appeared relating to hydrocarbon metabolism.[5] Aliphatic hydrocarbons are metabolized

---

[4] Buhler and Schindler 1984, Britton 1984, Singer and Finnerty 1984, Hartmans et al. 1989.

[5] Hou 1982, Singer and Finnerty 1984, Hartmans et al. 1989, Markovetz 1972, Ratledge 1978.

by a wide variety of microorganisms and fungi.[6] Surveys of hydrocarbon-metabolizing organisms in different soils demonstrated the presence of a large variety of microbes capable of oxidizing hydrocarbons (Perry and Scheld 1968, Jones and Edington 1968). In addition, soils obtained from oil fields contained a large proportion of hydrocarbonoclastic microbes (Brisbane and Ladd 1965). Similar observations have been made in marine environments (Colwell and Walker 1977). Ecological studies of hydrocarbon-metabolizing organisms have indicated that many hydrocarbons are degraded by a process called cooxidation. For example, *Pseudomonas methanica* grown on methane gas could oxidize ethane, propane, and butane to the corresponding alcohols, aldehydes, and acids (Singer and Finnerty 1984, Foster 1962). Cooxidation is an important phenomenon in the nature in degradation of hydrocarbons, especially recalcitrant molecules. Evidence indicates that the number of organisms involved in cooxidative metabolism of aliphatic hydrocarbons is quite large (Perry 1979).

1. *Terminal oxidation:* The most common n-alkane degradation proceeds via oxidation of the terminal methyl moiety. Several studies using bacteria, fungi, yeasts, and cell-free systems have shown that alkane is first oxidized at the C_1 position to primary alcohol and then to homologous fatty acid. Mainly three mechanisms of n-alkane oxidations to the corresponding alcohols exist. For more information on the mechanism, the reader is referred to earlier reviews by Buhler and Schindler (1984) and Britton (1984). The most widely studied mechanism is the hydroxylation of the ω-methyl group. This pathway involves the direct incorporation of molecular oxygen catalyzed by monooxygenases.
2. *Subterminal oxidation:* Even though alkane degradation occurs mainly via terminal oxidation, there are numerous microorganisms that metabolize alkanes via subterminal oxidation producing secondary alcohols and then corresponding ketones, which are further catabolized to fatty acids. It was indicated that the bacteria which utilize alkane as the sole source of carbon and energy, subterminal oxidation may be a minor pathway. For microorganisms which metabolize alkanes cooxidatively, the subterminal route seems to be a major pathway (Britton 1984, Buhler and Schindler 1984, Singer and Finnerty 1984).

Biotransformation of alkenes. Alkenes may be degraded by an initial attack either at unsaturated moiety, or elsewhere on the molecule re-

[6]Beerstecher 1954, Fuhs 1961, Klug and Markovetz 1971, Shennan and Levi 1974, Buhler and Schindler 1984, Singer and Finnerty 1984.

sulting into several types of products such as: (1) *w*-unsaturated alcohols or fatty acids, (2) primary or secondary alcohols or methyl ketones, (3) 1,2-epoxides, and (4) 1,2-diols depending on the mode of attack. Monooxygenases play an important role in oxidation of alkenes yielding primary and secondary alcohols. Metabolism of 1-alkenes with a chain length longer than C_5 generally proceed via methyl group oxidation (Hartmans et al. 1989). The mechanism of methyl group oxidation of alkenes would be similar to that of alkanes. In addition to the methyl group oxidation, many microorganisms were found to possess monooxygenases capable of epoxidating alkenes at the double bonds (May 1979). The first such report on the role of epoxide in the metabolism of alkenes was for ethene metabolism in *Mycobacterium* E20. For more information regarding the role played by epoxide, see the recent review by Hartmans et al. (1989).

Halogenated aliphatic compounds. The widespread distribution and contamination of halogenated aliphatic compounds and their apparent toxicological effects in species ranging from humans to microbes has caused much concern. Chlorinated solvents used as degreasing agents are generally known to resist degradation by conventional biological wastewater treatment processes. The six most commonly found volatile organic chemicals are tetrachloroethylene (PCE), trichloroethylene (TCE) 1,1,1-trichloroethane (TCA), cis-1,2-dichloroethylene (cis-1,2-DCE), 1,2-dichloroethane (1,2-DCA), and 1,1-dichloroethylene (1,1-DCE). These chemicals persist in the environment and are readily transported to subsurface waters. Current practice of removal of these volatile compounds include pumping the water to the surface and stripping the components in aeration towers or removing the chemicals by carbon adsorption procedures (Clark et al. 1988). These methods are expensive and simply transfer the pollutants from one phase (water) to another phase (air).

Halogenated organic compounds are often recalcitrant to microbial attack; for example, TCE was observed to exhibit a half-life of about 300 days (Wackett et al. 1989). However, halogenated compounds are biodegraded under specialized conditions; for example, Wilson and Wilson (1985) reported TCE oxidation by soil bacteria exposed to natural gas. Recently, several authors have shown that chlorinated ethenes are degraded by mixed and pure aerobic microorganisms that also metabolize either methane,[7] propane (Wackett et al. 1989), phenol or toluene (Nelson et al. 1986, 1987, 1988), and ammonia (Arciero et al. 1989).

Bacteria that grow on hydrocarbons usually introduce molecular O_2

[7]Hartmans et al. 1989, Fogel et al. 1986, Little et al. 1988, McCarty 1988.

into an organic molecule by oxygenases. Both types of oxygenases, namely mono- and dioxygenases, have been implicated in the oxidation of halogenated hydrocarbons. Epoxidation or the oxidation of carbon-carbon double bond was suggested to be the primary step in the oxidation of halogenated ethenes. The epoxide intermediate spontaneously undergoes several reactions, yielding products such as dichloroacetic acid, glyoxylic acid, or C_1 compounds. Studies suggested that under acidic conditions, TCE epoxide was converted to glyoxylic and dichloroacetic acid, while under basic conditions, carbon monoxide and formate were formed (Little et al. 1988, McCarty 1988, Vogel et al. 1987). Methanotrophs and other aerobes appear to have the capability to cometabolize several chlorinated methanes, ethanes, and ethylenes employing either mono- or dioxygenases. In general, it was observed that the rate of oxidation is inversely proportional to the degree of chlorination of a chemical. The highly chlorinated compounds, such as PCE, were not oxidized; conversely, vinyl chloride is metabolized much faster than DCEs.

Another mode of oxidation of halogenated aliphatic hydrocarbons occurs in some bacteria that can utilize these chemicals as the sole source of carbon and energy.[8] Many species of *Pseudomonas* and *Hyphomicrobium* have the capability to metabolize chlorinated alkanes as their primary substrates. Studies showed that complete metabolism of chloroalkanes proceeds in three different ways, as shown in Fig. 11.4. In the case of dichloromethane, removal of chlorine is accomplished by an enzyme, glutathione-dependent dehalogenase. Glutathione-mediated dehalogenation is generally initiated by nucleophilic attack of glutathione on electrophilic carbon that is bound to a halogen (Leisinger and Brunner 1986, Galli and Leisinger 1985). Degradation of 1,2-dichloroethane proceeded in two ways. Soil bacterium strain DE_2 transformed 1,2-dichloroethane by oxidative dehalogenation to 1,2-dichloroethanol, which was further converted to 2-chloroacetaldehyde and 2-chloroacetic acid (Fig. 11.4), whereas *Xanthobacter autotrophicus* degraded 1,2-dichloroethane to 1,2-chloroethanol by a hydrolytic dehalogenase (Janssen et al. 1985) (Fig. 11.4).

Degradative ecosystems for waste treatment

Liquid-waste reactors. Several biological waste-treatment processes can be employed to treat contaminated groundwater and leachate

---

[8]Leisinger and Brunner 1986, McCarty 1988, Hartmans et al. 1985, Stucki et al. 1983, Kohler-Staub et al. 1986, LaPat-Polasko et al. 1984, Galli and Leisinger 1985, Galli 1987.

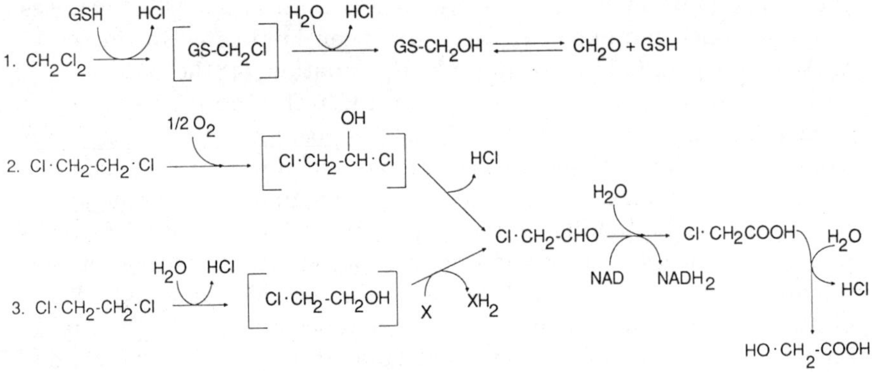

1. CH_2Cl_2 (GSH, HCl) → [$GS\text{-}CH_2Cl$] (H_2O, HCl) → $GS\text{-}CH_2OH$ ⇌ CH_2O + GSH

2. $Cl \cdot CH_2\text{-}CH_2 \cdot Cl$ (1/2 O_2) → [$Cl \cdot CH_2\text{-}CH(OH) \cdot Cl$] (HCl) → $Cl \cdot CH_2\text{-}CHO$

3. $Cl \cdot CH_2\text{-}CH_2 \cdot Cl$ (H_2O, HCl) → [$Cl \cdot CH_2\text{-}CH_2OH$] (X, XH_2) → $Cl \cdot CH_2\text{-}CHO$

$Cl \cdot CH_2\text{-}CHO$ (H_2O, NAD, $NADH_2$) → $Cl \cdot CH_2COOH$ (H_2O, HCl) → $HO \cdot CH_2\text{-}COOH$

Figure 11.4 Microbial degradation of chlorinated aliphatic compounds. Dehalogenation by (1) reduced glutathione (GSH), (2) oxidative attack, and (3) hydrolysis. *(Adapted from Leisinger and Brunner 1986, Kohler-Staub et al. 1986, Stucki et al. 1983, Janssen et al. 1985.)*

from hazardous waste sites that include (1) suspended growth processes such as activated sludge, sequencing batch reactors, aerated lagoons, waste stabilization ponds, and (2) fixed-film processes such as trickling filters, rotating biological contactors, activated biofilters, submerged filters and fluidized-bed reactors. For detailed information of these processes the reader may refer to the following publications: Shuckrow et al. (1980), Nyer (1985), Canter and Knox (1985), Rittmann, (1987), Bishop and Kinner (1986), and Shieh and Keenan (1986). Most of these treatment processes have traditionally been used to treat domestic and industrial discharge. However, treatment of the wastes contaminated with anthropogenic compounds of recent origin such as chlorinated and nonchlorinated aromatic and aliphatic hydrocarbons, chlorinated and nonchlorinated polynuclear aromatic compounds, pesticides, and other toxic and recalcitrant chemicals warrant innovative treatment techniques or improvements and modifications of the existing treatment technologies (Sutton 1985).

Innovative biotreatment techniques should economically treat hazardous wastes and comply with environmental legislation. Three new bioreactor systems that used microbial mixed cultures have been used for hazardous waste treatment: fluidized-bed biofilm reactors, fluidized-bed reactors with granular activated carbon, and one sludge reactor for denitrification and oxidation (Shieh and Keenan, 1986, Rittman, 1987).

Solid-waste composting

Composting is used to degrade residual solid wastes by mixed cultures of aerobic bacteria and fungi which degrade organic matter into CO_2,

H_2O, NH_3, and giving off heat. With regard to bioreclamation of recalcitrant and toxic wastes, composting technology has significant merits but it has not been widely applied. This technology requires addition of residual organic matter (i.e., wood chips, straw, paper) as the energy source for soil microbes that can degrade the hazardous waste along with the added substrates.

Rose and Mercer (1968) investigated the composting of insecticides in agricultural wastes and showed that the concentrations of diazinon and parathion rapidly decreased when composted under thermophilic conditions. The concentration of diazinon decreased from 3.3 ppm to less than 0.002 ppm in 42 days of composting. The concentration of parathion decreased by 50 percent within 12 days. Similarly, pp-DDT was reduced to 65 percent from its initial amount when composted for 50 days (Savage et al. 1989). Composting at 35 to 50°C provides higher degradation rates than 4 to 35°C, but higher temperatures are often not used because of the emission of volatile organics (McKinley et al. 1989).

In situ biorestoration. In situ biorestoration (or bioreclamation) of the subsurface environments contained with anthropogenic compounds is relatively new but promising technology. Biorestoration involves either biostimulation or bioaugmentation. In situ biorestoration of groundwater involves the stimulation of the indigenous mixed microflora to degrade contaminants in place (Lee et al. 1988, Thomas and Ward 1989, Harvey et al. 1984). Biostimulation involves injecting the nutrients necessary to increase the activity of in situ microbial mixed cultures in the subsurface. Bioaugmentation involves adding a pure or mixed culture of degradative organisms to the contaminated site. Transformation of organic compounds in the subsurface environment can occur through the action of microorganisms that are either present in the sediment pore water or are attached to sediment particles. Relatively high bacterial populations ranging from 10^6 to 10^7 cells per gram sediment (Wilson et al. 1987, Balkwill and Ghiorse 1985, Webster et al. 1985) and diverse kinds of bacteria (Lee et al. 1988; Smith et al. 1985, 1986) have been detected in the subsurface environments contaminated with toxicants.

Microorganisms in subsurface water are metabolically active and transform many commonly occurring groundwater pollutants. These include benzene, toluene, xylenes, chlorinated benzenes, chlorinated phenols, methylene chloride, naphthalene, methylnaphthalenes, and phenanthrene.[9] Other studies indicate that in situ populations de-

[9]Kuhn et al. 1988, Suflita and Miller 1985, Ihaveri and Mazzacca 1983, Wilson et al. 1985, Lee and Ward 1985, Barker and Patrick 1986.

grade gasoline-based hydrocarbons in contaminated aquifers that are oxygenated (Morgan and Watkinson 1989, Borden et al. 1986, Bedient et al. 1984). The extent of biodegradation of many organic pollutants in subsurface may be limited by the concentration of dissolved oxygen (Lee et al. 1988, Morgan and Watkinson 1989). Raymond et al. (1976) showed that removal of petroleum contaminants from groundwater was achieved by stimulating the native microbial population through the addition of oxygen and nutrients.

In situ biorestoration of aquifers contaminated with halogenated compounds requires a different approach since most of the halogenated compounds cannot serve as the primary carbon and energy source for growth of subsurface microorganisms (Roberts et al. 1989a,b). Laboratory research has shown that incubation of soil or aquifer materials with methane or natural gas will enrich bacteria (i.e., methanotrophs) that cometabolize a variety of halogenated compounds. These microorganisms possess methane monooxygenase that oxidizes methane as well as many halogenated methanes, ethanes, and ethylenes.[10] Recently, Wackett et al. (1989) employed propane as the growth substrate to obtain TCE degradation by a new class of bacteria. Propane monooxygenase was implicated in TCE oxidation. Similarly, researchers at the EPA laboratory in Gulf Breeze, Florida, found TCE-degrading activity in organisms that belonged to the genus *Pseudomonas* (Nelsen et al. 1986, 1987, 1988). These organisms transformed the added TCE while growing on an aromatic compound such as phenol or toluene. Subsequent studies demonstrated that the enzyme, toluene-dioxygenase was responsible for TCE degradation (Wackett and Gibson 1988).

Detoxification by Anaerobic Mixed Cultures

Recent understanding on the diversity and metabolism of anaerobic microorganisms has shown that anaerobic population can also detoxify and degrade various hazardous wastes. Notably, anaerobes also perform some unique detoxification reactions which are not known under aerobic conditions (e.g., reductive dechlorination of highly chlorinated aliphatic and aromatic hydrocarbons, reduction of aromatic ring to alicyclic ring structures, and ring fission). Furthermore, anaerobic biodegradation methods and protocols have been recently developed for assessment, monitoring, and control of residual and toxic organic removal by mixed-culture systems (Hickey 1987a,b; Bhatnagar and Zeikus 1989; Wu et al. 1989).

---

[10]Wilson and Wilson 1985, Fogel et al. 1986, McCarty 1988, Strand and Shippert 1986, Hensen et al. 1986, 1987, 1988.

Anaerobic degradation of aromatic compounds was first demonstrated by mixed cultures in 1934 (Tarvin and Buswell 1934). These authors showed that sewage sludge mixed cultures completely degraded benzoate, phenylacetate, phenyl propionate (hydrocinnamate), and cinnamate to methane and CO_2 in the absence of O_2. The mixed population responsible for benzoate conversion to methane was further studied by Ferry and Wolfe (1976). Evans (1977) reported that the initial anoxic transformation mechanism used by phototrophs for benzoate degradation was reduction of the ring. The reduced aromatic ring gives rise to an alicyclic ring, which is then hydrolytically cleaved to C_1-C_5 carboxylic acids that are eventually mineralized to CH_4 and CO_2 via acetate (Fig. 11.5). A pure culture of *Rhodopseudomonas palustris* degrades a wide range of oxygenated aromatic compounds under anoxic conditions (Harwood and Gibson 1988). Anaerobic transformation of certain nonoxygenated aromatic hydrocarbons was first demonstrated in 1980 (Ward et al. 1980). Conclusive evidence for an initial oxidative mechanism was provided re-

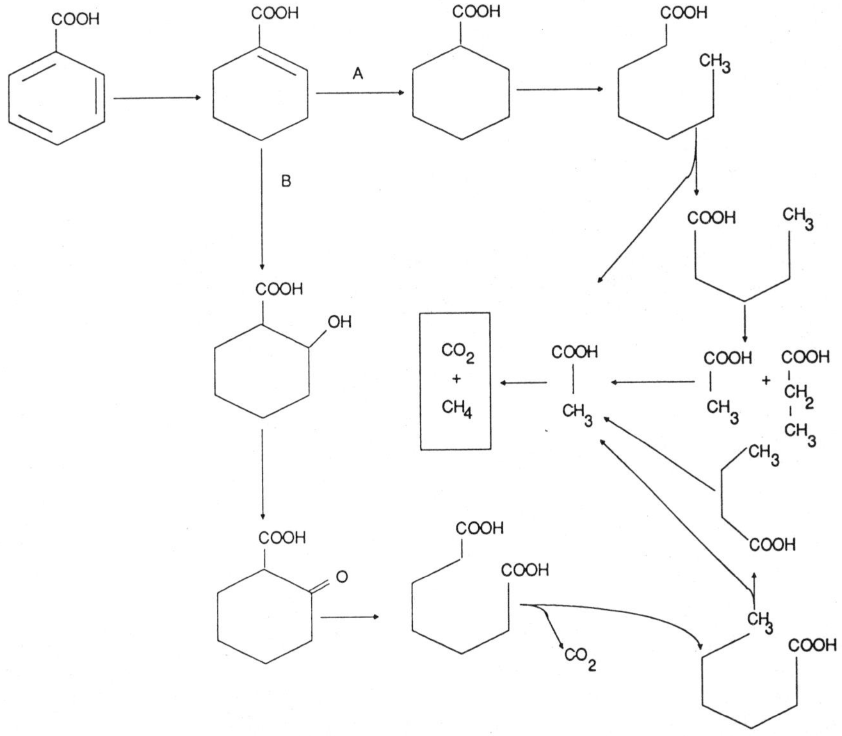

Figure 11.5 Proposed pathways for the anaerobic degradation of benzoate to methane by (A) a microbial consortium and (B) a *Rhodopseudomonas* sp. as well as a mixed culture. (*Adapted from Guyer and Hegeman 1969, Keith et al. 1978, Shlomi et al. 1978.*)

cently (Vogel and Grbic-Galic 1986) when the incorporation of oxygen from water into toulene and benzene was demonstrated under anoxic conditions.

Degradative mechanisms

Aromatic and aliphatic hydrocarbons

Aromatic hydrocarbons. Anaerobic degradation of toxic substituted monoaromatic compounds has been reported under different metabolic situations such as denitrification (Braun and Gibson 1984) sulfate reduction (Pfennig et al. 1981), fermentation (Schink and Pfennig 1982), methanogenesis (Ferry and Wolfe 1976, Kaiser and Hanselmann 1982, Young 1984) and anaerobic photosynthesis (Evans 1977). The catabolic pathways have not been elucidated in detail, though it has been strongly suggested that in the oxygen-substituted aromatics, the initial transformation step is the reductive ring cleavage scheme or its modification (Evans 1977). The aromatic nucleus is reduced to the corresponding cyclohexane, which is then cleaved by hydrolysis to yield aliphatic acids which are readily metabolized anaerobically.

Under methanogenic conditions aromatic compounds serve as substrates for microbial multispecies communities or consortia whose components mediate a sequence of coupled reactions resulting in complete degradation of the aromatic substrate to CO_2 and methane. A wide range of monoaromatic compounds are degraded under anaerobic conditions (Young 1984, Leisinger and Brunner 1986). At least two physiologically different anaerobic microbial communities are involved in the degradation of aromatic compounds such as benzoate: (1) organisms responsible for reductive cleavage of the aromatic ring to aliphatic acids and then to acetic acid, CO_2, H_2, etc.; (2) methanogens which utilize these substrates to produce methane. The proposed pathways of benzoate degradation are presented in Fig. 11.5.

The conversion of aromatics such as benzoate, phenol or *p*-hydroxybenzoate to acetate, H_2, and formate is thermodynamically unfavorable. Thus, to enable ring cleavage to proceed it is necessary to continuously remove these intermediary products. This is effectively achieved by methanogens. Ring cleavage seems to be closely linked to product utilization, resulting in an obligate syntrophic association between the ring cleavage and methanogenic organisms. Transformation of substituted benzenes with two or three substituents involves a negative standard free energy change. In these cases, the interactions between fermentative ring cleavage organisms and methanogens is not obligatory and thus the degradative organism(s) can be isolated in pure culture (Schink and Pfennig 1982, Kaiser and Hanselmann 1982), even though for complete degradation microbial communities are more effective. The anaerobic microbial transformation of homocyclic aromatic hydrocarbons was first reported by Ward and coworkers (1980). However, it was not until 1986 that conclusive results for an oxygen insertion reaction via water were obtained (Vogel and Grbic-Galic 1986).

Oil-polluted sediments incubated over 200 days under anaerobic conditions showed that up to 2 percent of the label from [^{14}C] benzene, 2.9 percent from the [^{14}C] ring-toluene, and 5 percent from the [^{14}C] methyl-toluene was transformed into CO_2 and CH_4 (Ward et al. 1980). Naphthalene and benzo(a)pyrene were not degraded. Later Schwarzenbach et al. (1983) and Reinhard et al. (1984) reported the selective removal of toluene and xylene from the anaerobic zones of two different groundwater aquifers contaminated with landfill leachate. They suggested that the biological transformation did not involve an initial reductive step, since they did not observe an increase in the concentration of reduced alicyclic rings.

A series of studies were conducted (Vogel and Grbic-Galic 1986, Grbic-Galic and Vogel 1987) using a microbial community enriched from anaerobic sewage sludge with ferulic acid (3-methoxy-4-hydroxy-cinnamic acid) as the sole energy and carbon source (Grbic-Galic and Young 1985). Pathways were proposed for degradation of ferrulic acid and other related compounds to CH_4 and CO_2. Ring-labeled [^{14}C] toluene or benzene were used along with $H_2{}^{18}O$ for biodegradation studies involving substrate disappearance, formation of intermediates and products, and the initial transformation mechanism. It was observed that benzene and toluene as sole carbon and energy source or together with methanol can be completely mineralized by the acclimated anaerobic community under methanogenic conditions. Toluene and benzene disappeared from the culture medium in 34 and 64 days, respectively. More than 50 percent was converted to CO_2 and CH_4 in 60 days and the rest was incorporated into cells or persisted in the culture fluid as various intermediates (alicyclic rings and aliphatic acids). After prolonged incubations complete mineralization was observed.

The first transformation intermediates involving ring oxidation were identified for both benzene and toluene (Grbic-Galic and Vogel 1987, Kuhn et al. 1988). Benzene was transformed to phenol and toluene was converted to *p*-cresol or *o*-cresol. In addition, toluene underwent a methyl-group oxidation to give benzyl alcohol. Benzoic acid was the most important intermediate of the oxidative pathways.

In general, after the first oxidative step, the degradation of nonoxygenated aromatics (benzene, toluene, etc.) follows similar pathway(s) as described for anaerobic transformation of oxygenated aromatics (Evans 1977, Berry et al. 1987). After modification of the ring substituents, the aromatic ring is oxidized followed by reduction to alicyclic intermediates which are then hydrolytically cleaved to aliphatic alcohols and acids leading to final conversion of these alcohols and acids to mineralization products (e.g., CO_2 and CH_4 under methanogenic conditions).

The most important observation in support of initial oxidation of aromatic ring came from the $H_2{}^{18}O$ studies, where it was demonstrated that the source of oxygen for initial oxidation was water (Vogel and

TABLE 11.1 Homocyclic Aromatic Hydrocarbons Transformed under Anaerobic Conditions with Mixed Microbial Cultures

Compound	Conditions	Mineralization	Reference[d]
Benzene	Denitrifying	ND	1
	Methanogenic	+[a]	2,3,7,9
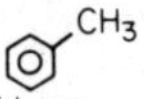 Toluene (CH_3)	Denitrifying	+	1,4,5
	Methanogenic	+	2,3,7,9
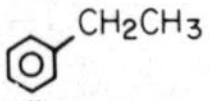 Ethylbenzene (CH_2CH_3)	Methanogenic	ND	2
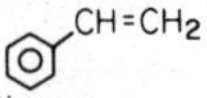 Styrene ($CH{=}CH_2$)	Fermentative[b]	—[c]	8
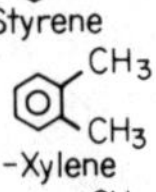 *o*-Xylene (CH_3, CH_3)	Denitrifying	ND	1
	Methanogenic	ND	2
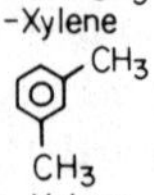 *m*-Xylene (CH_3, CH_3)	Denitrifying	+	1,4,5
	Methanogenic	ND	9
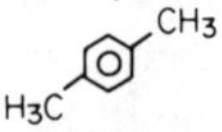 *p*-Xylene (CH_3, H_3C)	Denitrifying	ND	1
	Methanogenic	ND	2
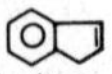 Indene	Methanogenic	+	Godsy and Grbic-Galic (unpublished)
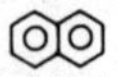 Naphthalene	Methanogenic	+	Godsy and Grbic-Galic (unpublished)
 Acenaphthene	Denitrifying	ND	10

ND = not determined

[a]Mineralization of the compound

[b]Methanogenic inoculum, but no methane produced

[c]Multicarbon compounds persist till the end of incubation (in addition to CO_2)

[d]References: 1—Major et al. 1988, 2—Wilson et al. 1986, 3—Vogel and Grbic-Galic 1986, 4—Kuhn et al. 1988, 5—Zeyer et al. 1986, 7—Grbic-Galic and Vogel 1987, 8—Churchman and Grbic-Galic 1987, 9—Wilson et al. 1987, 10—Michelcic and Luthy 1988 a,b.

SOURCE: Adapted from Grbic-Galic 1989.

Grbic-Galic 1986) as ^{18}O-*p*-cresol and ^{18}O-phenol were detected as products from toluene and benzene, respectively. Most of the studies on the anaerobic transformation of nonoxygenated homocyclic and heterocyclic aromatics are summarized in Tables 11.1 and 11.2.

Following the demonstration of the anaerobic biodegradation of monoaromatic hydrocarbons, Michelic and Luthy (1988a,b) reported degradation of polynuclear aromatic hydrocarbons including naphtha-

TABLE 11.2 **Heterocyclic Aromatic Hydrocarbons Transformed under Anaerobic Conditions with Mixed Microbial Cultures**

Compound	Conditions	Mineralization	Reference[d]
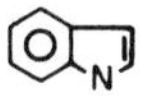 Indole	Denitrifying	ND	1
	Methanogenic	+	1,2
	Sulfate-reducing	+	3
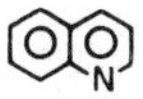 Quinoline	Methanogenic	± a	4 5,6
	Sulfate-reducing	ND	3
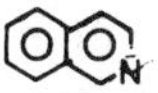 Isoquinoline	Methanogenic	+ —	4 6
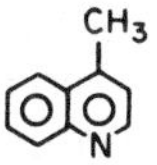 4-Methylquinoline	Methanogenic	—	6
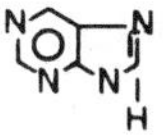 Purine	Fermentative[b]	+	7
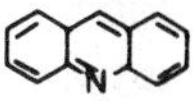 Acridine	Methanogenic	+	DeWitt and Grbic-Galic (unpublished)
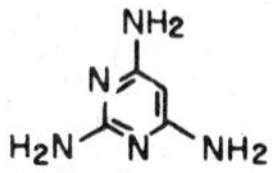 Melamine	Denitrifying	+	8
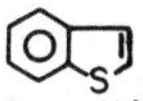 Benzothiophene	Methanogenic	+	9
	Fermentative[c]	+	10
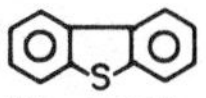 Dibenzothiophene	Fermentative[c]	+	10

[a]No mineralization products only intermediates detected
[b]Pure cultures of obligately or facultatively anaerobic fermentative bacteria
[c]Fermentative enrichments
[d]References: 1—Madsen et al. 1988, 2—Berry et al. 1987, 3—Bak and Widdel 1986, 4—Godsy et al. 1987, 5—Pereira et al. 1988, 6—Pereira et al. 1987, 7—Durre and Andreesen 1983, 8—Jutzi et al. 1982, 9—Godsy and Grbic-Galic 1988, 10—Maka et al. 1987.
SOURCE: Adapted from Grbic-Galic 1989.

lene and acenaphthene under denitrifying conditions in saturated soil microcosms. Naphthalene (7 mg/L) and acenaphthene (0.4 mg/L) were degraded below detectable levels in 45 and 40 days, respectively, after about 2 weeks of acclimation. Recently, Grbic-Galic (unpublished results) showed degradation of naphthalene under anaerobic conditions in the absence of nitrate with methanogenic consortia acclimated to benzene as the sole carbon and energy source (Grbic-Galic and Vogel 1987). After about one month of acclimation, the transformation of naphthalene resulted in aromatic intermediates. The degradation was complete after 30 weeks at 35°C. Recently, Godsy and Grbic-Galic (unpublished) also showed the transformation of indene in saturated laboratory microcosms derived from an aquifer contaminated with creosote. The microcosms degraded 12 mg/L of indene to undetectable levels in two months at 25°C, and stoichiometric quantities of CO_2 and CH_4 were produced (Table 11.1). The degradation pathways of polynuclear aromatic hydrocarbons (PAHs) under anaerobic conditions are not yet known.

Transformation of nitrogen- and sulfur-containing heterocyclic hydrocarbons has been observed by mixed bacterial cultures, and the important results are summarized in Table 11.2. Anaerobic degradation of different nitrogen heterocyclic compounds was observed under denitrifying, sulfate-reducing, fermentative, and methanogenic conditions. These transformations were demonstrated in mixed and pure cultures obtained from sewage sludge, anaerobic sediments, or anaerobic aquifer materials (see Table 11.2). In almost all these cases the initial anaerobic transformation step involves introduction of oxygen from the hydroxyl group of water. Pereira et al. (1988) demonstrated that water was the source of oxygen for initial oxidation of quinoline under methanogenic conditions employing $H_2{}^{18}O$.

Higher anaerobic degradation rates have been suggested for heterocyclic rather than homocyclic aromatic hydrocarbons. Indole (1 m*M*) was completely degraded by liquid cultures of *Desulfobacterium inodolicum* in 8 days (Bak and Widdel 1986); by sewage sludge–derived methanogenic enrichments in 60 days; and by freshwater sediment or soil slurries under denitrifying conditions in 144 days (Madsen et al. 1988). Quinoline and isoquinoline were completely mineralized to CO_2 and CH_4 in less than 30 days by anaerobic microcosms obtained from a creosote-contaminated groundwater aquifer (Godsy et al. 1987).

Very little information is available on the anaerobic biodegradation of heterocyclic sulfur compounds. Maka et al. (1987) reported anaerobic transformation of benzothiophene and dibenzothiophene by mixed microbial cultures enriched from anaerobic sewage sludge and from coal-storage site soils. Disappearance of 50 to 80 percent of the substrate was observed within 2 weeks at 37°C. Godsy and Grbic-Galic (1988) reported the degradation of benzothiophene (10 mg/L) by methanogenic microcosm enrichments after a 12-day lag period. Degradation was complete after 20 days, and CO_2 and CH_4 production was observed. Early oxidation intermediates were not detected, but a broad range of mononuclear aromatic and alicyclic compounds were formed which indicated that the initial step in degradation could have been oxidative in both the heterocyclic and homocyclic portion of benzothiophene molecule. The cyclic compounds were then broken down to aliphatic alcohols and acids, leading to further metabolism by known pathways.

Aliphatic hydrocarbons. Anaerobic transformation of aliphatic hydrocarbons was observed as early as 1960 (Muller 1957, Zobell 1946). Careful interpretation of anaerobic transformations is necessary because of the following problems: incomplete anaerobic conditions, degradation of added emulsifier (Bull 1980), or substrate impurities (Zobell and Prokop 1966). *Desulfovibrio desulfuricans* was reported to degrade methane, ethane, and *n*-octadecane using glucose as cosubstrate (Davis and Yarbrough 1966); it was later discovered that the cultures were impure. The mineralization of saturated hydrocarbons to methane and CO_2 under anoxic conditions is thermodynamically feasible (Thauer et al. 1977) as illustrated for ethane:

$$8C_2H_6 + 6H_2O \rightarrow 14CH_4 + 2HCO_3^- + 2H^+ \qquad (\Delta G^{\circ\prime} = -34.4 \text{ kJ mol ethane}^{-1})$$

Generally, the microbial degradation of saturated hydrocarbons depends on the presence of molecular oxygen involving oxygenases (Gibson 1975, Atlas 1981, Perry 1979).

The situation is different if there is at least one double bond in the hydrocarbon molecule. At this double bond hydration may give rise to an alcohol which could be oxidized to a ketone or via an aldehyde to the respective fatty acids. Thermodynamically, anaerobic degradation of ethene under methanogenic conditions, is more exergonic than ethane:

$$2C_2H_4 + 3H_2O \rightarrow 3CH_4 + HCO_3^- + H^+ \qquad (\Delta G^{\circ\prime} = -102.1 \text{ kJ mol ethene}^{-1})$$

It was demonstrated that 1-hexadecene was completely transformed to methane and CO_2 by anaerobic enrichment cultures (Schink 1985). Thus, a single double bond is sufficient for complete mineralization of an aliphatic linear hydrocarbon. The anoxic degradative pathway is different from that used by aerobes which would either epoxide the double bond (Huybergtse and van der Linden 1964) or start from the saturated end with a monooxygenase reaction (Ishikura and Foster 1961).

Halogenated aromatic and aliphatic hydrocarbons

Halogenated aromatics. The concept of reductive dehalogenation of the halogenated aromatics by anaerobic mixed cultures was recognized in 1982 (Suflita et al. 1982). Halogenated benzoates were partially or completely degraded to methane, CO_2, and hydrochloric acid by mixed methanogenic enrichment cultures. Several studies on anaerobic dehalogenation of various halogenated benzoates and phenols derivatives suggest that dehalogenation is an oxygen-sensitive process.[11] Several halogens are removed sequentially from the aromatic ring reductively (Bartels et al. 1984); no intermediates more toxic than the original substrate are formed during anaerobic dehalogenation.

Chlorobenzoates have long been used as model compounds for degradation studies because they are a toxic constituent of hazardous wastes, an intermediary product of several pollutants (e.g., PCBs, chlorophenols), and components of some herbicides. It was shown that 85 percent of 3-chlorobenzoate was mineralized to CH_4 and CO_2 using mixed cultures in anaerobic sludge (Shelton and Tiedje 1984) with benzoate as intermediate (Suflita et al. 1982). A series of mono-, di-, and trihalobenzoates have been shown to be dehalogenated by the sediment and sludge mixed microbial communities (Tiedje et al. 1987). Interestingly, all chlorines in the *meta* position were specifically removed.

The mixed microbial community in sludge that carried out the dechlorination of 3-chlorobenzoate was characterized (Shelton and Tiedje 1984). Nine different organisms were isolated out of which only five were thought to have major roles in the carbon and electron flow. Different partners of this consortia which degraded chlorobenzoate to CO_2 and CH_4 were: (1) the dechlorinating organism, DCB-1, that produces benzoate; (2) the benzoate-degrading strain, BZ-2, that produces H_2 and acetate; (3) the hydrogen-consuming methanogens *Methanospirillum* and *Methanobacterium* and the acetoclastic methanogen *Methanothrix* which uses acetate to make CO_2 and CH_4. At present strain DCB-1 is the only anaerobic organism in pure culture which dechlorinates. It has been named recently as *Desulfomonile tiedjei* (DeWeerd et al. 1990) and further studies on the mechanism and basis of dechlorination are under way.

Pentachlorophenol (PCP) is among the most widely studied chlorinated phenols, perhaps because PCP is widely used as a pesticide in

[11] Boyd and Shelton 1984, Boyd et al. 1983, Horowitz et al. 1983, Suflita et al. 1983.

wood preservation and as a herbicide for rice and sugar cane fields. Reductive dechlorination of PCP was suggested (Ide et al. 1972) during PCP degradation studies in flooded rice paddy fields. It seemed that the *ortho* and *para* chlorines were dechlorinated more easily than the *meta* isomer.

Chlorinated aromatics, including chlorophenols, chlorocatechols, and chloroguaiacols, in the paper and pulp mill effluents were shown to be transformed and degraded in an anaerobic fluidized-bed reactor (Hakulinen and Salkinoja-Salonen 1982, Salkinoja-Salonen et al. 1984). Reductive dechlorination of chlorinated catechols such as 4-chlororesorcinol was shown in fresh sludge as well as an enriched microbial community (Fathepure et al. 1987). Addition of yeast extract or trypticase increased the dechlorination rates from 1.1 to 2.6 μmol/mg/day and the lag time was decreased from 3 weeks to 2 days. Extensive degradation of PCP was reported by Guthrie et al. (1984) in anaerobic sludge. Chlorophenol removal was more rapid in the anaerobic reactor than in either an aerated lagoon or an activated sludge process.

Studies have demonstrated that chlorophenol degradation in fresh sludge proceeded via the reductive dechlorination at the *ortho* position. In batch culture studies PCP was sequentially converted to 3,4,5-trichlorophenol (3,4,5-TCP), 3,5-dichlorophenol, 3-chlorophenol which was finally metabolized to CH_4 and CO_2 in sewage sludges (Mikesell and Boyd 1986). Thus, *ortho* positions of PCP were dechlorinated. 2,4,6-TCP followed the same pattern yielding 4-CP.

We have recently developed granular anaerobic mixed cultures from specific enrichments on PCP (Bhatnagar et al. 1989a) which dechlorinate PCP in an upflow anaerobic sludge blanket (UASB) reactor via 2,4,6-TCP and 4-CP leading to the mineralization of PCP to CH_4 and CO_2 (Bhatnagar et al. 1989b) as shown in Fig. 11.6. These data suggest *meta* dechlorination is an initial dechlorination step. After the initial phase of adaptation to increasing levels of PCP in an upflow anaerobic sludge bed bioreactor, the granules dechlorinated and degraded PCP at the rate of 0.4 g/L reactor bed volume per day. The reactor was operated continuously with a recycle loop at an hydraulic retention time (HRT) of 14 h at 28°C. The synthetic feed contained 60 ppm PCP. PCP was never detected in the effluent after anaerobic treatment. Performance of this anaerobic bioreactor system was monitored by the rates of volatile fatty acids (VFA) degradation (Wu et al. 1989) over a period of more than 200 days. When the system was inhibited, several intermediates were detected and identified: 2,4,6-trichlorophenol, 2,4-dichlorophenol, and 4-chlorophenol, demonstrating that reductive dechlorination occurred initially at *meta* position, suggesting a novel pathway catalyzed by this anaerobic consortium. In batch culture studies 2,4,6-trichlorophenol and 2,4- and 2,6-

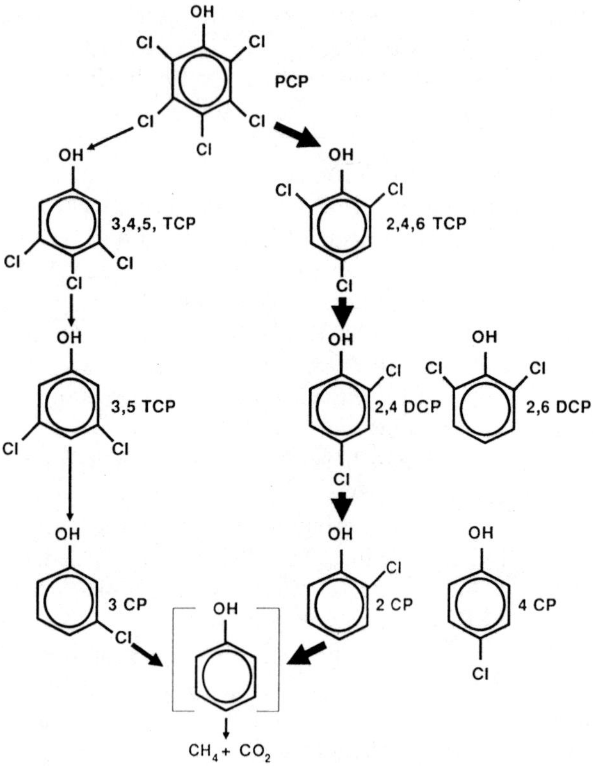

Figure 11.6 Proposed pathways of PCP dechlorination and degradation by mixed anaerobic cultures. Bold arrows indicate major pathway in a continuous UASB type reactor system. (*Bhatnagar et al. 1989b.*)

dichlorophenols were not inhibitory to the anaerobic granular biocatalysts, whereas 3,4,5-trichlorophenol was very toxic and 3,5-dichlorophenol was toxic too. Using radiolabeled (^{14}C) PCP, we demonstrated that over 75 percent of PCP was mineralized into $^{14}CO_2$ and $^{14}CH_4$. This type of PCP waste-treatment reactor system with specially developed anaerobic biomethanation granules has practical potential for commercial application in treatment of chloroorganic-contaminated groundwaters and landfill leachates.

Anaerobic mixed communities catalyzing reductive dechlorination of chlorinated benzenes and polychlorinated biphenyls (PCBs) could provide a biotreatment solution to environmental pollution problems in aquatic sediments. It is now known that highly chlorinated nonpolar compounds such as hexachlorobenzene (HCB), trichlorobenzene, and a variety of PCBs can undergo reductive dechlorination. The rates of dechlorination in case of HCB and PCBs is much lower than that for chlorobenzoates and chlorophenols (Tiedje et al. 1987,

Fathepure et al. 1988a, Quensen et al. 1988). Bailey (1983) suggested that dehalogenation of HCB in anaerobic sediments can be inferred from the data on the distribution of chlorobenzenes in the Great Lakes (Oliver and Nicol 1982). The analysis of these data indicated a slow dehalogenation of HCB in the anaerobic sediment. Hexachlorobenzene was almost completely dechlorinated within 3 weeks to tri- and dichlorobenzenes when incubated with sewage sludge under anaerobic conditions. Dechlorination proceeded via two routes, both involving the sequential elimination of chlorine from the aromatic ring. Most of the added HCB accumulated as 1,3,5-trichlorobenzene, which remained unchanged. The isomer 1,2,4-trichlorobenzene formed via minor route further dechlorinated to all isomers of dichlorobenzenes (Fathepure et al. 1988a). Monochlorobenzene was formed from trichlorobenzenes via dichlorobenzenes, and this activity was observed under both sulfate-reducing and methanogenic conditions (Holliger et al. 1988). Similarly, employing anaerobic biofilm column reactor, Vogel et al. (1989) and Fathepure and Vogel (1990) have shown the formation of 1,3- and 1,2-dichlorobenzene from HCB. These studies indicated acetate as a better primary carbon source for maximum dechlorination of HCB.

Highly chlorinated PCBs (i.e., penta, hexa, biphenyls) have been shown to be reductively dechlorinated by sediment enrichment cultures. Over 60 percent of the monochloro- and 20 percent of the dichlorobiophenyls were obtained when highly chlorinated biophenyls (Aroclor 1242, 1260) were treated with mixed microbial populations from the Hudson River (Quensen et al. 1988). This is of special interest because aerobic organisms do not degrade or dechlorinate penta-, hexa-, and higher chlorinated biphenyls.

PCBs were reductively dechlorinated by several populations of anaerobic bacteria (Brown et al. 1984) in the Hudson River sediments. Sediments containing >50 ppm PCBs showed losses of up to 30 percent of the chlorine originally present. High levels of individual mono-, di-, and trichlorobiphenyls in the samples now can be explained from the results obtained (Quensen et al. 1988) after reductive dechlorination of highly chlorinated PCBs. Apparently the reductive dechlorination of PCBs occurred in stepwise fashion until lower chlorinated PCBs were formed which were more difficult to reduce due to their increasing redox potential. The rate of reductive dechlorination under anaerobic conditions decreases with the decrease in the number of chlorines.

Halogenated aliphatics. The initial step in the degradation of halogenated aliphatic compounds is reductive dehalogenation. Unlike halogenated aromatic compounds, halogenated aliphatics can lose one or two halides during the reductive process. The mechanism for the loss of two halogens involves the removal of two vicinal halogens with the formation of double bond (Vogel et al. 1987). Reductive removal of a single halogen with hydrogen plays a major role in the degradation of halogenated aliphatic compounds. Of particular interest is the fate of chlorinated methanes, ethanes, and ethylenes, since they represent an important class of pollutants in many surface and subsurface environments. In 1981, biological transformation of halogenated methanes under methanogenic conditions was demonstrated (Bouwer et al. 1981). Subsequent studies showed dechlorination of chlorinated ethylenes and ethanes (Bouwer and McCarty 1983a).

Tetrachloroethylene (PCE) along with other chlorinated ethylenes is one of the major pollutants in groundwater, industrial wastewater, and landfill leachate. No aerobic organism(s) is known to date to dechlorinate and to degrade PCE. Anaerobic dechlorination of PCE by pure and mixed cultures has been recently demonstrated (Fathepure and Boyd 1988, Fathepure et al. 1988b). Anaerobic degradation of chloroform and tetrachloromethane in sediment samples probably go through an initial reductive dehalogenation step before leading to CO_2. Halogenated aliphatics which have been shown to be reductively dechlorinated in methanogenic mixed communities from sewage sludge and anoxic groundwater include carbon tetrachloride, chloroform, tri- and tetrachloroethane, tetrachloroethylene and trichloroethylene[12] (Fig. 11.7). Reduced iron porphyrins were also shown to dechlorinate some of these compounds (Klecka and Gonsior, 1984). Reductive dehalogenation was also reported with 1,2,2,2-tetrachloroethane leading to 1,1,2-trichloroethane (Bouwer and McCarty 1983b). Dehalogenation of 2-bromoethanesulfonic acid and 2-chloroethanesulfonic acid (Bouwer and McCarty 1983a) is possibly the cause of inactivation of these inhibitors (of methanogenic metabolism) in anaerobic digestors.

Biodegradative ecosystems for waste treatment

It is now known that anaerobic ecosystems for waste treatment have inherent advantages over aerobic ones, including no oxygen requirement; low sludge yields and disposal costs; and the process produces energy (CH_4) in lieu of consuming it.

A number of reactor configurations have been developed to maintain high levels of anaerobic mixed cultures, including the anaerobic contact reactor, filter reactor, UASB upflow reactor, and fluidized-bed

[12]Bouwer et al. 1981; Bouwer and McCarty 1983a,b; Parsons et al. 1984; Vogel and McCarty, 1985; Freedmen and Gosset, 1989.

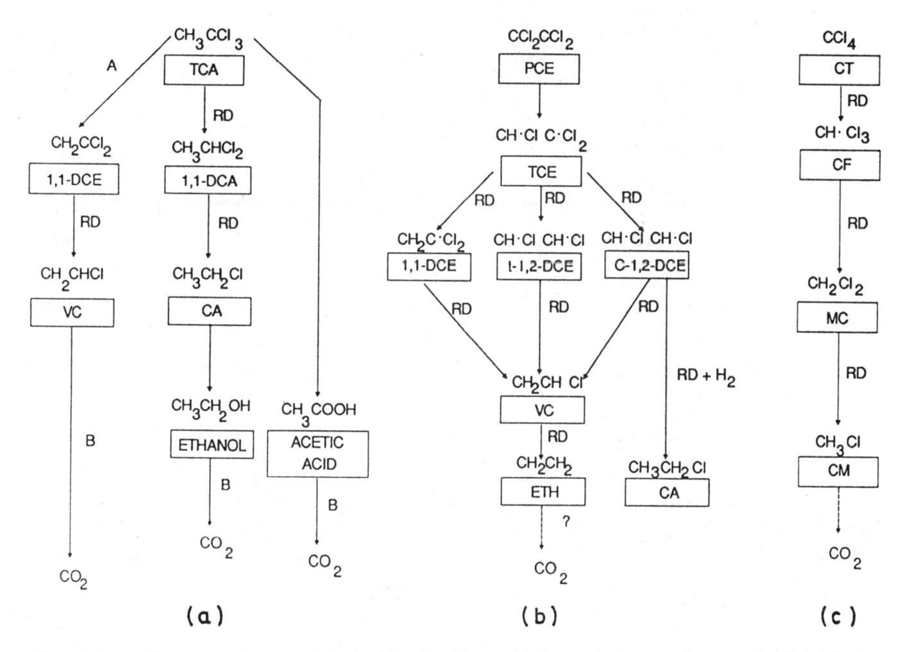

Figure 11.7 Suggested anaerobic dechlorination and degradation pathways for (*a*) 1,1,1-trichloroethane (TCA), (*b*) tetrachloroethylene (PCE), and (*c*) carbon tetrachloride (CT). RD = reductive dehalogenation; A = abiotic; b = biotic; CA = chloroethane; VC = vinyl chloride; ETH = ethylene; CM = chloromethane; CF = chloroform. (*Adapted from Vogel et al. 1987, Bario-Lage et al. 1986, Freedman and Gossett 1989.*)

systems. These reactor systems enable high biodegradation rates to occur. An anaerobic contact reactor was combined with aerobic polishing step (ANAMET Process, Huss, 1981) for treatment of landfill leachates (Bull et al. 1983) and sulfite evaporator condensates (Frostell 1984). Anaerobic filter reactors and fluidized-bed systems hold promise in treatment of industrial wastewater and landfill leachates containing halogenated organics. Lettinga et al. (1980) developed a granular mixed-culture upflow anaerobic sludge blanket (UASB). The UASB-type reactor system has been used to degrade pentachlorophenol (Bhatnagar et al. 1989b).

Biodetoxification by Novel Mixed-Culture Systems

Chlorinated chemicals, especially heavily chlorinated hydrocarbons, are toxic and persist for longer periods of time in oxygenated environments. However, these same chemicals can be readily dechlorinated to lesser toxic chlorinated products by anaerobic mixed-culture processes. The anaerobic rate of dechlorination decreases as the degree of chlorination of a compound is reduced (Bouwer and McCarty 1983a, Tiedje et al. 1987, Barrio-Lage et al. 1986); whereas, the rate of aerobic dechlorination increases as degree of chlorination decreases (Vogel et al. 1987, 1989; Kuhn et al. 1985; McCarty 1988). Thus a sequential two-stage anaerobic-aerobic bioreactor system has utility for complete destruction of harmful and persistent chemicals.

A method that utilized the concept of the sequential anaerobic and aerobic mixed-culture system to degrade toxic chemicals was documented with methoxychlor. This insecticide was only slightly degraded in soils under either aerobic or anaerobic conditions during a 3-month incubation period. However, when anaerobically incubated, methoxychlor was subsequently transferred (at the end of a 100-day incubation) to aerobic conditions, significant mineralization of methoxychlor occurred amounting to as much as 70-fold greater than that observed in soils maintained aerobically throughout the incubation period (Fogel et al. 1982). Recent studies (Fathepure and Vogel 1990, Vogel et al. 1989) employing anaerobic and aerobic biofilm reactors of mixed culture connected in series showed complete removal of polychlorinated compounds such as hexachlorobenzene, PCE, and CF. These chemicals were reductively transformed to dechlorinated products in an anaerobic biofilm reactor supported on acetate as the primary carbon. The effluent from the anaerobic biofilm containing dechlorinated products of HCB, PCE and CF was subsequently passed through an aerobic biofilm reactor. Analysis of effluent from the aerobic biofilm reactor indicated an effective removal of dechlorinated products present in the effluent from anaerobic reactor. In a similar study, the research conducted by Cambridge Analytical Associates Bioremediation Systems, Boston, Massachusetts (Dooley-Danna et al. 1989), has demonstrated the feasibility of in situ biodegradation of PCE and TCE. In this study, a laboratory aquifer simulator was filled with soil and amended groundwater was circulated through the soil. Naturally occurring bacteria were stimulated by controlled addition of nutrients. Methanogenic conditions were established within 2 weeks and PCE and TCE were rapidly dechlorinated to dichloroethylenes. Oxygen was then introduced to initiate the oxidative degradation of dichloroethylenes by methanotrophs. This sequential anaerobic-aerobic mixed-culture treatment scheme may be useful in biorestoration of aquifers contaminated with halogenated compounds.

It has been proposed that the biodegradation of toxicants could be enhanced by bioaugmentation, which involves the introduction of selected microbial populations in the contaminated environments. Specialized microorganisms can be developed by techniques such as adaptation or enrichment, and natural genetic exchange or genetic manipulation in the laboratory (Reineke and Knackmuss 1979,

Chakrabarty 1986). However, the ability of introduced microbes to remain viable, competitive, and genetically stable in the natural environments depends on many factors. These include (1) the concentration of target compound, whether it can support the growth of added organisms; (2) local environmental conditions such as the presence of inhibitory substances or native organisms that suppress the growth including the presence of predators; and (3) inaccessibility of pollutant itself for biodegradation.

In the past, inoculation of specialized organisms to the contaminated soils, surface water, and subsurface has been met with mixed success (Thomas and Ward 1989). Research has indicated that the biodegradation of toxic compounds such as phenols, pesticides, and several chlorinated compounds can be enhanced by applying specialized microorganisms (Mohn and Crawford 1985, Kilbane et al. 1982, Daughton and Hsieh 1977a,b). Most bioaugmentation studies have dealt with the addition of pure cultures of bacteria on chlorinated organics (Martinson et al. 1984, Focht and Brunner 1985) or fungi (Bumpus et al. 1985). Daughton and Hsieh (1977b) reported that 85 percent of parathion in soil was degraded when inoculated by adapted bacterial mixed cultures.

Genetic manipulation of microorganisms to produce engineered bacterial strains to degrade recalcitrant compounds is relatively new technology. Kellogg et al. (1982) introduced the term "molecular breeding" to describe the process of genetic exchange among mixed bacterial populations resulting into the selection of novel genotypes in a chemostat. Using this concept, Kilbane et al. (1982) obtained a *Pseudomonas cepacia* from a mixed culture that was fed 2,4,5-trichlorophenoxy acetic acid (2,4,5-T) as the sole carbon and energy source. The application of this bacterium to soil amended with 2,4,5-T showed successful biodegradation of the added chemical (Chatterjee et al. 1982). Similarly, Reineke and Knackmuss (1984) have constructed a chlorobenzene-degrading strain from benzene degraders by applying selection pressures using chlorobenzene.

One of the reasons why particular trophic groups get selected and exist as microbial mixed-culture communities on xenobiotic toxic compounds is that the degradation potential to metabolize these compounds does not exist in the genetic makeup of a single species. Also, there are possibilities of genetic exchange and rearrangement of genetic elements from different gene pools in microbial communities where complete mineralization of the toxicant is distributed between two or more different microbial species. Growth and stability of mixed cultures provide suitable conditions and opportunity for exchange of genetic information leading to the evolution of novel degradation pathways.

The relative importance of genetic transfer mechanisms such as transduction, transformation, and conjugation, including plasmid-mediated transfer, in microbial mixed cultures in their natural environment or in vitro laboratory systems is not well studied. Plasmid-mediated gene transfer may be significant because of the great versatility of plasmids contributing to the genetic flexibility and rearrangement of genes from different gene pools (Anderson 1968, Reanney 1982). Knackmuss and other groups (Minney et al. 1980, Reineke and Knackmuss 1979) have done pioneering work in demonstrating the role of plasmids in the genetic transfer within pseudomonads growing on chlorinated benzoic acid leading to the evolution of novel pathways. Using continuous culture enrichment procedures with a mixed culture containing the two pseudomonads (strain mt-2 and B13), a mutant was isolated which could grow on 4-chlorobenzoate, a substrate which initially did not support the growth of either of pseudomonad alone (Hartmann et al. 1979). By conventional mating experiments, the TOL plasmid was transferred from strain B13 to mt-2, then transconjugants were obtained which were similar to 4-chlorobenzoate utilizer isolated from continuous enrichments, indicating that plasmid transfer did occur to produce the new strain. Several promiscuous plasmids are known that code for a variety of catabolic functions (Focht and Alexander 1970, Wheelis 1975) including those involved in the degradation of herbicide 2,4-D (Pemberton and Fisher 1977).

Conclusion

Although the potential role and application of a mixed microbial community in the biodetoxification of xenobiotics is apparent, most bioaugmentation processes have used pure cultures. In many cases the performance of a pure culture in the degradation of a toxic organic xenobiotic compound is very poor when compared to that of an adapted mixed culture. Apparently, individual microbes do not have the genetic information to code for all the enzymes required for biodegradation because they have not been exposed to xenobiotics long enough to expect the evolution of a complete catabolic pathway. Therefore, it is thought that genetic manipulation to construct a complete metabolic pathway in one organism from different organisms by forced genetic transfer over a short period of time (directed evolution) with strong selective pressure may provide a "superbug" which can effectively meet the toxicity and environmental challenge to detoxify a given xenobiotic. Since these superbugs would still have to survive and compete in the environment as a mixed culture, it would seem that emphasis should be placed on the design, feeding, and use of "super mixed cultures" to ensure the job is done right.

Acknowledgments

The authors would like to express their sincere thanks to all scientists who provided pre-prints and other information for preparation of this chapter. We thank J. G. Zeikus for his comments in preparing this chapter. Also, we greatly appreciate the help and patience of Sue Ann Walker and Dianna Laverdiere who put together the manuscript. Support provided by the Michigan Biotechnology Institute is duly acknowledged.

References

Alexander, M. 1979. Role of cometabolism. Page 67 in A. W. Bourquin and P. H. Pritchard (eds.) U.S. E.P.A., Gulf Breeze, Florida.

Anderson, E. S. 1968. The ecology of transferable drug resistance in the Enterobacteria. *Ann. Rev. Microbiol.* 22:131.

Apajalahti, J. H. A., and M. S. Salkinoja-Salonen. 1987. Dechlorination and *para* hydroxylation of polychlorinated phenols by *Rhodococcus chlorophenolicus*. *J. Bacteriol.* 169:675.

Arciero, D., T. Vannelli, M. Logan, and A. B. Hooper. 1989. Degradation of trichloroethylene by ammonia-oxidizing bacterium *Nitrosomonas europaea*. *Biochem. Biophys. Res. Commun.* 159:640.

Atlas, R. M. 1981. Microbial degradation of petroleum hydrocarbons: an environmental perspective. *Microbial Rev.* 45:180.

Baggi, G., D. Catelani, E. Galli, and V. Treccani. 1972. The microbial degradation of phenylalkanes: 2-phenyl butane, 3-phenylpentane, 3-phenyldodecane, and 4-phenylheptane. *Biochem. J.* 126:1091.

Bailey, R. E. 1983. Comment on "Chlorobenzenes in sediments, water, and selected fish from Lakes Superior, Huron, Erie, and Ontario." *Environ. Sci. Technol.* 17:504.

Bak, F., and F. Widdel. 1986. Anaerobic degradation of indolic compounds by sulfate-reducing enrichment cultures and description of *Desulfobacterium indolicum* gen. nov. sp. nov. *Arch. Microbiol.* 146:170–176.

Bakker, A. 1977. Anaerobic degradation of aromatic compounds in the presence of nitrate. *FEMS Microbiol. Lett.* 1:103.

Balba, M. T., and W. C. Evans. 1979. The methanogenic fermentation of w-phenyl alkane carboxylic acids. *Biochem. Soc. Trans.* 7:403.

Balkwill, D. L., and W. C. Ghiorse. 1985. Characterization of subsurface bacteria associated with two shallow aquifers in Oklahoma. *Appl. Environ. Microbiol.* 50:580.

Barker, J. G., and G. C. Patrick. 1986. Natural altenuation of aromatic hydrocarbons in a shallow sand aquifer. In *Proceedings NWWA/API Conf. on Petroleum Hydrocarbons and Organic Chemicals in Groundwater, Prevention, Detection, and Restoration*. National Water Well Association, Worthington, Ohio.

Barrio-Lage, G., F. Z. Parsons, R. S. Nassar, and P. A. Lorenzo. 1986. Sequential dehalogenation of chlorinated ethenes. *Environ. Sci. Technol.* 20:96.

Bartels, I., H.-J. Knackmuss, and W. Reinecke. 1984. Suicide inactivation of catechol, 2,3-dioxygenase from *Pseudomonas putida* mt-2 by halocatechols. *Appl. Environ. Microbiol.* 47:500–505.

Bartha, R., and R. M. Atlas. 1977. The microbiology of aquatic oil spills. *Adv. Appl. Microbiol.* 22:225.

Bayly, R. C., and S. Dagley. 1969. Oxoenoic acids as metabolites in the bacterial degradation of catechols. *Biochem. J.* 111:303.

Beam, H. W., and J. J. Perry. 1974. Microbial degradation of cycloparaffinic hydrocarbons via co-metabolism and commensalism. *J. Gen. Microbiol.* 82:163.

Bedient, P. B., A. C. Rodgers, T. C. Bouvette, M. B. Tomson, and T. H. Wang. 1984. Groundwater quality at a creosote waste site. *Groundwater* 22:318.

Beerstecher, E. 1954. *Petroleum Microbiology*. Elsevier, New York.

Berry, D. L., E. L. Madsen, and J.-M. Bollag. 1987. Conversion of indole to oxindole under methanogenic conditions. *Appl. Environ. Microbiol.* 53:180–182.

Bhatnagar, L., and J. G. Zeikus. 1989. Assessment methods of anaerobic biodegradation in aquatic environment. In *Proc. International Congress on Degradation Assessment of Organic Substances in the Environment.* Paris, France.

Bhatnagar, L., S.-P. Li, M. K. Jain, and J. G. Zeikus. 1989a. Growth of methanogenic and acidogenic bacteria with pentachlorophenol as co-substrate. Pages 383–393 in G. Lewandowski, A. Armenante, and B. Baltzis (eds.), *Biotechnology Applications in Hazardous Waste Treatment*, New Jersey Institute of Technology, Newark, N.J.

Bhatnagar, L., W. Wu, M. K. Jain, and J. G. Zeikus. 1989b. Anaerobic biodetoxification of pentachlorophenol using syntrophic biomethanation granules in an upflow bioreactor system. Abstr. ACS National Meeting, Dallas, Texas.

Bishop, P. L., and N. E. Kinner. 1986. Aerobic fixed-film processes, p. 113. In H.-J. Rehm G. Reed (eds.), *Biotechnology* Vol. 8, VCH Verlagsgesellschaft mbH, D-6940, Weinheim, Fed. Rep. Germany.

Borden, R. C., P. B. Bedient, M. D. Lee, C. H. Ward, and J. T. Wilson. 1986. Transport of dissolved hydrocarbons influenced by oxygen-limited biodegradation. II. Field application. *Water Res.* 22:1986.

Bouwer, E. J., and P. L. McCarty. 1983a. Transformations of 1- and 2-carbon halogenated aliphatic organic compounds under methanogenic conditions. *Appl. Environ. Microbiol.* 45:1286–1294.

Bouwer, E. J., and P. L. McCarty. 1983b. Transformation of halogenated organic compounds under denitrification conditions. *Appl. Environ. Microbiol.* 45:1295–1299.

Bouwer, E. J., B. E. Rittmann, and P. L. McCarty. 1981. Anaerobic degradation of halogenated 1- and 2-carbon organic compounds. *Environ. Sci. Technol.* 15:596–599.

Boyd, S. A., D. R. Shelton, D. Berry, and J. M. Tiedje. 1983. Anaerobic biodegradation of phenolic compounds in digested sludge. *Appl. Environ. Microbiol.* 46:50–54.

Boyd, S. A., D. R. Shelton. 1984. Anaerobic biodegradation of chlorophenols in fresh and acclimated sludge. *Appl. Environ. Microbiol.* 47:272–277.

Braun, K., and D. T. Gibson. 1984. Anaerobic degradation of L-aminobenzoate (anthranilic acid) by denitrifying bacteria. *Appl. Environ. Microbiol.* 48:102–107.

Brisbane, P. G., and J. N. Ladd. 1965. The role of microorganisms in petroleum exploration. *Ann. Rev. Microbiol.* 19:351.

Britton, L. N. "Microbial degradation of aliphatic hydrocarbons. Page 89 in D. T. Gibson, (ed.) *Microbial Degradation of Organic Compounds* Marcel Dekker, New York, 1984.

Brown, J. G., R. E. Wagner, D. L. Bedford, M. J. Brennan, J. C. Carnahan, R. J. May, and T. J. Tofflemire. 1984. PCB transformations in Upper Hudson sediments. *Northeastern Environ. Sci. Issues* 34:167–179.

Bryant, M. P., L. L. Campbell, C. A. Reddy, and M. R. Crabhill. 1977. Growth of *Desulfovibrio* in lactate or ethanol media low in sulfate in association with H_2-utilizing methanogenic bacteria. *Appl. Environ. Microbiol.* 33:1162.

Bryant, M. P., E. A. Wolin, M. J. Wolin, and R. S. Wolfe. 1967. *Methanobacillus omelianski*, a symbiotic association of two species of bacteria. *Arch. Microbiol.* 59:20.

Buhler, M., and J. Schindler. 1984. Aliphatic hydrocarbons. Page 370. In K. Kieslich (ed.) *Biotransformations*, vol. 6A, Chemie Verlag, Weinheim, F.R.G.

Bull, A. T. 1980. Biodegradation: Some attitudes and strategies of microorganisms and microbiologists. Pages 107–136 Ellwood (ed.), *Contemporary Microbial Ecology*, Academic Press, London.

Bull, A. T., and C. M. Brown. 1979. Continuous culture applications to microbial biochemistry. Page 177 in J. R. Quayle (ed.), *Microbial Biochemistry*, University Park Press, Baltimore, Maryland.

Bull, P. S., J. V. Evans, R. M. Wechsler, and K. J. Cleland. 1983. Biological technology of the treatment of leachate from sanitary landfills. *Water Res.* 17:1473–1481.

Bumpus, J. A., M. Tien, D. Wright, and S. D. Aust. 1985. Oxidation of persistent environmental pollutants by white rot fungus. *Science* 228:1435.

Canter, L. W., and R. C. Knox. 1985. *Groundwater Pollution Control*, Lewis Publishing, Chelsea, Michigan.

Chakrabarty, A. M. 1986. Genetic engineering and problems of environmental pollution. Pages 515–530 in H. J. Rehm and C. Reed (eds.), *Biotechnology: Microbial Degradations*, vol. 8. VCH Publishers, Weinheim, FRG.

Chapman, P. J. 1975. Bacterial metabolism of 4-chlorobenzoic acid. *Abstr. Ann. Meet. Am. Soc. Microbiol.* 192.

Chartrain, M., L. Bhatnagar, and J. G. Zeikus. 1987. Microbial ecophysiology of whey biomethanation: Comparison of carbon transformation parameters, species composition, and starter culture performance in continuous culture. *Appl. Environ. Microbiol.* 53:1147–1156.

Chatterjee, D. K., J. J. Kilbane, and A. M. Chakrabarty. 1982. Biodegradation of 2,4,5-trichlorophenoxy acetic acid in soil by a pure culture of *Pseudomonas cepacia*. *Appl. Environ. Microbiol.* 44:514.

Churchman, J., and D. Grbic-Galic. 1987. Anaerobic transformation of styrene in methanogenic consortia and isolation of pure cultures. *Abstr. Annu. Meet. Am. Soc. Microbiol.* Q32, p. 287.

Clark, R. M., C. A. Frank, and B. W. Lykins, Jr. 1988. Removing organic contaminants from groundwater. *Environ. Sci. Technol.* 10:1126.

Colwell, R. R., and J. D. Walker. 1977. Ecological aspects of microbial degradation of petroleum in the marine environment. *Crit. Rev. Microbiol.* 5:423.

Cossar, D., C. M. Brown, and R. J. Watkinson. 1981. Some properties of a marine microbial community utilizing benzoate. *Soc. Gen. Microbiol.* 8:147.

Dagley, S. 1971. Catabolism of aromatic compounds by microorganisms. *Adv. Microb. Physiol.* 6:1.

Dagley, S. 1975. A biochemical approach to some problems of environmental pollution. *Essays Biochem.* 11:81.

Daughton, C. G., and D. P. Hsieh. 1977a. Parathion utilization by bacterial symbionts in a chemostat. *Appl. Environ. Microbiol.* 34:175.

Daughton, C. G., and D. P. Hsieh. 1977b. Accelerated parathion degradation in soil by inoculation with parathion degrading bacteria. *Bull. Environ. Contam. Toxicol.* 18:48.

Davis, J. B., and H. F. Yarbrough. 1966. Anaerobic oxidation of hydrocarbons by *Desulfovibrio desulfuricans*. *Chem. Geol.* 1:137.

DeWeerd, K. A., L. Mandelco, R. S. Tanner, C. R. Woese, and J. M. Suflita. 1990. *Desulfomonile tiedjei* gen. nov. and sp. nov., a novel anaerobic dehalogenating sulfate reducing bacterium. *Arch. Microbiol.* 154:23.

Dooley-Danna, M., S. Fogel, M. Findlay, The sequential anaerobic/aerobic biodegradation of chlorinated ethenes in an aquifer simulator. Page 5, A14, in *Abstr. Int. Symp. on Processes Governing the Movement and Fate of Contaminants in the Subsurface Environment*, Int. Assoc. Water Pollut. Res. Control., Stanford Univ., Stanford, 1989.

Durre, P., and J. R. Andreesen. 1983. Purine and glycine metabolism by purinolytic clostridia. *J. Bacteriol.* 154:192–199.

Evans, W. C. 1977. Biochemistry of the bacterial catabolism of aromatic compounds in anaerobic environments. *Nature* 270:17–22.

Fathepure, B. Z., and S. A. Boyd. 1988. Reductive dechlorination of perchloroethylene and the role of methanogens. *FEMS Microbiol. Lett.* 49:149.

Fathepure, B. Z., J. P. Nengu, and S. A. Boyd. 1988. Anaerobic bacteria that dechlorinate perchloroethylene. *Appl. Environ. Microbiol.* 53:2671.

Fathepure, B. Z., J. M. Tiedje, and S. A. Boyd. 1987. Reductive dechlorination of 4-chlororesorcinol by anaerobic microorganisms. *Environ. Toxicol. Chem.* 6:929–934.

Fathepure, B. Z., J. M. Tiedje, and S. A. Boyd. 1988. Reductive dechlorination of hexachlorobenzene to tri- and dichlorobenzenes in anaerobic sewage sludge. *Appl. Environ. Microbiol.* 54:327.

Fathepure, B. Z., and T. M. Vogel. 1990. The sequential anaerobic/aerobic biodegradation of chlorinated hydrocarbons, Q38, p. 294. Abstr. Ann. Mtg. Am. Soc. Microbiol., Anaheim, California.

Ferry, J. G., and R. S. Wolfe. 1976. Anaerobic degradation of benzoate to methane by a microbial consortium. *Arch. Microbiol.* 107:33.

Fewson, C. A. Biodegradation of aromatics with industrial relevance. Page 141 in T. Leisinger, A. M. Cook, R. Hutter, and J. Nuesch, (eds.) *Microbial Degradation of Xenobiotics and Recalcitrant Compounds*, Academic Press, London-New York, 1981.

Focht, D.-D., and M. Alexander. 1970. DDT metabolites and analogs: ring fission by *Hydrogenomonas*. *Science* 170:91.

Focht, D. D., and W. Brunner. 1985. Kinetics of biphenyl and polychlorinated biphenyl metabolism in soil. *Appl. Environ. Microbiol.* 50:1058.

Fogel, S., R. L. Lancione, and A. F. Sewall. 1982. Enhanced biodegradation of methoxychlor in soil under sequential environmental conditions. *Appl. Environ. Microbiol.* 44:113.

Fogel, M. M., A. R. Taddeo, and S. Fogel. 1986. Biodegradation of ethenes by a methene utilizing mixed culture. *Appl. Environ. Microbiol.* 51:720.

Foster, J. W. 1962. Bacterial oxidation of hydrocarbons. Page 241 in O. Hayaishi (ed.) *Oxygenases*. Academic Press, New York.

Freedman, D. L., and J. M. Gossett. 1989. Biological reductive dechlorination of tetrachloroethylene and trichloroethylene to ethylene under methanogenic conditions. *Appl. Environ. Microbiol.* 55:2144.

Frostell, B. 1984. Anaerobic-aerobic pilot-scale treatment of a sulfite evaporator condensate. *Pulp Pap. Mag. Can.* 85:T57–62.

Fuhs, G. W. 1961. Der mikrobielle abbau von kohlenwasserst offen. *Arch. Mikrobiol.* 39:374.

Galli, R. 1987. Biodegradation of dichloromethane in waste water using a fluidized bed reactor. *Appl. Microbiol. Biotechnol.* 27:206.

Galli, R., and T. Leisinger. 1985. Specialized bacterial strains for the removal of dichloromethane from industrial waste. *Conservation. Recyl.* 8:91.

Gibson, D. T., J. R. Koch, and R. E. Kallio. 1968. Oxidative degradation of aromatic hydrocarbons by microorganisms I. Enzymatic formation of catechol from benzene. *Biochemistry* 7:2653.

Gibson, D. T. 1975. Microbial degradation of hydrocarbons. Page 667 in E. D. Goldberd (ed.) *The Nature of Seawater*. Springer-Verlag, Heidelberg.

Gibson, D. T., and V. Subramanian. 1984. Microbial degradation of aromatic hydrocarbons. Page 181 in D. T. Gibson (ed.) *Microbial Degradation of Organic Compounds*. Marcel Dekker, New York.

Godsy, E. M., and D. Grbic-Galic. 1988. Anaerobic degradation pathways for benzothiophene in aquifier-derived methanogenic microcosms. *Abstr. Annu. Meet. Am. Soc. Microbiol.* A111, p. 301.

Godsy, E. M., D. F. Goerlitz, and D. Grbic-Galic. 1987. Anaerobic biodegradation of creosote contaminants in natural and simulated ground water ecosystems. Pages A17–A19 in B. J. Franks (ed.) *U.S. Geological Survey Toxic Waste-Ground Water Contamination Program: Proceedings of the Third Technical Meeting*, Pensacola, Florida, March 1987. U.S. Geological Survey Open File Report 87-109, Tallahassee, Florida.

Goldman, P., G. W. A. Milne, and M. T. Pignataro. 1967. Fluorine containing metabolites formed from 2-Fluorobenzoic acid by *Pseudomonas species. Arch. Biochem. Biophys.* 118:178.

Grbic-Galic, D. 1989. Microbial degradation of homocyclic and heterocyclic aromatic hydrocarbons under anaerobic conditions. *Dev. Ind. Microbiol.* 30:237.

Grbic-Galic, D., and T. M. Vogel. 1987. Transformation of toluene and benzene by mixed methanogenic cultures. *Appl. Environ. Microbiol.* 53:254–260.

Grbic-Galic, D., and L. Y. Young. 1985. Methane fermentation of ferulate and benzoate: anaerobic degradation pathways. *Appl. Environ. Microbiol.* 50:292–297.
Gunner, H. B., and B. M. Zuckerman. 1968. Degradation of diazinion by synergistic microbial action. *Nature* (London) 217:1183.
Guthrie, M. A., E. J. Kirsch, R. F. Wukasch, and C. P. L. Grady. 1984. Pentachlorophenol biodegradation-II. Anaerobic. *Water Res.* 18:451–461.
Hakulinen, R., and M. Salkinoja-Salonen. 1982. Treatment of pulp and paper industry wastewaters in an anaerobic fluidized bed reactor. *Process Biochem.* 17:18–22.
Hartmann, J., W. Reineke, and H.-J. Knackmuss. 1979. Metabolism of 3-chloro-, 4-chloro-, and 3,5-dichlorobenzoate by a pseudomonad. *Appl. Environ. Microbiol.* 37: 421–428.
Hartmans, S., J. A. M. de Bont, and W. Harder. 1989. Microbial metabolism of short chain unsaturated hydrocarbons. *FEMS Microbiol. Rev.* 63:235.
Hartmans, S., J. A. M. de Bont, J. Tramper, K. C. A. M. Luyben. 1985. Bacterial degradation of vinyl chloride. *Biotechnol. Lett.* 7:383.
Harvey, R. W., R. L. Smith, and L. George. 1984. Effect of organic contamination upon microbial distribution and heterotrophic uptake in a Cape Cod, Massachusetts aquifer. *Appl. Environ. Microbiol.* 48:1197.
Harwood, C. S., and J. Gibson. 1988. Anaerobic and aerobic metabolism of diverse aromatic compounds by photosynthetic bacterium *Rhodopseudomonas palustris. Appl. Environ. Microbiol.* 54:712–717.
Healey, J. B., Jr. and L. Y. Young. 1978. Catechol and phenol degradation by a methanogenic population of bacteria. *Appl. Environ. Microbiol.* 35:216.
Healy, J. B., Jr. and L. Y. Young. 1979. Anaerobic biodegradation of eleven aromatic compounds to methane. *Appl. Environ. Microbiol.* 38:84.
Henson, J. M., M. V. Yates, J. W. Cochran, and D. L. Shackleford. 1988. Microbial removal of halogenated methanes, ethanes, and ethylenes in an aerobic soil exposed to methane. *FEMS Microbiol. Ecol.* 53:193.
Henson, J. M., P. D. Nichols, and J. T. Wilson. 1986. Degradation of halogenated aliphatic hydrocarbons by a biochemically defined microbial community. 86th Ann. Mtg. Am. Soc. Microbiol., Washington, D.C.
Henson, J. M., M. V. Yates, and J. W. Cochran. 1987. Metabolism of chlorinated aliphatic hydrocarbons by a mixed bacteria culture growing on methane. Page 298 in Abst. 2-97, Ann. Mtg. Am. Soc. Microbiol.
Hickey, R. F., J. Vanderwielen, and M. S. Switzenbaum. 1987a. Production of trace levels of carbon monoxide during methanogenesis on acetate and methanol. *Biotechnol. Lett.* 9:63.
Hickey, R. F., J. Vanderwielen, and M. S. Switzenbaum. 1987b. The effect of organic toxicants on methane production and hydrogen gas levels during the anaerobic digestion of waste activated sludge. *Water Res.* 21:1417.
Holliger, C., A. J. M. Stams, and A. J. B. Zehnder. 1988. Anaerobic degradation of recalcitrant compounds. Page 211. in E. R. Hall and P. N. Hobson (eds.) *Anaerobic Digestion*, Pergamon Press, New York.
Horvath, R. S. 1972. Microbial cometabolism and the degradation of organic compounds in nature. *Bacteriol. Rev.* 36:146.
Horowitz, A., J. M. Suflita, and J. M. Tiedje. 1983. Reductive dehalogenations of halobenzoates by anaerobic lake sediment microorganisms. *Appl. Environ. Microbiol.* 45:1459–1465.
Hou, C. T. 1982. Microbial transformation of important industrial hydrocarbons. Page 81 in J. P. Rosazza (ed.) *Microbial Transformations of Bioactive Compounds*. CRC Press, Boca Raton, Florida.
Huss, L. 1981. Application of the Anamet Process for wastewater treatment. Pages 137–150 in D. E. Hughes, D. A. Stafford, B. J. Wheatley, W. Badder, G. Lettinga, E. J. Nyns, and W. Verstraeten (eds.), *Anaerobic Digestion 1981*. Elsevier, Amsterdam.
Huybergtse, R., and A. C. van der Linden. 1964. The oxidation of α-olefins by a *Pseudomonas* reaction involving double bonds. *Ant. V. Leeuwenhoek. J. Microbiol. Serol.* 30:185.

Ide, A., Y. Niki, F. Sakamoto, I. Wantanabe, and H. Wantanabe. 1972. Decomposition of pentachlorophenol in paddy soil. *Argric. Biol. Chem.* 36:1937–1944.

Ishikura, T., and J. W. Foster. 1961. Incorporation of molecular oxygen during microbial utilization of olefins. *Nature* (London) 192:892.

Janssen, D. B., A. Scheper, L. Dijkhuizen, and B. Witholt. 1985. Degradation of halogenated aliphatic compounds by xanthobacter autotrophicus DJ10. *Appl. Environ. Microbiol.* 49:673.

Jhaveri, V., and A. J. Mazzacca. 1983. Bioreclamation of groundwater: a case history. Page 242 in Proc. 4th Natl. Conf. on Management of Uncontrolled Hazardous Waste Sites, Washington, D.C.

Jigami, Y., T. Omori, and Y. Minoda. 1975. The degradation of isopropylbenzene and isobutylbenzene by *Pseudomonas* sp. *Agr. Biol. Chem.* 39:1781.

Johnston, H. W., G. G. Briggs, and M. Alexander. 1972. Metabolism of 3-chlorobenzoic acid by a *Pseudomonad. Soil-Biol. Biochem.* 4:187.

Jones, J. G., and M. A. Edington. 1968. An ecological survey of hydrocarbon-oxidizing microorganisms. *J. Gen. Microbiol.* 52:381.

Jutzi, K., A. M. Cook, and R. Hutter. 1982. The degradative pathway of the s-triazine Melamine. *Biochem. J.* 208:679–694.

Kaiser, J. P., and K. W. Hanselmann. 1982. Fermentative metabolism of substituted monoaromatic compounds by a bacterial community from anaerobic sediments. *Arch. Microbiol.* 133:185.

Keith, C. L., R. L. Bridges, L. R. Fina, K. L. Iverson, and J. A. Cloran. 1978. The anaerobic decomposition of benzoic acid during methane fermentation. IV. Dearomatization of the ring and volatile fatty acids formed on ring rupture. *Arch. Microbiol.* 118:113.

Kellogg, S. T., D. K. Chatterjee, and A. M. Chakrabarty. 1982. Plasmid-assisted molecular breeding: New technique for enhanced biodegradation of persistent toxic chemicals. *Science* 214:1133.

Kilbane, J. J., D. K. Chatterjee, J. S. Karns, S. T. Kellog, and A. M. Chakrabarty. 1982. Biodegradation of 2,4,5-trichlorophenoxy acetic acid by a pure culture of *Pseudomonas cepacia. Appl. Environ. Microbiol.* 44:72.

Kirk, T. K., W. J. Connors, R. D. Bleam, W. F. Hackett, and J. G. Zeikus. 1975. Preparation and microbial decomposition of synthetic [^{14}C] lignins. *Proc. Nat. Acad. Sci.* 72:2515–2519.

Kitagawa, M. 1956. Studies on the oxidation mechanism of methyl group. *J. Biochem.* (Tokyo) 43:553.

Klages, U., and F. Lingens. 1979. Degradation of 4-chlorobenzoic acid by a *Nocardia species. FEMS Microbiol. Lett.* 6:201.

Klecka, G. M., and S. J. Gonsior. 1984. Reductive dechlorination of chlorinated methanes and ethanes by reduced iron (II) porphyrins. *Chemosphere* 13:391–402.

Klug, M. J., and A. J. Markovetz. 1971. Utilization of aliphatic hydrocarbons by microorganisms. *Adv. Microbiol. Physiol.* 5:1.

Knackmuss, H.-J. 1981. Degradation of halogenated and sulfonated hydrocarbons. Page 189 in T. Leisinger, A. M. Cook, R. Hutter and J. Nuesch (eds.) *Microbial Degradation of Xenobiotics and Recalcitrant Compounds*. Academic Press, New York.

Kohler-Staub, D., S. Hartmans, R. Galli, F. Suter, and T. Leisinger. 1986. Evidence for identical dichloromethane dehalogenases in different methylotrophic bacteria. *J. Gen. Microbiol.* 132:2837.

Kuhn, E. P., P. J. Colberg, J. J. Schnoor, O. Warner, A. J. B. Zehnder, and R. P. Schwarzenbach. 1985. Microbial transformation of substituted benzenes during infiltration of river water to groundwater-laboratory column studies. *Environ. Sci. Technol.* 19:961.

Kuhn, E. P., J. Zeyer, P. Eicher, and R. P. Schwarzenbach. 1988. Anaerobic degradation of alkylated benzenes in denitrifying laboratory columns. *Appl. Environ. Microbiol.* 54:490.

LaPat-Polasko, L. T., P. L. McCarty, and A. J. B. Zehnder. 1984. Secondary substrate utilization of methylene chloride by an isolated strain of *Pseudomonas* sp. *Appl. Environ. Microbiol.* 47:825.

Lee, M. D., J. M. Thomas, R. C. Borden, P. B. Bedient, J. T. Wilson, and C. H. Ward. 1988. Biorestoration of aquifers contaminated with organic compounds. *CRC Critical Rev. Environ. Control* 18:29.

Lee, M. D., and C. H. Ward. 1985. Biological methods for the restoration of contaminated aquifers. *Environ. Toxicol. Chem.* 4:743.

Leisinger, T., and W. Brunner. 1986. Poorly degradable substances. Page 475 in H.-J. Rehm and G. Reed (eds.) *Biotechnology*, vol. 8, VCH Publishers, New York.

Lettinga, G., A. F. M. van Velsen, S. W. Hobma, W. de Zeeuw and A. Klapwijk. 1980. Use of upflow sludge blanket (USB) reactor concept for biological wastewater treatment, especially for anaerobic treatment. *Biotech. Bioeng.* 32:699.

Little, C. D., A. V. Palumbo, S. E. Herbes, M. E. Lidstrom, R. L. Tyndall, and P. J. Gilmer. 1988. Trichloroethylene biodegradation by a methane-oxidizing bacterium. *Appl. Environ. Microbiol.* 54:951.

Madsen, E. L., A. J. Francis, and J.-M. Bollag. 1988. Environmental factors affecting indole metabolism under anaerobic conditions. *Appl. Environ. Microbiol.* 54:74–78.

Major, D. W., C. I. Mayfield, and J. F. Barker. 1988. Biotransformation of benzene by denitrification in aquifer sand. *Ground Water* 26:8–14.

Maka, A., V. L. McKinley, J. R. Conrad, and K. F. Frannin. 1987. Degradation of benzothiophene and dibenzothiophene under anaerobic conditions by mixed cultures. *Abstr. Annu. Meet. Am. Soc. Microbiol.* 054, p. 269.

Malmstrom, B. G. 1982. Enzymology of oxygen. *Ann. Rev. Biochemistry*, 51:21.

Markovetz, A. J. 1972. Subterminal oxidation of aliphatic hydrocarbons by microorganisms. *Crit. Rev. Microbiol.* 1:225.

Marks, T. S., R. Wait, A. R. W. Smith, and A. V. Quirk. 1984. The origin of oxygen incorporated during the dehalogenation/hydroxylation of 4-chlorobenzoate by an *Arthrobacter* sp. *Biochem. Biophys. Res. Commun.* 124:669.

Martinson, M. M., J. G. Steirt, D. L. Saber, and R. L. Crawford. 1984. Microbiological degradation of pentachlorophenol in natural waters. In E. E. O'Rear (ed.) *Biodegradation*, 6th Proc. Int. Symp. Commonwealth Agric. Bureau, Washington, D.C.

May, S. W. 1979. Enzymatic epoxidation reactions. *Enzyme Microbiol. Technol.* 1:15.

McCarty, P. L. 1988. Bioengineering issues related to in-situ remediation of contaminated soils and groundwater. Page 143 in G. S. Omenn (ed.) *Environmental Biotechnology*, Plenum Press, New York.

McKinley, V. L., J. R. Vestal, and A. E. Eralp. 1989. Microbial activity in composting. Page 55 in *The Biocycle Guide to Composting Municipal Wastes*. The JG Press, Emmaus, Pennsylvania.

Merian, E., and M. Zander. 1982. Volatile aromatics. Page 117 in O. Hutzinger (ed.) *The Handbook of Environmental Chemistry*, vol. 3, part B. Springer-Verlag, Berlin.

Mihelcic, J. R., and R. G. Luthy. 1988a. Degradation of polycyclic aromatic hydrocarbon compounds under various redox conditions in soil-water systems. *Appl. Environ. Microbiol.* 54:1182–1187.

Mihelcic, J. R., and R. G. Luthy. 1988b. Microbial degradation of acenaphthene and naphthalene under denitrification conditions in soil-water systems. *Appl. Environ. Microbiol.* 54:1188.

Mikesell, M. D., and S. A. Boyd. 1986. Complete reductive dechlorination of pentachlorophenol by anaerobic microorganisms. *Appl. Environ. Microbiol.* 52:861–865.

Minney, S., R. J. Parkes, J. H. Slater, and A. T. Bull. 1980. The degradation of Fenuron by estuarine microbial communities in laboratory microcosms. Pages 117–118 in *Abstracts of the Second International Symposium on Microbial Ecology.*

Miyoshi, M. 1985. Die durehbohrung von membranen durch Pilzbaden *Jahrb. Wiss. Bot.* 28:269.

Mohn, W. W., and R. L. Crawford. 1985. Microbial removal of pentachlorophenol from soil using a Flavobacterium. *Enzyme Microb. Technol.* 7:617.

Morgan, P., and R. J. Watkinson. 1989. Hydrocarbon degradation in soils and methods for soil biotreatment. *CRC Critical Rev. Biotechnol.* 8:305.

Muller, F. M. 1957. On methane formation from higher alkanes. *Ant. V. Leeuwenhoek. J. Microbiol. Serol.* 23:369.

Muller, R., J. Thiele, U. Klages, and F. Lingens. 1984. Incorporation of [^{18}O] water into 4-hydroxybenzoic acid in the reaction of 4-chlorobenzoate dehalogenase from *Pseudomonas* species CBS3. *Biochem. Biophys. Res. Commun.* 124:178.

Nelson, M. J. K., S. O. Montgomery, W. R. Mahaffey, and P. H. Pritchard. 1987. Biodegradation of trichloroethylene and involvement of an aromatic biodegradative pathway. *Appl. Environ. Microbiol.* 52:949.

Nelson, M. J. K., S. O. Montgomery, E. J. O'Neill, and P. H. Pritchard. 1986. Aerobic metabolism of trichloroethylene by a bacterial isolate. *Appl. Environ. Microbiol.* 52: 383.

Nelson, M. J. K., S. O. Montgomery, and P. H. Pritchard. 1988. Trichloroethylene metabolism by microorganisms that degrade aromatic compounds. *Appl. Environ. Microbiol.* 54:604.

Nyer, E. K. 1985. *Groundwater Treatment Technology.* Van Nostrand Reinhold, New York.

Oliver, B. G., and K. D. Nicol. 1982. Chlorobenzenes in sediments, water, and selected fish from Lakes Superior, Huron, Erie, and Ontario. *Environ. Sci. Technol.* 16:532–536.

Olsen, R. H., J. J. Kukor, and M. D. Mikesell. 1989. Metabolic diversity of microbial BTEX degradation under aerobic and anoxic conditions. Pages 11–14 in *Biotreatment: The Use of Microorganisms in the Treatment of Hazardous Wastes.* HMCRI, Silver Springs, Maryland.

Omori, T., Y. Jigami, and Y. Minoda. 1975. Isolation, identification and substrate assimilation specificity of some aromatic hydrocarbon utilizing bacteria. *Agr. Biol. Chem.* 39:1775.

Parsons, F., P. R. Wood, and J. DeMarco. 1984. Transformations of tetrachloroethylene and trichloroethylene in microcosms and groundwater. *J. Am. Water Works Assoc.* 76:56–59.

Pemberton, J. M., and P. R. Fisher. 1977. 2,4-D plasmids and persistence. *Nature* 268: 732.

Pereira, W. E., C. E. Rostad, D. M. Updegraff, and J. L. Bennett. 1987. Fate and movement of azaarenes in an aquifer contaminated by wood-treatment chemicals. *Environ. Toxicol. Chem.* 6:163–176.

Pereira, W. E., C. E. Rostad, T. J. Leiker, D. M. Updegraff, and J. L. Bennett. 1988. Microbial hydroxylation of quinoline in contaminated groundwater: evidence for incorporation of the oxygen atom of water. *Appl. Environ. Microbiol.* 54:827–829.

Perry, J. J. 1979. Microbial cooxidations involving hydrocarbons. *Microbiol. Rev.* 43:59.

Perry, J. J., and H. W. Scheld. 1968. Oxidation of hydrocarbons by microorganisms isolated from soil. *Can. J. Microbiol.* 14:403.

Pfennig, N. 1978. Syntrophic associations and consortia with phototrophic bacteria. Page 16 in *Abstracts of the XII International Microbiology Congress of Microbiology.*

Pfennig, N., F. Widdell, and H. G. Truper. 1981. The dissimilatory sulfate-reducing bacteria. Pages 926–940 in M. P. Starr, H. Stolp, H. G. Truper, A. Balows, and H. G. Schlegel (eds.) *The Prokaryotes*, vol. 1. Springer-Verlag, Heidelberg.

Quensen, III, J. F., J.-M. Tiedje, and S. A. Boyd. 1988. Reductive dechlorination of polychlorinated biphenyls by anaerobic microorganisms from sediments. *Science* 242: 752.

Ratledge, C. 1978. Degradation of aliphatic hydrocarbons. In R. J. Watkinson (ed.) *Developments in Biodegradation of Hydrocarbons—I.* Applied Science Publisher, London.

Raymond, R. L., V. W. Jamison, and J. O. Hudson. 1976. Beneficial simulation of bacterial activity in groundwater containing petroleum products. *AICHE Symp. Ser.* 73: 390.

Reanney, D. C., W. P. Roberts, and W. J. Kelley. 1982. Genetic interactions among microbial communities. Pages 287–322, in A. J. Bull and J. H. Slater (eds.) *Microbial Interactions and Communities*, vol. I. Academic Press, London.

Reineke, W. 1984. Microbial degradation of halogenated aromatic compounds. Page 319 in D. T. Gibson (ed.) *Microbial Degradation of Organic Compounds.* Marcel Dekker, New York.

Reineke, W., and H.-J. Knackmuss. 1979. Construction of haloaromatic utilizing bacteria. *Nature* (London) 277:385–386.
Reineke, W., and H.-J. Knackmuss. 1984. Microbial metabolism of haloaromatics: Isolation and properties of chlorobenzene degrading bacterium. *Appl. Environ. Microbiol.* 47:395.
Reineke, W., and H.-J. Knackmuss. 1988. Microbial degradation of haloaromatics. *Ann. Rev. Microbiol.* 42:263.
Reinhard, M., N. L. Goodman, and J. F. Barker. 1984. Occurrence and distribution of organic chemicals in two landfill leachate plumes. *Environ. Sci. Technol.* 18:953–961.
Ribbons, D. W., and R. W. Eaton. 1982. Chemical transformations of aromatic hydrocarbons that support the growth of microorganisms. Page 59 in A. M. Chakrabarty (ed.) *Biodegradation and Detoxification of Environmental Pollutants*. CRC Press, Boca Raton, Florida.
Rittmann, B. E. 1987. Aerobic biological treatment. *Environ. Sci. Technol.* 21:128.
Roberts, P., L. Semprini, G. Hopkins, and P. McCarty. 1989a. Biostimulation of methanotrophic bacteria to transform halogenated alkenes for aquifer restoration. Page 203 in *Proc. Conf. on Petroleum Hydrocarbons, and Organic Chemicals in Groundwater—Prevention, Detection, and Restoration*. National Water Well Association/API. Houston, Texas.
Roberts, P. V., L. Semprini, G. D. Hopkins, D. Grbic-Galic, P. L. McCarty, and M. Reinhard. 1989b. In situ aquifer restoration of chlorinated aliphatics by methanotrophic bacteria. EPA-600/2-89/033. U.S. EPA, Ada, Oklahoma.
Rochkind-Dubinsky, M. L., G. S. Sayler, and J. W. Blackburn (eds.). 1987. *Microbial Decomposition of Chlorinated Aromatic Compounds*. Marcel Dekker, New York.
Rose, W. W., and W. A. Mercer. 1968. Fate of insecticides in composted agricultural wastes, Progress report, Part I. Natl. Canners Assoc., Washington, D.C.
Salkinoja-Salonen, M., R. Valo, J. Apajalahti, R. Hakulinen, L. Silakoski, and T. Jaakkola. 1984. Biodegradation of chlorophenolic compounds in wastes from wood processing industry. Pages 668–676 in M. J. Klug and C. A. Reddy (eds.) *Current Perspective in Microbial Ecology*, American Society for Microbiology, Washington, D.C.
Savage, G. M., L. F. Diaz, and C. G. Golueke. 1989. Biological treatment of organic toxic wastes. Page 156 in *The Biocycle Guide to Composting Municipal Wastes*. The JG Press, Emmaus, Pennsylvania.
Schink, B. 1985. Degradation of unsaturated hydrocarbons by methanogenic enrichment cultures. *FEMS Microbiol. Ecol.* 31:69.
Schink, B. 1988. Principles and limits of anaerobic degradation: environmental and technological aspects. Pages 771–846 in A. J. B. Zehnder (ed.), *Biology of Anaerobic Microorganisms*. John Wiley and Sons, New York.
Schink, B., and N. Pfennig. 1982. Fermentation of trihydroxybenzenes by *Pelobacter acidigallici* gen. nov. sp. nov., a new strictly anaerobic, non-sporeforming bacterium. *Arch. Microbiol.* 133:195.
Schink, B., and J. G. Zeikus. 1982. Microbial ecology of pectin decomposition in anoxic lake sediments. *J. Gen. Microbiol.* 128:393–404.
Schwarzenbach, R. P., W. Giger, E. Hoehn, and J. K. Schneider. 1983. Behavior of organic compounds during infiltration of river water to ground water. Field studies. *Environ. Sci. Technol.* 17:472–479.
Senior, E., A. T. Bull, and J. H. Slater. 1976. Enzyme evolution in a microbial community growing on the herbicide Dalapon. *Nature* (London) 263:476.
Shelton, D. R., and J. M. Tiedje. 1984. Isolation and partial characterization of bacteria in an anaerobic consortium that mineralizes 3-chlorobenzoic acid. *Appl. Environ. Microbiol.* 48:840–848.
Shelton, D. R., and J. M. Tiedje. 1984. General method for determining anaerobic biodegradation potential. *Appl. Environ. Microbiol.* 47:850–857.
Shennan, J. L., and J. D. Levi. 1974. The growth of yeast on hydrocarbons. *Prog. Ind. Microbiol.* 13:1.
Shieh, W. K., and J. D. Keenan. 1986. Fluidized bed biofilm reactors for wastewater treatment. Page 131, in A. Fiechter (ed.) *Advances in Biochemical Engineering/ Biotechnology*, vol. 33. Springer-Verlag, New York.

Shlomi, E. R., A. Lankhorst, and R. A. Prins. 1978. Methanogenic fermentation of benzoate in an enrichment culture. *Microbial Ecol.* 4:249.

Shukrow, A. J., A. P. Pajak, and C. J. Touhill. 1980. Management of hazardous waste leachate, SW 871. U.S. EPA, Washington, D.C.

Sielicki, M., D. D. Focht, and J. P. Martin. 1978. Microbial transformation of styrene and (^{14}C)-styrene in soil and enrichment cultures. *Appl. Environ. Microbiol.* 35:124.

Singer, M. E., and W. R. Finnerty. 1984. Microbial metabolism of straight chain and branched alkanes. Page 1 in R. M. Atlas (ed.) *Petroleum Microbiology*. Macmillan, New York.

Slater, J. H. 1978. The role of microbial communities in the natural environments. Pages 137–154 in K. W. A. Shatter and M. J. Somerville (eds.), *The Oil Industrial and Microbial Ecosystems*. Heydon and Sons, London.

Slater, J. H., and D. Lovatt. 1984. Biodegradation and significance of microbial communities. Pages 440–485 in D. T. Gibson (ed.), *Microbial Degradation of Organic Compounds*. Marcel Dekker, New York.

Smith, G. A., J. S. Nickels, J. D. Davis, R. H. Findlay, P. S. Yashio, J. T. Wilson, and D. C. White. 1985. Indices identifying subsurface microbial communities that adapted to organic pollution. Page 210 in N. N. Durham and A. E. Redelfs (eds.) *Second Int. Conf. on Groundwater Quality Res.*, Oklahoma State Univ. Print Services, Stillwater, Oklahoma.

Smith, G. A., J. S. Nickels, B. D. Kerger, J. D. Davis, S. P. Collins, J. T. Wilson, J. F. McNabb, and D. C. White. 1986. Quantitative characterization of microbial biomass and community structure in subsurface material: A prokaryotic consortium responsive to organic contamination. *Can. J. Microbiol.* 32:104.

Steiert, J. G., and R. L. Crawford. 1986. Catabolism of pentachlorophenol by a *Flavobacterium* sp. *Biochem. Biophys. Res. Commun.* 141:825.

Stirling, L. A., R. J. Watkinson, and I. J. Higgins. 1976. The microbial utilization of cyclohexane. *Proc. Soc. Gen. Microbiol.* 4:28.

Strand, S. E., and L. Shippert. 1986. Oxidation of chloroform in an aerobic soil exposed to natural gas. *Appl. Environ. Microbiol.* 52:203.

Stucki, G., U. Krebser, and T. Leisinger. 1983. Bacterial growth on 1,2-dichloroethane, *Experimentia.* 39:1271.

Suflita, J. M., A. Horowitz, D. R. Shelton, and J. M. Tiedje. 1982. Dehalogenation: A novel pathway for the anaerobic biodegradation of haloaromatic compounds. *Science* 218:1115–1117.

Suflita, J. M., and G. D. Miller. 1985. Microbial metabolism of chlorophenolic compounds in groundwater aquifers. *Environ. Toxicol. Chem.* 4:751.

Suflita, J. M., J. A. Robinson, and J. M. Tiedje. 1983. Kinetics of microbial dehalogenation of haloaromatic substrates in methanogenic environments. *Appl. Environ. Microbiol.* 45:1466–1473.

Sutton, P. M. 1985. Innovative engineered systems for biological treatment of contaminated groundwater. Page 7 in *Contaminated Soil Treatment Practice*, Silver Springs, Maryland.

Swisher, R. D. 1970. *Surfactant Biodegradation*. Marcel Dekker, New York.

Tarvin, D., and A. M. Buswell. 1934. The methane fermentation of organic acids and carbohydrates. *J. Am. Chem. Soc.* 56:1751.

Thauer, R. K., K. Jungermann, and K. Decker. 1977. Energy conservation in chemotrophic anaerobic bacteria. *Bacteriol. Rev.* 41:100.

Thiele, J. H., and J. G. Zeikus. 1988a. Control of interspecies electron flow during anaerobic digestion: Significance of formate transfer versus hydrogen transfer during syntrophic methanogenesis in flocs. *Appl. Environ. Microbiol.* 54:20–29.

Thiele, J. H., and J. G. Zeikus. 1988b. Interactions between hydrogen/formate producing bacteria and methanogens during anaerobic digestion. Pages 537–595 in L. E. Erickson and D. Y. C. Fung (eds.). *Handbook on Anaerobic Fermentation*, Marcel Dekker, New York.

Thiele, J., M. Chartrain, and J. G. Zeikus. 1988. Control of interspecies electron flow during anaerobic digestion: The role of floc formation in syntrophic methanogenesis. *Appl. Environ. Microbiol.* 54:10–19.

Thomas, J. M., and C. H. Ward. 1989. In situ biorestoration of organic contaminants in the subsurface. *Environ. Sci. Technol.* 23:760.

Tiedje, J. M., and T. O. Stevens. 1988. The ecology of an anaerobic dechlorinating consortium. Pages 3–14 in G. S. Owen (ed.) *Environmental Biotechnology*. Plenum Press, New York.

Tiedje, J. M., S. A. Boyd, and B. Z. Fathepure. 1987. Anaerobic degradation of chlorinated aromatic hydrocarbons. *Dev. Ind. Microbiol.* 27:117.

Vogel, T. M., C. S. Criddle, and P. L. McCarty. 1987. Transformation of halogenated aliphatic compounds, *Environ. Sci. Technol.* 21:722.

Vogel, T. M., B. Z. Fathepure, and H. Selig. 1989. Sequential anaerobic/aerobic degradation of chlorinated organic compounds. Page 1, A10 in *Abstr. Int. Symp. on Process Governing the Movement and Fate of Contaminants in the Subsurface Environment*. Assoc. Water Pollut. Res. Control, Stanford Univ., Stanford, California.

Vogel, T. M., and D. Grbic-Galic. 1986. Incorporation of oxygen from water into toluene and benzene during anaerobic fermentative transformation. *Appl. Environ. Microbiol.* 52:200–202.

Vogel, T. M., and P. L. McCarty. 1985. Biotransformation of tetrachloroethylene to trichloroethylene, dichloroethylene, vinyl chloride and carbon dioxide under methanogenic conditions. *Appl. Environ. Microbiol.* 49:1080–1083.

Wackett, L. P., and D. T. Gibson. 1988. Degradation of trichloroethylene by toluene dioxygenase in whole cell studies with *Pseudomonas putida* F1. *Appl. Environ. Microbiol.* 54:1703.

Wackett, L. P., G. A. Brusseau, S. R. Householder, and R. S. Hanson. 1989. Survey of Microbiol Oxygenases: Trichloroethylene degradation by propane oxidizing bacteria. *Appl. Environ. Microbiol.* 55:2960.

Ward, D. M., R. M. Atlas, P. D. Boehm, and J. A. Calder. 1980. Microbial biodegradation and chemical evolution of oil from the Amoco spill. AMBIO, *J. Human Environ. Res. Manag.*, Royal Swedish/Acad. Sci. 9:277–283.

Webster, J. J., G. J. Hampton, J. T. Wilson, W. C. Ghiorse, and F. R. Leach. 1985. Determination of microbial cell numbers in subsurface samples. *Groundwater* 23:17.

Wheelis, M. L. 1975. The genetics of dissimilarity pathways in *Pseudomonas. Ann. Rev. Microbiol.* 29:505.

Whittenbury, R. 1978. Bacterial nutrition. Pages 1–32 in J. R. Norris and M. H. Richmond (eds.) *Essays in Microbiology*. John Wiley and Sons, London.

Widdel, F. 1988. Microbiology and ecology of sulfate and sulfur reducing bacteria. Pages 469–586 in A. J. B. Zehnder (ed.) *Biology of Anaerobic Microorganisms*. John Wiley and Sons, New York.

Wilkinson, T. G., H. H. Topiwala, and G. Hamer. 1974. Interaction in a mixed bacterial population growing on methane in continuous culture. *Biotech. Bioeng.* 16:41.

Wilson, B. H., B. Bledsoe, and D. Kampbell. 1987. Biological processes occurring at an aviation gasoline spill site. Pages 125–137 in R. C. Averett and D. M. McKnight (eds.) *Chemical Quality of Water and the Hydrologic Cycle*. Lewis Publishers, Chelsea, Michigan.

Wilson, B. H., G. B. Smith, and J. F. Rees. 1986. Biotransformations of selected alkylbenzenes and halogenated aliphatic hydrocarbons in methanogenic aquifer material: a microcosm study. *Environ. Sci. Technol.* 20:997–1002.

Wilson, J. T., J. F. McNabb, J. W. Cochran, T. H. Wang, M. B. Tomson, and P. B. Bedient. 1985. Influence of microbial adaptation on the fate of organic pollutants in groundwater. *Environ. Toxicol. Chem.* 4:721.

Wilson, J. T., and B. H. Wilson. 1985. Biotransformation of trichloroethene in soil. *Appl. Environ. Microbiol.* 49:242.

Wolin, M. J. 1982. Hydrogen transfer in microbial communities. Pages 323–356 in A. T. Bull and J. H. Slater (eds.), *Microbial Interactions and Communities*, vol. I. Academic Press, New York.

Wu, W., R. Hickey, L. Bhatnagar, M. K. Jain, and J. G. Zeikus. 1989. Fatty acids degradation as a tool to monitor anaerobic sludge activity and toxicity. In *Proc. 44th Ann. Purdue Industrial Waste Conference*, West Lafayette, Indiana.

Young, L. Y. 1984. Anaerobic degradation of aromatic compounds. Pages 487–519 in D. T. Gibson (ed.) *Microbial Degradation of Organic Compounds*. Marcel Dekker, New York.

Zaitsev, G. M., and Y. N. Karasevich. 1980a. Utilization of 4-chlorobenzoic acid by *Arthrobacter globiformis*. *Mikrobiologiya* 50:35.

Zaitsev, G. M., and Y. N. Karasevich. 1980b. Preparative metabolism of 4-chlorobenzoic acid in *Arthrobacter globiformis*. *Mikrobiologiya* 50:423.

Zeikus, J. G. 1983a. Metabolism of one-carbon compounds by chemotrophic anaerobes. Pages 218–229 in *Advances in Microbial Physiology*, vol. 24, Academic Press, New York.

Zeikus, J. G. 1983b. Metabolic communication between biodegradative populations in nature. Pages 423–462 in Slater, Whittenbury, and Wimpenny (eds.), *Microbes in Their Natural Environments*, vol. 34. Cambridge University Press, Cambridge, England.

Zeikus, J. G., A. L. Wellstein, and T. K. Kirk. 1982. Molecular basis for the biodegradative recalcitrance of lignin in anaerobic environments. *FEMS* 15:193–197.

Zeyer, J., E. P. Kuhn, and R. P. Schwarzenbach. 1986. Rapid microbial mineralization of toluene and 1,3-dimethyl benzene in the absence of molecular oxygen. *Appl. Environ. Microbiol.* 52:944.

Zobell, C. E. 1946. Action of microorganisms on hydrocarbons. *Bacteriol. Rev.* 10:1, 1946.

Zobell, C. E., and J. F. Prokop. 1966. Microbial oxidation of mineral oils in Barataria Bay bottom deposits. *Z. Allg. Mikrobiol.* 6:143.

Chapter

12

The Role of Consortia in Microbially Influenced Corrosion

Nicholas J. E. Dowling, Marc W. Mittelman, and David C. White

Institute for Applied Microbiology
University of Tennessee
Knoxville, Tennessee

Introduction

Microbially influenced corrosion (MIC) is not a recently recognized problem (Houghton et al. 1988, Tiller 1986, Miller 1981). Despite adequate descriptions of microbial corrosion early in this century, little progress has been made in the elucidation of mechanisms and reasonable solutions to the problem. One notable exception was an article by Von Wolzogen Kuhr and Van der Vlugt, who in 1934 proposed a hypothesis for "cathodic depolarization" where certain sulfate-reducing bacteria increased the corrosion rate by accelerating the reductive, cathodic reactions (the consumption of hydrogen generated). Cathodically produced hydrogen is oxidized by these organisms and establishes a link between their well-being and the corrosion reaction itself. Other processes cannot be disregarded. Sulfate-reducing bacteria require the presence and activity of other organisms to exist in aerobic environments. Some of the characterized 10 or 11 genera of sulfate-reducing bacteria (Widdel 1988) do not exist exclusively as autotrophs but require organic substrates which are available from other organisms within biofilm communities. Large, complex molecules such as proteins and polysaccharides are commonly available and do not accumulate in biofilms for the same reasons that apply in sedimentary systems (Laanbroek and Veldkamp 1982). Indeed it is reasonable to assume that sediments have similar substrate dynamics as biofilms.

Partial degradation of substrates by primary degrading organisms channel monomeric products to fermentative bacteria, which in turn provide short-chain acids and alcohols to terminal oxidizing bacteria. The latter, including the sulfate-reducing bacteria, tend to be obligately anaerobic and are situated close to the solid substratum. This model, which has been described at length by Hamilton (1985), appears to fit well with the significant deterioration of metal structures observed in the environment. Clearly, the different physiological groups within the biofilm operate in concert. The obligately anaerobic bacteria, for example, are never found in biofilms without other organisms which scavenge oxygen. Under some conditions, new metal surfaces (produced by mechanical scouring for example) have a large oxygen requirement to maintain the surface oxide layer. This, however, is a short-term event and seems unlikely to be sustained long enough to provide the anaerobic environment required by obligate anaerobes. Thus other bacteria are required to not only provide the substrates required by sulfate-reducing bacteria, but also to scavenge the oxygen which prevents growth of these organisms. In this way, we see that interaction is imperative for some classes of microorganisms. The "closer" relationship of syntrophic organisms with their hydrogenophilic counterparts (Wolin 1982) in biofilms has not been examined.

From a practical standpoint, there appears to be no simple solution to the problem of MIC beyond materials selection, mechanical cleaning, and low-level biocide use. Few materials are truly impervious to attack in biological systems (Kuhn and Rae 1988) and all structural materials are considered to have a finite life in service. Studies are often instituted only when a catastrophic shortening of this service life is demonstrated and are specific in nature and limited in scope. Even in the event of a putative link between MIC and the observed failure, subsequent studies are likely to focus on short-term solutions. Perhaps the single biggest flaw of research into this topic has been the study of the corrosion properties of individual organisms in axenic cultures. While the relationship of a single microorganism in a biofilm to the substratum is important, there can be no reasonable extrapolation of material performance under environmental stress until the substratum is challenged with the in situ fouling community.

Engineering materials of most kinds are usually composed of complex arrangements, whether ordered by crystalline structure such as metals or aggregates such as concrete. Many of these materials have flaws which are exploitable either by virtue of their surface structure or composition. This chapter sets out to explore the deleterious effect of different microorganisms in consortium on these diverse materials from problematic and mechanistic points of view.

Economic Importance of MIC

The economic effects of MIC are as diverse and far-reaching as are the consortia responsible. Perhaps the best studied of these activities has been the corrosion of underground iron gas and water pipelines by sulfate-reducing bacteria. Corrosion of these mild steel and cast-iron pipe systems is classically viewed as dependent upon the availability of oxygen. At the end of the nineteenth century, however, it was noted that serious forms of pitting-type corrosion were associated with extremely anoxic soils. Apparently another type of corrosion process was involved. Von Wolzogen Kuhr and van der Vlugt (1934) coined the term "graphitization" to describe the action of sulfate-reducing bacteria (SRB) (obligately anaerobic organisms) on cast iron. Furthermore, they proposed that a hydrogenase enzyme system previously described by Stephenson and Strickland (1931) was responsible for the "cathodic depolarization" of the metal, and this resulted in accelerated corrosion.

It is clear that SRB are found in corrosive soil environments in association with a broad range of physiologically distinct microorganisms. Gaylarde and Johnston (1986) described accelerated corrosion of mild steel when cultures of *Desulfovibrio vulgaris* were grown with *Vibrio anguillarum*. Their results demonstrated the importance of conducting coculture experiments and explain, in part, the relatively low corrosion rates obtained with laboratory SRB monocultures compared with those observed in the environment. Since many estimates of environmental MIC rates are based upon monoculture experiments, it seems likely that most projections of the economic impact of anaerobic corrosion on buried pipelines are too conservative.

A number of workers have attempted to place a dollar figure on the economic effects of underground pipeline corrosion. In one of the first studies of MIC effects on pipeline corrosion, Greathouse and Wessel (1954) described U.S. economic losses in excess of $100 million per year. Allred et al. (1959) found that SRB were responsible for more than 77 percent of the corrosion of oil-well underground pipelines. Postgate (1984) has noted that despite the "obvious economic importance of anaerobic corrosion, useful estimates of its economic cost are difficult to make."

Biological fouling of heat exchanger surfaces, cooling tower structures, and attendant distribution systems accounts for over $4.5 billion in economic losses in the noncommunist world. Losses to U.S. refinery heat exchangers amount to greater than $1.4 billion per year

(Knudsen 1981). Purkiss (1971) has quoted losses in British industrial systems exceeding 300 million pounds per year in 1971 pounds. It is often difficult to separate the results of fouling phenomena such as mechanical blockages, heat transfer resistance, and corrosion. Aerobic, slime-forming consortia create differential oxygen cells as well as heat transfer resistance and decreased pipeline capacity. Open cooling towers are particularly susceptible to this type of multifaceted fouling problem. Eukaryotes such as algae and diatoms can produce significant slime deposits in addition to acidic metabolic by-products. Within an aerobic layer composed of such slime-forming isolates as *Bacillus subtilis, B. cereus, Flavobacterium* spp., *Aerobacter* spp., and *Pseudomonas* spp., conditions become ideal for the growth of SRB. Iron-oxidizing bacteria such as *Sphaerotilus* and *Creuothrix* spp. produce ferric hydroxide, a potent initiator of pitting corrosion in addition to forming differential oxygen cells (Purkiss 1971).

The fouling and corrosion of jet fuel storage tanks by microbial consortia—principally fungal contaminants—creates tremendous problems for military and civilian aircraft operations. Severe structural damage resulting from biogenic organic acid attack was commonplace in the 1950s and 1960s, prior to the introduction of biocides and improved measures to prevent water intrusion into the fuel systems. Cabral (1980) analyzed samples of kerosene A1-type jet fuel from aircraft fuel systems in Argentina; 65 percent of those tested were found to be contaminated with *Cladosporium resinae*, a ubiquitous, filamentous fungi. While a number of other fungi were isolated, most notably *Penicillium* spp., only *C. resinae* was capable of utilizing the fuel as a sole carbon source. Various consortia were tested on a kerosene medium; however, only those mixtures containing *C. resinae* were able to utilize the kerosene as a sole carbon source. Schiapparelli and Meybaum (1980) have identified dodecanoic acid (a kerosene breakdown product) as a major organic acid involved in the corrosion of aluminum fuel tanks. Other corrosion mechanisms associated with aluminum fuel tanks include the differential oxygen cells created by bacteria as a result of oxygen uptake in localized areas (Hill 1984).

Estimates of costs associated with abiological and biological corrosion in the pulp and paper industry have ranged from $5 to $9 per ton of product or up to 20 percent of maintenance costs including capital expenditure (Davies 1984, Safade 1988). Approximately 50 percent of these costs are associated with the Kraft process alone. Tatnall (1981) described severe microbial corrosion in a paper mill closed-water ("white water") system. He found that corroded areas were covered by aerobic slime consisting of *Pseudomonas, Aerobacter, Flavobacterium*, and *Bacillus* spp. A brown, gritty deposit containing *Desulfovibrio, Desulfotomaculum*, and *Clostridium* spp. was found in association with the aerobic slime layer. Corrosion of paper manufacturing equipment used in the Kraft process by thermophilic *Desulfovibrio* and *Desulfotomaculum* spp. has been described by Grimm et al. (1984).

The growth and activity of *Thiobacillus* spp. in concrete pipes is an interesting example of consortial involvement in corrosion accelerated by physically separated members. MIC of concrete has resulted in catastrophic failures of the upper (aerated) regions of concrete sewage pipes (Sand and Bock 1984). Early investigators attributed the rapid corrosion of sewage transport pipelines to the production of hydrogen sulfide in the sewage followed by a "catalytic oxidation" to sulfuric acid vapor. It was thought that this process was abiological in nature, resulting from a chemical reaction catalyzed at the concrete surface (Thistlethwayte 1972). Milde et al. (1983) and Parker (1947) have since described a number of species of thiobacilli which are involved in concrete corrosion.

Dissimilatory SRB present in the sewerage and sludge at the bottom of the concrete pipes produce volatile reduced sulfur compounds which dissolve in the thin water film on the upper regions of the pipe. Here the water film is relatively aerated and thiobacilli can use the sulfide and mercaptans as electron sources. The resulting sulfate and lowered pH leaches out the calcium binder and serious structural loss results. At the initiation of the corrosion process (at neutral pH), facultatively chemolithotrophic thiobacilli such as *T. novellus* and *T. intermedius* predominated. As the pH at the concrete surface decreased below 6, *T. neopolitanus* numbers increased. At pH values below 5, *T. thiooxidans* predominated, further lowering the pH. It appears that elemental sulfur, a product of abiological sulfide oxidation, can also be used as an energy source for the thiobacilli associated with the Hamburg, FRG concrete pipes (Milde et al. 1983).

Treatment by forced aeration of the sewerage decreased SRB activity and lowered volatile sulfide production. A decrease in the number of thiobacilli and the associated rate of concrete corrosion then resulted (Bielecki and Schremmer 1987). With the advent of the National Pollution Discharge Elimination Standards (NPDES) in 1972 controlling discharge of pollutants into waterways, the level of heavy metals in sewage has declined significantly. This change, while having salutary benefits for the environment, has resulted in an exacerbation of the MIC problem in concrete sewage systems. By removing metals involved in suppressing bacteria, concrete corrosion increased several orders of magnitude. Interestingly, the heavy metals appeared to have been effective against the SRB and not against the thiobacilli (personal communication A. Walker, L.A. Sanitation Dept.). Test systems and detection methods for thiobacilli have been used successfully (Kerger et al. 1987).

Asphalt, which is the residue from petroleum crude oil distillates, is used as cement in the construction of roads. Pendrys (1989) has described its degradation by a consortia of seven gram-negative aerobic bacteria. The organisms belonged to the *Pseudomonas, Acinetobacter, Alcaligenes, Flavimonas*, and *Flavobacterium* genera; the consortia was capable of degrading the saturate and naphthene aromatic fractions of an asphalt cement-20. The primary mechanism for this corrosion process appears to be the production of an emulsifier which is produced by *A. calcoaceticus*, one of the consortial members. However, consortial members which cannot produce this emulsifier can corrode the asphalt and remain viable by utilizing the water-soluble asphalt fraction as a carbon source. Several million tons of asphalt are used annually in the United States; however, there are no figures available for the contribution of MIC to asphalt degradation (Ramamurti et al. 1984).

The corrosion of containers for high-level radioactive wastes has implications for all of humanity. The materials of construction and design of these containers in geological disposal sites must be sufficient to survive for at least 1000 years (Day et al. 1985). Marsh and Taylor (1988) have considered the theoretical contribution of microbial consortia to the corrosion of these containers. Based upon environmental conditions within the repository, the authors surmised that only SRB (of the categories examined) would be capable of growth in the progressively anaerobic environment. Employing calculations which assume lactate and/or acetate as sole carbon sources, a "worst-case" loss of about 13 mm mild steel during the 1000-year lifespan was predicted in addition to the 216 mm lost as a result of "abiological" localized corrosion. Unfortunately, the influence of biofilms as mediators of corrosion was ignored in these calculations, as was the potential autotrophic growth of SRB on H_2 and CO_2. In order to form long-term projections on the behavior of such materials, some estimation must be made for the water activity of surrounding rock formations. The activity of many microorganisms is severely curtailed in dessicated environments. Most problematic is projection of local weather trends 1000 years in the future and the effect of the spent fuel thermal output for the first 500 years, which will tend to draw water to the repository. Obvious measures such as siting these repositories away from gypsum formations (sulfate sources) must also be considered.

General Microbiological Considerations

MIC testing

Currently, there are no standards for microbiological testing of metallic, ceramic, vitreous, or other engineering materials. The American

Society for Testing and Materials has yet to define specific microorganisms that may be used in a laboratory test to characterize specific materials similar, for example, to the regimes used for textiles and wood resistance (Onions, 1975). This is probably due to the fact that microbiological attack of such materials by definition is slow and nonuniform.

Another interesting requirement is that a "standard test system" must be uniformly applicable. For MIC this will present a problem since the corrosion products of some alloys are toxic. A good example of this are the brasses and copper-nickel alloys where the adhesion of organisms is controlled by the alloying distribution (Chamberlain and Garner 1988). Any MIC test system will have to address these not insignificant points, and as yet no such system has been proposed.

Isolates of specific bacteria that are obtained from a corrosion site and reintroduced to the same alloy in the laboratory often demonstrate little attack. From this we conclude that either microorganisms were isolated that were not responsible for the field corrosion situation or that the isolates are indeed responsible but did not perform the same way in the laboratory. This problem is directly influenced by Winogradsky's (1949) observations on the enrichment of "zymogenous" organisms in batch culture. These observations demonstrate that isolates obtained by such batch cultural methods have a high probability of not being the dominant contributors to a turnover (or effect) in biofilms or sediments. Furthermore if "environmentally relevant" organisms are in fact isolated and used as a test system in batch culture, their performance will undoubtedly be very different in such an arrangement from that experienced in the environment. Dowling et al. (1988) have showed that batch culture is very different from continuous-flow situations: the worst corrosion rate associated with carbon steel C1020 was demonstrated to be one in which continuous flow was interrupted after a few days with a period of batch (stagnant) culture. This also appears to be the industrial experience: systems left in batch (stagnant) culture that are untreated tend to be very susceptible to MIC (Pollock 1989).

Modification of the in situ environment

Considerable time and effort have gone into investigations of the formation of the initial biofilm. Baier (1981) has shown that "conditioning films" are laid down which consist of polymeric macromolecules which chemically adhere very quickly to the "new" surface presented. This is followed by recruitment to the surface of bacteria, perhaps predominantly *Pseudomonas* spp. or other similar organisms which produce copious quantities of extracellular glycocalyx (Costerton 1985).

The actual attachment of a single bacterium to an uncolonized site appears to follow a specific sequence of events: first, a reversible phase of electrostatic attachment followed by an irreversible phase mediated by extracellular polymer production (Marshall 1976, McCoy 1987, Haydon 1961). At some undetermined time these oxygen-scavenging organisms are joined by facultative and obligately anaerobic bacteria among which are the SRB.

The problem is that research has only recently produced detailed information on the nature of adhesion and corrosion and that relationship is far from clear. An example of the disarray in this topic can be illustrated by reading the proceedings of two symposia on the subject edited by Dexter (1986) and Licina (1988). For example, it is entirely possible to have a pipe that is heavily fouled, but after the mixture of detritus and biofilm is removed, no corrosion is observed. In practice, many different environments present themselves which vary in temperature, salinity, pH, and so on, and therefore generalizations about MIC fit badly when conceived as a single set of circumstances. For example, the thickness of the electrical double layer is very dependent upon the salinity. In microbiological terms, temperature and the availability of carbon and energy are probably the most important concerns in MIC. Seasonable variability in MIC attack may follow not just temperature but also the effects on heterotrophic organisms of algal blooms in source water which supplies piping systems. If such a correlation exists, then a lake with high agricultural runoff of nitrates and phosphates and the associated algal bloom may well contribute to the deterioration of metal superstructures in contact with that water. An interesting note is that chemical corrosion of ferrous alloys is generally impeded in the presence of nitrate and phosphate.

The nature of a "mature biofilm" can vary considerably. Biofilms have been described as stratified layers (Hamilton 1985) subject to planar diffusion of oxidized and reduced molecular species to and from the metal substratum (Fig. 12.1). For the purposes of this discussion, planar diffusion describes those mass transport processes which operate perpendicular to the surface. In the lower, reduced-layer SRB occupy an anaerobic niche where low-molecular-weight acids may be oxidized at the expense of sulfate. The product, sulfide, may diffuse away or precipitate with the metal close to the substratum. In the upper layers, oxygen is scavenged by aerobic organisms which force fermentative processes in lower anoxic strata. The fermentative products of this region are fed to the SRB in the reduced zone. This description is quite reasonable for waters with high organic loading such as in the pulp and paper industry (Safade 1988, Morgan and May 1970, Cloete and Gray 1985).

Alternatively, in environments where the organic loading is not so

Figure 12.1 A stratified biofilm in high-organic-loading wastes.

high, such as an alpine-type freshwater lake, the biofilms may be discontinuous (Fig. 12.2) and form pockets of reduced-anoxic areas in oxic plains (Lappin-Scott and Costerton 1989). As with the stratified model, the discontinuous biofilm model requires the interaction of aerobic, facultatively anaerobic, and obligately anaerobic bacteria. In this latter model, however, there are distinct areas where corrosion may be facilitated by differential oxygen cells (NACE 1970). This occurs when adjacent areas of a metal surface are exposed to different oxygen concentrations and redox potentials. The areas under reduced oxygen tension will tend to recruit the anodic (dissolutive) processes, while the areas of high oxygen tension will recruit the cathodic (re-

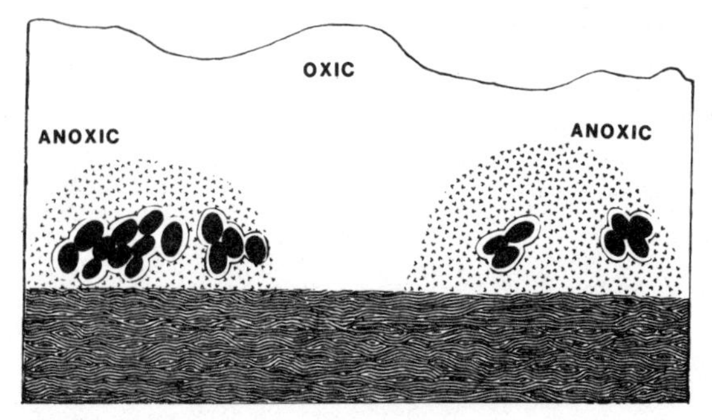

Figure 12.2 A "nonuniform" biofilm in low organic-loading waters.

ductive) processes. Thus corrosion will tend to occur at sites of maximum bacterial consumption of oxygen (the anoxic zones) and be of a distinctly "localized" nature.

Corrosion activity as a measure of consortial production

There are few studies attempting to correlate bacterial activity with corrosion rate. Partly this is because in the complex slime layers that occur on metal substrata there is no pure "biofilm" per se, but a mixture of bacterial mass, extracellular polymer, and corrosion-deposition products. In this situation with a very complicated series of corrosion reactions, the contribution of a single species is difficult to ascertain. The Aberdeen group have used $^{35}SO_4$ reduction to gain some idea of the rates of sulfate reduction present in biofilms (Maxwell and Hamilton 1986, Hamilton et al. 1988). The method involves placing a metal coupon complete with an intact biofilm into a wide-bore Hungate roll tube. The tube is purged with nitrogen and a reduced mineral salts medium is added. Radiolabeled $^{35}SO_4$ is introduced for a contact time of approximately 5 h (Fig. 12.3). Subsequently, the mineral medium is acidified and the volatilized sulfide captured on zinc acetate–soaked filter paper. The filter paper may then be retrieved for scintillation counting in a suitable phosphor cocktail. This assay was developed from a similar one suitable for the study of sediment SO_4-reduction rates (Rosser and Hamilton 1983).

While the correlation of corrosion rate (by weight loss) with sulfate-reduction activity is hampered by the acidification of the mineral salts, useful information can still be obtained. This is particularly evident if low-grade alloys are abandoned in favor of stainless steels

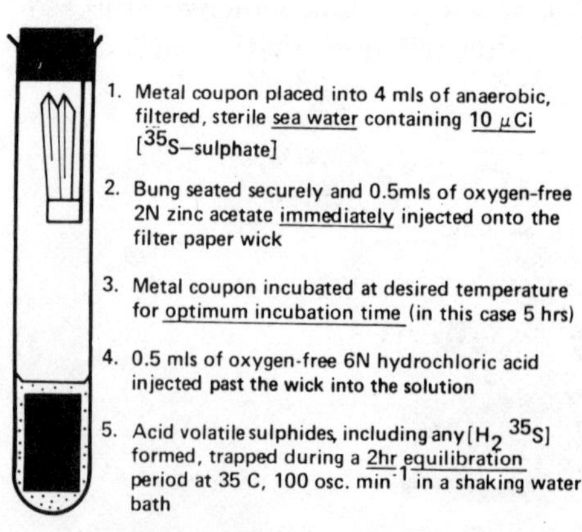

Figure 12.3 Radiolabel method to survey sulfate-reduction rates on metal coupons exposed to seawater. [*By kind permission of the authors: S. Maxwell and W. A. Hamilton (1986).*]

where the corrosion products (pyrite formation) do not interfere with volatilization of the sulfide. In such circumstances, however, weight loss must give way to electrochemical monitoring of the corrosion rate since weight loss is relatively insensitive.

Radiolabeling studies to determine aerobic incorporation rates into biofilms have also been conducted (Geesey et al. 1986) on inert substrates. In this case glass cover slips were placed into cultures where the bacteria were encouraged to adhere. The slips were recovered, washed to remove loosely attached cells, then exposed to ^{3}H-glucose for varying periods of time. The results showed that adherent cells were significantly more "active" than planktonic cells. This kind of arrangement is certainly amenable to work in the corrosion field.

The significance of the data as obtained above needs some qualification. A large part of the biomass associated with such films are not obligately anaerobic. Neither can they be considered aerobic. Both measures of estimating heterotrophic "production" have serious flaws. In reality the organisms interact together and not as subsets to catabolize incoming substrate, and in so doing create mass transfer discontinuities (Rittmann 1982). These discontinuities must also take into account "pockets" of organisms which are subject to radial diffusion (i.e., not just perpendicular to the surface). In this way it is easy to see that MIC can be a dynamic process where the corrosion rate can oscillate as a function of the type of biofilm, which is itself affected by the versatility of the member species and the incoming substrate.

Interactions Among Consortial Members

It has been a widely held contention amongst corrosion specialists that corrosion appears to be worse when a wide variety of microorganisms is present. Certainly, isolations from fresh corrosion tubercles have yielded a wide variety of isolates that fall into a diverse number of physiological types (Franklin et al. 1989). Recent experiments have documented the enhancement of corrosion processes by consortia of bacteria in controlled experimental systems. Experiments utilizing a sterilizable flow-through test system after that reported by Franklin et al. (1989) have shown that the corrosion of mild steel measured nondestructively by electrochemical impedance spectroscopy for a consortium concentrated from corrosion tubercles in a freshwater system with a trace of sulfate was 7.5 times faster than the sterile control (White et al. 1990). Monocultures of aerobic heterotrophic bacteria resulted in corrosion rates which were three times faster than the sterile control. A biculture of any of several facultative heterotrophic bacteria and a sulfate-reducing bacterium resulted in a corrosion rate which was 4.2 times faster than that of the control (White et al. 1990).

Electrochemical impedance spectroscopy (EIS), a method of deriving corrosion information from low-amplitude sinusoidal voltage oscillations, has proved to be quite valuable in MIC studies due to its relatively nondestructive properties (Dowling et al. 1988). EIS data showed that the mechanism of corrosion for bicultures and environmental consortia is quite different from that with monocultures and sterile chemical corrosion.

The presence of fermentative bacteria that produce acetate stimulate the growth of the sulfate-reducing bacterium *Desulfobacter* sp. (Dowling et al. 1988). The presence of acid-producing heterotrophs in aerobic seawater systems also stimulates corrosion by these SRB on coupons of stainless steel. Presumably this occurs both by creating the metabolic niche and supplying the nutritional substrate (White et al. 1990). With the judicious utilization of ^{13}C and gas chromatographic–mass spectrometric (GC/MS) analysis, it should be possible to demonstrate the transfer of carbon between the heterotrophic bacteria and the obligate anaerobe by following the incorporation into the phospholipid ester-linked fatty acid 10-methyl palmitic acid which is characteristic of this sulfate-reducing bacterium (Dowling et al. 1986). In this system, the consortium could be fed low levels of a ^{13}C-labeled substrate such as glutamate that is not utilized by the anaerobe and the coupon containing the consortia extracted for GCMS analysis for the "heavy" fatty acid biomarkers.

Pope et al. (1984) have described corrosion by local reduction of the pH by organic acid production. While it can be established that the pH in propagating pits may be very low (less than 2.0), it seems unlikely that only acetic acid (pK_a 4.75), a very important and common fermentation product, could be the cause. Coupled with other organisms which utilize this short-chain acid at the expense of oxidizing other electron acceptors (perhaps ferric ions), corrosion rates are faster than observed with just the fermentative bacteria (Gaylarde and Johnston 1982). Further support against the specific acetate-mediated corrosion hypothesis is given by chemical corrosion studies: supplementation of mineral salts medium with acetic acid achieves only a low increase in corrosion rate which does not compare to the high rates observed with microorganisms.

Another common hypothesis is the effect of the differential oxygen cell mentioned earlier. This is an established mechanism in chemical corrosion. Oxygen-depleted areas tend to recruit the anodic processes. Since a large number of microorganisms respire at the expense of oxygen, the probability is that such microbes will easily enter crevices, create local anoxic conditions, and thereby promote corrosion. Further complicating this study has been the interaction of obligate anaerobic organisms which often produce significantly corrosive end products.

Thus not only is there an oxygen differential but also a reduced redox potential at the anodic site.

A significant flaw associated with almost all field studies has been the failure to identify sources of nutrient flux. It is axiomatic that microorganisms require carbon and energy. In grossly polluted systems such as busy, enclosed harbors, the biofilms generated are very thick and diverse in community structure (Desmukh et al. 1988) and the resulting MIC is quite extensive and rapid. It seems reasonable to assume that the problem would be mitigated by a simple reduction in the pollution level. Desmukh et al. (1988) demonstrated the high levels of corrosion associated with the presence of waste effluent. SRB were specifically implicated by interaction with other organisms which provided primary breakdown of the suspended waste. Unfortunately, while such experimentation on marine corrosion has shown that the corrosiveness of certain waters varies from place to place and is influenced by pollution loading, attempts to isolate MIC as the sole cause have not been successful.

Mechanisms

Some simple mechanisms have already been discussed for various types of MIC (organic acid production, differential oxygen cells, etc.). While it is clear that they have an important bearing on MIC, they cannot be the sole cause. Other mechanisms put forth include the "cathodic depolarization" hypothesis, where hydrogen generated in the reductive processes is rapidly removed to low partial pressures by hydrogenophilic bacteria. This has been shown for methanogenic bacteria (Daniels et al. 1987) as well as SRB (Pankhania et al. 1986) and a range of others (Mara and Williams 1971). The presence of "hydrogenase" enzymes has been taken as indicating potential for cathodic depolarization. Although organisms with hydrogenase enzymes may be responsible for higher corrosion rates than those without (Booth and Tiller 1968), there have been few attempts to separate the "effect" of the enzyme from other potentially corrosive and possibly never described activities. Microorganisms are involved in a great many processes which may affect the corrosion rate besides a hydrogenase-linked cathodic depolarization hypothesis. Supporting the depolarization hypothesis has been electrochemical data which describe cathodic polarization curve movements and current direction (Ringas and Robinson 1987, Daumas et al. 1988). Only recently has the technology been available to rigorously examine this phenomenon at the molecular level.

Hydrogen is available not only from corrosion reactions but also from certain fermentative bacteria. No estimations have been conducted to quantitate these different sources of hydrogen in biofilms and assess their effect on the corrosion rate. Further complicating this problem are SRB which contain more than one hydrogenase. These enzymes can be engaged in either import or export functions (Lupton et al. 1984). Only a few studies have been performed where the interactions of SRB with fermentative organisms have been examined with respect to the corrosion rate of their metal substratum. Gaylarde and Johnston (1982) showed that a coculture of *Vibrio anguillarum* and *Desulfovibrio vulgaris* accelerated the corrosion rate of a mild (carbon) steel over that exhibited by either of those organisms in pure culture. This study was carried out using a reduced medium and lactate. No mechanistic information was provided, however. Dowling et al. (1988) subsequently showed that a nominally *aerobic* coculture arrangement with *Vibrio natriegens* and two SRB substantially changed the impedance diagram associated with the overall corrosion process as compared to pure culture of vibrio alone. This study was different from the first in that no substrate was specifically incorporated into the medium for the SRB and anaerobiosis was provided by the activity of the vibrio alone. Electrochemical impedance analysis (Fig. 12.4) indicated that an unspecified colloidal corrosion product associated with a low-

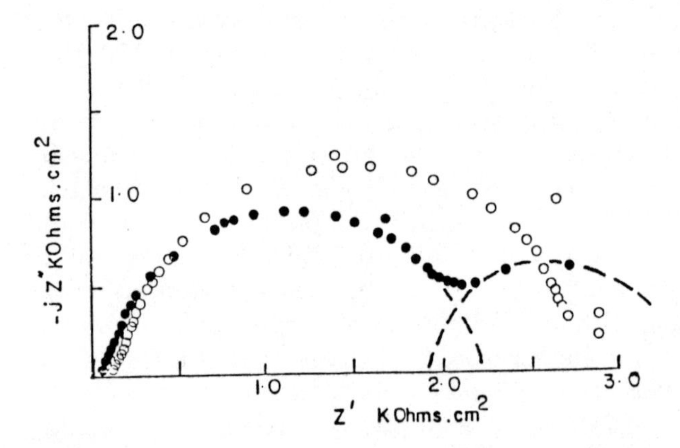

Figure 12.4 Electrochemical impedance spectra (*Z*′-real impedance vs. *Z*″-imaginary impedance) which shows the inception of a low-frequency capacitive loop (right side of the diagram) associated with the corrosion of carbon steel. This was only observed when mild steel coupons were in sterile conditions or in contact with *Vibrio natriegens* alone (closed circles). A coculture of the vibrio and two sulfate-reducing bacteria (open circles) apparently removed the low-frequency capacitive loop. Note that the occurrence of a secondary loop indicates diffusion through a particular corrosion product which was apparently removed by the sulfate-reducing bacteria. (*After Dowling et al. 1988.*)

frequency impedance capacitive loop may have been removed in the presence of the SRB. These data together indicate that not only can SRB accelerate the corrosion process in coculture with another organism but that the corrosion processes promoted by each organism are distinct.

Hydrogen sulfide, whether of chemical or biological origin, provides aggressive anions in solution (Assefpour-Dezfuly and Ferguson 1988). The generation of these anions by SRB contributes to the overall corrosion process. This effect has been separated from any other corrosive activity that these bacteria might possess in two ways: (1) Examination of the corrosion rates associated with nonhydrogenase SRB (no cathodic depolarization), and (2) corrosion rates associated with hydrogenase-containing bacteria with benzyl viologen (no sulfide production) replacing sulfate as the electron acceptor (Booth and Tiller 1968).

Another mechanism of metal corrosion associated with microorganisms is hydrogen embrittlement (Ford and Mitchell 1989). This is a process whereby microorganisms release diatomic hydrogen as a result of fermentation processes which then diffuses into the solid metal and concentrates in a particular area. This hydrogen coalesces to form a bubble which presses the metal out in a blister. While hydrogen embrittlement is not immediately a corrosion problem, there is the potential for stress corrosion cracking associated with blister formation. Catastrophic failure can occur without warning. While embrittlement is a well-known phenomenon, the contribution due to microorganisms is undetermined.

A serious but unquantified aspect of microbial corrosion has been the scavenging of chloride anions from service water systems and other raw water conduits. In order for corrosion to occur, electroneutrality must be observed at the anodic sites (Lin et al. 1981). Thus some anions must diffuse to the surface in order to account for the positive charges generated in the anodic process. Biofilms tend to recruit metal ions by the extracellular polysaccharide acting as an exchange resin (Rendleman 1978). These charges must be balanced by anions obtained from solution. The degree to which the polysaccharide retains aggressive anions such as chloride and affects the corrosion rate by this method is not understood. Jolley et al. (1988) have used Auger electron spectroscopy and X-ray photoelectron spectroscopy to examine the biocorrosion of copper by *Pseudomonas atlantica* polysaccharide. Figure 12.5 shows the Auger electron lines where bacteria and various organic materials and culture supernatant contributed to the dissolution of copper. This information shows that the extracellular polymer acts not just as a support matrix for the bacteria but also behaves as a trap for ions which promote corrosion.

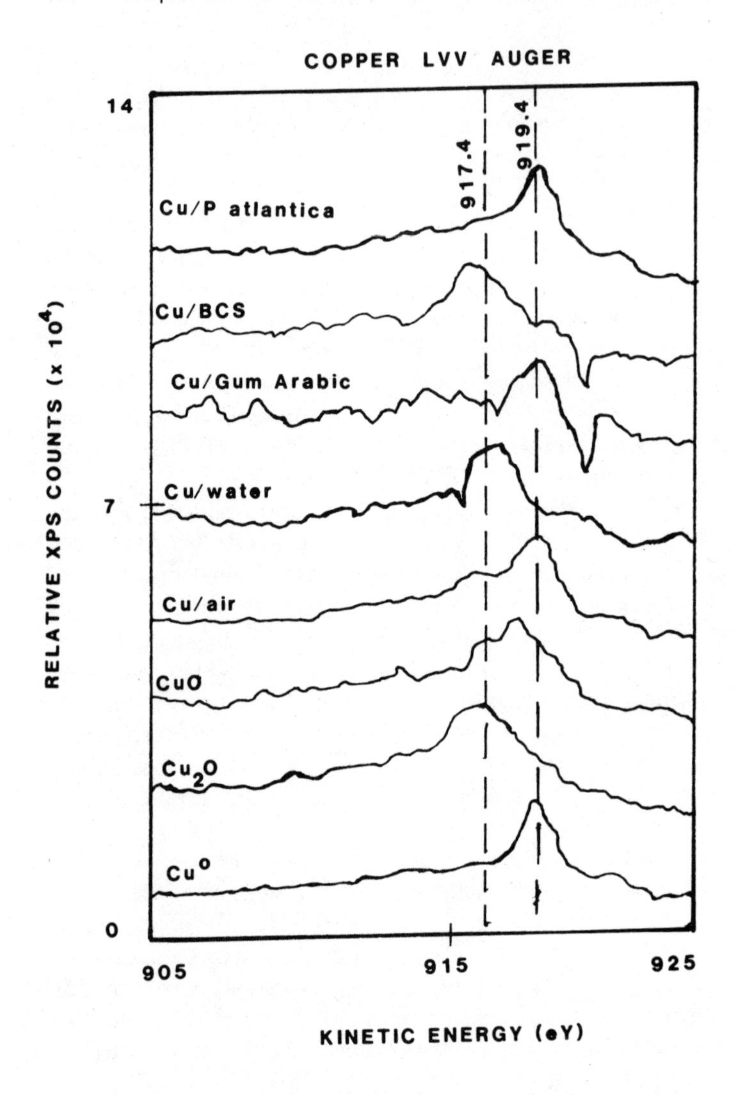

Figure 12.5 Auger electron spectral lines associated with various combinations of bacteria and biopolymers in contact with copper surfaces. (*From Jolley et al. 1988.*)

The Effects of Surface Perturbations on MIC Communities

Weld effects

Elements such as iron, copper, zinc, and aluminum all have extensive use as engineering materials. The more desirable properties, however, such as strength, elasticity, and corrosion resistance, generally improve with the incorporation of specific alloying elements. Alloys, unfortunately, are subject to types of corrosion not found in uniform, single-element materials. An example is "intergranular attack." This is a type of corrosion whereby adjacent grains of slightly differing composition form a galvanic couple. These discontinuities produce anodic and cathodic segregation and hence localized corrosion. In ordinary circumstances, stainless steels are manufactured to produce an even distribution of elements. Localized heat treatments such as welding cause a discontinuity in this arrangement. The grain size, elemental composition, and crystal structure all vary across a weld. Such areas are heavily attacked. Whether microorganisms are recruited to these sites (by magnetic or electric fields, for example) or organisms present merely exert a greater corrosive effect has not been established. While huge conglomerations of bacteria arranged in tubercles have been observed with many types of welds, their role in the actual deterioration of the weld is not understood (Borenstein 1988a, Tatnall 1981, Kobrin 1976).

Heat-affected zone (HAZ)

Welds can be divided into three areas: the base metal, which is unaffected by the weld adjacent to it; the weld metal, which contained the liquefied pool of metal behind the heat strike (a mixture of the rod and base metal); and the heat-affected zone (HAZ), an area where dislocations occur due to the proximity of the heat strike but where no rod material is incorporated.

While the weld metal itself can be attacked, the HAZ is an area that may be preferentially attacked. Figure 12.6 depicts a weld failure where the primary region of attack is the HAZ immediately adjacent to the fusion line. MIC in such cases is characterized by tiny perforations in the pipe wall which lead to very large subsurface cavities (Borenstein 1988b). Figure 12.7 shows a 304L stainless steel weld. A large pit was found immediately under the tubercle. So-called weeping welds are those where several of these subsurface cavities have joined at a single point and provided a continuous path through the wall of the pipe. While it is certain that these cavities are anaerobic, no knowledge of the microbiology of the interior of these regions has been obtained due in part to the kinetic heating produced when slicing with a saw (no cooling jets may be used which might contaminate the site). Dowling et al. (1989) have shown that MIC attack of AISI 316/E308 stainless steel weldments in seawater initiates in the HAZ. Rather interestingly, if the weld is autogenous (i.e., heat strike only, no weld metal), then corrosion initiates in the weld zone at places

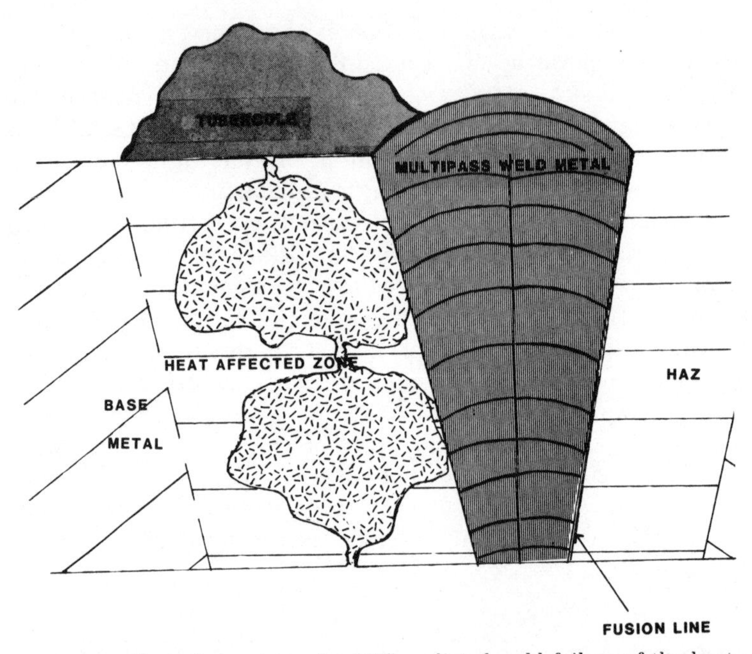

Figure 12.6 Typical structure of a MIC-mediated weld failure of the heat-affected zone (HAZ). Note that a tubercle covers a pin-hole entrance to a series of caverns within the heat affected zone. Attack can also occur in the weld metal, apparently without HAZ deterioration.

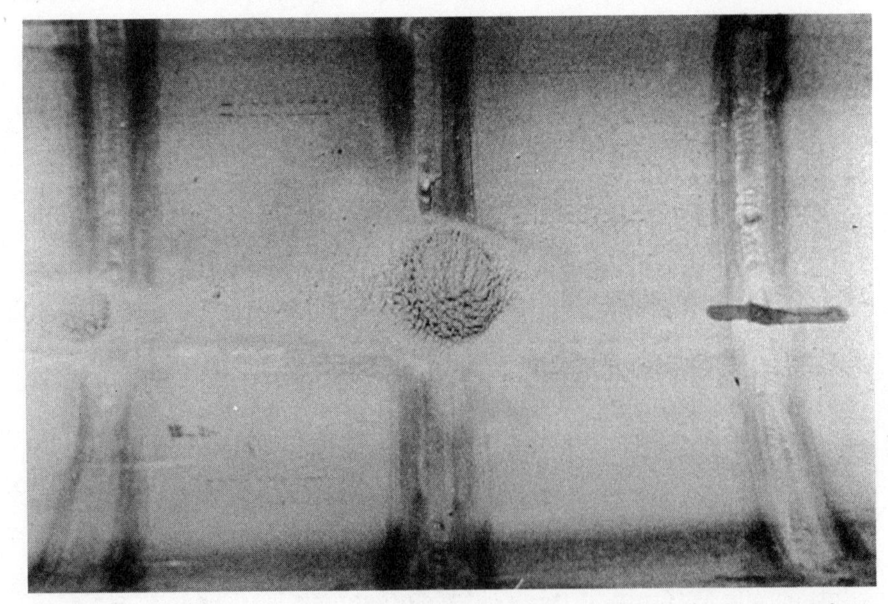

Figure 12.7 Tubercular mound covering a 304L weld in the bottom of a 304L pipe that had been exposed to untreated "freshwater." Note that these mounds may appear within a two-week exposure period. (*By kind permission of Susan W. Borenstein.*)

where slag accumulates. Slag is composed of manganese and silicon globules which are produced in the welding process.

Weld metal

The weld metal is a mixture of the melted composition from the rod or welding wire and the base metal. Depending upon the criteria required for that weld, the rod and base metal may or may not be identical. In certain stainless steel welds, the overriding importance will be for resistance to "hot cracking" and not corrosion resistance. Thus, these welds are manufactured without regard to their susceptibility to MIC (Borenstein 1988c). A duplex structure is characteristic of 300 series stainless steel welds. In this type of arrangement, two crystal structures, ferrite and austenite, are produced by differential cooling-solidification rates after the heat strike has passed, where previously the base metal was fully austenitic. MIC attack of such welds appears to selectively remove either one phase or the other, but not both. Figure 12.8 depicts the skeletal remains of the ferrite after removal of the surrounding austenite in the pit. Borenstein (1989) found that the

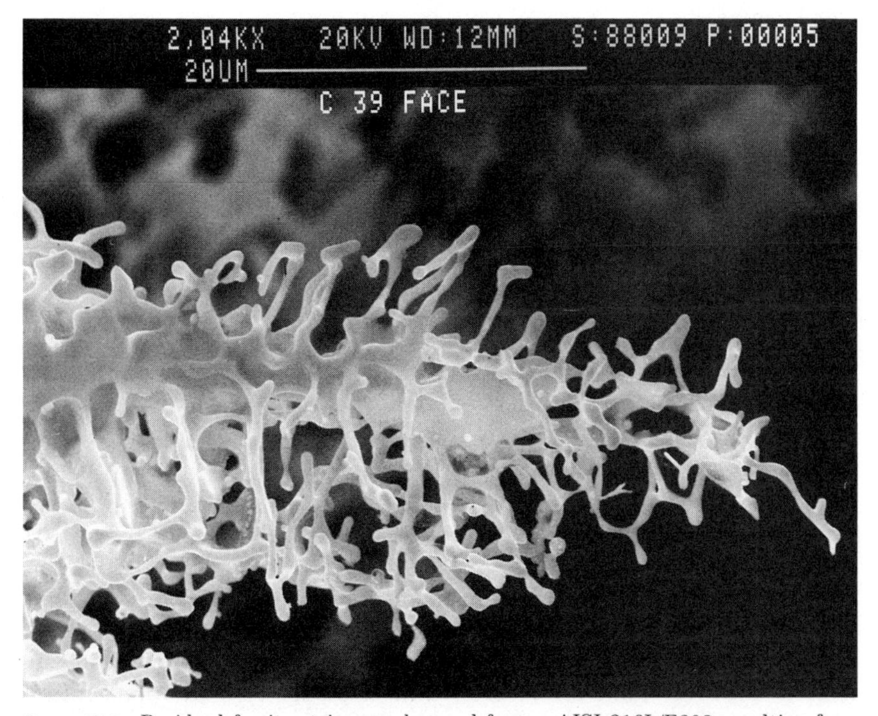

Figure 12.8 Residual ferrite stringers observed from a AISI 316L/E308 resulting from MIC. The austenite phase has been selectively removed leaving the chromium rich ferrite behind. This residual crystal structure has only been observed with MIC and ferric chloride attack. (*Photograph courtesy of Rebecca Wood, Singleton Laboratories.*)

content of ferrite in a weld apparently had no effect on MIC susceptibility, although attack of these welds apparently required a duplex structure. Tubercule formation was never observed on solution-annealed (heat-treated to produce single-phase austenitic) welds, whereas similar, untreated duplex welds were attacked. Thus it seems that MIC attack requires the presence of such dislocations. Examination of welded materials other than stainless steels tends to confirm that elemental segregation is an important factor in weld failures (Little et al. 1988).

The microbiology of stainless steel tubercules is complicated. SRB are frequently found associated with pseudomonads and *Bacillus* and *Alcaligenes* species. Whereas the distribution of organisms does not appear to change much, the total number of individuals varies a great deal. Dexter (1988) described a tubercule that had a structured arrangement with obligately anaerobic organisms on the interior and aerobic, iron-precipitating bacteria on the exterior. Extrapolation from the microbiological composition of a tubercule to that of the subsurface cavities, however, seems unreasonable.

Metal composition

The surface chemistry of the metal substratum exerts a significant effect on adhesion and, therefore, corrosion. An excellent example of this was work by Chamberlain and Garner (1988), which involved exposure of 90/10 copper-nickel alloys to varying amounts of iron in the sea. The alloy containing no iron corroded rapidly and demonstrated little bacterial adhesion. It appears that the rapid (chemical) dissolution of toxic metals may therefore have been involved in the prevention of colonization similar to an antifouling paint which releases copper. Other studies have shown that copper-nickel alloys are particularly affected when voracious oxygen-consuming bacteria are present (Schriffrin and De Sanchez 1985). The same authors also described a significant increase in corrosion rate when mixed cultures of *Pseudomonas* spp., *Micrococcus* spp., and *Corynebacterium* spp. were introduced into the test system. The dèterioration of copper and copper-based alloys have been described as associated with the formation of exopolymers (Geesey et al. 1986) which would tend to support the data of Schiffrin and De Sanchez (1985). The NACE (1970) basic corrosion course lists SRB as significantly corrosive to brasses (copper-zinc alloys) due to the presence of aggressive HS anions.

A large number of commonly used engineering alloys contain elements that are considered toxic. Stainless steels, for example, contain high levels of nickel, chromium, and molybdenum. Corrosion of these alloys poses the interesting question of why biofilm organisms are apparently immune to the consequent production of toxic cations. An ESCA survey, however, showed that significant quantities of these metals do not accumulate in the biofilm (Dowling et al. 1989). One possible explanation is that these cations are not retained by the bacterial extracellular polymer and diffuse rapidly away. The case for copper has been discussed above. Nickel-based alloys (where nickel is present in the highest quantity of any element) have also been implicated as susceptible to MIC. Stoecker (1984) discussed a case-study where a Monel tube (72% nickel) was severely attacked by selective leaching of the nickel, leaving a copper sponge. Unfortunately, follow-up microbiology is rarely conducted to determine elementary conditions such as total direct counts, viable counts, nickel or copper resistance of the isolates, numbers of anaerobes, and so on.

Cathodic protection

It is often uneconomical to use high-grade alloys to perform low-risk tasks that are not "safety-related." To this end, many construction companies select materials that are more susceptible to corrosion but are less expensive and have better mechanical or welding capabilities, for example. To offset the increased corrosion, interim measures such as painting with anticorrosion or antifouling coatings and cathodic protection may be employed. The latter is quite a common procedure in which the anodic sites associated with the corrosion process segregate to a separate electrode either by impressed current (use of an industrial potentiostat) or via sacrificial anodes (made of zinc or magnesium, for example). Further details may be obtained from McKenzie (1987).

The cathodic (reductive) reactions segregate to the "protected" structure and locally the pH increases due to production of hydroxyl ions. Two major effects spring from this that have the potential to affect and be affected by many different types of organisms: (1) deposition of a calcareous shell due to the elevated pH, and (2) production of cathodic hydrogen. These effects include elevating the numbers of hydrogenophilic bacteria (such as SRB) surrounding the protected structure (Guezennec et al. 1988) and removal of the calcareous shell (Edyvean et al. 1988). These processes do not, however, constitute immediate corrosion but rather economic loss due to the increased current required to maintain the displaced potential (Fiksdal and Guezennec 1988). Serious corrosion is not appreciated until accidental disruption of the cathodic protection (whether by complete dissolution of the sacrificial anode or interrupted power to the potentiostat) occurs. In this event, the "cloud" of organisms (including SRB), which have been encouraged by the evolution of cathodic hydrogen, will be available to accelerate corrosion. Microorganisms of different physiological types are therefore likely to have a synergistic effect on the deterioration of cathodic protection effectiveness. Factors influencing the effectiveness and efficiency of cathodic protection include the water activity in soils, salinity in water systems, and organic loading.

Dissecting MIC Community Structure

Problems with classical methods

The ability of microbiologists to adequately characterize physiologically distinct members of MIC communities has often been limited by the available cultural techniques. Isolation and enumeration of the unusual bacteria which make up MIC communities is frustrated by the diverse growth requirements of consortial members. Mittelman and Geesey (1987) and Pope (1986) have described classical cultural techniques for corrosion consortia. Unfortunately, media for enumeration of the sulfate-reducing, sulfur-oxidizing, and slime-forming consortial members are unacceptably selective. The presence of "viable, nonculturable" organisms in extreme environments and biocide-treated industrial systems adds a further complication to classical cultural techniques.

Lipid analyses

In order that the deficiencies associated with cultural methods may be addressed, techniques are now being employed which utilize signature biomarkers rather than the presence of viable cells, as indicators of community structure and biomass. Vestal and White (1989) have reviewed the use of lipid analyses to evaluate the biomass, community structure, metabolic status, and activity of consortial communities. These techniques exploit the ubiquitous nature and relatively short half-life of membrane-associated phospholipid fatty acids (PLFA). Since these compounds are present in all cells and they are turned over rapidly in the environment (White 1988), their presence is an indicator of viable biota.

Analysis of specific PLFA patterns provides a means by which physiologically distinct groups of bacteria may be identified and enumerated. With the exception of archaebacteria, which contain ether-linked rather than ester-linked fatty acids, these lipid-based methods can be used to assess the eukaryotic and prokaryotic biotic structure at a given point in time. Dowling et al. (1988) utilized PLFA analysis to characterize the biofilm community structure of corroded carbon steel coupon. The analysis showed the presence of large numbers of facultative anaerobes, with smaller numbers of SRB present in the corroded areas.

An assessment of community activity, an important determinant of corrosion potential, can be made using a measure of substrate incorporation into cellular lipids (White et al. 1977). ^{14}C-acetate, which can be readily metabolized by most eubacteria and fungi, is quickly incorporated into lipids. Following incubation with the labeled acetate, biofilms can be extracted and the amount and rate of lipid synthesis estimated. Franklin et al. (1989) used this uptake procedure to evaluate the effects of oxidizing biocides on biofilm activity and corrosion rates of mild steel coupons.

Molecular methods

The techniques of molecular biology provide new tools to examine the distribution and community structure of microbial consortia in biofilms. The techniques that have been developed for quantitative recovery of DNA and ribosomal RNA from sediments should work for the biofilms involved in corrosion. It is particularly difficult to recover nucleic acids from soils or sediments rich in clay, and the nucleic acids recovered are typically fragmented. Presumably these problems would be decreased in the analysis of corrosion biofilms. This emerging technology is enormously powerful, as oligonucleotides made with sequences of more than 10 to 20 bases provide a specificity that is virtually absolute. Once the sequence is known and the appropriate fragment defined, the synthetic DNA probe can be made in substantial quantities. This probe is then tagged (most often with ^{32}P) by nick-translation or polynucleotide kinase. The environmental samples are treated with detergents to lyse the bacteria and the nucleic acids recovered and purified. Once the nucleic acid is recovered from the environment, it is placed in buffers of appropriate ionic strength and heated to denature (form single strands). The single strands are then allowed to anneal (hybridize) with the probe and the unhybridized single-stranded nucleic acid, and excess probe removed by various techniques. The degree of homology can be controlled by the hybridization conditions. Under proper conditions it is possible to detect single gene copies if enzymatic amplification techniques, such as polymerase chain reaction, are utilized (Ogram and Sayler 1988).

Selection of the appropriate DNA probe is dependent on the gene of interest, augmented by the DNA sequence information available for that gene. With access to gene libraries it is possible to select for a wide variety of genes. Genes that are associated with certain groups of bacteria can be selected if the distribution of the enzyme is known. As more of the sequences of genetic determinants of specific processes become known, the probes can be modified or mixed probes utilized. The probe defines the presence of the specific gene, not the enzyme of the metabolic activity. It has been established that certain genes can be detected outside of bacteria adsorbed to the sediment (Ogram and Sayler 1988). Even with these provisos, the technique is powerful and can give insight into the enzyme distribution in a community. Problems may exist in referring the presence of the gene to the presence of a specific bacteria. Some genes are widely distributed; for example, the APS-reductase enzyme which activates sulfate and is essential to all the SRB also was found to occur in sulfur-oxidizing bacteria as well (Postgate 1984). DNA probes are particularly useful in the detection of genetically engineered microbes in the environment where the specificity of the probe can be controlled. The use of DNA gene probes has been reviewed (Jain et al. 1988).

An alternative or complementary method that allows detection of the ribosomal RNA (rRNA) with probes has some additional properties. The rRNA are remarkably conserved molecules that are involved in protein synthetic systems common to all life. Initial work concentrated on the 5S rRNA, which consists of about 120 nucleotides. This short sequence was sufficiently invariate that it showed a paucity of independently varying nucleotide positions. The 16S rRNA (~1600 nucleotides) was ideal for phylogenetic comparisons and with the advent of DNA cloning and sequencing methods, provides exciting possibilities. There are invariate sequences universal to all life, sequences common to the major kingdoms—eukaryotes (plants, animals, and microeukaryotes) and the prokaryotes (eubacteria and archaebacteria). There are sequences which can be used to define closely phylogenetically related groups (which may not be functionally obvious, such as the plant mitochondrial rRNA, the methane-oxidizing bacteria, and the plant-tumor-inducing *Agrobacterium*) and individual species or strains. The facts that there are 10^4 copies of the rRNA in each cell, that the analytical systems can detect 50 or so molecules, and that the probe nucleotides can penetrate the intact cells of bacteria in environmental samples provides a system in which appropriately labeled probes can be used to identify specific bacteria or groups of bacteria in biofilms (DeLong et al. 1989).

The disadvantages of the rRNA probe technology are that for maximum effectiveness the sequence of the 16SRNA must be known. This means the organism must be culturable and the sequencing performed. Determination of the sequence is an artful procedure not the least of which comes in defining the best positioning (placement of "skipped" bases) to use in phylogenetic matching. There is a constant problem of the almost ubiquitous RNAase contamination which can easily ruin the experiments. Nonetheless, the application of nucleic acid probe technology to environmental systems is a powerful means to defining the distribution of microbes in samples and their community structure.

Immunofluorescent techniques

When fluorescent probes are applied to biofilms, the localization of specific cells can be readily determined. This exciting new technology is reviewed by Olsen et al. (1986). An elegant use of this technology in examining the effects of a commonly used antibiotic on the community structure in the bovine rumen has been reported (Stahl et al. 1988). Pope (1986) has reviewed techniques for detecting members of MIC communities. Included in these discussions is a summary of immunofluorescent methods for detecting sulfate-reducing and sulfur-oxidizing bacteria in tubercles. As with other immunochemical techniques, however, the use of antibody probes for the detection of specific consortial members is somewhat limited by the ability to produce the probes and target them to specific environmental isolates. Despite their current technological limitations, these emerging biochemical, molecular, and immunological techniques could provide great insight into the activities of specific organisms in the corrosion consortium.

Treatment Considerations

The literature is replete with descriptions of biocide and physical treatments for the control of biofouling and MIC activities. Unfortunately, the majority of these studies are performed with monocultures and are designed to evaluate treatment efficacy with respect to planktonic populations. The phenomenon of "bacterial regrowth" in municipal water systems is an example of the failings of classical treatment evaluations (LeChevalier 1988). Biofouling and corrosion treatments which fail to control adherent populations cannot succeed in eliminating or controlling deleterious microbial activities. All too often, failures in control measures are attributed to "biocide-resistent" microbial populations. Indeed, this apparent resistance is sometimes used as a rationale for regular changes in the type(s) of biocides employed in the treatment regime. In reality, "biocide resistance" can nearly always be attributed to the presence of adherent biofilms (Costerton and Lashen 1984). Resistance to commonly employed biocides such as sodium hypochlorite, in the classical sense of antibiotic resistance, has not been shown. The multitude of extra- and intracellular target sites and relatively high concentrations utilized in industrial systems effectively precludes such resistance mechanisms.

Franklin et al. (1989) examined the effects of sodium hypochlorite and sodium hypochlorite–sodium bromide biocide combinations on a consortia of MIC putative bacteria. The test system employed mild steel coupons in a flowing freshwater medium maintained at pH 8.5. An inoculum of bacteria obtained from a tubercle containing actively respiring bacteria was used in the biocide challenge assays. The consortium, consisted of a spore-forming *Bacillus* sp.; an iron-reducer (Fe^{3+}); an iron-precipitating, slime-forming strict aerobe; an acetic acid–producing, facultative anaerobe; and a lactate-utilizing, sulfate-reducing bacterium. Both the corrosion rate as measured by electrochemical impedance spectroscopy and the metabolic activity of the community decreased following addition of 16 mg/L sodium hypochlorite and sodium bromide (Fig. 12.9). A three- to four-order-of-magnitude decrease in each of the community members was noted following the biocide dosage. However, within 24 h following cessation of

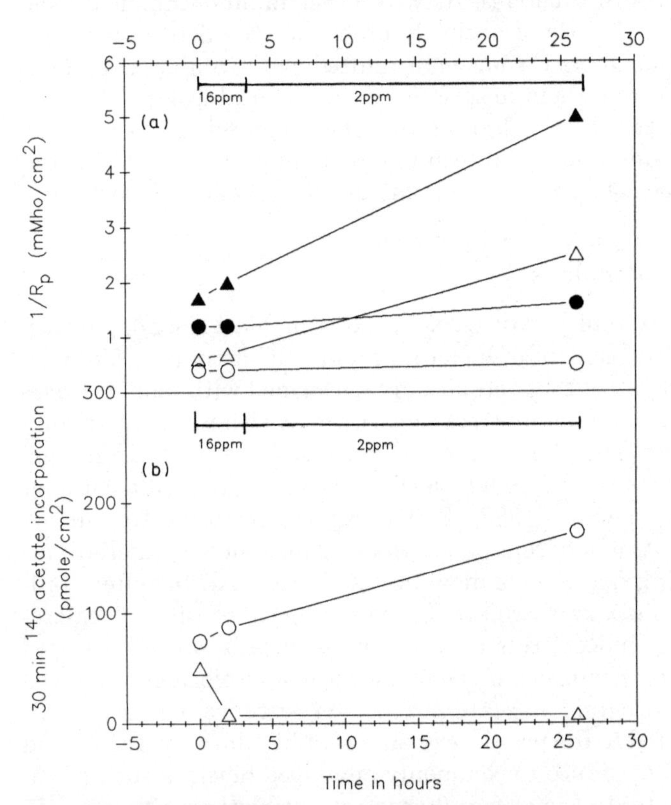

Figure 12.9 The effect of chlorine biocide treatment on corrosion rate (as measured by reciprocal polarization resistance) and metabolic activity (^{14}C-acetate incorporation into phospholipid fatty acids). (*a*) (○) HOCl, inoculated; (●) HOCL, sterile; (△) nontreated, inoculated; (▲) nontreated, sterile. (*b*) (△) Nontreated, inoculated; (○) HOCl, treated, inoculated.

biocide dosing, the community had recovered to its approximate prebiocide structure and number. The intrinsic biocide demand associated with adherent biofilms and their associated corrosion products can account for the failure of apparently high concentrations of oxidizing biocides to completely inactivate biofilm bacteria (Ruseska et al. 1982).

The corrosion products which result from MIC activities can interact with a number of commonly employed biocide treatments. For example, the oxidizing biocides, such as sodium hypochlorite, can oxidize ferrous iron, manganese, and hydrogen sulfide. The resulting deposits often lead to deleterious deposits within cooling towers and other industrial water systems (Trulear and Wiatr 1988). Other nonoxidizing biocides, such as acrolein, are rapidly inactivated by hydrogen sulfide. While differential sensitivity to biocide application has not been reported for MIC consortial bacteria, Ridgway (1987) has shown that fast-growing *Mycobacteria* spp. are significantly more resistant to inactivation by chlorine than are other aerobic bacteria present in consortial biofilms on wastewater-treatment reverse-osmosis membrane surfaces.

Clearly, there is justification for improving evaluation methodologies. It is apparent that the presence of consortial corrosion communities within adherent biofilms is the primary limiting factor in treatment efficacy.

Summary

1. MIC attack is usually associated with consortial activities. One caveat to this is that a single species within a community may exert a disproportionate effect. The corrosion of aluminum fuel tanks by *C. resinae* provides an example of such an effect.
2. Complicated consortia are involved in promoting corrosion in particular areas depending upon materials and their condition (e.g., welds). The relationship of these organisms to each other and to their substratum is not well understood.
3. Most significant economic effects of corrosion which have been identified as microbiological in origin are found to be a direct result of consortial, rather than monospecies, activities.
4. Novel methods for dissection of MIC consortia and their activities are currently available but have yet to be extensively used. These include nondestructive electrochemical techniques, signature biomarker analyses, and molecular genetic techniques.
5. Treatment and prevention maintenance strategies must consider the influence of consortial biofilms on their efficacy. Evaluations of

treatment efficacies should utilize biofilm communities. Treatments which do not consider their influence are, at best, ineffective. At their worst, they can often exacerbate preexisting corrosion conditions.

6. The development and use of new surface-specific biocides, MIC-resistant alloys (containing bactericidal elements perhaps), self-draining pipe systems with constant flow (no stagnant areas) and limiting zones where detritus may collect, will be more important in the design of efficient plant operations in the future.

Acknowledgments

Some of the work described in this chapter was supported by the following organizations: U.S. Office of Naval Research, no. N00014-87-K00012 (molecular biology of adhesion processes program); U.S. Office of Naval Research, no. N00014-89-J-3095 (biosurfaces program); Gas Research Institute, no. 5086-260-1303 (Anaerobic Biocorrosion of pipe-line steels). The assistance of Sandy Isbill and Stephanie Brooks in preparation of the manuscript is gratefully acknowledged.

References

Allred, R. C., J. D. Sudbury, and D. C. Olson. 1959. Corrosion is controlled by bactericide treatment. *World Oil* 149:111–112.

Assefpour, M., and W. G. Ferguson. 1988. Effect of low concentrations of hydrogen sulfide in seawater on fatigue crack growth in a CMn structural steel. *Corrosion* 44(7): 443–449.

Baier, R. 1981. Initial events in microbial film formation. In: Marine Biodeterioration: An interdisciplinary study. Proceedings of the symposium on marine biodeterioration. Uniformed Services, University of Health Sciences. 20–23, April, 1981. Naval Institute Press, Annapolis, Maryland.

Bielecki, R., and H. Schremmer. 1987. Leichtweiss-Institut fur Wasserbau der Technishen Universitat Braunschweig. Mitteilungen Heft 94/1987. ISSN 0343–1223.

Booth, G. H., and A. K. Tiller. 1968. Cathodic characteristics of mild steel in suspensions of sulphate-reducing bacteria. *Corrosion Science* 3:583–600.

Borenstein, S. W. 1988. Microbiologically influenced corrosion failures of austenitic stainless steel welds. *Materials Performance*, August, pp. 62–66.

Borenstein, S. W. 1988b. Guidelines for destructive examination of potential MIC-related failures. EPRI publication no. ER-6345. Microbial corrosion: 1988 workshop proceedings. Ed. G. J. Licina.

Borenstein, S. W. 1988c. Microbiologically influenced corrosion failure analyses. *Materials Performance*. March, pp. 51–54.

Borenstein, S. W. 1989. Influence of welding variables on microbiologically influenced corrosion of austenitic stainless steel weldments. MS thesis, University of Tennessee, Knoxville.

Cabral, D. 1980. Corrosion by microorganisms of jet aircraft integral fuel tanks. Int. *Biodeterion. Bull.* 16(1):23–27.

Chamberlain, A. H. L., and B. J. Garner. 1988. The influence of iron content on the biofouling resistance of 90/10 copper-nickel alloys. *Biofouling* 1(1):79–96.

Cloete, T. E., and F. Gray. 1985. Microbiological control in paper mills. Paper Southern Africa 51(4):26.

Costerton, J. W. 1985. The role of bacterial exopolysaccharides in nature and disease. Developments in Industrial Microbiology 26:249–261.

Costerton, J. W., and E. S. Lashen. 1984. Influence of biofilm on efficacy of biocides on corrosion-causing bacteria. *Materials Performance* 23(2):13–17.

Daniels, L., N. Belay, B. S. Rajagopal, and P. Weimer. 1987. Bacterial methanogenesis and growth from CO^2 with elemental iron as the sole source of electrons. *Science* 237: 509–511.

Daumas, S., Y. Massiani, and J. Crousier. 1988. Microbiological battery induced by sulphate-reducing bacteria. *Corrosion Science* 28(11):1041–1050.

Davies, R. D. 1984. Corrosion problems in the pulp and paper industry. *Paper Southern Africa* 4(4):18–27.

Day, D. H., A. E. Hughes, and G. P. Marsh. 1985. The management of radioactive wastes. *Reports on progress in physics* 48:101–169.

DeLong, E. F., G. S. Wickham, and N. R. Pace. 1989. Phylogenetic stains: Ribosomal RNA-based probes for the identification of single cells. *Science* 243:1360–1363.

Deshmukh, M. B., R. B. Srivastava, and A. A. Karande. 1988. Corrosion of mild steel and copper by microorganisms in polluted seawater. Pages 725–737 in M.-F. Thomson, R. Sarojini, and R. Nagabhushanam, (eds.), *Marine Biodeterioration: Advanced Techniques Applicable to the Indian Ocean*. Oxford & IBH Publishing Co., New Delhi, India.

Dexter, S. C. 1988. Role of microfouling organisms in marine corrosion. Plenary lecture: 7th International Congress on Marine Corrosion and Fouling. Valenica, Spain. November.

Dexter, S.: *Biologically Induced Corrosion*. Proceedings of the Conference on Biologically Induced Corrosion, June 10–12, 1985, Gaithersburg, Maryland. National Association of Corrosion Engineers (NACE), Houston, Texas 1986.

Dowling, N. J. E., J. Guezennec, M. L. Lemoine, A. Tunlid, and D. C. White. 1989. Analysis of carbon steels affected by bacteria using electrochemical impedance and direct current techniques. *Corrosion* 44 (12):869–874.

Dowling, N. J. E., M. J. Franklin, D. C. White, C. H. Lee, and C. Lundin. 1988. The effect of microbiologically influenced corrosion on stainless steel weldments in seawater. Page No. 187. NACE '89. New Orleans.

Dowling, N. J. E., F. Widdel, and D. C. White. 1986. Phospholipid ester-linked fatty acid biomarkers of aceate-oxidizing, sulphate-reducing, and other sulfide-forming bacteria. *J. General Microbiology*. 132:1815–1825.

Edyvean, R. G. L., C. J. Hutchinson, and N. J. Silk. 1988. Observations on the cathodic protection of steel in biologically active seawater. 7th International Congress for Marine Corrosion and Fouling. Valencia, Spain.

Fiksdal L., and J. Guezennec. 1988. Increased current demand required by cathodic protection of carbon steel in the presence of *Vibrio natriegens*. 7th International Congress for Marine Corrosion and Fouling. Valencia, Spain.

Ford, T. E., and R. Mitchell. 1989. Hydrogen embrittlement: a microbiological perspective. *Corrosion 89*. p. 189. NACE, New Orleans.

Franklin, M. J., D. E. Nivens, M. W. Mittelman, A. A. Vass, R. F. Jack, N. J. E. Dowling, R. P. Mackowski, S. L. Duncan, D. B. Ringleberg, and D. C. White. 1989. An analogue MIC system with specific bacterial consortia, to test effectiveness of materials selection and countermeasures. *Corrosion 89*. p. 513. NACE proceedings of conference. New Orleans.

Gaylarde, C. C., and J. M. Johnston. 1986. Anaerobic metal corrosion in cultures of bacteria from estuarine sediments. Pages 137–143 in S. C. Dexter (ed.), *Biologically Induced Corrosion*. NACE, Houston, Texas.

Gaylarde, C. C., and J. M. Johnston. 1982. The effect of *Vibrio anguillarum* on the anaerobic corrosion of mild steel by *Desulfovibrio vulgaris*. *Int. Biodeterioration Bull.* 18(4):111–116.

Geesey, G. G., S. D. Salas, and M. W. Mittelman. 1986. Effects of environmental conditions on the sessile existance of an estuarine sediment bacterium. GERBAM: Deuxieme Colloque International de Bacteriologie marine-CNRS, Brest. 1–5 Octobre 1984. IFREMER, Proceedings of conference 3:243–247.

Geesey, G. G., M. W. Mittelman, T. Iwaoka, P. R. Griffiths. 1986. Role of bacterial exopolymers in the deterioration of metallic copper surfaces. *Materials Performance*, February, pp. 37–40.
Greathouse, G. A., and C. J. Wessel. 1954. *Deterioration of Materials*. Reinhold, New York.
Grimm, D. T., C. G. Hollis, and S. T. Threlkeld. 1984. The nature of sulfate-reducing bacteria isolated from a high temperature Kraft liner board machine. Proceedings of the Technical Association of the Pulp & Paper Industry, pp. 111–113.
Guezennec, J., M. Therene, N. Dowling, and D. C. White. 1988. Influence de la protection cathodique sur la croissance des bacteries sulfato-reductrices et sur la corrosion d'aciers doux dans les sediments marins. 7th International Congress on Marine Corrosion and Fouling. Valencia, Spain.
Hamilton, W. A. 1985. Sulphate-reducing bacteria and anaerobic corrosion. *Ann. Rev. Microbiol.* 39:195–217.
Hamilton, W. A., A. N. Moosaavi, N. Pirrie. 1988. Mechanism of anaerobic microbial corrosion in the marine environment. Paper submitted to European Federation of Corrosion First Workshop on Microbial Corrosion at Sintra, Portugal, March 7–9, 1988.
Haydon, D. A. 1961. The surface charge of cells and some other small particles as indicated by electrophoresis. *Biochim. Biophys. Acta* 50:450–457.
Hill, H. C. 1984. Biodegradation of petroleum products. Pages 582–617 in R. M. Atlas (ed.) *Petroleum Microbiology*. Macmillan, New York.
Houghton, D. R., R. N. Smith, and H. O. W. Eggins. 1988. Biodeterioration. Proceedings of Conference of the Seventh International biodeterioration symposium, Cambridge, UK 6–11 September 1987.
Jain, R. K., R. S. Burlage, and G. S. Sayler. 1988. Methods for detecting recombinant DNA in the environment. *CRC Crit. Rev. Biotechnol.* 8:33–84.
Jolley, J. G., G. G. Geesey, M. R. Hankins, R. B. Wright, and P. L. Wichlacz. 1988. Auger electron spectroscopy and X-ray photoelectron spectroscopy of the biocorrosion of copper by gum arabic, bacterial culture supernatant and *Pseudomonas atlantica* exopolymer. *Sur. Int. Anal.* 11:371–376.
Kerger, B. D., P. D. Nichols, C. A. Antworth, W. Sand, E. Bock, J. C. Cox, T. A. Langworthy, and D. C. White. 1987. Signature fatty acids in the polar lipids of acid-producing *Thiobaccilli*: methoxy, cyclopropyl, α-hydroxy-cyclopropyl and branched and normal monoenoic fatty acids. *FEMS Microb. Ecol.* 38:67–77.
Knudsen, J. G. 1981. Fouling of heat transfer surfaces. Pages 57–82 in P. J. Marto and R. H. Nunn (ed.), *Power Condenser Heat Transfer Technology*. Hemispher Publishing Co., New York.
Kobrin, G. 1976. Corrosion by microbiological organisms in natural waters. *Materials Performance* 15(7):38.
Kuhn A. T., and T. Rae. 1988. Aqueous corrosion of Ni-Cr alloys in biological environments and implications for their biocompatability. *Br. Corrosion J.* 23(4):259–266.
Laanbroek, H. J., and H. Veldkamp. 1982. Microbial interactions in sediment communities. *Phil. Trans. R. Soc. Lond. B*, 297:533–550.
Lappin-Scott, H. M., and J. W. Costerton. 1989. Bacterial biofilms and surface fouling. *Biofouling* 4:323–342.
LeChevallier, M. W., C. D. Cawthon, and R. G. Lee. 1988. Inactivation of biofilm bacteria. *Appl. Environ. Microbiol.* 54:2492–2499.
Licina, G. 1988. *Microbial Corrosion: 1988 Workshop Proceedings*. Electric Power Research Institute (EPRI) publication ER-6345. October 18, 1988, Charlotte, N.C.
Lin, L. F., C. Y. Chao, and D. D. MacDonald. 1981. A point defect model for anodic passive films. II. Chemical breakdown and pit initiation. *J. Electrochemic. Soc.* 128(6): 1194–1198.
Little, B., P. Wagner, and J. Jacobus. 1988. The impact of sulfate-reducing bacteria on welded copper-nickel seawater piping systems. *Materials Performance*. 27(8):57–61.
Lupton, F. S., R. Conrad, J. G. Zeikus. 1984. Physiological function of hydrogen metabolism during growth of sulfidogenic bacteria on organic substrates. *J. Bacteriol.* 159(3):843–849.
Mara, D. D., and D. J. A. Williams. 1971. Polarization studies of pure Fe in the presence

of hydrogenase-positive microbes—I. Non-photosynthetic bacteria. *Corrosion Science* 11:895–900.

Marsh, G. P., and K. J. Taylor. 1988. An assessment of carbon steel containers for radioactive waste disposal. *Corrosion Science* 28:289–320.

Marshall, K. C. 1976. *Interfaces in Microbial Ecology*. Harvard University Press, Cambridge.

Maxwell, S., and W. A. Hamilton. 1986. Modified radiorespirometric assay for determining the sulfate reduction activity of biofilms on metal surfaces. *J. Microbiol. Meth.* 5:83–91.

McCoy, W. 1987. Fouling biofilm formation. Pages 24–25 in M. W. Mittelman and G. G. Geesey (eds.), *Biological Fouling of Industrial Water Systems: A Problem Solving Approach*, Water Micro Associates, San Diego, California.

McKenzie, S. G. 1987. Techniques for monitoring corrosion of steel in concrete. *Corrosion Prevention and Control*, February 11–17, 1987.

Milde, K., W. Sand, W. Bock. 1983. *Thiobacilli* of corroded concrete walls of the Hamburg sewer system. *J. Gen. Microbiol.* 129:1227–1333.

Miller, J. D. A., E. R. Schiapparelli, and D. E. Meybaum. 1981. *Metals*. In A. H. Rose (ed.), *Microbial Biodeterioration*. Academic Press, London.

Mittelman, M. W., and G. G. Geesey. 1987. Biological fouling of industrial water systems: a problem solving approach. *Water Micro Associates*, San Diego, California.

Morgan, F. L., and O. W. May. 1970. Scale control in white-water systems. *TAPPI*, 53(11):2096–2098.

NACE: NACE Basic Corrosion Course. 1970. NACE, Houston, Texas.

Ogram, A. V., and G. S. Sayler. 1988. The use of gene probes in the rapid analysis of natural microbial communities. *J. Indust. Microbiol.* 3:281–292.

Olsen, G. J., D. J. Lane, S. J. Giovanni, N. R. Pace, and D. A. Stahl. 1986. Microbial ecology and evolution: a ribosomal approach. *Ann. Rev. Microbiol.* 40:337–365.

Onions, A. 1975. Organisms for Biodeterioration Testing-moulds and fungi. In D. W. Lovelock and R. J. Gilobert (eds.), *Microbial Aspects of the Biodeterioration of Materials*. Academic Press, New York.

Pankanhania, I. P., A. N. Moosaavi, W. A. Hamilton. 1986. Utilization of cathodic hydrogen by *Desulfovibrio vulgaris* (Hildenborough). *J. Gen. Microbiol.* 132:3357–3365.

Parker, C. D. 1947. Species of sulphur bacteria associated with the corrosion of concrete. *Nature* 159:439.

Pendrys, J. P. 1989. Biodegradation of asphalt cement-20 by aerobic bacteria. *Appl. Environ. Microbiol.* 55:1357–1362.

Pollock, W. I. 1989. Hydrotest water can cause MIC pitting. *Materials Performance*. 28(7):68.

Pope, D. H., D. J. Duquette, A. H. Johannes, and P. C. Wayner. 1984. Microbiologically influenced corrosion of industrial alloys. *Materials Performance*, April.

Pope, D. H. 1986. Discussion of methods for the detection of microorganisms involved in microbiologically influenced corrosion. Pages 275–282 in S. C. Dexter (ed.), *Biologically Induced Corrosion*. NACE Publications, Houston, Texas.

Postgate, J. R. 1984. *The Sulphate-Reducing Bacteria*. Cambridge University Press, London.

Purkiss, B. E. 1971. Corrosion in industrial situations by mixed microbial floras. Pages 107–128 in J. D. A. Miller (ed.), *Microbial Aspects of Metallurgy*. Medical and Technical Publishing, Aylesbury, England.

Ramamurti, K., G. P. Jayaprakash, and C. F. Crumpton. 1984. Microbial biodegradation of asphalt and related hydrocarbons—a literature review. Pages 1–44 in Report no. FHWA-KS-84/1 Interim report. Division of Operations, Bureau of Material and Research, Kansas Department of Transportation, Topeka, Kansas.

Ratledge, C., and S. G. Wilkinson. 1988. *Microbial Lipids*, vol. 1. Academic Press, New York.

Rendleman, J. M. 1978. Metal-polysaccharide complexes—Part I. *Fundamental Chemistry* 3:47–79.

Ridgway, H. F. 1987. Microbial fouling of reverse osmosis membranes: genesis and control. Pages 138–193. In M. W. Mittelman and G. G. Geesey (eds.), *Biological Fouling of Industrial Water Systems: A Problem Solving Approach*. Water Micro. Associates, San Diego, California.

Ringas, C., and F. P. A. Robinson. 1987. Corrosion of stainless steel by sulfate-reducing bacteria-electrochemical techniques. *Corrosion* 44(6):386–396.

Rittman, B. E. 1982. Comparative performance of biofilm reactor types. *Biotech. Bioeng.* 24:1341–1370.

Rosser, H. R., and W. A. Hamilton. 1983. Simple assay for accurate determination of [^{35}S] sulfate reduction. *Appl. Environ. Microbiol.* 45(6):1956–1959.

Ruseska, I., J. Robbins, J. W. Costerton, and E. S. Lashen. 1982. Biocide testing against corrosion-causing oilfield bacteria helps control plugging. *Oil Gas J.* 80:253–264.

Safade, T. L. 1988. Tackling the slime problem in a paper-mill. *Paper Technology and Industry* 25(6):280–285.

Sand, W., and E. Bock. 1984. Concrete corrosion in the Hamburg sewer system. *Environ. Technol. Lett.* 5:517–528.

Schiapparelli, E. R., and B. R. Meybaum. 1980. The role of dodecanoic acid in the microbiological corrosion of jet aircraft integral fuel tanks. *Int. Biodet. Bull.* 16(3): 61–66.

Schiffrin, D. J., and S. R. De Sanchez. 1985. The effects of pollutants and bacterial microfouling on the corrosion of copper-base alloys in seawater. *Corrosion* 41(1): 31–38.

Stahl, D. A., B. Flesher, H. R. Mansfeld, L. Montgomery. 1988. Use of phylogenetically based hybridization probes for studies of ruminal microbial ecology. *Appl. Environ. Microbiol.* 54:1079–1084.

Steocker, J. G. 1984. Guide for the investigation of microbiologically induced corrosion. *Materials Performance* 23(8):48–55.

Stephensen, M., and L. H. Stickland. 1931. Hydrogenase. II. The reduction of sulfate-to-sulfide by molecular hydrogen. *Biochem. J.* 25:215–220.

Tatnall, R. 1981. Case histories: Bacteria induced corrosion. *Materials Performance* 20(9):32.

Thistlethwayte, D. K. B. 1972. *Control of Sulfides in Sewerage Systems*. Butterworths, Melbourne.

Tiller, A. K. 1986. A review of the European research effort on microbial corrosion between 1959 and 1984. In S. Dexter (ed.), *Biologically Induced Corrosion*. NACE, Houston, Texas.

Trulear, M. G., and C. L. Wiatr. 1988. Recent advances in halogen based biocontrol. Proceedings of the National Association of Corrosion Engineers Annual Meeting, March 21–25, St. Louis, Missouri. Paper no. 19.

Vestal, J. R., and D. C. White. 1989. Lipid analysis in microbial ecology. *Bioscience* 39: 535–541.

Von Wolzogen Kuhr, C. A. V., and L. S. Van der Vlugt. 1934. Water (The Hague), 18(16):147.

White, D. C., R. F. Jack, N. J. E. Dowling, M. J. Franklin, D. E. Nivens, S. Brooks, M. W. Mittelman, A. A. Vass, and H. W. Isaacs. 1990. Microbially influenced corrosion of carbon steels. Submitted to Corrosion/90, NACE, Las Vegas, Nevada.

White, D. C., R. J. Bobbie, S. J. Morrison, D. K. Oosterhof, C. W. Taylor, and D. A. Meeter. 1977. Determination of microbial activity of estuarine detritus by relative rates of lipid biosynthesis. *Limnol. Oceanogr.* 22:1089–1099.

White, D. C. 1988. Validation of quantitative analysis for microbial biomass, community structure, and metabolic activity. *Adv. Limnol.* 31:1–18.

Widdel, F. 1988. Microbiology and ecology of sulfate- and sulfur-reducing bacteria. In A. J. B. Zehnder (ed.), *Biology of Anaerobic Microorganisms*. Wiley-Interscience, New York.

Winogradsky, S. 1949. *Microbiologie du Sol. Oeuvres Completes*. Mason, Paris.

Wolin, M. J. 1982. Hydrogen transfer in microbial communities in biotechnology. In A. T. Bull and J. H. Slater (eds.), *Microbial Interactions and Communities*, vol. 1. Academic Press, London.

Chapter

13

Mixed Cultures in Biological Leaching Processes and Mineral Biotechnology

Olli H. Tuovinen

Department of Microbiology
The Ohio State University
Columbus, Ohio

Bruce C. Kelley

CRA
Roseville, NSW, Australia

Stoyan N. Groudev

Higher Institute of Mining and Geology
Research and Training Center of Mineral Biotechnology
Sofia, Bulgaria

Introduction

Biological leaching processes in mineral industry have been developed to commercial-scale applications in three major areas, namely the leaching of (1) copper ores, (2) uranium ores, and (3) the pretreatment of gold-bearing pyrite and arsenopyrite ore concentrates. Mineral biotechnology has potential for many other applications for processing of mineral resources, but the exploitation by the mining industry awaits further developments in research and economic evaluation.

Mineral materials that are amenable to solubilization in biohydrometallurgical processes are essentially sulfide mineralizations, although metals of interest may not necessarily be incorporated or locked in sulfide lattice, as in the case of uranium minerals. *Thiobacillus*-type bacteria used in mineral leaching processes require acid, sulfate-containing environments; acid solutions used in these reactions also enhance the solubility of metals. The same bacteria have application in the desulfurization of coal, but the advent of commercialization of coal biotechnology for sulfur removal is not imminent.

Bacteria in dump and underground leaching processes are indigenous on exposed ore surfaces and in pore spaces and mine water at the source. Strong selection pressure in microbial consortia at mine sites is brought about by inorganic electron donors; low concentrations of available organic carbon; and high concentrations of protons, metal ions, and sulfate. Successful attempts to modify bacterial cultures in commercial-scale copper or uranium leaching operations have not been reported, whereas efforts to enhance the activities of resident microorganisms have sometimes been successful. In concentrate leaching for precious metal recovery, stirred tank reactors are the method of choice also for large-scale processing under controlled conditions. In both approaches of applied mineral biotechnology, a good knowledge of the chemical and microbiological fundamentals of the leaching reactions is of key importance in attempts to develop suitable cultures for mineral leaching systems.

In this chapter, mineral biotechnology is shown to be unique in its problems due to the nature, heterogeneity, and bulkiness of the raw material to be treated and apparent lack of operational biological control in commercial-scale on-site applications. Sulfide minerals, that is, the primary substrates and electron donors, are typically insoluble and inorganic and their distribution and association with gangue minerals display heterogeneity. The concentrations of minerals and metals within ore deposits also display tremendous variation. Provision of bacterial contact with mineral surfaces and fluxes of reactants and products are kinetically important factors because the oxidative reactions occur on mineral surfaces. Several of these problems are also relevant in coal biotechnology, although coal desulfurization by bacteria is not within the scope of this chapter.

Microbial communities in sulfide or coal mines and mine waters are complex, site-specific, and in general not well understood. The indigenous bacteria in these environments display considerable heterogeneity at genetic and physiological levels. The descriptive and phenomenological studies reported in the literature have not yielded proven methods which could be employed in a predictable manner to control and modify microbial communities and their activities at mine sites where the biological leaching of metals is practiced. Efforts to mimic genetic, physiological, and taxonomic diversity have revealed that mixed cultures hold great promise in mineral biotechnology. However, the methodology to characterize mixed cultures and to control and stabilize their composition and activities must be developed before the full potential of mixed cultures can be better employed for mineral biotechnology.

Substrates

The acid, oxidative leaching of different sulfide minerals by microorganisms has been well documented, and the reader is directed to a number of excellent reviews on the subject.[1] Some of the commercially important sulfide minerals and examples of reactions representing their oxidative dissolution are listed in Table 13.1.

It should be noted that sulfide minerals display nonstoichiometric relationships and also contain varying amounts of other metals. Pyrrhotite (FeS) is an example of a mineral that shows varying nonstoichiometry with respect to the Fe:S ratio, signified with a formula $Fe_{1-x}S$ (e.g., $Fe_{0.95}S$). Cobalt, apart from occurring as a discreet sulfide mineral of its own (CoS), is usually present in varying quantities in iron sulfides pyrite (FeS_2) and pyrrhotite as well as in pentlandite (Ni_9S_8) phases. The formula of pentlandite, a nonstoichiometric sulfide mineral, is often written as $(Ni,Fe)_9S_8$ to indicate that it actually also contains varying amounts of iron. The distribution of metals in several mineral phases and varying nonstoichiometric relationships in natural sulfide minerals makes it extremely difficult to establish accurate stoichiometric reactions in bacterial leaching systems. Furthermore, with natural sulfide ore materials and concentrates, several concurrent oxidative reactions occur, involving molecular oxygen, ferric iron, and bacteria as oxidants, which cannot usually be attributed to a single reaction mechanism.

Iron, both in its divalent and trivalent state, plays a central role in sulfide mineral oxidation (Table 13.2). The major iron sulfide mineral pyrite tends to be ubiquitous in its distribution and is associated with many mineral deposits of economic significance. Pyrite also constitutes an abundant mineral in precious metal–containing ores and concentrates that are amenable to the bacterial leaching. In this application, the recovery of gold and silver, locked up in the sulfide matrix, can be considerably enhanced upon bacterial oxidation of pyrite and arsenopyrite (FeAsS). Biological oxidation of these sulfide minerals improves the accessibility of the trapped precious metals to standard extraction reagents such as cyanide.

[1]Brierley 1978; Kelly et al. 1979; Ralph 1979, 1985; Lundgren and Malouf 1983; Torma 1988.

TABLE 13.1 Examples of Oxidation Reactions of Some Industrially Important Sulfide Minerals

Metal	Mineral	Formula	Oxidative dissolution
Cobalt	Cobaltite	CoS	$CoS + 2O_2 \rightarrow Co^{2+} + SO_4^{2-}$
Copper	Covellite	CuS	$CuS + 2O_2 \rightarrow Cu^{2+} + SO_4^{2-}$
	Chalcocite	Cu_2S	$Cu_2S + 0.5O_2 + 2H^+ \rightarrow CuS + Cu^{2+} + H_2O$
			$Cu_2S + 2.5O_2 + 2H^+ \rightarrow 2Cu^{2+} + SO_4^{2-} + H_2O$
	Digenite	Cu_9S_5	$Cu_9S_5 + 2O_2 + 8H^+ \rightarrow 5CuS + 4Cu^{2+} + 4H_2O$
	Bornite	Cu_5FeS_4	$Cu_5FeS_4 + 9O_2 + 4H^+ \rightarrow 5Cu^{2+} + Fe^{2+} + 4SO_4^{2-} + 2H_2O$
			$2Cu_5FeS_4 + 18.5O_2 + 10H^+ \rightarrow 10Cu^{2+} + 2Fe^{3+} + 8SO_4^{2-} + 5H_2O$
	Chalcopyrite	$CuFeS_2$	$CuFeS_2 + O_2 + 4H^+ \rightarrow Cu^{2+} + Fe^{2+} + 2S^0 + 2H_2O$
			$2CuFeS_2 + 8.5O_2 + 2H^+ \rightarrow 2Cu^{2+} + 2Fe^{3+} + 4SO_4^{2-} + H_2O$
Gold	Arsenopyrite	$FeAsS$	$FeAsS + 3.5O_2 + H_2O \rightarrow Fe^{3+} + AsO_4^{3-} + SO_4^{2-} + 2H^+$
			$2FeAsS + 6.5O_2 + 3H_2O \rightarrow 2Fe^{2+} + 2AsO_4^{3-} + 2SO_4^{2-} + 6H^+$
Iron	Pyrite	FeS_2	$2FeS_2 + 7.5O_2 + H_2O \rightarrow 2Fe^{3+} + 4SO_4^{2-} + 2H^+$
	Pyrrhotite	FeS	$2FeS + 4.5O_2 + 2H^+ \rightarrow 2Fe^{3+} + 2SO_4^{2-} + H_2O$
			$2FeS + 1.5O_2 + 6H^+ \rightarrow 2Fe^{3+} + 2S^0 + 3H_2O$
Lead	Galena	PbS	$PbS + 2O_2 \rightarrow Pb^{2+} + SO_4^{2-} \rightarrow PbSO_4$
Molybdenum	Molybdenite	MoS_2	$2MoS_2 + 9.5O_2 + 5H_2O \rightarrow 2MoO_4^- + 4SO_4^{2-} + 10H^+$
Nickel	Pentlandite	Ni_9S_8	$Ni_9S_8 + 16.5O_2 + 2H^+ \rightarrow 9Ni^{2+} + 8SO_4^{2-} + H_2O$
	Millerite	NiS	$NiS + 2O_2 \rightarrow Ni^{2+} + SO_4^{2-}$
Zinc	Sphalerite	ZnS	$ZnS + 2O_2 \rightarrow Zn^{2+} + SO_4^{2-}$

TABLE 13.2 Examples of Ferric Iron–Mediated Leaching Reactions of Some Industrially Important Sulfide Minerals

Mineral	Ferric iron–mediated reaction
Covellite	$CuS + 2Fe^{3+} \rightarrow Cu^{2+} + 2Fe^{2+} + S^0$
Chalcocite	$Cu_2S + 4Fe^{3+} \rightarrow 2Cu^{2+} + 4Fe^{2+} + S^0$
Bornite	$Cu_5FeS_4 + 12Fe^{3+} \rightarrow 5Cu^{2+} + 13Fe^{2+} + 4S^0$
Chalcopyrite	$CuFeS_2 + 4Fe^{3+} \rightarrow Cu^{2+} + 5Fe^{2+} + 2S^0$
Pyrite	$FeS_2 + 14Fe^{3+} + 8H_2O \rightarrow 15Fe^{2+} + 2SO_4^{2-} + 16H^+$

The prerequisite for bacterial oxidation is the chemical dissociation of the insoluble metal sulfide from the mineral surface, as exemplified for copper sulfide covellite:

$$CuS \leftrightarrows Cu^{2+} + S^{2-}$$

The dissociation then provides the bacterial metabolic system with an oxidizable substrate (i.e., a source of electrons). From an industrial viewpoint, the rate of substrate provision is extremely important, because it potentially represents a rate-limiting step. A number of factors affect the dissolution step. Dislocations, imperfections, and the presence of impurities or inclusions all contribute to the susceptibility of sulfide minerals to bacterial attack. Indeed, such structural deformations in the mineral lattice may determine the sites of initial bacterial attachment (Rodriguez-Leiva and Tributsch 1988).

As a general rule, mineral sulfides with high solubility products tend to be good substrates for bacterial leaching (Tributsch and Bennett 1981a,b). However, the equilibrium conditions of sulfide mineral dissociation are also affected by other factors, notably ferric ion (Fe^{3+}) and protons (H^+). Electron extraction from the sulfide valence band by Fe^{3+}, together with the presence of an excess of holes in the valence band, increases the susceptibility of sulfide minerals to oxidative leaching process (Tributsch and Bennett 1981a,b; Vaughan 1984). The concentration of H^+ is also highly important. Protons can chemically break surface bonds, shifting electron levels of many sulfides into a range where they can more directly interact with the bacterial metabolic system.

Attachment of bacteria on mineral surfaces is a common phenomenon, but the interaction between the bacterium and the mineral surface remains poorly understood. The removal of weakly bonded —SH groups produced by the H^+-sulfide interaction, was postulated to involve an unidentified molecular carrier (Tributsch and Bennett 1981a,b). Rodriguez-Leiva and Tributsch (1988) suggested that a thin film exists between the attached bacterial cell outer membrane and the surface of sulfide mineral, and that the corrosion process occurs within this interfacial film. If the film is aqueous in nature, then a molecular carrier would be required to transport hydrophobic sulfur into the bacterial cell. Alternatively, if the film was to be entirely organic and therefore capable of dissolving sulfur, then a carrier system may not need to operate. Instead, soluble sulfur species would be presented for bacterial oxidation. While further experimental evaluation is required, some evidence exists to favor an organic corrosion film (Rodriguez-Leiva and Tributsch 1988). Following bacterial attachment and the initiation of a corrosion pit, the leaching process continues until conditions within the pit become detrimental for further activity. Thus pit depth may be limited by the mass transfer of oxygen and carbon dioxide or by buildup of inhibitory levels of reaction products (Rodriguez-Leiva and Tributsch 1988). Recent studies in exposing thin FeS_2 films to *Thiobacillus ferrooxidans* have confirmed the morphological features associated with this corrosion phenomenon (Bärtels et al. 1989).

Several concurrent reactions are involved in the oxidative, microbiological leaching of sulfide minerals. These are sometimes collectively termed as a direct, indirect, and electrochemical mechanism of biological leaching.

The direct mechanism initially relies on the dissociation of sulfide mineral followed by the bacterial oxidation of the sulfur moiety to sulfate. Iron is brought into solution by the oxidation of pyrite and other iron sulfides and thus the bacterium is presented with reduced iron and sulfur compounds as electron donors and energy sources for growth as presented in Fig. 13.1. The products of these reactions are ferric sulfate and sulfuric acid, which together make up a highly oxidizing, corrosive environment in which the chemical or indirect leaching can occur. The products of the indirect attack include ferrous ion and elemental sulfur (Table 13.2, Fig. 13.1), both of which can be directly oxidized by the bacterium to complete the leaching cycle. Secondary minerals, if produced during the leaching, are formed via dissolution and may include secondary sulfides such as covellite or Fe(III)-complexes such as jarosite. The proton concentration is critical, both in assisting with the chemical attack on the mineral sulfide and in maintaining the solubility of the Fe^{3+}.

Uranium leaching is another example of the indirect chemical action of acidic ferric sulfate which is produced by the biological oxidation of pyrite (Table 13.3). This requires a mineralogical association between pyrite, or some other iron sulfide, and the uranium mineral. The tetravalent uranium in minerals such as uraninite is chemically oxidized to the hexavalent form. With ferric iron as the oxidant (Fig. 13.2), the leaching rates can be considerably enhanced. The role of

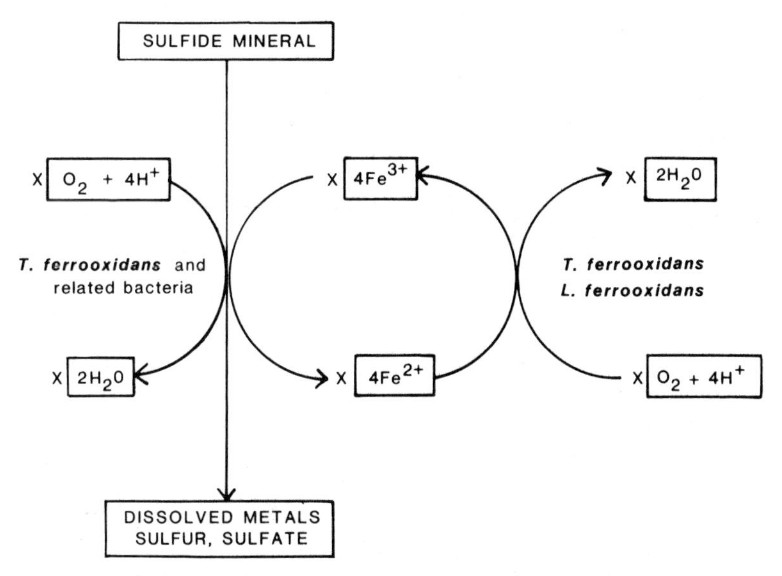

Figure 13.1 Schematic illustration of the bacterial leaching of sulfide minerals. The oxygen-dependent route on the left represents the direct oxidative dissolution; the ferric iron–dependent route on the right represents the indirect leaching in which bacteria regenerate the chemical oxidant Fe^{3+}. The X denotes multiples of electron equivalents and is characteristic to each mineral.

TABLE 13.3 Reactions in the Oxidative Dissolution and Leaching of Uranium

Component	Reaction	Equation
Tetravalent U^{IV}	Oxidation	$U^{4+} + 1.5O_2 + 2H^+ \rightarrow UO_2^{2+} + H_2O$
Hexavalent U^{VI}	Hydrolysis	$U^{6+} + 2H_2O \rightarrow UO_2^{2+} + 4H^+$
Uranium oxide UO_3	Dissolution	$UO_3 + 2H^+ \rightarrow UO_2^{2+} + H_2O$
Uraninite UO_2	Oxidation	$UO_2 + 2Fe^{3+} \rightarrow UO_2^{2+} + 2Fe^{2+}$
		$UO_2 + 0.5O_2 + 2H^+ \rightarrow UO_2^{2+} + H_2O$
Ferrous iron Fe^{2+}	Oxidation	$4Fe^{2+} + O_2 + 4H^+ \rightarrow 4Fe^{3+} + 2H_2O$
Pyrite FeS_2	Oxidation	$2FeS_2 + 7.5O_2 + H_2O \rightarrow 2Fe^{3+} + 4SO_4^{2-} + 2H^+$

bacteria is to regenerate ferric sulfate solutions by the reoxidation of the ferrous iron thus reduced and by continued oxidation of pyrite. While direct bacterial oxidation of hexavalent uranium (U^{4+}) has been demonstrated with washed cell suspensions (DiSpirito and Tuovinen 1982a,b), the significance of the bacterial participation in the direct leaching is doubtful because the predominant pathway in the presence of iron is via the redox cycle of Fe^{3+}/Fe^{2+}.

The electrochemical mechanism is based on the electrode potential of the participating mineral and the redox potential of the leach solu-

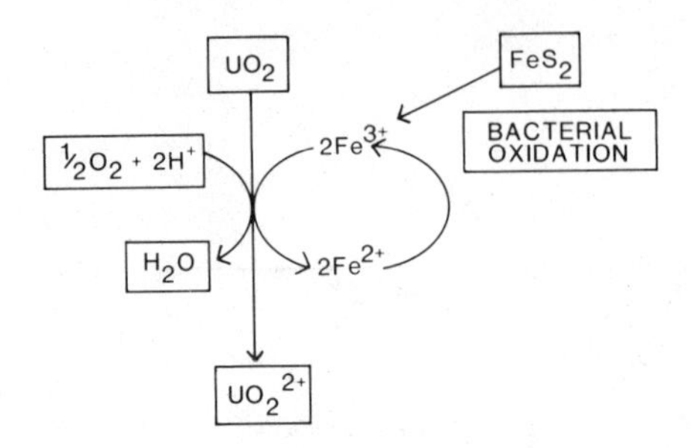

Figure 13.2 Schematic illustration of the bacterial leaching of pyritic uranium ores. UO_2 denotes a mineral with hexavalent uranium (e.g., uraninite or pitchblende). The oxidative leaching driven by molecular oxygen (*left*) in acid solutions is relatively slow compared with ferric iron–mediated oxidation (*right*) in which bacteria assist by renerating the lixiviant.

tion (Wadsworth 1979, 1984). Each mineral has a different electrode potential, and it is generally recognized that the electrochemical series of minerals also ranks their susceptibility to bacterial and chemical leaching. The disintegration of a single crystal is considerably enhanced when another mineral with a higher electrode potential is placed in contact to set up a galvanic couple. The galvanic couple will preferentially promote the oxidative leaching of the mineral with the lower electrode potential, that is, the anodic oxidation. The galvanic coupling effect is schematically illustrated in Fig. 13.3 for $CuFeS_2$ (anodic) and FeS_2 (cathodic). This enhancement of bacterial leaching has been demonstrated with mixtures of chalcopyrite and pyrite (Mehta and Murr 1982; Ahonen et al. 1986). The role of bacteria is to participate at least in the oxidation of Fe and S entities, but detailed studies of the bacterially assisted electrochemical dissolution have not been reported.

Under natural leaching conditions, given the ubiquitous nature of iron-containing sulfides such as pyrite, the three major mechanisms would concurrently contribute to mineral sulfide degradation. The primary substrates available to support the growth of the mixed cultures involved in microbiological leaching processes are, therefore, reduced compounds of iron and sulfur. The biochemical pathways of both iron and sulfur oxidation have been discussed elsewhere (Ingledew 1982, Kelly 1988b) and will only be briefly summarized here.

For commercialization, microbiological leaching processes are in need of improvement of the leaching rates. Kinetic aspects of substrate oxidation are complex and not well understood. Ferrous iron oxidation by growing cultures of *T. ferrooxidans* displays exponential kinetics of substrate utilization (Ahonen and Tuovinen 1989). Chemostat studies have shown that iron oxidation is subject to a dynamic complex of substrate and end-product inhibition (Jones and

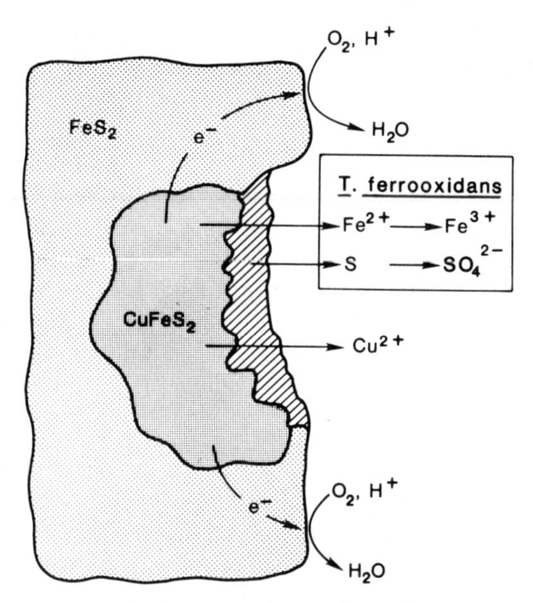

Figure 13.3 Galvanic coupling effect between chalcopyrite (anodic) and pyrite (cathodic). Electron exit is depicted via the pyrite phase to underscore the cathodic and semiconductive nature of the pyrite. Sulfur zone on the exposed surface of chalcopyrite may be a mixture of polysulfides or polythionates and S^0. Thiobacilli in this scheme oxidize ferrous iron to ferric iron and S to sulfate. The direct participation of thiobacilli as final acceptors of the e^- flow from chalcopyrite-pyrite is obscure.

Kelly 1983). Exponential kinetics should also apply to growth with soluble sulfur compounds (thiosulfate, polythionates).

With sulfide minerals the exponential phase of substrate utilization is usually short-lived. Two discreetly different models have been developed in attempts to determine kinetic equations and rate constants. A shrinking-particle model predicts that the leaching rate is directly proportional to the surface area of the mineral. Thus it is apparent that the rate is a function of the particle size of the ore material. A shrinking-core model predicts that the rate is inversely proportional to the extent of the dissolution process and thus to the concentration of the dissolved metal. The shrinking-core model is based on the assumption that a product zone is formed which slows down the fluxes of reactants and products; this problem is coupled with an increasing distance that the leach liquor has to penetrate before it reaches the contact surface. The theoretical discussion on development of the respective rate equations has been presented by Sohn (1979) and Wadsworth (1979). A common lack of knowledge in bacterial leaching systems is the poor understanding and defining of the mixed rate control of heterogeneous reactions, involving diffusion as well as rate-limiting biochemical and chemical steps which have yet to be defined for single-sulfide minerals as well as for natural, heterogeneous sulfide ores. To highlight this problem of mixed rate limitation in heterogeneous reactions, chalcopyrite and other sulfide minerals undergo a passivation effect due to the formation of a sulfur layer, although chemically poorly defined, on the mineral surface (Wadsworth 1984). This zone not only provides a diffusion barrier but also has an adverse effect on charge transfer. The passivation occurs particularly in ferric iron–mediated leaching systems and appears to be difficult to remove even with sulfur-oxidizing thiobacilli.

Sulfur

The biochemical pathway of sulfur oxidation, which is generally accepted to operate relatively uniformly in all of the thiobacilli, involves a complex series of intermediate valence state sulfur compounds (Kelly 1988b). Some sulfur compounds, including those central in the biochemical pathway of thiobacilli, and their respective oxidation reactions are listed in Table 13.4. Among these, sulfite (SO_3^{2-}) represents a central intermediate. Thiosulfate is cleaved to sulfite and polysulfide or sulfur; the reaction is mediated rhodanese activity in acidophilic thiobacilli. The oxidation of thiosulfate to tetrathionate ($S_4O_6^{2-}$), coupled to oxidative phosphorylation, has also been demonstrated in these bacteria (Tuovinen 1977, Lundgren et al. 1986). Not

TABLE 13.4 Reactions Involved in the Biochemical Pathway of the Oxidation of Inorganic Sulfur Compounds by Thiobacilli

Compound	Reaction	Enzyme
Sulfide	$S^{2-} + 3H_2O \rightarrow SO_3^{2-} + 6H^+ + 6e^-$	Sulfite reductase
Sulfur	$S^0 + O_2 + H_2O \rightarrow SO_3^{2-} + 2H^+$	Sulfur oxygenase
Sulfite	$SO_3^{2-} + H_2O \rightarrow SO_4^{2-} + 2H^+ + 2e^-$	Sulfite:cyto *c* oxidoreductase
	$SO_3^{2-} + AMP \rightarrow APS + 2e^-$	APS-reductase
Adenylyl sulfate	$APS + P_i \rightarrow SO_4^{2-} + ADP$	ADP-sulfurylase
	$APS + PP_i \rightarrow SO_4^{2-} + ATP$	ATP-sulfurylase
Thiosulfate	$S_2O_3^{2-} + CN^- \rightarrow SO_3^{2-} + SCN^-$	Thiosulfate sulfur transferase (rhodanese)
	$2S_2O_3^{2-} \rightarrow S_4O_6^{2-} + 2e^-$	Thiosulfate:cyto *c* oxidoreductase
	$S_2O_3^{2-} + 2e^- \rightarrow SO_3^{2-} + S^{2-}$	Thiosulfate reductase
Trithionate	$S_3O_6^{2-} + H_2O \rightarrow SO_4^{2-} + S_2O_3^{2-} + 2H^+$	Trithionate hydrolase

Note: P_i, inorganic *ortho*-phosphate; PP_i, inorganic pyrophosphate.

all the enzymes and electron transport components in the pathway have been characterized, and thus some reactions remain putative at the present time (Kelly 1988b). Half reactions are presented in Table 13.5 for those inorganic S compounds that are most commonly used as substrates for growing cultures of thiobacilli. Compared with the total yield of only $1e^-$ per $1Fe^{2+}$ oxidized, the number of electrons from the oxidation of inorganic sulfur compounds is considerably higher on a molar basis (Table 13.5).

The oxidation of sulfite to sulfate is an energy-linked process involving different reaction steps. Sulfite oxidation by the enzyme sulfite oxidase is linked to the respiratory chain at the level of cytochrome *c*. Alternatively, sulfite oxidation can proceed via the adenylyl phosphosulfate (APS) pathway (APS reductase) which is coupled to ADP sulfurylase and adenylate kinase and thus to energy generation via substrate level phosphorylation. Evidence also exists for the involvement of ATP sulfurylase activity preceded by APS reductase, but its role in conserving energy remains obscure.

A relationship between the rate of leaching of sulfide minerals and the bacterial metabolic system was suggested by Cwalina et al. (1988). The effectiveness of leaching was correlated with the activity of some key enzymes involved in the metabolism of inorganic sulfur compounds, including sulfide oxidase, thiosulfate oxidase, and rhodanese. These results are in keeping with previous demonstrations that strain-related differences in leaching rates of sulfide minerals could be attributed to specific activities of enzymes in the oxidation of iron and sulfur (Groudev 1980).

Iron

High ratios of Fe^{3+}/Fe^{2+} in mine effluents and leach liquors are often indicative of biological activities in acid leaching systems. Other natural mechanisms of iron oxidation in acid (pH 1–2) leach liquors are

TABLE 13.5 Electron Donor Half Reactions in the Oxidation of Inorganic Sulfur Compounds

Electron donor	Half reaction
Sulfide	$S^{2-} + 4H_2O \rightarrow SO_4^{2-} + 8H^+ + 8e^-$
Sulfur	$S^0 + 4H_2O \rightarrow SO_4^{2-} + 8H^+ + 6e^-$
Thiosulfate	$S_2O_3^{2-} + 5H_2O \rightarrow 2SO_4^{2-} + 10H^+ + 8e^-$
Tetrathionate	$S_4O_6^{2-} + 10H_2O \rightarrow 4SO_4^{2-} + 20H^+ + 14e^-$
Trithionate	$S_3O_6^{2-} + 6H_2O \rightarrow 3SO_4^{2-} + 12H^+ + 8e^-$
Pyrite	$FeS_2 + 8H_2O \rightarrow 2SO_4^{2-} + Fe^{2+} + 16H^+ + 14e^-$

kinetically so slow that their contribution to the overall balance of Fe^{3+}/Fe^{2+} is insignificant when contrasted with the bacterial oxidation. The oxidation of ferrous iron by *T. ferrooxidans* has been well documented (Ingledew 1982, Cox and Brand 1984). The half reactions involved in this oxidation are coupled by the electron transport chain in energy-linked reactions.

$$4Fe^{2+} \rightarrow 4Fe^{3+} + 4e^-$$

$$O_2 + 4H^+ + 4e^- \rightarrow 2H_2O$$

$$4Fe^{2+} + O_2 + 4H^+ \rightarrow 4Fe^{3+} + 2H_2O$$

T. ferrooxidans grows under optimum conditions at about pH 2. Iron oxidation activity spans a range of physiological conditions from the acidic environment encompassing the cell wall and periplasm through to the relative neutrality of the cell cytosol where oxygen is reduced. Ferrous ion oxidation external to the cell membrane is favored by high acidity, as this maintains the substrate in its reduced, divalent state and thus available for bacterial oxidation. Under less acidic conditions, as found in the cell cytosol, rapid chemical oxidation of the Fe^{2+} substrate and precipitation of the Fe^{3+} product would occur.

Ingledew (1982, 1986) proposed that electrons are transferred from the outer cell wall via a polynuclear Fe(III) layer to the periplasm. Rusticyanin, the copper protein present at significant levels within the cell (5 percent of total cell protein) was initially proposed as the primary electron acceptor. However, there exists doubt as to whether the rates of reduction of rusticyanin are sufficiently high to account for the observed rates of iron oxidation (Lappin et al. 1985, Blake and Shute 1987). The presence of other potential primary electron acceptors on the reducing side of rusticyanin has been proposed, including an iron-sulfur cluster (Fry et al. 1986, Fukumori et al. 1988) and *c*-type cytochromes (Blake et al. 1988, Sato et al. 1989).

Further characterization of the respiratory chain operating in *T. ferrooxidans* is required. Very little has been published on the composition of the respiratory chains operating in other iron-oxidizing members of the mineral leaching microbial consortium, notably *L. ferrooxidans*. It is likely that microorganisms will differ in electron transfer components and kinetic parameters for a range of substrates, a factor which in part will contribute to the make-up of the mixed cultures involved in the leaching process under a given set of environmental conditions.

Energy generation based on the chemiosmotic coupling has been discussed for acidophilic iron oxidizers (Ingledew 1982, Cox and Brand, 1984). A large gradient of H^+ naturally exists across the cell membrane as a result of the acidic growth environment coupled with the near-neutral conditions present within the cell cytosol. This facilitates the passage of H^+ into the cell via the reversible membrane-bound ATPase, coupled with ATP synthesis. The influx of H^+ must be balanced by removal of H^+ from within the cytosol to ensure that the H^+ concentration does not increase to levels where cytosolic acidification would eventually result in cell death. In the case of *T. ferrooxidans*, the H^+ concentration is controlled by internal proton consumption in the reduction of oxygen at the inner surface of the cell membrane, linked to ferrous iron oxidation. Cytochrome oxidase activity has also been reported to occur on the periplasmic side (Kai et al. 1989). If it is not due to an experimental artifact, then its ability to reduce molecular oxygen would warrant reinvestigation of other mechanisms of ATP formation.

The acidophilic leaching microorganisms are unique in their ability to exploit their harsh environment to their own advantage. The role of H^+ in assisting with mineral sulfide dissociation and therefore substrate provision, coupled with maintaining substrate solubility and availability and contributing to the energetic requirements of the cell exemplifies the unique adaptation of iron-oxidizing thiobacilli in highly selective environments.

Organisms

In studies of the contribution of iron-oxidizing bacteria to mineral sulfide dissolution, most experimental work has focused on the activities of the acidophilic chemolithotroph *T. ferrooxidans*. However, in the natural leaching environment, a diverse group of iron- and sulfur-oxidizing organisms is involved in a series of complex interactions. The ever-changing make-up of the microbial consortium is largely influenced by the prevailing physicochemical environmental conditions which tend to be site-specific. Variations in the physical and chemical characteristics of the mineral, climatic conditions, pH, and the availability of oxygen, carbon dioxide, water, and appropriate nutrients at the reaction site will affect the qualitative and quantitative nature and distribution of the microbial populations that develop. This highly interactive environment is continually changing, leading to a succession of microbial populations (Ralph 1979, 1985; Lundgren and Malouf 1983).

In general, the complex microbial associations that develop during the leaching of sulfide minerals are dominated by acidophiles with iron- and sulfur-oxidizing abilities. These organisms often display tolerance to high levels of acidity and high concentrations of several dissolved metals. The major groups of organisms of interest (Table 13.6)

TABLE 13.6 Partial Listing of Acidophilic Bacteria of Interest in Metal Leaching Processes

Type	Main mode of metabolism	Species designation
Mesophiles	Inorganic carbon and energy source	*Thiobacillus ferrooxidans*, *T. thiooxidans*, *T. prosperus*, *Leptospirillum ferrooxidans*
	Organic carbon and energy source	*Acidiphilium cryptum* and related heterotrophs
	Inorganic or organic carbon and energy source (facultative heterotrophs)	*T. acidophilus*, *T. organovorus*, *T. cuprinus*
Moderate thermo-acidophiles	Inorganic carbon and energy source	*Sulfobacillus thermosulfidooxidans*, *Metallosphaera sedula*, TH strains and several other unnamed isolates from thermal springs and from samples derived from sulfide ore and coal deposits
Extreme thermo-acidophiles	Inorganic carbon and energy source	*Sulfolobus solfaraticus*, *S. acidocaldarius*, *Acidianus brierleyi*, *A. infernus*

are only briefly introduced in this chapter. The taxonomic position of some of the more recently isolated organisms has yet to be fully resolved. For the purpose of the present discussion, the bacteria have been broadly separated, on the basis of temperature response, as mesophiles, moderate thermophiles, and extreme thermophiles.

Mesophiles

Thiobacillus ferrooxidans. By far the most studied organism in biological leaching systems to date, *T. ferrooxidans* is a gram-negative rod-shaped bacterium occurring generally as single rods and occasionally in pairs. While many strains have been isolated, the organism generally displays optimum growth with Fe^{2+} at pH 1.5 to 2.5 and over a temperature range of 28 to 32°C. Psychrotrophic iron-oxidizing thiobacilli have been enriched from mine waters, capable of growth at temperatures as low as 4°C (Ferroni et al. 1986; Ahonen and Tuovinen 1989, 1990). It is not known whether truly psychrophilic thiobacilli exist. *T. ferrooxidans* is a chemolithotroph, deriving its energy for growth from the oxidation of ferrous ion and inorganic sulfur compounds, including sulfides, sulfur, and various S-oxyanions such as thiosulfate and tetrathionate. The carbon metabolism of *T. ferrooxidans* is characteristic of a chemoautotroph, carbon dioxide being assimilated via the Calvin-Benson cycle catalyzed by the enzyme ribulose bisphosphate carboxylase. For nitrogen requirement, the preferred source is ammonium ion. The ability by some strains to fix dinitrogen has been reported (Mackintosh 1978, Stevens et al. 1986). It is generally recognized that the species *T. ferrooxidans* is a heterogeneous taxon and comprises strains of wide genomic diversity, some showing negligible DNA homology with one another (Harrison 1986).

Leptospirillum ferrooxidans. *L. ferrooxidans* was originally isolated from copper deposits in Armenia (Balashova et al. 1974). It is morphologically quite distinct from *T. ferrooxidans* because the cell shape varies from curved rods to spirals. The organisms possess a polar flagellum and generally tend to be more motile than *T. ferrooxidans*. *L. ferrooxidans* is an acidophilic iron-oxidizing bacterium but is incapable of oxidizing sulfur compounds. Unlike *T. ferrooxidans*, no strains of *L. ferrooxidans* have been reported to oxidize sulfur. Thus the ability of the organism to leach certain sulfide minerals, for example, chalcopyrite, is negligible unless it is accompanied in mixed culture by sulfur-oxidizing bacteria (Balashova et al. 1974, Norris 1983). *L. ferrooxidans* has been shown to oxidize pyrite (Norris and Kelly 1982, Merrettig et al. 1989) in spite of its inability to directly oxidize the sulfur entity of this sulfide mineral. Thus, in the absence of thiobacilli, a part of the sulfur entity accumulates as elemental sulfur during pyrite oxidation by *L. ferrooxidans*. In mixed culture with *T. ferrooxidans*, the oxidation proceeds to the level of sulfate (Merrettig et al. 1989).

In mixed batch cultures growing with pyrite, some strains of *L. ferrooxidans* have been shown to gradually dominate over *T. ferrooxidans*. Coupled with this, pyrite oxidation is more extensive when compared to that by *T. ferrooxidans* alone (Norris and Kelly 1982, Norris 1983). Under ferrous iron–limiting conditions in chemostat competition experiments, Norris et al. (1988) demonstrated that *L. ferrooxidans* effectively outcompeted and eventually displaced *T. ferrooxidans*. This phenomenon apparently resulted from differences in kinetic parameters of iron oxidation between the two organisms, *L. ferrooxidans* displaying a higher affinity for ferrous iron. This strain was also less sensitive to the end-product competitive inhibition of ferrous iron oxidation by ferric iron (Norris et al. 1988).

L. ferrooxidans is able to grow at low pH values which are either marginal or prohibitive to *T. ferrooxidans* (Norris 1983). The influence of pH on the microbial make-up of mixed cultures growing with pyrite was demonstrated by Helle and Onken (1988), who showed that the dominance of *L. ferrooxidans* over *T. ferrooxidans* at pH 1.5 was readily reversed by increasing the pH to 2.3. These examples indicate that *L. ferrooxidans* constitutes an important component of microbial communities involved the leaching of sulfide minerals.

Thiobacillus thiooxidans. *T. thiooxidans* shares morphological characteristics with *T. ferrooxidans*. However, this rod-shaped chemolithotroph cannot oxidize ferrous iron, nor can it degrade pyrite or chalcopyrite to any great extent in pure culture. It relies instead on the oxidation of S^0 and soluble sulfur compounds to provide the energy necessary for growth. *T. thiooxidans* has been shown to promote the leaching of chalcopyrite and nonferrous sulfide minerals (Kelly et al. 1979, Lizama and Suzuki 1989). In mixed culture comprising *T. thiooxidans* and *L. ferrooxidans*—two organisms which complement each other to be able to oxidize an iron sulfide—extensive degradation of chalcopyrite has been demonstrated (Norris 1983). Thus the bacterium is an important member of microbial consortia, complementing the activity of the iron oxidizers by oxidizing the sulfur passivation films that may form on mineral particle surfaces, particularly on chalcopyrite, during the leaching process.

As recently discussed by Kelly (1988a), the range of mesophilic, acidophilic bacteria contributing to mineral oxidations under acid conditions is far more extensive than the few major representatives listed above. Clearly the succession of microbial populations developing during the leaching process represents a highly mixed and interactive population at the mesophilic level, extending its activities to suboptimal temperatures that may continually or seasonally prevail in various parts of leach mines.

Moderate thermophiles

The moderately thermophilic, acidophilic iron- and sulfur-oxidizing bacteria represent a diverse group of organisms which, broadly speaking, display optimum growth temperatures of about 50°C. While several isolates have been well studied over a number of years, there has yet to be a formal taxonomic classification within this group.

The isolation of a number of moderate thermophiles, including strains TH1, ALV, and BC1, has been discussed by Norris and Barr (1985). Subsequent studies have shown that TH1-like organisms appear to be widespread. They have since been isolated from a number of geothermal environments as well as mine sites. A moderate thermoacidophile which was isolated from a pyritic deposit in western Australia showed close DNA:DNA homology with TH1, further extending the already broad distribution of this group (Holden et al. 1988). In addition, strain BC1 appears to be closely related to strain TH1 based on comparative whole-cell electrophoretic protein patterns (Norris and Barr 1985). The close DNA:DNA homology between the two strains indicates that BC1 represents another isolate of the same species.

Another moderate thermophile, *Sulfobacillus thermosulfidooxidans*, has been described which is capable of oxidizing iron, sulfur, and sulfide minerals (Golovacheva and Karavaiko 1978, Karavaiko 1988). It is distinctly different from the TH1 organisms on the basis of motility and its ability to form spores. The genus *Sulfobacillus* is gram-positive. Based primarily on spore formation, variants have been described which have been named as subspecies to *S. thermosulfidooxidans* (Vartanyan et al. 1988).

The morphology, mineral oxidizing capacity, and further characterization of the TH1, BC1, ALV, and other moderately thermophilic bacteria have been discussed by Norris and Barr (1985) and Norris et al. (1986a). Strain TH1 grows chemolithoheterophically on iron in the presence of yeast extract (Norris and Barr 1985). Strain ALV, on the other hand, displays no requirement of a reduced sulfur source when growing autotrophically on ferrous iron due to its ability to utilize sulfate as a sole source of sulfur (Norris and Barr 1985). All strains are capable of growing autotrophically.

The oxidation of mineral substrates by moderate thermophiles has been discussed by Norris et al. (1986a). In general, these bacteria can oxidize a range of commercially important sulfide minerals. Because the leaching rates are faster than those measured with mesophilic cultures, these organisms may avail themselves to tank leaching processes for treatment of sulfide concentrates. However, optimization studies are yet to be reported that contrast the operational parameters and cost factors between mesophiles and moderately thermophilic bacteria.

A number of strains of moderately thermophilic sulfur-oxidizing acidophilic bacteria have also been reported. These organisms lack the capacity to oxidize iron. Whilst the type strain BC13 (Norris et al. 1986b) is a gram-negative organism and similar in some respects to *T. thiooxidans*, its thermophilic and genomic characteristics necessitate its classification as a new species. Sulfur oxidation by strain BC13 is fast compared with the moderately thermophilic counterparts that can oxidize both iron and sulfur (Norris et al. 1986b). The taxonomic position of the moderately thermophilic acidophilic bacteria needs to be formally addressed.

As a group, both the iron and sulfur oxidizers undoubtedly play an important role in microbial communities oxidizing mineral sulfides at temperatures above the ambient surface temperature. It is not uncommon that elevated temperature zones and microcosms develop inside leach dumps and heaps, depending on geology, content of sulfide minerals, and heat transfer characteristics of the given deposits. It is thus to be expected that moderate thermoacidophiles are established in deeper, subsurface layers of heaps and dumps where temperature stratification provides a suitable thermal setting for their competitive advantage.

Extreme thermophiles

Brock et al. (1972) were the first to report *Sulfolobus* as a new genus of sulfur-oxidizing bacteria. Initially classified as *S. acidocaldarius*, these mainly spherical organisms were shown to be facultative autotrophs, capable of growing with sulfur or simple organic compounds at 55 to 80°C under acidic conditions. Subsequent studies have shown that *Sulfolobus* spp. are effective in leaching sulfide mineral substrates, including molybdenite and chalcopyrite-containing ore materials (Brierley and Brierley 1986) as well as pyrite and arsenopyrite concentrates (Lawrence and Marchant 1988, Lindström and Gunneriusson 1990). The interest in this group of organisms eventually led to the recognition of a diversity of types. Subsequently the genus *Sulfolobus* was divided into three distinct species: *S. acidocaldarius, S. solfataricus*, and *S. brierleyi* (Zillig et al. 1980). Another species, *S. ambivalens*, was described by Zillig et al. (1985), again from a thermal spring and later renamed as *Desulfurolobus ambivalens* (Zillig et al. 1986). Thus, all original strains *Sulfolobus* strains had been isolated from either hot springs and thermal acid soil or sediment samples. *Sulfolobus* spp. have since been isolated from a coal heap in the United Kingdom (Marsh and Norris 1983a). It still remains unclear whether these types of organisms are commonly present in environments altered by mining of coal and mineral resources. If present, these environments must provide thermally elevated habitats because the extreme thermophiles are inactivated at the mesophilic range of temperatures.

The metabolic diversity of the extremely thermophilic iron and sulfur oxidizers was further demonstrated in growth studies in which *S. brierleyi*, subsequently classified as *Acidianus brierleyi*, was shown to be capable of (1) anaerobic growth coupled with the reduction of elemental sulfur, and (2) aerobic growth coupled with the oxidation of sulfur (Segerer et al. 1985, 1986). Growth of *D. ambivalens* (*S. ambivalens*) under anaerobic conditions by reducing sulfur with hydrogen was also reported by Zillig et al. (1985).

In general, the extreme thermophiles of the *Sulfolobus* type have an optimum growth temperature on mineral sulfides of about 65 to 70°C, although there is evidence that other isolates are capable of growth at even higher temperatures (Stetter 1986, Kelly and Deming 1988). The metabolic versatility extends from iron- and sulfur-oxidizing isolates to organisms capable of oxidizing sulfur only. All extreme thermoacidophiles of interest in mineral biotechnology are archaebacteria, including *Metallosphaera sedula*, which was recently isolated from a geothermal site and shown to leach sulfide ore samples with an optimum temperature of 75°C (Huber et al. 1989).

These facultative autotrophs are of significant commercial interest, given their ability to effectively degrade recalcitrant sulfide minerals (e.g., chalcopyrite) at high temperatures. It is particularly the rapid leaching rates observed at elevated temperatures (Marsh et al. 1983b, LeRoux and Wakerley 1988) that appear attractive for commercial application because they are exceedingly superior to those obtained with the mesophilic organisms at ambient temperatures. The oxidative leaching reactions are exothermic and may involve a considerable amount of heat evolution during the active phase of the process. This problem necessitates the installation of a cooling system when mesophilic organisms are used in stirred-tank leaching processes. Heat evolution has not been reported to reach levels prohibitive to *Sulfolobus* spp. Because these thermophiles may undergo irreversible thermal inactivation at ambient (mesophilic) temperatures, temperature control appears warranted also for thermophilic leaching. Strictly comparable studies of mesophilic and thermophilic leaching which would also address heat and mass transfer and material balance of the system as well as problems caused by changes in the solubility of reactants (e.g., O_2 and CO_2) and products (e.g., complexes of ferric iron) are yet to be reported.

Mixed Cultures

It is generally recognized that mixed cultures offer several advantages over pure cultures in mineral leaching processes. A range of heterotrophic bacteria, fungi, and yeasts are acidophilic and persist in leaching environments. However, their contribution to the leaching process remains poorly understood. Indeed, there have been relatively few systematic investigations of the benefits arising from the use of mixed cultures during mineral sulfide oxidation. It is now generally recognized that cultures of acidophilic thiobacilli commonly carry heterotrophic counterparts. The resolution of the various members to pure cultures is tedious and difficult because the heterotrophic satellites readily survive in association with *T. ferrooxidans* colonies, and vice versa (Johnson et al. 1987).

A role for the heterotrophic bacteria in the leaching process was indicated by the increase in the rate of leaching of a copper-nickel sulfide concentrate by *T. ferrooxidans* in the presence of *Beijerinckia lacticogenes* (Tsuchiya et al. 1974, Tsuchiya 1977). Although thought to result from microbial mutualistic associations involving the removal of inhibitory organic compounds and the fixation of atmospheric nitrogen by the heterotroph, independent duplication of these results has not been reported and exact mechanism remains obscure.

Acid mine waters and leach liquors constitute a diverse source of facultatively and obligately heterotrophic acidophiles. Facultatively heterotrophic *T. cuprinus* isolates have been described which can oxidize sulfide minerals and sulfur at pH 3 to 4 (Huber and Stetter 1990), apparently without the involvement of the Fe^{3+}/Fe^{2+} redox couple. These bacteria can utilize a range of organic compounds for energy and carbon. Other thiobacilli, e.g., *T. acidophilus* and *T. organovorus*, have also been described which can be maintained either in sulfur or heterotrophic media at low pH values.

It is probable that acidophilic heterotrophs play an important role in mixed cultures during mineral sulfide leaching. Heterotrophs such as *Acidiphilium* spp. have been shown to enhance the growth of *T. ferrooxidans* when present in mixed culture (Harrison 1984, Wichlacz and Thompson 1988). Indeed, the leaching of cobalt sulfide sample by *T. ferrooxidans* was enhanced in the presence of the acidophilic heterotroph, provided that ferrous iron was present as substrate for the autotroph (Wichlacz and Thompson 1988). In the absence of ferrous iron, however, the presence of the heterotroph led to a reduction in cobalt sulfide leaching. Acidophilic heterotrophs, present in varying levels in cultures of *T. ferrooxidans* (Harrison 1981, Johnson et al. 1987), are presumed to survive by scavenging reduced organic compounds excreted by *T. ferrooxidans*; thus the level of organic compounds is maintained at noninhibitory levels to the autotroph.

T. ferrooxidans is sensitive to many organic acids. This may relate to the large transmembrane pH gradient operating in organisms with an internal pH near neutrality and growing in acidic environments. Such a pH gradient may permit the accumulation of weak acids in to the cell cytosol (Matin 1978, Ingledew 1982, Alexander et al. 1987), resulting in cytosolic acidification and cell death. The scavenging role of the heterotrophs may therefore be invaluable for the survival of organisms such as *T. ferrooxidans*. Although not validated experimentally, it is possible that the heterotrophic oxidation of organic compounds may also lead to an increase in the level of CO_2 available for fixation by *T. ferrooxidans* (Wichlacz and Thompson 1988).

The role of the heterotrophic organisms in the mixed culture may extend to other areas in addition to the provision or utilization of organic metabolites. Mineral sulfide leaching may be enhanced by some vitamins, cofactors, chelating agents, and surfactants either added directly to the leaching system or produced by the heterotrophic population. On the other hand, the heterotrophs may also adversely affect the leaching rate through the production of inhibitors or by physically restricting fluxes of reactants and products on mineral surfaces. It must also be remembered that the heterotrophic organisms compete with the autotrophic organisms for key inorganic nutrients and, most importantly, for dissolved oxygen.

Johnson et al. (1990) described several isolates of heterotrophic organisms from abandoned mine waters which were able to utilize glucose and reduce ferric iron when growing on solidified, mixed-substrate media. Several other novel types of organisms were found from the same sources, including (1) autotrophic iron oxidizers that were differentiated from *T. ferrooxidans* on the basis of their filamentous shape, and (2) iron-oxidizing heterotrophs with characteristically filamentous cell shape. Mixed cultures were described which recycled iron between the oxidized and reduced states. In the absence of supplementary organic carbon, pyrite oxidation by a *Leptospirillum*-type chemolithotrophic isolate was considerably enhanced in the presence of a heterotroph, presumably due to scavenging of inhibitory organics produced by the autotrophic *Leptospirillum*. In the presence of glucose, the same mixed culture displayed reduced oxidation of pyrite, presumed to result from heterotrophic iron reduction which was counterproductive to ferric-iron mediated pyrite oxidation by *Leptospirillum* (Johnson et al. 1990).

In addition to the role of heterotrophs in mixed cultures, microbial growth on mineral sulfides also involves mixed populations of autotrophs. Pyrite was shown to be more extensively degraded by mixed cultures dominated by *L. ferrooxidans* (Norris and Kelly 1982, Norris 1983) when compared to *T. ferrooxidans* alone. The ability of *L. ferrooxidans* to outcompete *T. ferrooxidans* at low ferrous iron concentrations is coupled with the greater tolerance of *L. ferrooxidans* to increasing levels of acidity. *L. ferrooxidans* therefore represents an important component of the mixed culture, as the presence of this organism extends the range of conditions under which pyrite can be degraded to much lower pH values than can be tolerated by *T. ferrooxidans* (Norris 1983, Helle and Onken 1988). No strains of *L. ferrooxidans* have been shown to oxidize sulfur. Therefore, in mixed cultures the cooperative degradation of mineral sulfides by *L. ferrooxidans* and *T. ferrooxidans*, together with a range of sulfur-oxidizing acidophiles such as *T. thiooxidans* occurs. Enhanced rates of chalcopyrite and sphalerite leaching by a defined mixed culture of *T. ferrooxidans* and *T. thiooxidans*, compared with the respective rates in single cultures, were demonstrated in column leaching studies by Lizama and Suzuki (1989). In order to leach samples of chalcopyrite concentrates, *L. ferrooxidans* required the presence of sulfur-oxidizing bacteria in mixed culture (Balashova et al. 1974, Norris 1983). Similarly, the role of the moderately thermophilic sulfur oxidizers, as typified by strain BC13, in sulfide mineral oxidation is related to cooperative effects.

Microbial communities in leach mines display considerable strain variation within species such as *T. ferrooxidans*. Strains may greatly differ with respect to several phenotypic traits, such as the specific enzyme activities, diazotrophy, resistance to metal ions and pH, to mention a few. Genotypic differences in terms of plasmid profiles and phenotypic differences in tolerance to toxic metals have been demonstrated among strains of *T. ferrooxidans* isolated from a leach mine (Tuovinen et al. 1980). Strain variation among *Thiobacillus, Leptospirillum*, and thermoacidophilic species may well extend to their iron- and sulfur-oxidizing capabilities.

It is therefore apparent that the mineral sulfide leaching process involves a highly mixed and diverse group of organisms. In a natural leaching system, this community is continually changing in response to physicochemical changes in its environment. In fact, quite different strains may be isolated even from a single small lump of ore. The presence of mixed populations is in many ways beneficial to the leaching process, as it greatly extends the range of microbial metabolic capabilities required for effective oxidation of sulfide minerals. However, the complex interactions that occur in such environments remain to be elucidated.

Environmental Parameters

In both the natural and the industrial leaching process environments, the oxidative leaching of sulfide minerals is controlled by a complex interaction of physicochemical parameters. Tailings dams, dump leaching operations, and waste rock dumps represent extremely variable environments for the microbiological oxidation of sulfide minerals and ferrous iron to take place. The diverse nature of the parameters that make up the leaching environment have been previously discussed at more length (Ralph 1979, 1985; Lundgren and Malouf 1983; Karavaiko 1985).

The global environment is subjected to climatic conditions, notably the effects of rainfall and temperature. The geology and mineralogical composition of the ore, concentrate, or waste material are also critical. The nature and abundance of the sulfide mineralization vary considerably, generally reflecting the complex nature of the ore body. The ore may also contain metals such as arsenic, silver, and molybdenum which, once solubilized, may prove toxic to members of the microbial consortium. At the reaction site, the activity of the mixed population is further dependent on the provision of oxidizable substrate, oxygen, carbon dioxide, water, and appropriate inorganic nutrient ions. Similarly, the pH is a critical factor, not only because it sets the limits of microbial activity but also because the solubility of various metal species is pH-dependent.

pH

The presence of acid-consuming gangue elements such as silicates and carbonates affects both bacterial establishment and the rate of sulfide mineral oxidation. Excessive acid consumption by alkaline materials (e.g., calcite, dolomite) may elevate the pH beyond the physiological requirements for the major members of the microbiological leaching consortia. Calcium as well as magnesium are often present in excessive concentrations because of the dissolution of alkaline materials. Calcium in sulfate environments readily forms poorly soluble $CaSO_4$ complexes (gypsum) which coat mineral surfaces and thus amplify diffusion-related problems.

$$CaCO_3 + 2H^+ + SO_4^{2-} + H_2O \rightarrow CaSO_4 \cdot 2H_2O + CO_2$$

Magnesium is problematic in metal recovery by solvent extraction. The dissolution of silicates is undesired because silicates form altered, colloidal complexes which adversely influence the permeability and interfacial characteristics of ore material, as presented for K-feldspar with the following equations.

$$3KAlSi_3O_8 + 2H^+ + 12H_2O \rightarrow KAl_3Si_3O_{10}(OH)_2 + 2K^+ + 6H_4SiO_4$$

$$2KAl_3Si_3O_{10}(OH)_2 + 2H^+ \rightarrow 3Al_2Si_2O_5(OH)_4 + 2K^+$$

Poorly soluble products establish barriers that may physically restrict the access of gases, solutions, and microorganisms to reaction sites and may lead to a build-up of inhibitory levels of reaction products.

The eventual establishment of a favorable leaching environment depends on the development of a microbial succession. Microbial communities vary quantitatively and qualitatively as a function of the changing environment, particularly following the initial colonization of sulfidic materials. The ability to produce acid conditions will partly depend on activities of the S- and Fe-oxidizing bacteria. In contrast to sulfuric acid production when sulfur is oxidized to sulfate, ferrous iron oxidation is an acid-consuming reaction (1 H^+ consumed per 1 Fe^{2+} oxidized) but the subsequent hydrolysis of Fe^{3+} to various aquo-coordinated complexes is a series of acid-producing reactions, as shown below:

$$2Fe^{3+} + 2H_2O \rightarrow 2FeOH^{2+} + 2H^+$$

$$2FeOH^{2+} + 2H_2O \rightarrow 2Fe(OH)_2{}^+ + 2H^+$$

$$2Fe(OH)_2{}^+ + 2H_2O \rightarrow 2Fe(OH)_3 + 2H^+$$

Hydroxyspecies of Fe(III) can undergo several solution-phase reactions to form a range of hydroxysulfate and oxyhydroxide complexes. The formation of an ordered structure such as jarosite is typical in these solutions. These reactions also influence the pH.

$$2Fe_2(SO_4)_3 + 6H_2O \rightarrow 2Fe(OH)_3 + 3H_2SO_4$$

$$3Fe^{3+} + K^+ + 2SO_4^{2-} + 6H_2O \rightarrow KFe_3(SO_4)_2(OH)_6 + 6H^+$$

Not all sulfide minerals produce acid upon oxidation. As for Fe^{2+}, the oxidative dissolution of pyrrhotite and various nonferrous monosulfides such as ZnS, Cu_2S, and NiS constitutes net acid-consuming reactions. The major acid-producing sulfide mineral is pyrite.

Mixed populations are very important in initially establishing a suitable setting for the microbiological leaching. The sulfur-oxidizing capacity of organisms such as *T. ferrooxidans* and *T. thiooxidans* provides for sulfuric acid production and therefore conditions acidic enough to promote substrate dissolution and iron-oxidation. *T. ferrooxidans* has an optimum pH range for growth of 1.5 to 2.5, although the lower limit for growth during sulfide mineral degradation may be as low as pH 1.2. *T. thiooxidans* tolerates pH values in the range of 0.5 to 1.0. In this context, *L. ferrooxidans* should be recognized as an important iron-oxidizing partner capable of growing at pH values that may be prohibitively low to *T. ferrooxidans*.

The pH is also an important factor in influencing the solubility of substrates for bacterial oxidation. Prior to oxidation, the substrates must be rendered soluble by a chemical dissociation from the parent mineral. A number of factors contribute to this, including high solubility products of the metal sulfides, an excess of holes in the sulfide valence band, and electron extraction by ferric iron. Proton-catalyzed reactions also break chemical bonds in the sulfide valence band, inducing a shift in electronic states favoring interaction with the bacterial metabolic system (Tributsch and Bennett 1981a,b).

In this context it is interesting to note the description of a recent *Thiobacillus* isolate, *T. prosperus*, from a geothermally influenced marine environment (Huber and Stetter 1989). This bacterium grows with ferrous iron, elemental sulfur, or sulfide minerals at pH range 1 to 4 and is only partially inhibited by NaCl at concentrations up to 35 g/L. In contrast, *T. ferrooxidans* is subject to strong inhibition by 10 g NaCl per liter. The use of high-chloride environments in bacterial leaching processes warrants further experimental evaluation because ferric chloride is an efficient oxidant. A bacterial leaching process in a mixed sulfate-chloride environment might be particularly amenable in areas with saline groundwater or those in proximity of coastal regions but with limited freshwater supplies. Chloride ion would alleviate some precipitation problems owing to the increased solubility of metals as chloride species. However, acid chloride solutions will also enhance the corrosiveness of the leach solutions.

The rate of the chemical oxidation of ferrous ion is pH-dependent, being practically negligible at pH values optimal for *T. ferrooxidans* and *L. ferrooxidans*. At pH values in excess of 3, the chemical oxidation of ferrous ion becomes increasingly rapid (Ackermann and Kleinmann 1984), effectively reducing the level of substrate available for biological oxidation. Coupled with this, the low solubility of the ferric iron product at a pH above 2.5 would rule out the important role of ferric ion as a chemical leaching species.

Temperature

Just as a microbial succession is established as a function of the prevailing pH, temperature is also an important determinant contributing to the continuing development of microbial mixed populations. For the lower temperature scale supporting microbiological activity, Ferroni et al. (1986) reported the isolation of psychrotrophic *T. ferrooxidans* strains which grew over the range 12 to 25°C with a minimum growth temperature of 2°C. Similarly, Chashchina and Kukharchuk (1988) tentatively reported on active iron oxidation at 8°C by enrichment cultures from a Pb-Zn sulfide deposit in Siberia; a culture was shown to readily oxidize ferrous iron at 8°C, whereas the same inoculum displayed an extensive (>30 days) lag period before growth commenced at 28°C. Arrhenius plots of rate constants for Fe^{2+} and S^0 oxidation, tested with growing cultures originally enriched from a copper mine composite water sample, displayed linearity between 4 and 28°C (Ahonen and Tuovinen 1989, 1990). The temperature coefficient Q_{10} for iron and sulfur oxidation was about 2. The activation energies were approximately 80 kJ/mol for Fe^{2+} and 65 kJ/mol for S^0 oxidation, suggesting that the rate of substrate oxidation was limited by a chemical or biochemical reaction rather than diffusion under the experimental conditions.

Suboptimal temperatures have been encountered in many underground leaching situations, in spite of the fact that the oxidations of sulfide minerals are exothermic reactions. In some leaching environments such as tailings dams and dump operations, heat dissipation may be minimized by engineering constraints and the low conductivity of the mining materials (Ralph 1979, Lundgren and Malouf 1983). Limited dissipation may lead to elevated temperature zones, prohibitive even to thermophiles, with chemical oxidations prevailing and with the possibility of spontaneous combustion in the case of fine-grained sulfide minerals.

Moderately elevated temperatures within the range 50 to 60°C have been recorded in waste rock dumps associated with uranium (Harries and Ritchie 1981) and copper (Murr and Berry 1979, Murr and Brierley 1978) operations. Such temperatures are well in excess of the upper growth limits of about 40°C for the mesophilic thiobacilli. Indeed, moderate thermoacidophiles appear to be readily found in a variety of leaching environments as well as geothermally influenced settings. These include copper leach dumps (Brierley 1978), a large-scale experimental copper waste leaching facility (Murr and Brierley 1978), coal spoil heaps (Marsh and Norris 1983a,b), hot springs (LeRoux et al. 1977, Marsh and Norris 1983a), and pyrite tailings with spontaneous combustion of interior layers (Holden et al. 1988). There is little doubt, therefore, that mixed cultures of moderately thermophilic, acidophilic bacteria are present under appropriate environmental conditions within many leaching operations.

The extent of colonization of leaching and waste rock dumps by extreme thermophiles is far less clear. Temperatures in excess of 60°C favoring extreme thermophiles have been reported in a number of mining operations. Temperatures in excess of 80°C have been observed in low-grade copper dumps in Bulgaria (Groudev et al. 1978) and in the United States (Beck 1967), but the presence of extreme thermoacidophiles in these operations has not been reported. The first demonstration of extreme thermoacidophiles from a mine site was reported in the United Kingdom where *Sulfolobus* spp. were isolated from a coal heap (Marsh and Norris 1983a). Most other reported isolations of extreme thermophiles of the genus *Sulfolobus* have been made from samples of acid hot springs. Future studies are required to ascertain the presence and role of such extreme thermoacidophiles in waste rock dumps and leaching operations.

The environment within large-scale waste rock dumps is extremely variable and is therefore inducive to the presence of highly mixed populations. Apart from the qualitative and quantitative differences in sulfide and gangue components, physical parameters such as temperature and oxygen vary widely throughout the dump. This has been elegantly demonstrated in an ongoing monitoring project of temperature and oxygen profiles in a uranium waste rock dump at Rum Jungle (Northern Territory, Australia) (Harries and Ritchie 1986, Harries et al. 1988). The diverse range of temperatures observed within the dump at any time would provide appropriate settings for a spectrum of mesophilic, moderately thermophilic, and extremely thermophilic iron- and sulfur-oxidizing organisms. Changes in temperature within the dump over longer time periods are bound to also lead to major population shifts. The gradual decrease in temperature observed in some zones of the Rum Jungle waste dump would eventually favor proliferation of mesophilic organisms. It is apparent that the contribution of mixed cultures in mineral leaching operations should be viewed not only in terms of population variability at any one time, but also in terms of population dynamics over considerably longer time periods.

Nitrogen

There is no clear large-scale demonstration whether the supplementation of leach liquors with ammonia-N in leaching operations enhances microbial activities. Natural, but relatively low-level sources of inorganic nitrogen (ammonium, nitrate) are process water supply, rainwater, and leaching from surrounding soils. In all leach mines there is a background residual level of ammonium and nitrate as a result of use of explosives for blasting and fracturing of the ore material. Ammonium incorporates into basic ferric sulfates:

$$NH_4{}^+ + 3Fe^{3+} + 2SO_4^{2-} + 6H_2O \rightarrow NH_4Fe_3(SO_4)_2(OH)_6 + 6H^+$$

$$NH_4{}^+ + 3Fe^{3+} + 2HSO_4^{2-} + 6H_2O \rightarrow NH_4Fe_3(SO_4)_2(OH)_6 + 8H^+$$

The reaction is dependent on the pH and ionic composition of the bulk solution. The formation of ammoniojarosite is undesired because it contributes to the loss of available nitrogen and may create diffusion barriers of interfacial fluxes. Similarly, phosphate also forms poorly soluble complexes with iron (Hoffmann et al. 1985), which may be problematic in leaching situations:

$$Fe^{3+} + H_2PO_4{}^- \rightarrow FePO_4 + 2H^+$$

$$FePO_4 + Fe^{2+} \rightarrow FePO_4Fe^{2+}$$

$$FePO_4 + FeH_3PO_4{}^0 \rightarrow FePO_4FeH_2PO_4{}^0$$

Particularly in reactor leaching of sulfide concentrates, the provision of mineral nutrients is important because there is no extraneous source of phosphate or nitrogen besides the water and atmospheric NH_3 and N_2. The ability of some strains of *T. ferrooxidans* to fix atmospheric dinitrogen has been substantiated at the molecular level by the isolation and sequencing of the nitrogenase structural genes (*nif* HDK) from *T. ferrooxidans* (Pretorius et al. 1986, 1987). Moderately thermophilic iron-oxidizing bacteria of the TH1/BC group, designated as NMW-6 have been found to have DNA sequences which hybridize with the *nif* HDK genes of *T. ferrooxidans* (Holden et al. 1988). Nitrogenase activity, however, could not be demonstrated by the acetylene reduction method in the thermophile (Holden 1989).

Oxygen and carbon dioxide

Generally speaking, the organisms active in mineral leaching processes are aerobic and grow autotrophically; i.e., these organisms are dependent on O_2 as an electron acceptor and CO_2 as a carbon source. A continuous supply of oxygen and carbon dioxide is therefore essential. Acevedo and Gentina (1989) pointed out that growth of these acidophiles requires a balance of CO_2 and O_2 which is inadequately provided by the CO_2/O_2 ratio of ca. 0.0015 in air. Therefore, carbon dioxide limitation is to be expected under ambient conditions.

Gas transport from the surface of a dump to the site of oxidation at depth occurs by diffusion, convection, and advection (Jaynes et al. 1983), the relative contribution of these mechanisms being dependent on a number of parameters as discussed by Harries and Ritchie (1985). The transport of oxygen to the sites of oxidation may be the rate-limiting process in some dump leaching operations. Low levels of oxygen have been found in pore spaces associated with a reclaimed coal strip mine (Jaynes et al. 1983) and waste rock dumps (Harries and Ritchie 1985).

The concentration of oxygen greatly influences the nature of the microbial population present. The acidophilic species *T. ferrooxidans* and *T. thiooxidans* are obligate aerobes and require oxygen for growth to occur. Low concentrations of oxygen will impose constraints on the rate of substrate oxidation. *T. ferrooxidans* has been shown to couple the oxidation of sulfur to ferric ion reduction under anaerobic conditions (Brock and Gustafson 1976, Sugio et al. 1985), but this anaerobic oxidation-reduction activity is not coupled to growth. Goodman et al. (1983) reported on an experiment in which *T. ferrooxidans* solubilized sulfide minerals under apparently anaerobic conditions, but the mechanism of this leaching has not been resolved.

Microaerophilic conditions are required for nitrogen fixation by *T. ferrooxidans* (Mackintosh 1978, Stevens and Tuovinen 1986, Stevens et al. 1986). The oxygen sensitivity of the nitrogenase system on one hand and the requirement of oxygen as the terminal electron acceptor for iron oxidation coupled with energy transduction and growth on the other hand are in apparent contradiction to each other. The bacterium does not possess specialized structures that would protect against oxygen. A possible way for the bacterium to overcome these opposing requirements may be by rapid rate of respiration and oxygen uptake. This would necessitate a high content of electron transport components in *T. ferrooxidans*, which is in agreement with reports that rusticyanin makes up to 5 percent of cellular protein.

In a mixed culture, in its natural environment at a leach mine, there is competition for available oxygen as the electron acceptor, not only among the microorganisms but also against concurrent chemical reactions. While a range of mesophilic organisms may be present in dump operations that may be restricted by the concentration of available oxygen, it is envisaged that organisms are present that are capable of metabolic activity under microaerophilic or anaerobic conditions, although such activity may not necessarily be growth-related. Denitrifying thiobacilli (*T. denitrificans*) have been found in water samples at some leach mines but their significance is questionable. Anaerobic oxidation of sulfur coupled with nitrate may have little relevance in leach mines because nitrate is in limited supply and the denitrifying S oxidizers are restricted to near-neutral environments. It is characteristic to sulfide ore materials that they deplete electron acceptors (Fe^{3+}, O_2) also in the absence of biological catalysts. For example, monosulfide minerals undergo nonoxidative dissolution which results in the formation of H_2S which is noted for its oxygen-scavenging property.

Many sites within dump leaching operations contain elevated levels of carbon dioxide (Jaynes et al. 1983; Harries and Ritchie 1983, 1985), thus satisfying the inorganic carbon requirements of these organisms. Elevated levels of carbon dioxide typically result from the chemical interaction of carbonate-containing gangue with acid leach solutions (Jaynes et al. 1983).

At elevated temperatures, such as those favoring moderate and extreme thermoacidophiles, the solubility of oxygen, carbon dioxide, and other gases is reduced. The moderately thermophilic iron-oxidizing bacteria, unlike their mesophilic and extremely thermophilic counterparts, are relatively inefficient in their ability to assimilate carbon dioxide and require CO_2 enrichment for optimum growth. Microbial communities at leach mines would also be likely to contain sulfur-oxidizing moderate thermophiles which can assimilate carbon dioxide with similar efficiency to that of the mesophilic *T. ferrooxidans*.

The metabolic diversity of the acidophilic chemolithotrophs is further evidenced by the demonstration that *S. brierleyi* (*Acidianus brierleyi*) grows either anaerobically by reducing molecular sulfur or aerobically by oxidizing sulfur (Segerer et al. 1985). These findings were further supported by the observations of Zillig et al. (1985). While *Sulfolobus* relies on a different biochemical pathway for carbon dioxide assimilation compared with its lower-temperature leaching counterparts (Kandler and Stetter 1981, Norris et al. 1989), the efficiency of carbon dioxide reduction by the bacterium more closely resembles that of the mesophilic *T. ferrooxidans*.

Toxic metals

The complex microbial associations which develop on minerals are dominated by organisms with sulfur- and iron-oxidizing abilities which tolerate high levels of acidity and dissolved metals. These organisms are subject to inhibition by a range of soluble metal cations or oxyanions (Norris 1989). Metal toxicity varies between different strains of the same species and as a function of previous exposure to the metal in question, pH, growth substrate, stage of growth, chemical composition of growth medium, and experimental design (Norris 1989).

Generally speaking, *T. ferrooxidans*, moderately thermophilic iron-oxidizing bacteria, and *Sulfolobus* spp. are similarly affected by uranium, copper, and aluminum (Norris et al. 1986b). Silver and mercury tend to be the most toxic metals to these organisms. However, in sulfide mineral leaching, the solubility of silver is extremely limited (usually <1 mg/L) because of the formation of poorly soluble complexes (e.g., Ag_2S). Therefore, metal toxicity testing in ferrous sulfate media may generate entirely different inhibition profiles as compared with bacterial responses in sulfide mineral–containing media. In terms of leaching operations, uranium is an important target metal. Compared with *T. ferrooxidans, L. ferrooxidans* tends to display a higher level of tolerance to uranium. However, *T. ferrooxidans* can be adapted to higher concentrations of uranium.

In the case of another industrially important metal, copper, *L. ferrooxidans* is clearly far more sensitive to copper when compared with *T. ferrooxidans* (Norris et al. 1986b, Norris 1989). This may have important implications in the make-up of the mixed population during copper dump leaching operations, particularly if the pH falls below pH 1.5 where *L. ferrooxidans* would be expected to dominate (Norris 1983, Helle and Onken, 1988). However, it is extremely unlikely that rate-limiting steps in leach mines are related to problems of metal toxicity. In concentrate leaching in stirred tank reactors, tolerance may constitute a problem because the concentrations of toxic metals attain considerably higher levels. The toxicity of arsenic is of particular interest in view of the rapidly approaching full commercial status of reactor leaching of refractory gold-containing arsenopyrite flotation concentrates. In general, the arsenite species is more toxic than the arsenate species to microorganisms. The construction of arsenic-resistant *T. ferrooxidans* recombinant plasmids has been reported (Rawlings et al. 1984) as a first step in the development of arsenic-resistant strains. The introduction of plasmids specifying resistance to metal ions such as silver, mercury, and cadmium may also be considered in the long term, although the ability of recombinant organisms to survive and compete effectively in large-scale leaching operations remains questionable. At present, one would also be faced with environmental regulatory requirements concerning the deliberate release and on-site testing of genetically modified thiobacilli at leach mines.

The choice of organisms for a leaching operation may in part be based on the selection of appropriately resistant microorganisms. As an example, the iron-oxidizing thiobacilli show a low tolerance to molybdenum (Tuovinen et al. 1971). The extreme thermophiles, however, display greater tolerance to molybdenum (Brierley and Brierley, 1986) and have been shown to enhance molybdenite solubilization.

Thiobacilli have acquired properties similar to those of other bacteria to cope with certain toxic metals. Some strains of *T. ferrooxidans* have an NADPH-dependent mercuric reductase enzyme, a flavoprotein, which reduces Hg^{2+} to Hg^0 (Booth and Williams 1984, Shiratori et al. 1989) and thus confers resistance to mercury. *T. thiooxidans* has been found to produce metal-binding proteins upon acquiring resistance to Zn^{2+} and Cd^{2+} (Sakamoto et al. 1989) and these resemble metallothionein proteins that have been characterized from a number of prokaryotic and eukaryotic sources. Besides specific proteins, other mechanisms for metal tolerance may be related to a general property of the cytoplasmic membrane to act as a barrier to exclude metal ions from the cell interior.

Applications for Mineral Deposits and Ores

Dump and heap leaching processes

At present, biological dump and heap leaching processes are applied on an industrial scale for treatment of low-grade copper and pyrite-bearing uranium ores. Operations of this type are located in more than twenty countries in five continents. Accurate estimates are not available to evaluate the relative contribution of the mineral biotechnology to the world market. Torma (1987) estimated that the copper produced by dump leaching amounts to about 25 percent of the total copper production in the United States, with largest operations being located in the western states.

Industrial dump leaching is carried out usually in the vicinity of the relevant mine site to minimize transportation costs. The low-grade, run-of-the-mine ores are brought by trucks or by conveyor belts and deposited on impermeable ground, usually in valleys or on moderately steep hills. Many of the dumps have the shape of a truncated cone. Large dumps may be higher than 100 m with two or more high lifts. The surface of each separate step of such dumps is several hectares and the total amount of the ore mass is in the hundreds of thousands or even millions of tons.

Simplified flowcharts of bacterial leaching operations are presented in Figs. 13.4 and 13.5. In a typical operation, leach solution containing different agents which participate in copper and uranium solubilization, namely chemolithotrophic bacteria, ferric ions, dissolved oxygen, and sulfuric acid, is pumped to the top of the dumps. The solution is introduced into the dump by spraying, flooding, or injecting

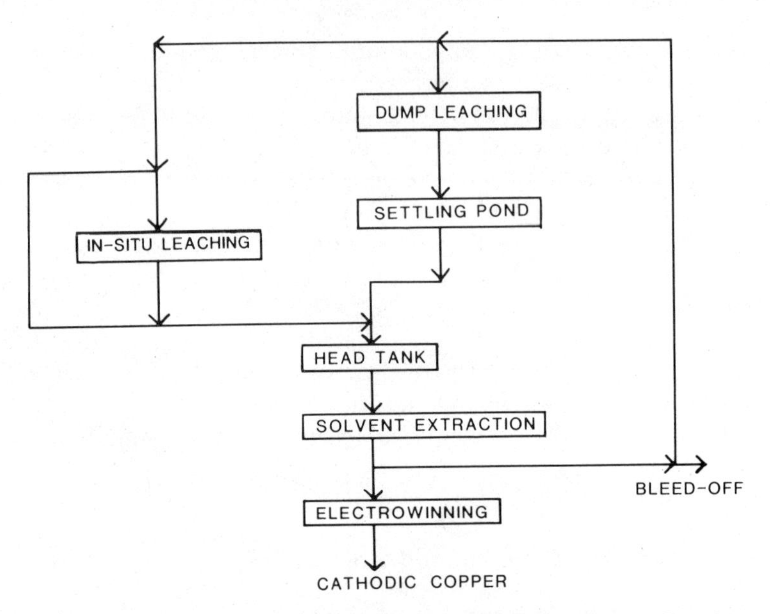

Figure 13.4 A simplified flowchart of a dump leaching operation based on the bacterially assisted oxidation of copper minerals.

through vertical pipes. The solution percolates through the ore mass and is channeled to a collection (holding) pond and then to the metal recovery plant. Copper is recovered from the pregnant solution by solvent extraction and electrowinning (Fig. 13.4) or by cementation with iron. Uranium is recovered by ion exchange. The barren solutions are supplemented with make-up water and, if needed, with sulfuric acid to adjust the pH to the desired level for recycle to the dumps. In some operations, the ferrous iron in the barren solutions may be separately oxidized to the ferric state before recycling of the solution. Some ferric iron is hydrolyzed and precipitated as insoluble salts outside the dumps and acid is regenerated as a result of the hydrolysis.

Leaching dumps represent extremely complex and variable environments. The metal solubilization is a result of various biological and chemical reactions; their rates and relative contributions to the total metal extraction are quite different and vary from ore to ore in leaching operations. Large variations exist even within an ore dump, both in different sections and within one and the same section. Mixed microbial communities associated with the leach mines are similarly bound to vary from site to site. A uniform feature, however, is the ubiquity of iron- and sulfur-oxidizing thiobacilli, which have been found in all leach mines of pyritic ores investigated thus far.

Some important factors affecting both the microbial activity and metal leaching are the following:

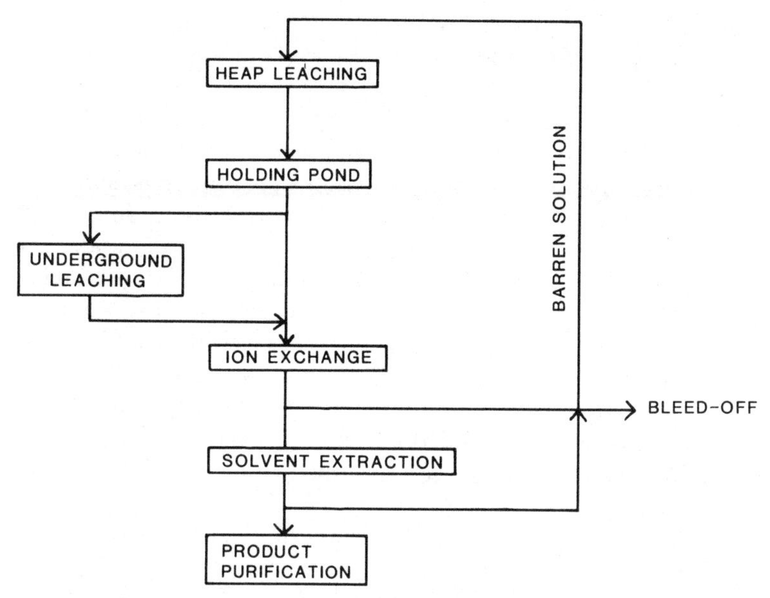

Figure 13.5 A simplified flowchart of a heap and underground stope leaching operation for the bacterially assisted recovery of uranium from pyritic ore materials.

- Physical characteristics of the ore mass (size of the mineral surface exposed to air and leach solution, permeability to leach solution, thermal conductivity, shape, and size of the dump)
- Chemical and mineralogical composition of the ore, content of copper or uranium minerals, phase composition of the copper minerals, primary (e.g., chalcopyrite) and secondary (e.g., covellite, chalcocite) sulfides, oxidic materials, presence and distribution of pyrite, lattice imperfections and impurities in their structure, character of the gangue material, including content of clays and carbonates which are highly-acid consuming
- Physicochemical conditions within the dump as well as composition and characteristics of influent leach solution (temperature, pH, redox potential, moisture, availability of oxygen, carbon dioxide and nutrients, microbial population)

Many of these factors cannot be controlled or can be controlled only to a limited degree. This is particularly valid for dumps which have been constructed with little or no control as to the size, shape, and particle size of the ore material. The physical and chemical conditions within the dumps are quite different from those optimal for the microbial growth and activity. Furthermore, the operational conditions cannot be easily controlled even in dumps which have been built with appropriate size and shape and contain ore having appropriate composition and particle size.

An active dump leaching operation represents a perturbed environment that is radically different from the natural situation, that is, from a native ore body or a waste pile where spontaneous leaching proceeds at a low rate after rainfall. It has sometimes been suggested that the most efficient microorganisms for iron and sulfide oxidation and metal solubilization for on-site application may not actually be indigenous to the ore material. However, the introduction of strains, either laboratory-bred cultures or wild types isolated from other sources and possessing improved oxidatives toward the substrates of interest, may not lead to improvement of the leaching rate. The main rate-limiting factors in leach mines are related to mass transfer of oxygen and fluxes of reactants and products within the ore material being leached.

Oxygen availability can be considerably increased by proper design and construction of dumps with optimum shape and size. Furthermore, the method of leach solution application to the surface of the dump, the amount of fines present in the ore, and the degree of compaction greatly influence the amount of oxygen that can penetrate into the dump. For these reasons, the near-future progress in this area of raw material processing is likely to involve improvements in dump design to permit improved control of biological leaching rather than the utilization of genetically improved bacteria.

Until now, industrial dump leaching operations have been based on exploiting bacteria indigenously associated with ore material. A resident community in a leaching operation is envisioned to contain a spectrum of various bacteria which differ in their physiological and genetic traits (Tuovinen et al. 1981, DiSpirito et al. 1982). The number and the oxidative activity of the indigenous bacteria in acid mine waters and on surfaces of ore materials can be manipulated only to a limited degree, mainly with changes in environmental and operational parameters such as pH control and flow rate.

Acidophiles capable of oxidizing ferrous iron, elemental sulfur, and sulfide minerals constitute the most significant and widely distributed population in leach dumps. Regardless of the selective environmental conditions, the microbial populations in industrial dump leaching operations are characterized by considerable variation and heterogeneity. *T. ferrooxidans* is of major importance in these engineered environments. *T. thiooxidans* and *L. ferrooxidans* are invariably found but in lower numbers than *T. ferrooxidans*. Acidophilic heterotrophs that represent bacteria such as *T. acidophilus, T. organoparus*, and *Acidiphilium* spp. are also typically present, usually in lower numbers than

T. thiooxidans. Moderately thermophilic chemolithotrophs related to *Sulfobacillus thermosulfidooxidans* and thiobacilli are found in some rich-in-pyrite dumps or dump regions. These moderate thermophiles sometimes occur in mixed populations with the mesophilic chemolithotrophs in temperature zones at 20 to 40°C, that is, below their temperature optimum. In these mixed populations the number of the moderate thermophiles is usually lower than that of the mesophilic chemolithotrophs, especially *T. ferrooxidans*. However, in zones with temperatures higher than 40°C, the moderate thermophilic chemolithotrophs are the prevalent microorganisms. At present, there is little evidence to suggest that extreme thermophiles related to the *Sulfolobus* and *Acidianus* spp. would be typical inhabitants in dump leaching operations.

Many heterotrophic microorganisms as well as some phototrophic sulfur bacteria and algae are found in dump leaching operations (Groudev et al. 1978) but their role is more or less incidental and has little relevance to the progress of leaching. Some heterotrophs, mainly of the acidophilic *Acidiphilium* spp. type, are close associates of *T. ferrooxidans* and reported to utilize organics formed by the chemolithotrophic bacteria. These heterotrophs are typical in microbial consortia found in acid mine waters and dump leaching operations. Other heterotrophic microbes have little significance in the leaching process: they are found sporadically in these environments and always in lower numbers than the chemolithotrophic bacteria.

It is now commonly recognized for dump leaching processes that the name *T. ferrooxidans* has been used in a generic fashion to refer to bacteria that oxidize sulfide minerals and ferrous iron under acidic conditions. The bacterial distribution is mostly confined to the upper layers (the top 0.5 to 1 m) of dumps with densities as high as in excess of 10^8 bacteria per gram of ore. In properly constructed, highly permeable dumps, the bacterial numbers typically decrease with increasing depth. In deeper layers (8 to 10 m), bacterial counts tend to be negligible (Harries and Ritchie 1983, Karavaiko 1985, Groudeva et al. 1991). In some cases, iron oxidizers have been found in zones as deep as 30 m, mainly in association with void spaces, which are formed as a result of the channeling of the leach solutions. The presence of impervious clay layers, formed within some dumps as a result of compaction by trucks during dump construction, considerably influences the distribution of bacteria. Compacted layers tend to retain bacterial cells as leach solutions infiltrate through these zones. This effect causes stratification and zonation of bacteria, with a total oxidizing activity which may be much higher than that achieved under simulated laboratory conditions (Beck 1967).

There is no clear relationship between the numbers of bacteria in dump effluents and those within the dumps. Usually, *T. ferrooxidans* counts in effluents are lower than those in the upper layers of the dump. This is presumed to result from the infiltration effect during the percolation of the leach solutions through the ore material. Dumps, with chemolithotrophic bacteria attached on ore particles, may be regarded as continuous-flow systems which, once in a steady state, display little change in population levels over long periods. As in other fixed-film steady-state systems, an equilibrium exists between the cells attached to the ore particles and those in solution. Although this equilibrium is subject to transitory changes upon rainfall, bacterial counts in dump effluents usually reflect the progress of the leaching process.

High numbers of *T. ferrooxidans* in dump effluents have sometimes been construed to index active bacterial leaching. This association has not been quantitated and its general validity can be questioned in view of the tremendous variation, spatially and temporally, in physical and chemical factors affecting the microbial growth and activity at dump sites. It is not uncommon that bacterial numbers in dump effluents display fluctuation which has no distinct or predictable pattern and the cause of which is unknown. Thus, there is no apparent relationship between the factors affecting the efficiency of leaching and the number of *T. ferrooxidans* in dump effluents. Interior layers in large dumps tend to be naturally thermostated at relatively constant temperatures which may greatly exceed the ambient surface conditions. Therefore, seasonal fluctuations in bacterial densities in dump effluents are not particularly evident.

Changes in pH and flow rate affect the number of *T. ferrooxidans* in dump effluents (Groudev et al. 1978). After a heavy rainfall, or after a sharp increase in the flow rate, bacterial numbers in effluents decrease and the total amount of bacteria washed out from the dump increase.

Usually, bacterial numbers are elevated after rest periods in dump irrigation. Rest periods are practiced in the industry with the intention of increasing metal yields upon subsequent resumption of active irrigation schedule. Although the beneficial effects of rest periods are generally recognized, the underlying mechanisms promoting better yields are not known. This may be related to improved mass transfer of O_2 and CO_2 in the interior layers. Rest period is, of course, also a cost-saving feature due to lack of pumping and reagent consumption, which may partially explain why it is widely practiced in leaching operations that recover copper from marginal ore materials.

The number of iron oxidizers sharply decreases upon cementation of copper with scrap iron (Groudev et al. 1978), presumably due to coprecipitation of bacteria.

$$Cu^{2+} + Fe \rightarrow Fe^{2+} + Cu$$

Contact with solvent extraction reagents is generally detrimental to thiobacilli. On the other hand, bacterial numbers increase in holding ponds, which provide contact time for the bacterial oxidation of ferrous iron. In some industrial operations, the residence time in these ponds is too short for efficient iron oxidation. As a result, and combined with subsequent dilution of leach liquor with make-up water, bacterial numbers in barren solution are lower compared with pregnant effluents (Groudeva et al. 1991).

The determination of bacterial activities in these situations would be important, but adequate methodology has not been developed for this purpose. As in any created or natural environments, there are instances when there is no relationship between the number and activity of bacteria. The measurement of in situ rates of bacterial iron oxidation and $^{14}CO_2$ fixation has been used to evaluate bacterial activity in acid leaching environments (Karavaiko and Moshniakova 1971, Groudeva et al. 1991). However, these measurements are difficult to standardize and interpret because the activity depends on multiple interactive factors which vary from sample to sample.

There is only scanty information on the distribution of acidophiles other than *T. ferrooxidans* in different stages of leaching operations. The isolation of moderate thermoacidophiles from ore material and leach liquor samples suggests that they are important in dump leaching processes at elevated temperature zones. Relative rates of chemical and microbiological oxidations of sulfide minerals at the different temperature zones have yet to be evaluated.

Some *Thiobacillus* species, capable of oxidizing S^0, thiosulfate, and polythionates at neutral pH, have been sporadically isolated from samples of leach liquors and ore materials (Groudeva et al. 1991). These bacteria are more numerous in initial phases of dump leaching processes when the prevailing pH of bulk solution is close to neutral. Their low numbers and inability to oxidize sulfide minerals suggest that they do not play an important role in industrial dump and heap leaching operations. Sulfate-reducing bacteria have been found in some deep layers of leach dumps, characterized by low redox potential. Their presence would signify inadequate aeration and pH control. Amoebae, potentially preying on bacteria for food, have been detected in leach liquor samples representing near-neutral pH zones at leach mines (Groudev et al. 1978). Predation is not considered to represent a threat to active bacterial leaching because acid solutions in active leach mines characteristically have low pH values which are prohibitive to amoebae. Moreover, colonization and attachment of bacteria on ore surfaces and other particulates would also negate amoebal predation.

Underground leaching processes

The underground leaching of ores (sometimes also in situ leaching or solution mining) is similar in many respects to dump leaching. It usually applied for small, high-grade, and large, low-grade ore deposits. It has utility when either the total amount or the concentration of the metal is too low to justify a conventional mining operation. To date, underground leaching has been practiced for copper and uranium ores at mines which had been first exploited by conventional methods. Solution mining has also been practiced for virgin ore bodies. In both cases, ore material is fractured by blasting with explosives in order to produce the desired particle size and to facilitate lixiviant penetration in the leaching process. Leach solutions may be introduced into fractured ore bodies by spraying, flooding, or injecting. No significant progress has been made in the evaluation of microbiological events in solution mining.

Some uranium ore bodies consisting of sandstone layers impregnated with uranium minerals are sometimes sufficiently permeable without additional blasting. Leach solutions are introduced into the mineralization through injection wells by using positive pressures greater than the hydrostatic pressure in the deposit. Pregnant solutions are collected through production wells which are drilled to create a low-pressure dump.

The flowsheets of the industrial underground leaching operations are essentially similar to those adopted for dump leaching operations. One of the major considerations for the underground leaching operation is the selection of a suitable oxidant to initiate and maintain the leaching process. The solutions used for the acid leaching of fractured copper sulfide or pyrite-bearing uranium ores have a similar composition with the lixiviants used in the dump leaching operations. However, the regeneration of the lixiviant is a critical, separate stage in the process flowsheet for the leaching of deposits situated below the natural water table and too deep for economic mining by the conventional methods, as well as for the leaching of unfractured virgin uranium ore bodies. This regeneration, similar to that involved in the dump leaching, comprises (1) the oxidation of acidic ferrous sulfate, a process which is difficult to control under deep solution mining conditions and impossible to carry out within the unfractured ore bodies; and (2) pH adjustment to the desired level by addition of sulfuric acid.

The microbiological considerations for underground leaching processes are much the same as for dump leaching, although some physical factors, especially under deep solution mining conditions, are different. Additional important considerations for underground leaching operations are the wide range of prevailing temperatures depending on the geology of the ore body, elevated hydrostatic pressure, and varying oxygen concentrations depending upon the oxygen partial pressure in the gas and the total pressure prevailing in the system. Temperatures in underground sections in leach mines show little seasonal fluctuation and, with few exceptions, are usually in the suboptimum range. Elevated hydrostatic pressures do not appear to be prohibitive to the oxidation of iron and sulfide minerals by *T. ferrooxidans* (Bosecker et al. 1979, Davidson et al. 1981). As for low oxygen tension, it may be postulated that the Fe^{3+}/Fe^{2+} pair constitutes a major redox component in subsurface environments, with Fe^{3+} serving as a sink for electrons for the anaerobic oxidation of sulfur demonstrated in thiobacilli. Further studies are needed to elucidate anaerobic oxidation and nonoxidative dissolution reactions mediated by bacteria in the leaching of sulfide minerals. In general, little is also known about the distribution, number, and activity of mesophilic and thermophilic acidophiles in virgin ore bodies after injection of leach solutions.

Underground leaching is not practiced as commonly as dump leaching and therefore less information is available about the microbial diversity in these leaching processes. Available data from a few site-specific studies (Burton et al. 1983, Brauckmann et al. 1988, Rossi 1988) suggest that the active microorganisms in underground mine waters are similar to those described in dump leaching operations (Table 13.7). *T. ferrooxidans* is again prevalent. Numbers of *T. ferrooxidans* in leach solution samples from the now defunct demonstration-scale copper leaching experiment at the Avoca Mine, in Ireland, were in the range of 10^1 to 10^4 cells/mL (Burton et al. 1983). In a uranium leaching mine, bacterial concentrations in leach liquor samples in the underground leach circuit varied between <1

TABLE 13.7 Microorganisms in Industrial-Scale Dump Leaching Operations at Vlaikov Vrah and Tsar Assen, Bulgaria

Microorganisms	Range of cell concentration per mL	
	Vlaikov Vrah	Tsar Assen
T. ferrooxidans and *L. ferrooxidans*	10^1–10^7	10^1–10^5
Acidophilic sulfur oxidizers	10^1–10^7	1–10^5
Neutrophilic sulfur oxidizers	<1–10^3	<1–10^2
Thermoacidophilic (50°C) Fe and S oxidizers	<1–10^3	<1
Acidophilic heterotrophs	1–10^3	<1–10^4

and $>10^6$ per milliliter of leach solution (Tuovinen et al. 1981, DiSpirito et al. 1982). Considerable variation, of a similar order of magnitude, was apparent in the numbers of heterotrophic glucose oxidizers enumerated in the same samples. Densities in the range of 10^5 to 10^7 of *T. ferrooxidans*-type bacteria have been reported in dump effluents in the Nikolaevsk and Volkov mines in the U.S.S.R. It should be emphasized again that the numbers of bacteria in leach solution may index little or none of the bacterial activities on ore surfaces in surface heaps and underground stopes. However, due to the generally unfavorable environmental conditions, mainly to the insufficient aeration and low temperatures in ore deposits located above or slightly below the water table, bacterial numbers may be relatively low both in the ore being leached and in the effluents.

The BACFOX (*bac*terial *f*ilm *ox*idation) process has been used to produce acidified ferric sulfate leach solutions for the leaching of uranium ores (Livesey-Goldblatt et al. 1977). In this process, ferrous iron solution is contacted with a solid, porous matrix that contains a biofilm of *T. ferrooxidans* and *L. ferrooxidans* and encrustations of Fe(III) complexes such as jarosites. Rotating biological contactors (RBC) have also been described which employ a fixed biofilm for Fe^{2+} oxidation, but their application, in industrial scale, for regeneration of acid ferric sulfate solutions has not been pursued. Bacterial iron oxidation in the treatment of acid mine waters has been practiced on a large scale with *T. ferrooxidans* immobilized on diatomaceous earth, which yields a large surface area (Murayama et al. 1987). Fixed-film bioreactors, employing high surface area to support colonization and with continuous-flow mode to maintain a steady state, have in bench-scale studies yielded iron oxidation rates superior to other bioreactor techniques reported thus far (Grishin and Tuovinen 1988).

Tank leaching

Bacterial leaching in stirred tank reactors has been widely used in laboratory and pilot-scale studies for oxidative dissolution of various sulfide mineral concentrates as well as for coals containing sulfur impurities. Stirred tank reactors can be equipped with instrumentation for monitoring all important leaching parameters. Data derived from these studies are valuable in efforts of scaling up of leaching processes and developing preliminary economical evaluation.

In contrast to dump and underground leaching, tank leaching is technically a relatively well-defined system. Tank leaching can be carried out under conditions which are very close to the optimum conditions for the growth and activity of the microorganisms of interest. It is necessary to differentiate between (1) the small-scale tank leaching experiments which are carried out under aseptic conditions by using presterilized mineral substrates and mineral salt solutions which are inoculated with pure strains or mixed enrichment cultures, and (2) the large-scale tank leaching experiments which are carried out under nonaseptic, semienclosed conditions.

Small-scale experiments are important for understanding the role of defined test cultures, pure or mixed, in leaching processes. Large-scale leaching experiments are closer to the real industrial conditions where it would be impossible to maintain the leach system free from extraneous sources of microorganisms. Flowcharts of tank leaching processes are presented in Figs. 13.6 and 13.7.

It is natural that sulfide concentrates are sources of several nonferrous metals in major, minor, and trace quantities, and these may be problematic in conventional metallurgical processes. The use of the bacterial leaching as a pretreatment procedure to dissolve metals prior to conventional concentrate treatment has been proposed. Several pretreatment options may be feasible. The upgrading of sulfide concentrates for conventional metallurgical processes has been evaluated by employing bacterial leaching to selectively remove copper, zinc, and cadmium from off-grade complex lead concentrates (Torma 1978) and by leaching of unwanted elements such as arsenic to make the concentrate more acceptable for conventional processing (Groudeva and Groudev 1984). The renewed interest in gold during the 1980s, coupled with increased activities in exploration and mining of this precious metal, has focused attention on the need to develop more efficient technologies to exploit the large global reserves of refractory gold. This type of gold is generally bound in sulfide minerals,

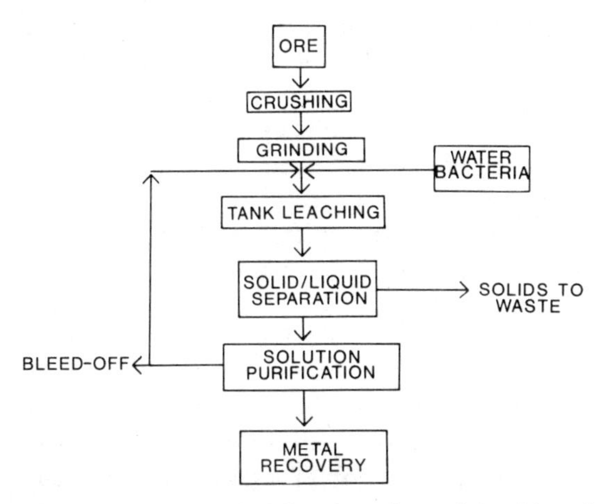

Figure 13.6 A simplified flowchart for tank leaching of a high-grade ore material.

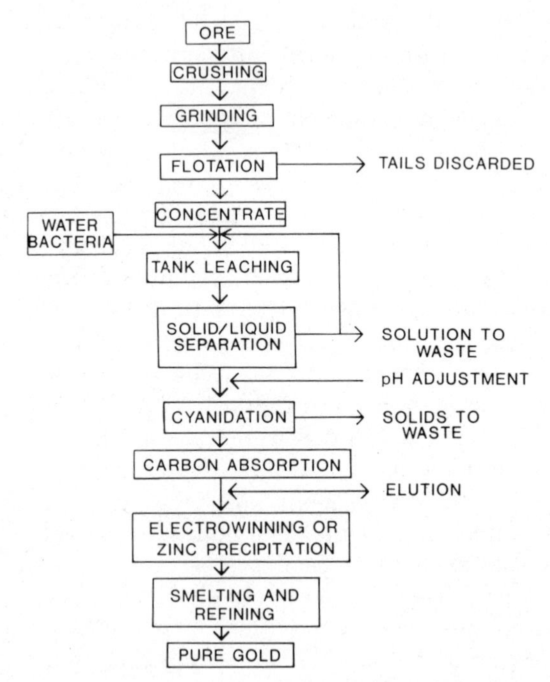

Figure 13.7 A simplified flowchart for tank leaching of gold-containing arsenopyrite and pyrite concentrates.

notably arsenopyrite and pyrite. The gold is often fine-grained, submicroscopic, and sometimes tightly locked within the sulfide mineral. Refractory gold ores tend to be site-specific and therefore the choice of processing technology must be made on consideration of a number of factors from mineralogy through to process capital and operating costs (Haines 1986).

An emerging technology is the bacterial leaching of refractory gold-containing sulfide concentrates in continuously stirred tank reactors. The bacterial oxidation of gold- and silver-bearing sulfide minerals to expose finely disseminated Au and Ag in the sulfide matrix and to facilitate its subsequent chemical extraction has rapidly approached commercial-scale demonstration. The higher capital and operating costs associated with stirred tanks generally require that the feed materials are high-value pyrite and arsenopyrite concentrates. However, the process is also being considered for the treatment of high-grade run-of-mine ores. The choice of feed material can only be made after careful metallurgical and economic evaluation of a number of critical process parameters, including process residence time, reactor volumes, and oxidation level required for optimum gold recovery.

While the process has been tested on numerous ores and concentrates at laboratory scale (e.g., Bruynesteyn et al. 1986, Lawrence and Marchant 1988, Nobar et al. 1988), the closest demonstration to a commercial process has been conducted by the General Mining Union Corporation Limited (GENCOR) at the Fairview Mine, South Africa. A 10-ton/day plant was commissioned during October 1986, with an estimated contact time of about 7 days for the bacterial pretreatment, and has been operating successfully since then. A review of the early operation was presented by van Aswegen et al. (1988). By 1989, the full-scale continuous-flow process had been expanded to treat 18 tons/day with an average contact time of 3 days for bacterial leaching before cyanidation.

Despite the large number of studies conducted on reactor leaching of gold concentrates,[2] there has been a lack of information on the mixed cultures operating during the leaching process. The likely commercial process configuration will involve a train of continuously stirred tank reactors in series. Each tank along the train will therefore present a different leaching environment as the bioleached pulp is progressively transferred from tank to tank. As the level of oxidation of the sulfide material progresses to a higher level in each tank, the availability of substrate will vary as a function of particle size, the relationship to gangue components, and the buildup of passivation layers. The leaching rates vary from tank to tank, and this will in turn affect the demand for oxygen, carbon dioxide, and inorganic nutrients. Depending on the pH control and the mineral being oxidized, the level of acidity may increase along the reactor train, with the pH being considerably lower in the last tank when compared to the first. Such differences in the leaching environment would influence the nature of the mixed culture. For example, in the first tank at a pH above 1.5, *T. ferrooxidans* strains might be expected to dominate. As the pH along the train decreases below pH 1.5, a mixed culture of *L. ferrooxidans* and strains of sulfur-oxidizing bacteria might well be anticipated. The make-up of the mixed culture is related to the selective nature of the leaching process. Continuous operation under limiting conditions of O_2, CO_2, and nutrients would create a high selection pressure on the mixed population, selecting for organisms with higher affinities for the limiting nutritional factors. Eventually, a stable mixed population would develop in each tank. Any perturbation to the process, such as a power loss, change of feed, or variation in temperature may alter the make-up of the mixed culture with a potential loss in productivity.

Microbiological leaching systems are complex and therefore difficult

[2]Livesey-Goldblatt et al. 1983, Haines 1986, Southwood and Southwood 1986, Pinches et al. 1988, Hansford and Drossou 1988.

to model. There are insufficient data at the fundamental research level and little experience at the operational level to predict how mixed cultures in continuous tank leaching processes change in their genetic make-up and phenotypic expression. Reports to date indicate that the leaching bacteria are relatively insensitive to transitory perturbations, as long-term process stability has been reported under a range of operating conditions. Care must be taken to ensure that acidophilic heterotrophic contaminants do not proliferate to the point where they effectively compete with the leaching bacteria for oxygen and nutrients. Tank leaching may eventually offer the advantage of using selected mixed cultures for particular ore types since the degree of operator control over the process allows for some selectivity. The use of genetically engineered organisms, tolerant of high levels of soluble arsenic generated during the leaching of arsenopyrite concentrates, can therefore be envisaged. To date, the process has successfully relied on the adaptation of bacteria to elevated levels of arsenic typically found in a plant situation. In the case of the GENCOR process at the Fairview Mine, the gradual development of bacterial resistance by adaptation is implicated by virtue of the fact that leach solutions contain >10 g/L arsenic, which is about tenfold excess of the As concentration that tends to be inhibitory to *T. ferrooxidans* upon first exposure. Future use of high-As-level arsenopyrite concentrates that require almost complete oxidation may require organisms capable of tolerating much higher levels of arsenic. Alternatively, the introduction of plasmids specifying resistance to arsenic as well as to a range of other toxic metal ions (e.g., Ag, Hg, Mo) may eventually find application. This could apply across a number of leaching microorganisms, as they differ in their sensitivities to these metals. In addition, strains carrying other desirable traits, including enhanced iron oxidation and improved carbon dioxide assimilation, could also be envisaged for use in tank leaching operations.

Compared with *T. ferrooxidans*, the thermophiles usually oxidize the pyrite and arsenopyrite at faster rates. However, their tolerance to high pulp density is lower than that of *T. ferrooxidans* and they require CO_2-enriched aeration to support favorable growth and leaching rates (Norris 1988). In tank leaching, the selection pressure afforded by temperatures of 50 to 60°C could be coupled to the observed improvements in leaching rates using thermophilic bacteria. If the apparent inhibition of thermophiles caused by high pulp density can be alleviated, decreased residence times could lead to small plant size with a consequent reduction in capital costs. Reduced cooling requirements could also lead to a lowering of operating costs depending on the location of the leach plant. To operate at high temperatures, thermophilic organisms with high affinities for O_2 and CO_2 may be required in view of the reduced solubilities of these gases at high temperatures and at low pH.

A major problem associated with the full utility of tank leaching to date has been the lack of suitable techniques for enumerating or monitoring biomass levels during the leaching processes. Enumeration of *T. ferrooxidans* by colony counting has been adequately resolved with various formulations of solid media-containing mineral salts (e.g., Butler and Kempton 1987, Schrader and Holmes 1988, Visca et al. 1989). Solid media are now also available for *Sulfolobus* spp. (Lindström and Sehlin 1989). While these methods avail themselves to quantitative enumeration, problems associated with representative biomass sampling of the bulk solid and liquid phases have not been resolved. Enumeration techniques based on cultural methods of recovery are, of course, limited by the medium of choice and they have little utility in providing insight into mixed-culture systems in terms of genetic potential, diversity, and interactions. The use of genetic probes, for instance by Southern blot hybridization, has been tentatively evaluated for *T. ferrooxidans* (Yates and Holmes 1986) and may greatly help to identify and quantitate members and to look for genetic traits, such as mercury resistance (Barkay et al. 1985, Shiratori et al. 1989) in mixed populations. Molecular probes have yet to be better explored in the microbiological characterization of leaching processes. Detection methodology based on immunoassays is another specific and sensitive diagnostic approach to selectively look for and quantitate defined members in mixed cultures in samples from bacterial leaching operations (Muyzer et al. 1987, Arredondo and Jerez 1989), but is limited by the source of the antigen and has other limitations due to its inability to directly reveal any specific phenotypic markers in mixed communities. However, combined with other detection techniques such DNA fluorescence and respiratory dyes (Baker and Mills 1982, Muyzer et al. 1987), the methodology should yield useful information on community changes in leaching processes.

The yield of gold extraction has been found to be proportional to the degree of the preceding bacterial oxidation of pyrite or arsenopyrite (Lawrence and Bruynesteyn 1983, Lindström and Gunneriusson 1990). Even a partial oxidation of the sulfide matrix can be sufficient for achieving a high yield of gold extraction (Southwood and Southwood 1986). In pyrite and arsenopyrite ores, gold occurs mainly in imperfections in sulfide crystal lattice. In the case of pyrite, these sites are also the preferred regions of bacterial oxidative attack. For arsenopyrite, such relationships are yet to be reported. Arsenopyrite has a lower electrode potential relative to pyrite and therefore preferential oxidation of arsenopyrite should occur in complex pyrite-arsenopyrite concentrates, provided that the process is only electrochemically controlled, which rarely is the case. If gold is associated with the arsenopyrite phase, some selectivity of arsenopyrite oxidation over that of pyrite is afforded because of the differences in the respective electrode potentials, thereby exposing gold for cyanidation without complete dissolution of pyrite (Lawrence and Marchant 1987). However, the selectivity, stoichiometric relationships, and kinetics of the bacterial leaching of arsenopyrite-pyrite concentrates have yet to be worked out. Similarly, the redox reactions involved in the oxidation of As to the level of arsenite and its subsequent oxidation to arsenate have not been elucidated.

After bacterial leaching, the leach residue is separated from the liquid phase and is either neutralized with lime and leached with cyanide or is subjected to thiourea leaching to extract gold. The leach solution containing arsenate is neutralized with lime to pH 3.5 to precipitate this toxic oxyanion. Safe disposal of arsenic is via the formation of poorly soluble ferric arsenate complexes ($FeAsO_4 \cdot xH_2O$) which may vary in the stoichiometry, solubility, and stability (Krause and Ettel 1989). In leach liquors from the bacterial oxidation of pyrite-arsenopyrite concentrates, both jarosite and scorodite ($FeAsO_4 \cdot x2H_2O$) have been identified as secondary crystalline products (Fig. 13.8). The precipitation of ferric arsenate complexes reduces the effective available concentration of arsenate and thus decreases the toxicity of these solutions to leach bacteria. After the ferric arsenate precipitate is discarded, the pH and nutrient concentrations of the solution are adjusted and the treated solution is returned to the leaching step. Block-Bolten et al. (1985) demonstrated that the extraction of gold from pyrite and arsenopyrite ore materials could be reduced to a single-step leach process with the use of *T. ferrooxidans*, which was adapted to oxidize sulfide minerals in the presence of high concentrations of thiourea. The utility of this process may be limited because of the instability of thiourea in acid leach solutions.

Concluding Remarks

At present, phenotypic improvements of acidophilic thiobacilli can be readily obtained by means of selection via adaptation and by classical genetic methods based on both natural and induced mutation. Several traits can be targeted for improvement in genetically modified strains. These characteristics are concerned with improved growth rates, higher iron- and sulfur-oxidizing activities, modified attachment properties to mineral particles, increased resistance to metal ions, reduced sensitivity to organic compounds, changes in kinetic parameters such as a higher affinity for dissolved oxygen, and enhanced ability to fix nitrogen. The modified strain must be stable and able to compete and retain the activity of desirable traits in the presence of native and

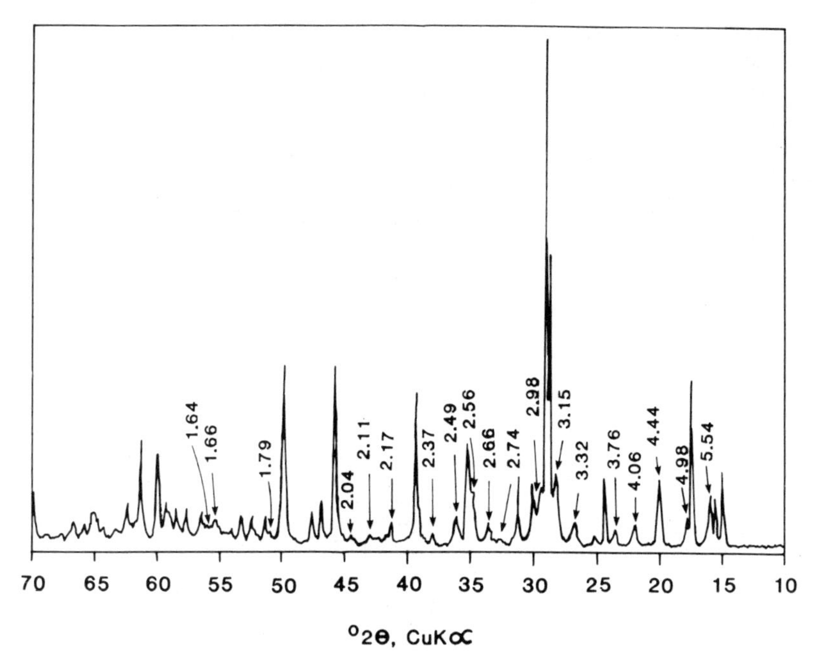

Figure 13.8 X-ray diffractogram of a sample of precipitates formed in a *Sulfolobus* culture growing with a pyrite-arsenopyrite concentrate (unpublished data, courtesy of J.M. Bigham and E.B. Lindström). Scorodite peak assignments (arrows) are indicated in Å; the unassigned peaks represent well-ordered jarosite.

wild-type microorganisms. At present, a major objective of the molecular genetics of thiobacilli and of thermoacidophiles is to facilitate the understanding of their biochemistry and regulatory responses. While genes of *T. ferrooxidans* have been successfully transformed and expressed in *Escherichia coli*, transformation systems are only now being developed for *T. ferrooxidans* and related organisms. Thus, near-future objectives must include the finding of suitable markers, possibly metal resistance traits, and experimenting with transformation conditions for setting up genetic systems for acidophilic chemolithotrophs. In view of recent improvements in methods such as electroporation for gene transfer, breakthroughs in the transformation of iron-oxidizing acidophiles, both mesophilic and thermophilic, are imminent. Future studies need also to address the introduction of suitable genetic markers into laboratory strains to facilitate strain identification and to provide selection mechanisms to enhance the competitiveness, ultimately in a bioleaching operation. As the tank bioleaching process nears full-scale commercialization in many countries, being also applicable to the biological processing of high-sulfur coals, it is obvious that we will need to improve our understanding of the complex microbiology involved. This need will take on greater urgency as the sophistication of tank leaching improves, particularly in view of the projected use of selected mixed cultures.

Acknowledgments

O.H.T. gratefully acknowledges research support from Outokumpu Research Ltd. and the Nordisk Industrifond. We thank Laurie Haldeman for typing this manuscript.

References

Acevedo, F., and Gentina, J. C. 1989. Process engineering aspects of the bioleaching of copper ores. *Bioproc. Eng.* 4:223–229.

Ackerman, T. E., and Kleinmann, R. L. P. 1984. In-line aeration and treatment of acid-mine drainage. *U.S. Bureau of Mines Rep. Invest.* 8866, U.S. Dept. Interior.

Ahonen, L., Hiltunen, P., and Tuovinen, O. H. 1986. The role of pyrrhotite and pyrite in the bacterial leaching of chalcopyrite ores. In *Fundamental and Applied Biohydrometallurgy* (R. W. Lawrence, R. M. P. Branion, and H. G. Ebner, eds.), pp. 13–23. Elsevier, Amsterdam.

Ahonen, L., and Tuovinen, O. H. 1989. Microbiological oxidation of ferrous iron at low temperatures. *Appl. Environ. Microbiol.* 55:312–316.

Ahonen, L., and Tuovinen, O. H. 1990. Kinetics of sulfur oxidation at suboptimal temperatures. *Appl. Environ. Microbiol.* 56:560–562.

Alexander, B., Leach, S., and Ingledew, W. J. 1987. The relationship between chemiosmotic parameters and sensitivity to anions and organic acids in the acidophile *Thiobacillus ferrooxidans*. *J. Gen. Microbiol.* 133:1171–1179.

Arredondo, R., and Jerez, C. A. 1989. Specific dot-immunobinding assay for detection and enumeration of *Thiobacillus ferrooxidans*. *Appl. Environ. Microbiol.* 55:2025–2029.

Baker, K. H., and Mills, A. L. 1982. Determination of the number of respiring *Thiobacillus ferrooxidans* cells in water samples by using combined fluorescent antibody 2-(*p*-iodophenyl)-3-(*p*-nitrophenyl)-5-phenyltetrazolium chloride staining. *Appl. Environ. Microbiol.* 43:338–344.

Balashova, V. V., Vedenina, I. Y., Markosyan, G. E., and Zavarzin, G. A. 1974. The auxotrophic growth of *Leptospirillum ferrooxidans*. *Mikrobiologiya* 43:581–585.

Barkay, T., Fouts, D. L., and Olson, B. H. 1985. Preparation of a DNA gene probe for detection of mercury resistance genes in gram-negative bacterial communities. *Appl. Environ. Microbiol.* 49:686–692.

Bärtels, C.-C., Chatzitheodorou, G., Rodriguez-Leiva, M., and Tributsch, H. 1989. Novel technique for investigation and quantification of bacterial leaching by *Thiobacillus ferrooxidans*. *Biotechnol. Bioeng.* 33:1196–1204.

Beck, J. V. 1967. The role of bacteria in copper mining operations. *Biotechnol. Bioeng.* 9:487–497.

Blake, R. C., and Shute, E. A. 1987. Respiratory enzymes of *Thiobacillus ferrooxidans*. A kinetic study of electron transport between iron and rusticyanin in sulfate media. *J. Biol. Chem.* 262:14983–14989.

Blake, R. C., White, K. J., and Shute, E. A. 1988. Electron transfer from Fe(II) to rusticyanin is catalyzed by an acid-stable cytochrome. In *Biohydrometallurgy* (P. R. Norris and D. P. Kelly, eds.), pp. 103–110. Science and Technology Letters, Kew, Surrey, U.K.

Block-Bolten, A., Dalta, M. S., Torma, A. E., and Steensma, R. 1985. New possibilities in the extraction of gold and silver from zinc and lead sulfide flotation waste. In *Complex Sulfides Processing of Ores, Concentrates, and By-Products* (A. D. Zunkel, R. S. Boorman, A. E. Morris, and R. J. Wesley, eds.), pp. 149–166, The Metallurgical Society, Warrendale, Penn.

Booth, J. E., and Williams, J. W. 1984. The isolation of a mercuric ion-reducing flavoprotein from *Thiobacillus ferrooxidans*. *J. Gen. Microbiol.* 130:725–730.
Bosecker, K., Torma, A. E., and Brierley, J. A. 1979. Microbiological leaching of a chalcopyrite concentrate and the influence of hydrostatic pressure on the activity of *Thiobacillus ferrooxidans*. *Eur. J. Appl. Microbiol. Biotechnol.* 7:85–90.
Brauckmann, B., Boppe, W., Beyer, W., Lerche, W., and Steppke, H.-D. 1988. Investigations of increased biological in situ leaching of the "Old Deposit" of the Preussag Rammelsberg ore mine. In *Biohydrometallurgy* (P. R. Norris and D. P. Kelly, eds.), pp. 521–523. Science and Technology Letters, Kew, Surrey, U.K.
Brierley, C. L. 1978. Bacterial leaching. *Crit. Rev. Microbiol.* 6:207–262.
Brierley, J. A., and Brierley, C. L. 1986. Microbial mining using thermophilic microorganisms. In *Thermophiles: General, Molecular, and Applied Microbiology* (T. D. Brock, ed.), pp. 279–305. John Wiley, New York.
Brock, T. D., Brock, K. M., Belly, R. T., and Weiss, R. L. 1972. *Sulfolobus*: a new genus of sulfur-oxidizing bacteria living at low pH and high temperature. *Arch. Mikrobiol.* 84:54–68.
Brock, T. D., and Gustafson, J. 1976. Ferric iron reduction by sulfur- and iron-oxidizing bacteria. *Appl. Environ. Microbiol.* 32:567–571.
Bruynesteyn, A., Hackl, R. P., and Wright, F. 1986. The Biotankleach process. In *Gold 100. Proceedings of the International Conference on Gold*, vol. 2: *Extractive Metallurgy of Gold* (C. E. Fivaz and R. P. King, eds.), pp. 353–365. South African Institute of Mining and Metallurgy, Johannesburg.
Burton, C., Cowman, S., Heffernan, J., and Thorne, B. 1983. In situ bioleaching of sulphide ores at Avoca, Ireland. Part I—Development, characterization and operation of a medium scale (6000 t) experimental leach site. In *Recent Progress in Biohydrometallurgy* (G. Rossi and A. E. Torma, eds.), pp. 213–241. Associazione Mineraria Sarda, Iglesias, Italy.
Butler, B. J., and Kempton, A. G. 1987. Growth of *Thiobacillus ferrooxidans* on solid media containing heterotrophic bacteria. *J. Ind. Microbiol.* 2:41–45.
Chashchina, N. M., and Kukharchuk, L. E. 1988. Effect of low temperatures on distribution and activity of certain microorganisms in a zinc-lead deposit. *Mikrobiologiya* 57:152–157.
Cox, J. C., and Brand, M. D. 1984. Iron oxidation and energy conservation in the chemoautotroph *Thiobacillus ferrooxidans*. In *Microbial Chemoautotrophy* (W. R. Strohl and O. H. Tuovinen, eds.), pp. 31–46. The Ohio State University Press, Columbus, Ohio.
Cwalina, B., Weglarz, L., Dzierzewicz, Z., and Wilczok, T. 1988. Dependence of effectiveness of leaching of metallic sulphides on enzymes involved in inorganic sulphur metabolism in *Thiobacillus ferrooxidans*. *Appl. Microbiol. Biotechnol.* 28:100–102.
Davidson, M. S., Torma, A. E., Brierley, J. A., and Brierley, C. L. 1981. Effects of elevated pressures on iron- and sulfur-oxidizing bacteria. *Biotechnol. Bioeng. Symp.* 11: 603–618.
DiSpirito, A. A., Silver, M., Voss, L., and Tuovinen, O. H. 1982. Flagella and pili of iron-oxidizing thiobacilli isolated from a uranium mine in northern Ontario, Canada. *Appl. Environ. Microbiol.* 43:1096–2000.
DiSpirito, A. A., and Tuovinen, O. H. 1982a. Uranous ion oxidation and carbon dioxide fixation by *Thiobacillus ferrooxidans*. *Arch. Microbiol.* 133:28–32.
DiSpirito, A. A., and Tuovinen, O. H. 1982b. Kinetics of uranous ion and ferrous ion oxidation by *Thiobacillus ferrooxidans*. *Arch. Microbiol.* 133:33–37.
Ferroni, G. D., Leduc, L. G., and Todd, M. 1986. Isolation and temperature characterization of psychrotrophic strains of *Thiobacillus ferrooxidans* from the environment of a uranium mine. *J. Gen. Appl. Microbiol.* 32:169–175.
Fry, I. V., Lazaroff, N., and Packer, L. 1986. Sulfate-dependent iron oxidation by *Thiobacillus ferrooxidans*: characterization of a new EPR detectable electron transport component on the reducing side of rusticyanin. *Arch. Biochem. Biophys.* 246:650–654.
Fukumori, Y., Yano, T., Sato, A., and Yamanaka, T. 1988. Fe(II)-oxidizing enzyme purified from *Thiobacillus ferrooxidans*. *FEMS Microbiol. Lett.* 50:169–170.
Golovacheva, R. S., and Karavaiko, G. I. 1978. A new genus of thermophilic spore-forming bacteria, *Sulfobacillus*. *Mikrobiologiya* 47:658–665.

Goodman, A. E., Babij, T., and Ritchie, A. I. M. 1983. Leaching of a sulphide ore by *Thiobacillus ferrooxidans* under anaerobic conditions. In *Recent Progress in Biohydrometallurgy* (G. Rossi and A. E. Torma, eds.), pp. 361–376. Associazione Mineraria Sarda, Iglesias, Italy.

Grishin, S. I., and Tuovinen, O. H. 1988. Fast kinetics of Fe^{2+} oxidation in packed-bed reactors. *Appl. Environ. Microbiol.* 54:3092–3100.

Groudev, S. 1980. Leaching of copper-bearing mineral substrates with wild microflora and with laboratory-bred strains of *Thiobacillus ferrooxidans*. In *Biogeochemistry of Ancient and Modern Environments* (P. A. Trudinger, M. R. Walter, and B. J. Ralph, eds.), pp. 485–503. Springer-Verlag, Berlin.

Groudev, S. N., Genchev, F. N., and Gaidarjiev, S. S. 1978. Observations on the microflora in an industrial copper dump leaching operation. In *Metallurgical Applications of Bacterial Leaching and Related Microbiological Phenomena* (L. E. Murr, A. E. Torma, and J. A. Brierley, eds.), pp. 253–274. Academic Press, New York.

Groudeva, V. I., and Groudev, S. N. 1984. Removal of arsenic from sulphide concentrates by means of microorganisms. In *Biotech 84 USA*, pp. A57–A65. Online Publications, Pinner, U.K.

Groudeva, V. I., Groudev, S. N., and Vassilev, D. V. 1991. Microflora of two industrial copper dump leaching operations. In *Dump and Underground Biological Leaching of Ores* (G. I. Karavaiko, S. N. Groudev, and G. Rossi, eds.), Centre of International Projects GKNT, Moscow, U.S.S.R., in press.

Haines, A. K. 1986. Factors affecting the choice of technology for the recovery of gold from refractory arsenical ores. In *Gold 100. Proceedings of the International Conference on Gold*, vol. 2: *Extractive Metallurgy of Gold* (C. E. Fivaz and R. P. King, eds.), pp. 227–233. South African Institute of Mining and Metallurgy, Johannesburg.

Hansford, G. S., and Drossou, M. 1988. A propagating-pore model for the batch bioleach kinetics of refractory gold-bearing pyrite. In *Biohydrometallurgy* (P. R. Norris and D. P. Kelly, eds.), pp. 345–358. Science and Technology Letters, Kew, Surrey, U.K.

Harries, J. R., Hendy, N., and Ritchie, A. I. M. 1988. Rate controls on leaching in pyritic mine wastes. In *Biohydrometallurgy* (P. R. Norris and D. P. Kelly, eds.), pp. 233–241. Science and Technology Letters, Kew, Surrey, U.K.

Harries, J. R., and Ritchie, A. I. M. 1981. The use of temperature profiles to estimate the pyritic oxidation rate in a waste rock dump from an open cut mine. *Water Air Soil Pollut.* 15:405–423.

Harries, J. R., and Ritchie, A. I. M. 1985. Pore gas composition in waste rock dumps undergoing pyritic oxidation. *Soil Sci.* 140:143–152.

Harries, J. R., and Ritchie, A. I. M. 1986. The impact of rehabilitation measures on the physicochemical conditions within mine wastes undergoing pyritic oxidation. In *Fundamental and Applied Biohydrometallurgy* (R. W. Lawrence, R. M. R. Branion, and H. G. Ebner, eds.), pp. 341–351. Elsevier, Amsterdam.

Harrison, A. P. 1981. *Acidiphilium cryptum* gen. nov., sp. nov.: heterotrophic bacteria from acidic mineral environments. *Int. J. Syst. Bacteriol.* 31:327–332.

Harrison, A. P. 1984. The acidophilic thiobacilli and other acidophilic bacteria that share their habitat. *Ann. Rev. Microbiol.* 38:265–292.

Harrison, A. P. 1986. Characteristics of *Thiobacillus ferrooxidans* and other iron-oxidizing bacteria, with emphasis on nucleic acid analyses. *Biotechnol. Appl. Biochem.* 8:249–257.

Helle, U., and Onken, U. 1988. Continuous bacterial leaching of a pyritic flotation concentrate by mixed cultures. In *Biohydrometallurgy* (P. R. Norris and D. P. Kelly, eds.), pp. 61–75. Science and Technology Letters, Kew, Surrey, U.K.

Hoffmann, M. R., Hiltunen, P., and Tuovinen, O. H. 1985. Inhibition of ferrous ion oxidation by *Thiobacillus ferrooxidans* in the presence of oxyanions of sulfur and phosphorus. In *Processing and Utilization of High Sulfur Coals* (Y. A. Attia, ed.), pp. 683–698. Elsevier, Amsterdam.

Holden, P. J. 1989. Characteristics of a moderately thermophilic iron-oxidizing bacterium: aspects of nitrogen metabolism. In *Proceedings of the 8th Australian Biotechnology Conference*, pp. 590–593. Sydney, Australia.

Holden, P. J., Kelley, B. C., and Madgwick, J. C. 1988. Isolation of an iron-oxidizing moderate thermophile. *Austral. Microbiol.* 9(2):208.

Huber, G., Spinnler, C., Gambacorta, A., and Stetter, K. O. 1989. *Metallosphaera sedula* gen. and sp. nov. represents a new genus of aerobic, metal-mobilizing, thermoacidophilic archaebacteria. *Syst. Appl. Microbiol.* 12:38–47.

Huber, G., and Stetter, K. O. 1989. *Thiobacillus prosperus* sp. nov., represents a new group of halotolerant metal-mobilizing bacteria isolated from a marine geothermal field. *Arch. Microbiol.* 151:479–485.

Huber, G., and Stetter, K. O. 1990. *Thiobacillus cuprinus* sp. nov., a novel facultatively organotrophic metal-mobilizing bacterium. *Appl. Environ. Microbiol.* 56:315–322.

Ingledew, W. J. 1982. *Thiobacillus ferrooxidans*. The bioenergetics of an acidophilic chemolithotroph. *Biochim. Biophys. Acta* 683:89–117.

Ingledew, W. J. 1986. Ferrous iron oxidation by *Thiobacillus ferrooxidans*. *Biotechnol. Bioeng. Symp.* 16:23–33.

Jaynes, D. B., Rogowski, A. S., Pionke, H. B., and Jacoby, E. K. 1983. Atmosphere and temperature changes within a reclaimed coal strip mine. *Soil Sci.* 136:164–177.

Johnson, D. B., Macvicar, J. H. M., and Rolfe, S. 1987. A new solid medium for the isolation and enumeration of *Thiobacillus ferrooxidans* and acidophilic heterotrophic bacteria. *J. Microbiol. Meth.* 7:9–18.

Johnson, D. B., Said, M. F., Chauri, M. A., and McGinness, S. 1990. Isolation of novel acidophiles and their potential use in bioleaching operations. In *Biohydrometallurgy 89* (J. Salley, R. G. L. McCready, and P. L. Wichlacz, eds), pp. 403–414. Department of Supplies and Services, Government of Canada, Ottawa, Ont.

Jones, C. A., and Kelly, D. P. 1983. Growth of *Thiobacillus ferrooxidans* on ferrous iron in chemostat culture: influence of product and substrate inhibition. *J. Chem. Technol. Biotechnol.* 33B:241–261.

Kai, M., Yano, T., Fukumori, Y., and Yamanaka, T. 1989. Cytochrome oxidase of an acidophilic iron-oxidizing bacterium, *Thiobacillus ferrooxidans*, functions at pH 3.5. *Biochem. Biophys. Res. Commun.* 160:839–843.

Kandler, O., and Stetter, K. O. 1981. Evidence for autotrophic CO_2 assimilation in *Sulfolobus brierleyi* via a reductive carboxylic acid pathway. *Zbl. Bakt. Hyg., I Abt. Orig.* C 2:111–121.

Karavaiko, G. I. 1988. Methods of isolation, evaluation and studying of microorganisms. In *Biogeotechnology of Metals—Manual* (G. I. Karavaiko, G. Rossi, A. D. Agate, S. N. Groudev, and Z. A. Avakyan, eds.), pp. 47–86. Centre for International Projects GKNT, Moscow, U.S.S.R.

Karavaiko, G. I. 1985. *Microbiological Processes for the Leaching of Metals from Ores—State-of-the-Art Review*. Centre of International Projects GKNT, Moscow, U.S.S.R.

Karavaiko, G. I., and Moshniakova, S. A. 1971. A study on chemosynthesis and rate of bacterial and chemical oxidative processes under conditions of copper-nickel ore deposits of Kolsky Peninsula. *Mikrobiologiya* 40:551–557.

Kelly, D. P. 1988a. Evolution of the understanding of the microbiology and biochemistry of the mineral leaching habitat. In *Biohydrometallurgy* (P. R. Norris and D. P. Kelly, eds.), pp. 3–14. Science and Technology Letters, Kew, Surrey, U.K.

Kelly, D. P. 1988b. Oxidation of sulphur compounds. *Symp. Soc. Gen. Microbiol.* 42:66–98.

Kelly, D. P., Norris, P. R., and Brierley, C. L. 1979. Microbiological methods for the extraction and recovery of metals. *Symp. Soc. Gen. Microbiol.* 29:263–308.

Kelly, R. M., and Deming, J. W. 1988. Extremely thermophilic archaebacteria: biological and engineering considerations. *Biotechnol. Progr.* 4:47–62.

Krause, E., and Ettel, V. A. 1989. Solubilities and stabilities of ferric arsenate compounds. *Hydrometallurgy* 22:311–337.

Lappin, A. G., Lewis, C. A., and Ingledew, W. J. 1985. Kinetics and mechanism of reduction of rusticyanin, a blue copper protein from *Thiobacillus ferrooxidans*, by inorganic cations. *Inorg. Chem.* 24:1446–1450.

Lawrence, R. W., and Bruynesteyn, A. 1983. Biological pre-oxidation to enhance gold and silver recovery from refractory pyritic ores and concentrates. *CIM Bull.* 76:107–110.

Lawrence, R. W., and Marchant, P. B. 1987. Biochemical pretreatment in arsenical gold ore processing. In *Arsenic Metallurgy Fundamentals and Applications* (R. G. Reddy, J. L. Hendrix, and P. B. Queneau, eds.), pp. 199–211. The Metallurgical Society, Warrendale, Pennsylvania.

Lawrence, R. W., and Marchant, P. B. 1988. Comparison of mesophilic and thermophilic oxidation systems for the treatment of refractory gold ores and concentrates. In *Biohydrometallurgy* (P. R. Norris and D. P. Kelly, eds.), pp. 359–374. Science and Technology Letters, Kew, Surrey, U.K.

Le Roux, N. W., and Wakerley, D. S. 1988. Leaching of chalcopyrite ($CuFeS_2$) at 70°C using *Sulfolobus*. In *Biohydrometallurgy* (P. R. Norris and D. P. Kelly, eds.), pp. 305–317. Science and Technology Letters, Kew, Surrey, U.K.

Le Roux, N. W., Wakerley, D. S., and Hunt, S. D. 1977. Thermophilic thiobacillus-type bacteria from Icelandic thermal areas. *J. Gen. Microbiol.* 100:197–201.

Le Roux, N. W., Wakerley, D. S., and Perry, V. F. 1978. Leaching of minerals using bacteria other than thiobacilli. In *Metallurgical Applications of Bacterial Leaching and Related Microbiological Phenomena* (L. E. Murr, A. E. Torma, and J. A. Brierley, eds.), pp. 167–191. Academic Press, New York.

Lindström, E. B., and Gunneriusson, L. 1990. Thermophilic bioleaching of arsenopyrite using *Sulfolobus* and a semi-continuous laboratory procedure. *J. Ind. Microbiol.*, 5: 375–382.

Lindström, E. B., and Sehlin, H. M. 1989. High efficiency of the plating of the thermophilic sulfur-dependent archaebacterium *Sulfolobus acidocaldarius. Appl. Environ. Microbiol.* 55:3020–3021.

Livesey-Goldblatt, E., Norman, P., and Livesey-Goldblatt, D. R. 1983. Gold recovery from arsenopyrite/pyrite ore by bacterial leaching and cyanidation. In *Recent Progress in Biohydrometallurgy* (G. Rossi and A. E. Torma, eds.), pp. 627–641. Associazione Mineraria Sarda, Iglesias, Italy.

Livesey-Goldblatt, E., Tunley, T. H., and Nagy, I. F. 1977. Pilot plant bacterial film oxidation (BACFOX process) of recycled acidified uranium plant ferrous sulphate leach solution. In *Conference Bacterial Leaching* (W. Schwartz, ed.), pp. 175–190. Verlag Chemie, Weinheim, FRG.

Lizama, H. M., and Suzuki, I. 1989. Bacterial leaching of a sulfide ore by *Thiobacillus ferrooxidans* and *Thiobacillus thiooxidans*. Part II: Column leaching studies. *Hydrometallurgy* 22:301–310.

Lundgren, D. G., and Malouf, E. E. 1983. Microbial extraction and concentration of metals. *Advan. Biotechnol. Process.* 1:223–249.

Lundgren, D. G., Valkova-Valchanova, M., and Reed, R. 1986. Chemical reactions important in bioleaching and bioaccumulation. *Biotechnol. Bioeng. Symp.* 16:7–22.

Mackintosh, M. E. 1978. Nitrogen fixation by *Thiobacillus ferrooxidans. J. Gen. Microbiol.* 105:215–218.

Marsh, R. M., and Norris, P. R. 1983a. The isolation of some thermophilic, autotrophic, iron- and sulphur-oxidizing bacteria. *FEMS Microbiol. Lett.* 17:311–315.

Marsh, R. M., and Norris, P. R. 1983b. Mineral sulphide oxidation by moderately thermophilic acidophilic bacteria. *Biotechnol. Lett.* 5:585–590.

Matin, A. 1978. Organic nutrition of chemolithotrophic bacteria. *Ann. Rev. Microbiol.* 32:433–468.

Mehta, A. P., and Murr, L. E. 1982. Kinetic study of sulfide leaching by galvanic interaction between chalcopyrite, pyrite, and sphalerite in the presence of *Thiobacillus ferrooxidans* (30°C) and a thermophilic microorganism (55°C). *Biotechnol. Bioeng.* 24: 919–924.

Merrettig, U., Wlotzka, P., and Onken, U. 1989. The removal of pyritic sulphur from coal by *Leptospirillum*-like bacteria. *Appl. Microbiol. Biotechnol.* 31:626–628.

Murayama, T., Konno, Y., Sakata, T., and Imaizumi, I. 1987. Application of immobilized *Thiobacillus ferrooxidans* for large-scale treatment of acid mine drainage. *Meth. Enzymol.* 136:53–540.

Murr, L. E., and Brierley, J. A. 1978. The use of large-scale testing facilities in studies of the role of microorganisms in commercial leaching operations. In *Metallurgical Applications of Bacterial Leaching and Related Microbiological Phenomena* (L. E. Murr, A. E. Torma, and J. A. Brierley, eds.), pp. 491–520. Academic Press, New York.

Murr, L. E., and Berry, V. K. 1979. Observations of a natural thermophilic microorganism in the leaching of a large scale experimental copper-bearing waste body. *Metall. Trans.* B10:523–531.

Muyzer, G., de Bruyn, A. C., Schmedding, D. J. M., Bos, P., Westbroek, P., and Kuenen, G. J. 1987. A combined immunofluorescence-DNA-fluorescence staining technique for enumeration of *Thiobacillus ferrooxidans* in a population of acidophilic bacteria. *Appl. Environ. Microbiol.* 53:66–664.

Nobar, A. M., Ewart, D. K., Alsaffar, L., Barrett, J., Hughes, M. N., and Poole, R. K. 1988. Isolation and characterisation of a mixed microbial community from an Australian mine. In *Biohydrometallurgy* (P. R. Norris and D. P. Kelly, eds.), pp. 530–531. Science and Technology Letters, Kew, Surrey, U.K.

Norris, P. R. 1983. Iron and mineral oxidation with *Leptospirillum*-like bacteria. In *Recent Progress in Biohydrometallurgy* (G. Rossi and A. E. Torma, eds.), pp. 83–96. Associazione Mineraria Sarda, Iglesias, Italy.

Norris, P. R. 1988. Bacterial diversity in reactor mineral leaching. In *Proceedings of the 8th International Biotechnology Symposium* (G. Durand, L. Bobichon, and J. Florent, eds.), pp. 1119–1130. Société Française de Microbiologie, Paris.

Norris, P. R. 1989. Mineral-oxidizing bacteria: metal-organism interactions. In *Metal-Microbe Interactions* (R. K. Poole and G. A. Godd, eds.), pp. 99–116. IRL Press, Oxford.

Norris, P. R., and Barr, D. W. 1985. Growth and iron oxidation by acidophilic moderate thermophiles. *FEMS Microbiol. Lett.* 28:221–224.

Norris, P. R., Barr, D. W., and Hinson, D. 1988. Iron and mineral oxidation by acidophilic bacteria: affinities for iron and attachment to pyrite. In *Biohydrometallurgy* (P. R. Norris and D. P. Kelly, eds.), pp. 43–59. Science and Technology Letters, Kew, Surrey, U.K.

Norris, P. R., and Kelly, D. P. 1982. The use of mixed microbial cultures in metal recovery. In *Microbial Interactions and Communities* (A. T. Bull and J. H. Slater, eds.), pp. 443–474. Academic Press, London.

Norris, P. R., Marsh, R. M., and Lindström, E. B. 1986a. Growth of mesophilic and thermophilic acidophilic bacteria on sulfur and tetrathionate. *Biotechnol. Appl. Biochem.* 8:318–329.

Norris, P. R., Nixon, A., and Hart, A. 1989. Acidophilic, mineral-oxidizing bacteria: the utilization of carbon dioxide with particular reference to autotrophy in *Sulfolobus*. In *Microbiology of Extreme Environments and Its Potential in Biotechnology* (M. S. Da Costa, J. C. Duarte, and R. A. D. Williams, eds.), pp. 24–43. Elsevier Applied Science, London, U.K.

Norris, P. R., Parrott, L., and Marsh, R. M. 1986b. Moderately thermophilic mineral-oxidizing bacteria. *Biotechnol. Bioeng. Symp.* 16:253–262.

Pinches, A., Chapman, J. T., te Riele, W. A. M., and van Staden, M. 1988. The performance of bacterial leach reactors for the preoxidation of refractory gold-bearing sulphide concentrates. In *Biohydrometallurgy* (P. R. Norris and D. P. Kelly, eds.), pp. 329–344. Science and Technology Letters, Kew, Surrey, U.K.

Pretorius, I.-M., Rawlings, D. E., and Woods, D. R. 1986. Identification and cloning of *Thiobacillus ferrooxidans* structural *nif* genes in *Escherichia coli*. *Gene* 45:59–65.

Pretorius, I.-M., Rawlings, D. E., O'Neill, E. G. Jones, W. A., Kirby, R., and Woods, D. R. 1987. Nucleotide sequence of the gene encoding the nitrogenase iron protein of *Thiobacillus ferrooxidans*. *J. Bacteriol.* 169:367–370.

Ralph, B. J. 1979. Oxidative reactions in the sulfur cycle. In *Biogeochemical Cycling of Mineral-Forming Elements* (P. A. Trudinger and D. J. Swaine, eds.), pp. 369–400. Elsevier, Amsterdam.

Ralph, B. J. 1985. Biotechnology applied to raw minerals processing. In *Comprehensive Biotechnology*, vol. 4 (C. W. Robinson and J. A. Howell, eds.), pp. 201–234. Pergamon Press, Oxford.

Rawlings, D. E., Pretorius, I.-M., and Woods, D. R. 1984. Construction of arsenic-resistant *Thiobacillus ferrooxidans* recombinant plasmids and the expression of autotrophic plasmid genes in a heterotrophic cell-free system. *J. Biotechnol.* 1:129–133.

Rodriguez-Leiva, M., and Tributsch, H. 1988. Morphology of bacterial leaching patterns by *Thiobacillus ferrooxidans* on synthetic pyrite. *Arch. Microbiol.* 149:401–405.

Sakamoto, K., Yagasaki, M., Kirimura, K., and Usami, S. 1989. Resistance acquisition of *Thiobacillus thiooxidans* upon cadmium and zinc ion addition and formation of cadmium ion-binding and zinc ion-binding proteins exhibiting metallothionein-like properties. *J. Ferment. Bioeng.* 67:266–273.

Sato, A., Fukumori, Y. Yano, T., Kai, M., and Yamanaka, T. 1989. *Thiobacillus ferrooxidans* cytochrome *c*-552: purification and some of its molecular features. *Biochim. Biophys. Acta* 976:129–134.

Schrader, J. A., and Holmes, D. S. 1988. Phenotypic switching of *Thiobacillus ferrooxidans*. *J. Bacteriol.* 170:3915–3923.

Segerer, A., Neuner, A., Kristjansson, J. K., and Stetter, K. O. 1986. *Acidianus infernus* gen. nov., sp. nov., and *Acidianus brierleyi* comb. nov.: facultatively aerobic, extremely acidophilic sulfur-metabolizing archaebacteria. *Int. J. Syst. Bacteriol.* 36: 559–564.

Segerer, A., Stetter, K. O., and Klink, F. 1985. Two contrary modes of chemolithotrophy in the same archaebacterium. *Nature* (London) 313:787–789.

Shiratori, T., Inoue, C., Sugawara, K., Kusano, T., and Kitagawa, Y. 1989. Cloning and expression of *Thiobacillus ferrooxidans* mercury ion resistance genes in *Escherichia coli*. *J. Bacteriol.* 171:3458–3464.

Sohn, H. Y. 1979. Fundamentals of the kinetics of heterogeneous reaction systems. In *Rate Processes of Extractive Metallurgy* (H. Y. Sohn and M. E. Wadsworth, eds.), pp. 1–42. Plenum Press, New York.

Southwood, M. J., and Southwood, A. J. 1986. Mineralogical observations on the bacterial leaching of auriferous pyrite: a new mathematical model and implications for the release of gold. In *Fundamental and Applied Biohydrometallurgy* (R. W. Lawrence, R. M. R. Branion, and H. G. Ebner, eds.), pp. 98–113. Elsevier, Amsterdam.

Stetter, K. O. 1986. Diversity of extremely thermophilic archaebacteria. In *Thermophiles: General, Molecular, and Applied Microbiology* (T. D. Brock, ed.), pp. 39–74. John Wiley & Sons, New York.

Stevens, C. J., Dugan, P. R., and Tuovinen, O. H. 1986. Acetylene reduction (nitrogen fixation) by *Thiobacillus ferrooxidans*. *Biotechnol. Appl. Biochem.* 8:351–359.

Stevens, C. J., and Tuovinen, O. H. 1986. Ferrous ion oxidation, nitrogen fixation (acetylene reduction), and nitrate reductase activity by *Thiobacillus ferrooxidans*. In *Proceedings of the 2nd Annual Meeting of BIOMINET* (R. G. L. McCready, ed.), pp. 37–45. CANMET, Ottawa.

Sugio, T., Domatsu, C., Munakata, O., Tano, T., and Imai, K. 1985. Role of ferric ion-reducing system in sulfur oxidation of *Thiobacillus ferrooxidans*. *Appl. Environ. Microbiol.* 49:1401–1406.

Torma, A. E. 1978. Complex lead sulfide leaching by microorganisms. In *Metallurgical Applications of Bacterial Leaching and Related Microbiological Phenomena* (L. E. Murr, A. E. Torma, and J. A. Brierley, eds.), pp. 375–387. Academic Press, New York.

Torma, A. E. 1987. Impact of biotechnology on metal extractions. *Miner. Proc. Extr. Metall. Rev.* 2:289–330.

Torma, A. E. 1988. Leaching of metals. In *Biotechnology*, vol. 6B: *Special Microbial Processes* (H.-J. Rehm and G. Reed, eds.), pp. 367–399. Verlag Chemie, Weinheim, FRG.

Tributsch, H., and Bennett, J. C. 1981a. Semiconductor-electrochemical aspects of bacterial leaching 1. Oxidation of metal sulfides with large energy gaps. *J. Chem. Technol. Biotechnol.* 31:565–577.

Tributsch, H., and Bennett, J. C. 1981b. Semiconductor-electrochemical aspects of bacterial leaching 2. Survey of rate-controlling sulfide properties. *J. Chem. Technol. Biotechnol.* 31:627–635.

Tsuchiya, H. M. 1977. Leaching of Cu-Ni sulphide concentrate from the Duluth gabbro. In *Conference Bacterial Leaching* (W. Schwartz, ed.), pp. 101–106. Verlag Chemie, Weinheim, FRG.

Tsuchiya, H. M., Trivedi, N. C., and Schuler, M. L. 1974. Microbial mutualism in ore leaching. *Biotechnol. Bioeng.* 16:199–995.

Tuovinen, O. H. 1977. Pathways of the utilization of inorganic sulphur compounds by *Thiobacillus ferrooxidans*. In *Conference Bacterial Leaching* (W. Schwartz, ed.), pp. 9–20. Verlag Chemie, Weinheim, FRG.

Tuovinen, O. H., Silver, M., Martin, P. A. W., and Dugan, P. R. 1981. The Agnew Lake uranium mine leach liquors: chemical examinations, bacterial enumeration, and composition of plasmid DNA of iron-oxidizing thiobacilli. In *Proceedings of the Inter-*

national Conference on Use of Microorganisms in Hydrometallurgy, pp. 59–69. Hungarian Academy of Sciences, Pećs, Hungary.
Tuovinen, O. H., Niemelä, S. I., and Gyllenberg, H. G. 1971. Tolerance of *Thiobacillus ferrooxidans* to some metals. *Antonie van Leeuwenhoek* 37:489–496.
van Aswegen, P. C., Haines, A. K., and Marias, H. J. 1988. Design and operation of a commercial bacterial oxidation plant at Fairview. In *Randol International Gold Conference*, pp. 144–147. Perth, Western Australia.
Vartanyan, N. S., Pivovarova, T. A., Tsaplina, I. A., Lysenko, A. M., and Karavaiko, G. I. 1988. New thermoacidophilic bacterium of the genus *Sulfobacillus*. Mikrobiologiya 57:268–274.
Vaughan, D. J. 1984. Electronic structures of sulfides and leaching behavior. In *Hydrometallurgical Process Fundamentals* (R. G. Bautista, ed.), pp. 23–40. Plenum Press, New York.
Visca, P., Bianchi, E., Polidoro, M., Buonfiglio, V., Valenti, P., and Orsi, N. 1989. A new solid medium for isolating and enumerating *Thiobacillus ferrooxidans*. *J. Gen. Appl. Microbiol.* 35:71–81.
Wadsworth, M. E. 1979. Hydrometallurgical processes. In *Rate Processes of Extractive Metallurgy* (H. Y. Sohn and M. E. Wadsworth, eds.), pp. 133–241. Plenum Press, New York.
Wadsworth, M. E. 1984. Heterogeneous rate processes in the leaching of base metal sulfides. In *Hydrometallurgical Process Fundamentals* (R. G. Bautista, ed.), pp. 41–76. Plenum Press, New York.
Wichlacz, P. L., and Thompson, D. L. 1988. The effect of acidophilic bacteria on the leaching of cobalt by *Thiobacillus ferrooxidans*. In *Biohydrometallurgy* (P. R. Norris and D. P. Kelly, eds.), pp. 77–86. Science and Technology Letters, Kew, Surrey, U.K.
Yates, J. R., Lobos, J. H., and Holmes, D. S. 1986. The use of genetic probes to detect microorganisms in biomining operations. *J. Ind. Microbiol.* 1:129–135.
Yates, J. R., and Holmes, D. S. 1986. Molecular probes for the identification and quantitation of microorganisms found in mines and mine tailings. *Biotechnol. Bioeng. Symp.* 16:301–309.
Zillig, W., Stetter, K. O., Wunderl, S., Schulz, W., Priess, H., and Scholz, I. 1980. The *Sulfolobus-"Caldariella"* group: taxonomy on the basis of the structure of DNA-dependent RNA polymerases. *Arch. Microbiol.* 125:259–269.
Zillig, W., Yeats, S., Holz, I., Böck, A., Gropp, F., Rettenberger, M., and Lutz, S. 1985. Plasmid-related anaerobic autotrophy of the novel archaebacterium *Sulfolobus ambivalens*. *Nature* (London) 313:789–791.
Zillig, W., Yeats, S., Holz, I., Böck, A., Rettenberger, M., Gropp, F., and Simon, G. 1986. *Desulfurolobus ambivalens*, gen. nov., sp. nov., an autotrophic archaebacterium facultatively oxidizing or reducing sulfur. *Syst. Appl. Microbiol.* 8:197–203.

Index

A. fumigatus, 247
Acetobacter, 7, 57, 111, 209
  aceti, 57
Acetogenium, 210
  kivui, 244
Acidianus, 407
  brierleyi, 390, 401
Acidiphilium, 392, 406–407
Acid tolerance, 162–163
Acinetobacter, 296, 346
Actinomucer, 10
  elegans, 8
Actinomyces, 236
  levoris, 222
Aerobacter, 344
Aflatoxin, 158–159
Agrobacterium, 364
Alcaligenes, 300, 346, 360
  calcoaceticus, 346
  faecalis, 239–240
Amensalism, 119–123, 124, 207, 262–263, 270, 277
Amylomyces, 8
Antibiotics:
  in bovine rumen, 365
  commercial production, 222–223
  as inhibitors, 10
Arthrobacter, 223, 297
  globiformis, 297
  simplex, 11, 224, 226–227
Ashbya gossypi, 221
Aspergillus, 41, 158, 223, 234, 236
  awamori, 213, 215, 237, 249–251
  flavus, 141, 158, 159, 222
  nidulans, 222
  niger, 10, 249
  oryzae, 4, 8, 14, 156, 158, 161
  parasiticus, 141, 158
  phoenicis, 235–236
  soyae, 14, 158
  wentii, 235–236
Azospirillum, 238

BACFOX (bacterial film oxidation) process, 412
Bacillus, 14, 33, 41, 59, 178–179, 182, 223, 244, 247, 344, 360, 366
  cereus, 124, 155, 344
  coagulans, 178, 213
  lichenoformis, 154, 178
  macerans, 238
  polymyxa, 178
Bacillus (*Cont.*):
  sphaericus, 224
  subtilis, 154, 344
Bacteria:
  acetic acid, 52, 57
  anaerobic thermophilic, 216–217
  chitinoclastic, 252
  hydrolytic fermentative, 268–270
  lactic acid (*see* Lactic acid bacteria)
  methanogenic, 268, 271
  in secondary fermentation, 93–95
  sulfidogenic (*see* Sulfate-reducing bacteria)
  syntrophic acetogenic, 270–272, 274–278, 281–284
  and yeasts, 14, 20, 22–24, 52–55, 124
  (*See also specific bacteria*)
Bacteriophages, 56
  phage control methods, 118
  phage-host interactions, 117–118
  phage infections, 4–5, 109, 117–118, 126
  phage-mediated transductions, 108, 116
  phage-resistant mutants, 109, 113–114
Bacterium, 41
Bacteroides cellulosolvens, 245, 269
Barm (bread leaven), 20
Basidiomycetes, 11
Beauvaria, 10
Beijerinckia lactinogenes, 391
Bifidobacterium bifidum, 223
Bioaugmentation, 311, 326–327, 328
Biocide resistance, 365
Biological fouling, 343–344, 365
Biological leaching processes (*see* Mineral biotechnology)
Biomethanation:
  ecosystems, 261–262, 272
  rate-limiting steps in, 277–278, 285
  syntrophic, 279–280, 281
Bioreactors, 297–280, 324–325
  biofilm, 326
  fixed-film, 412
  UASB (upflow anaerobic sludge blanket), 279, 280, 321, 324–325
Biotechnology, mixed cultures in (*see* Vegetable fermentations)
Biotin, 222, 296
Blakeslea trispora, 220–221
Botrytis cinerea, 41, 43, 57, 306
Botulinum toxin, 137, 147–148, 150–151

Botulism, 150–151, 153, 155
Breadmaking:
  baker's yeast, *Saccharomyces cerevisae*, 19–20, 23, 26–27, 30–33
  flat breads, 29–33
  historical development, 17–18
  lactic acid bacteria, 19–20, 22–25, 27–29, 31–33
  microbial leavens used, 19
  microflora and flavor, 23
  microflora of bread doughs, 20–33
  sourdough breads, 24–29
  white and variety breads, 20–24
Brettanomyces, 41, 47, 59
  clausenii, 238
Brevibacterium, 111, 298
  linens, 296
Byssochlamys, 60

Cabbage:
  concentration of salt, 76
  distribution of LAB, 75
  growth of microorganisms, 76
  (*See also* Sauerkraut)
Campylobacter jejuni, 136
Candida, 10, 33, 41, 45, 51
  colliculosa, 45
  krusei, 26, 50
  lipolytica, 209
  metalondinsis, 222
  milleri, 19, 26–27, 28–29
  parapsilosis, 222
  pulcherrima, 41, 45–46, 50–51
  stellata, 45, 50, 51
  tropicalis, 222
  utilis, 222, 237, 240, 248, 254
  wickerhamii, 12, 238
Cassava, 33, 150, 163
"Cathode depolarization," 341, 343, 353
Cellulomonas, 239–240
  flavigena, 240
  gelida, 238
Cellulose, fermentation of, 11, 213, 215–218
  advantages of mixed cultures, 234–236
  enzymatic degradation, 234–247
  microorganisms involved, 234–247
  mono- versus cocultures, 237, 241–242
  mutualism of mixed culture, 239, 245
  and nitrogen fixation, 238–239
  pretreated substrates, 241, 244
  role of cellulase, 234–235, 237, 247
Cellvibrio, 236
Cereal products, fermented, 154–158
Cheesemaking:
  genera used, 111–112
  raw versus pasteurized milk, 119
  safety of, 119, 159
Chemical oxygen demand (*See* COD)
Chlorella, 222
Chromobacterium, 41
Cladosporium resinae, 344
Climax community, 161
Clostridium, 210, 244–245, 245–254, 344
  acetobutylicum, 217–218, 253
  bifermentans, 175
  botulinum, 141, 146–149, 151, 153–156, 163
  butyricum, 93, 137, 175, 239, 255
  cellulyticum, 217
  formicoaceticum, 212
  paraputrificum, 175
  perfingens, 149
  saccharolyticum, 245, 269
  sphenoides, 176
  sporogenes, 150, 175
  tertium, 94
  thermoaceticum, 212
  thermocellum, 210, 213, 216, 218, 240–241, 244
  thermohydrosulfuricum, 216, 240–241, 250
  thermosaccharolyticum, 216, 240–241, 250
  thermosulfurogenes, 216, 250
  tyrobutyricum, 175–176
COD conversion rate, 278–279, 281, 284–285
Cometabolism, 296, 297–298
Commensalism, 123–124, 207, 248–249, 262, 274
Commercial chemical production by mixed cultures, 205–232
  acetic acid and its salts, 209–212
  acetone-butanol, 217–218
  antibiotics, 222–223
  biotin, 222
  carotenoids, 220–221
  citric acid, 209
  cocultures, 206, 210, 215–217, 220, 224–225
  commodity chemicals, 208–218
  ethanol, 214–217
  lactic acid, 212–214
  malic acid, 208–209
  organic acids, 208–214, 219
  propionic acid, 219
  relative costs, 214, 219
  required purity, 208
  riboflavin, 221–222
  specialty chemicals, 218–227
  steroid transformations, 11–12, 223–227
  theoretical aspects, 207–208
  vitamins, 219–222
  "wild" fermentations, 217
Competition, 115–116, 207, 248–249, 262–263
Conjugation, 107–108, 113, 116, 328
Cooxidation, 307

Corynebacterium, 33, 223, 360
  equi, 227
Crenotrix, 344
Cryptococcus, 41, 51
Cucumbers:
  bacteria in interior, 73
  bloater damage, 80, 83–84, 93
  brining, 72–74
  controlled fermentation, 81, 83–85
  fermentation for preservation, 72
  microbial load on surface, 72
  natural fermentation, 80
  open versus closed tanks, 84, 95–97
  pasteurization, 71–72
  primary fermentation, 80–85
  problem of softening, 72–73
  pure culture fermentation, 80–81
  purging of CO_2 from brines, 83
  significance of the flower, 72
  spoilage in secondary fermentation, 94
Curtobacterium, 296
Curvularia, 223
  lunata, 224, 226

Dairy fermentations, 105–133
  acetaldehyde and "green" flavor, 112
  amensalism, 119–123
  antipathogenic action, 123
  centralized preparation of starters, 116
  cheesemaking, 111–112, 119, 159
  commensalism, 123–124
  competition, 115–116
  genera, 111–112
  historical pespective, 105–111
  interactions in starter cultures, 114–126
  isolation of new strains, 109
  lactic cultures as preservatives, 123
  microbiology of, 105–106
  mixed dairy cultures, 111–114
  mixed starter systems, 126–127
  mutualism (symbiosis), 124–126
  parasitism, 117–118
  phage-control methods, 118, 126–127
  phage-host interaction, 117–118
  phage-resistant mutants, 109, 113–114
  postfermentation contamination, 152
  safety of, 119, 121, 151–153
  species, subspecies used, 112–113
  starter culture development, 109–111
  strain evolution, 107–108
  strains used, 113
  undefined or defined cultures, 126–127
  yeast and *Lactobacillus*, 124
Debaryomyces, 51
Desulfobacter, 352
  postgatei, 266
Desulfobacterium indolicum, 318
Desulfomonile tiedjei, 320
Desulfotomaculum, 344–345
Desulfovibrio, 344–345
  desulfuricans, 253–254, 319
  vulgaris, 184, 265, 275–276, 281, 343, 354
Desulfurolobus ambivalens, 390
Detoxification of hazardous waste, 12–13, 293–340
  advantages of mixed cultures, 293–294
  by aerobic mixed cultures, 300–312
  aliphatic hydrocarbons, 306–307, 319–320
  by anaerobic mixed cultures, 312–325
  aromatic hydrocarbons, 301–302, 314–319
  aromatic ring cleavage, 302–303
  bioaugmentation, 311, 326–327, 328
  biorestoration of ground water, 311–312, 315, 326
  biostimulation, 311
  biotransformation of alkenes, 307–308
  degradation of benzene derivatives, 13, 299, 301–305, 314–323
  degradation of styrene, 297
  degradative mechanisms, aerobic: 301–309
    anaerobic, 314–324
  dehalogenation, 304–305, 305
  ecological role of microbial communities, 295–300
  ecosystems for waste treatment, 309–312, 324–325
  fixed-film processes, 310
  genetically engineered bacteria, 327–328
  halogenated aliphatics, 307–308, 323–324
  halogenated aromatics, 303–304, 320–323
  insecticides and herbicides, 297–300, 311, 320–321, 327–328
  liquid-waste reactors, 309–310
  mechanisms of reductive dehalogenation, 320–323
  by novel mixed-culture systems, 325–328
  sequential anaerobic-aerobic mixed culture, 326
  solid-waste composting, 310–311
  suspended growth processes, 310
  treatment of landfill leachates, 324, 325
  types of microbial communities, 295–300
Directed evolution, 328
DNA:
  conjugation, 107, 116
  fluorescence, 417
  probes, 363–364, 417
  recombinant, 109

Edwardsiella tarda, 149
E.lactis, 162

Enterobacteriaceae, 73, 91, 178
Environmental problems:
  aliphatic hydrocarbons, 306
  salt disposal, 72, 77
  (*See also* Detoxification of hazardous waste)
Eremothecium ashbyii, 221
Erwinia herbicola, 178
Erysipelothrix rhusiopathiae, 149
Escherichia coli, 15, 119, 124, 136, 150, 152, 178, 253, 419
Ethanol:
  from cellulose, 237–238, 241, 244
  commercial production, 11, 47, 214–217
  fermentation versus chemical synthesis, 214
  from sorghum, 11
  from starch, 248–250
  in wine production, 37–38

Fish fermentations, 8, 149–151
  botulism, 150–151
  pathogens, 149–151
  problem of polluted waters, 150
Flavimonas, 346
Flavobacterium, 294, 296, 304, 344, 346
  farinofermentans, 141, 157, 158
Fruit fermentation, 153–154
Fungi, 1–2, 10, 163, 237
  bacteria and, 11, 13
  in benzene degradation, 13
  in commercial chemical production, 208–209, 220–221, 223
  on grapes, 41, 43
  in silage fermentation, 177–178
  tea fungus, 7
Fusarium, 236
  oxysporum, 222
  solani, 222
Fusidium, 236

Gene transfer, 107–108, 109, 116, 137
Geotrichum, 33, 111
Gluconobacter, 57, 209
"Graphitization," 343

Hafnia alvei, 178
Hanseniaspora, 43
Hansenula, 51
  anomala, 25, 50, 55, 61
Hedonic scale, 29
Hyphomicrobium, 296, 309

Initiation of fermentation, 71–77
Insulin production, 15
Interspecies hydrogen transfer, 271

Klebsiella, 178
  aerogenes, 253
  pneumoniae, 244
Kloeckera, 39, 43, 45
  apiculata, 41, 45–47, 50–51, 55
Kluyveromyces, 51
  fragilis, 50, 214–215, 218

LAB (*see* Lactic acid bacteria)
Lactic acid bacteria:
  bacteriocins produced by, 121–122
  in breadmaking, 19–20, 22–25, 27–31
  conjugation of strains, 107–108
  in dairy industry, 107–127*passim*, 152
  genetic systems in, 107
  heterofermentative, 88, 172, 181, 182
  homofermentative, 88, 172, 181, 183, 186, 212
  inhibitors of, 76–77
  insertion sequences in, 108
  in lactic acid production, 212–214
  pathogen inhibition by, 119, 152, 160, 163
  salt and acid tolerance, 163
  in silage fermentation, 171, 172–175, 179–183, 185–187, 193–195
  in vegetable fermentation, 71, 73, 75–81, 83–88, 90–93
  in winemaking, 53–54, 56, 57–58
  (*See also specific bacteria*)
Lactobacillus, 37, 58, 78, 107–108, 111–112, 123–124, 153–154, 172, 173, 219
  acidophilus, 8, 112, 124, 149, 212
  bavaricus, 87
  brevis, 25, 27, 29, 33, 85, 112, 172, 174, 181
  buchneri, 172
  bulgaricus, 3, 113, 124, 125, 212, 223
  casei, 25, 32, 108, 112–113, 172, 183, 212, 219
  cellobiosus, 84
  curvatus, 145
  delbrueckii, 27, 112, 212–213
  fermentum, 25, 27, 31, 32, 108, 112
  fructivorans, 39
  helveticus, 107, 112–113
  kefir, 112, 124
  koumiss, 124
  murinus, 108
  plantarum, 25, 27, 32, 77, 83–85, 90, 93, 95, 145, 172, 174, 181, 183, 186
  reuteri, 108
  sake, 145
  sanfrancisco, 28–29
Lactococcus, 107–108, 111, 124
  cremoris, 107, 112, 119, 121, 124
  diacetylactis, 123
  fermentum, 108
  helveticus, 107, 112–113
  lactis, 108, 111–112, 119, 121
  murinus, 108

Lactococcus (*Cont.*):
reuteri, 108
Legumes, fermented, 154–158
miso, 3–4, 8, 14, 156, 161
pathogens, 155
tempeh, 7, 155
tofu, 156
Leptospirillum ferrooxidans, 384, 387–388, 393, 396–397, 402, 406, 412, 415
Leuconostoc, 33, 37, 78, 107, 108, 111, 172, 174, 181
buchneri, 181
cellobiosis, 181
cremoris, 111
diacetylactis, 111–113, 123
fermentum, 181
lactis, 124
mesenteroides, 31, 33, 85–86, 88, 90, 92, 111, 153, 172, 174, 181
oenos, 41, 56
Listeria, 119, 121
monocytogenes, 136, 146–149, 152, 178
Listeriosis, 148
"Lysis-from-without," 117

Malolactic fermentation, 37–39, 41, 53–54, 56–57
Meats, fermented:
Clostridium botulinum hazard, 147–148
factors favoring *Staphylococcus aureus*, 146–147
pathogens, 146–149
Salmonella contamination, 147
salt and pH requirements, 148
sausage fermentation, 145–149
starter cultures used, 145
Metallosphaera sedula, 390
Methanobacillus omelianski, 298
Methanobacterium, 298, 320
formicicum, 275–276, 281–282, 284
ruminatium, 245
thermoautotropicum, 241
Methanobrevibacterium arboriphilum, 275
Methanogenesis, 261–292
amenalistic interaction, 277
anaerobic biodegradation, 261, 285
anaerobic ecological niches, 263, 284
biomethanation ecosystems, 261–262, 272
carbon flow pattern, 268
COD conversion rate, 278–279, 281, 284–285
commensal interaction, 274–275
concentration gradients of electron acceptors, 266
ecoengineering anaerobic pathway design, 284–285
environmental metabolic stratification, 267

Methanogenesis (*Cont.*):
examples of physical barriers, 263
formate:bicarbonate electron cycle model, 282–284
geochemical parameters, 263–267
hydrogen threshold concept, 264–266
hydrolytic fermentative bacteria, 268–270
metabolic and ecophysiological context, 268
metabolic compartmentalization on biocatalyst level, 281–284
metabolic compartmentalization on bioreactor level, 279–280
methanogenic bacteria, 268, 272–275
methanogenic carbon flow coordination, 277–279
methanogen-SRB competition, 275–277
microbial interactions defined, 262–263
mutual exclusion of oxidation reactions, 264
mutualistic interaction, 274, 278
redox potentials of related reactions, 263–264, 266
species juxtapositioning, 280–281
sulfidogenic bacteria, 275–277
syntrophic acetogenic bacteria, 270–272, 274–276, 278, 281–284
Methanogens, 268, 271, 272–275
acetoclastic, 272–274, 277, 284
Methanosarcina:
acetivorans, 274
barkeri, 255, 265, 274
mazei, 274
Methanospirillum, 299, 320
hungatei, 265
Methanothrix, 320
sohngenii, 274
Methanotrophs, 309, 326
Metschnikowia, 46
pulcherrima, 41, 46, 47
MIC (*see* Microbially influenced corrosion)
Microbial detoxification, definition, 294
Microbial interactions, definitions, 114–126, 262–263
Microbially influenced corrosion, 341–372
adhesion and, 348, 360, 365, 367
in alloys, 357, 360, 361
batch culture versus continuous flow, 347
biofilm models, 348–349
"cathode depolarization" mechanism, 341, 343, 353
cathodic protection, 361–362
of concrete sewage pipes, 345
"conditioning films," 347
of containers for radioactive waste, 346
differential oxygen cell, 349, 352–353
dissecting MIC community structure, 362–365

Microbially influenced corrosion (*Cont.*):
economic effects, 343–346
effects of surface perturbations, 356–362
energy requirements, 348
fungal contaminants, 343
heat-affected zone (HAZ) of weld, 357–359
hydrogen embrittlement, 355
influence of biofilms, 346, 347–349, 365, 367
interactions among consortial members, 351–353
intergranular attack, 357
lipid analyses, 362–363
"mature biofilm," 348
as measure of consortial production, 350–351
mechanisms, 353–356
MIC testing, 346–347
modification of in situ environment, 347–350
molecular methods to identify microbes, 363–364
organic acid production, 352–353
possible solutions, 342
radiolabeled sulfur studies, 350–351
role of consortia, 341–372
role of fermentative bacteria, 342
sulfate-reducing bacteria, 341–342, 345, 346, 352, 354–355, 360
sulfate reduction and corrosion rate, 350–351
temperature effects, 348
toxic corrosion products, 347
treatment considerations, 365–367
of underground iron pipelines by SRB, 343
weld effects, 356–360
Micrococcus, 41, 111, 304, 360
varians, 145
Mineral biotechnology, 373–427
accessing refractory gold, 375, 413–418
acidophilic organisms, 384, 385, 386–387, 391, 398, 406, 407
bacteria on mineral surfaces, 377–378
biological oxidation of pyrite, 375, 378–380, 387, 393, 414, 416–418
dump and heap leaching processes, 403–410
effect of nitrogen, 399
effect of O_2 and CO_2, 399–401
environmental parameters, 394–403
extreme thermophiles, 390–391, 398
(*See also Sulfolobus* species)
factors in microbial activity and metal leaching, 404–406
galvanic coupling effect, 380
Mineral biotechnology (*Cont.*):
heterotrophic versus autotrophic organisms, 392–393
iron substrate, 375, 380, 383–385, 412
mesophiles, 386–388
metal toxicity, 401–403
mixed cultures, 391–394
moderate thermophiles, 388–389, 398, 401
need for genetically engineered strains, 416–419
problems of, 374
reactions influencing pH, 394–397
role of thiobacilli, 374, 382, 392, 403
shrinking-core model, 381
shrinking-particle model, 381
substrates, 375–385
sulfur primary substrate, 373–374, 375, 380, 382–383
tank leaching, 412–418
temperature effects, 397–398
underground leaching processes, 410–412
uranium leaching, 375, 378–379, 403–404, 410–411
Mixed-culture fermentations:
advantages of, 3–5, 10–11
classification, 7–8
control of, 8–10
definition, 1, 206
disadvantages of, 6–7
future of, 10–15
in nature, 2–3, 11
sequential transformations versus, 223
use of inhibitors, 9–10
Molds, 123, 135, 154, 163
antibiotic-like action, 155
in *koji* production, 156–157
mold starter cultures, 149
"Molecular breeding," 327
Monascus, 158
purpureus, 7
Monilia nivea, 222
Monoculture, 7–8
Mucor, 41
Multiculture, 8
Mutant cultures, 113–114, 328
Mutualism, 124–126, 207, 239, 245, 262, 274, 278
Mycobacterium, 308, 367
phlei, 224
Mycotoxin formation, 158–159

National Pollution Discharge Elimination Standards (NPDES), 345
Neurospora, 158, 159
intermedia, 8
Neutralism, 207

Nisin, 121
Nocardia, 223, 226, 296
  restrictus, 225
Norwalk agent, 136

Obligately anaerobic organisms, 342, 343, 348, 352
  (*See also* Sulfate-reducing bacteria)
Oleuropein, 76–77
Olives:
  air purging of brines, 91–92
  anaerobic, buried tanks, 97–98
  controlled fermentation, 92–93
  fermentation by yeasts versus LAB, 77, 91
  gas pocket formation, 91
  natural fermentation, 90–92
  primary fermentation, 90–93
  role of oleuropein in fermentation, 76–77
  spoilage by clostridia, 93
  stuck fermentations, 92
  treatment with alkali, 76–77, 91
  "yeast spots," 90–91
  zapatera spoilage, 94
Oriental food fermentations, 1–16
  koji, 2, 3–4, 9, 14, 156–157, 161
  miso, 3–4, 8, 14, 156, 161
  ontjom, 7
  ragi, 2, 4, 8, 9
  sake, 8, 15
  shoyu, 3, 8–9, 15
  sufu (Chinese cheese), 8
  tempeh, 7, 155
  tofu, 156
Osmotic stress, 162

Pacyysolen tannophilus, 247
Parasitism, 117–118, 207
Pasteurization, 59, 71–72, 119, 151
Pediococcus, 8, 37, 78, 108, 111, 121, 148, 172, 173, 186
  acidilactici, 145, 172
  cerevisiae, 31, 33, 85, 90, 93, 147
  halophilus, 9, 156–157, 161
  pentosaceus, 145, 172
Penicillium, 41, 111, 158, 344
  camemberti, 159
  corylophilum, 239
  roqueforti, 159
Phage (*see* Bacteriophage)
Pichia, 45, 51
  membranefacieus, 60
  stipitis, 238
Polyculture, 8
Polyporus, 240
Polysaccharides, enzymatic degradation of, 233–259
  and biomass conversion, 234–255

Polysaccharides, enzymatic degradation of (*Cont.*):
  cellulose materials, 234–247
  chitin, 252–254
  mechanisms, 234–255
  pectin, 254–255
  starch-into-ethanol conversion, 248–250
  starchy materials, 247–252
  (*See also* Cellulose, fermentation of)
Postfermentation, 95–98, 152
Predation, 207, 262–263
Preservatives, 123, 149
Primary fermentation, 77–93
  cucumbers, 80–85
  genera of lactic acid bacteria, 77–78
  kimchi, 88, 90
  olives, 90–93
  sauerkraut, 85–88
  yeasts, 78–80
Propionibacterium, 111, 219
  shermani, 123, 219
Pseudomonas, 41, 294, 296, 298, 302, 304–305, 309, 312, 328, 344, 346, 347, 360
  alcaligenes, 297
  atlantica, 355
  cepacia, 327
  cocovenenans, 141, 157
  methanica, 306–307
  putida, 297

Rhizobium, 2
Rhizopus, 8, 12, 41, 251–252
  arrhizus, 208–209
  oligosporus, 7, 157
Rhodococcus chlorophenolicus, 305
Rhodopseudomonas palustris, 313
Rhodotorula, 41
  mucilaginosum, 222
RNA:
  dsRNA and killer activity, 51
  ribosomal, in MIC research, 363–364
Ruminicoccus flavefaciens, 245

SAB (*see* Syntrophic acetogenic bacteria)
Saccharomyces, 38–39, 41, 44, 47–51, 55
  carlsbergensis, 222
  cerevisiae, 84, 222, 237–238, 240, 249
  dairensis, 27
  diastaticus, 249
  rouxii, 156–157, 161
  uvarum, 214–215
Saccharomycopsis, 251–252
  fibuligera, 9, 248–249, 254
Safety of fermented foods, 135–169
  adaptation to osmotic and acid stress, 162–163

Safety of fermented foods (*Cont.*):
  bacterial pathogens, 137–141
  bongkrek poisoning, 157–158
  botulism, 150, 153, 155
  control of pathogens and toxins, 145–158
  dairy products, 151–153
  diseases from fermented foods, 137–141
  exclusion of pathogens, 160–161
  factors affecting safety, 143
  fish, 149–151
  fruits and vegetables, 153–154
  importance of foodborne disease, 136–137
  incidence of disease, 141, 143–145
  legume and cereal products, 154–158
  meats, 145–149
    (*See also* Meats, fermented)
  mold-yeast-LAB connections, 163
  mycotoxin formation, 158–159
  role of salt and pH, 148, 153–154, 156, 160
  spoilage and pathogenic organisms inhibited, 136
  viruses, 137, 141, 159
Salmonella, 136, 141, 146–149, 152, 156
  oranienburg, 162
  typhimurium, 150, 155
Salt-tolerant organisms, 156, 161, 162–163
Sauerkraut:
  American versus European processes, 86
  controlled fermentation, 86–88
  gaseous and nongaseous stages, 87–88
  natural fermentation, 85–86
  primary fermentation, 85–88
  purging of CO_2 from brine, 88
  role of temperature and salt concentration, 85
  spoilage problems, 85–86
  succession of LAB species active, 85, 153
Schizophyllum commune, 209
Schizosaccharomyces romys, 222
Schwanniomyces:
  alluvius, 249
  castellii, 249
  uvarum, 249
Scytalidium lignicola, 237
Secondary fermentation, 93–95
Serratia marcescens, 297
Sewage digestion, 277, 283
Sewage sludge, cultures from, 210, 222, 245, 313, 315, 318, 324
Shigella, 150
Silage fermentation, 171–204
  acid detergent fiber (ADF), 184
  additives, 190, 197
  anaerobic environment, 171–172
Silage fermentation (*Cont.*):
  Bacillus, 178–179
  bacterial inoculant, 186
  buffering capacity of crop, 189, 197
  changes in the crop, 183–184
  chop length, effect of, 192
  clostridia, 175–177, 180, 188–189, 191, 193, 195–197
  effect of wilting, 188
  enterobacteria, 178
  factors affecting dynamics, 184–192
  fermentation products, 181–183
  fungi, 177–178
  initial microbial populations, 186–187
  lactic acid bacteria, 171, 172–175, 179–183, 185–187, 193–195
  Listeria, 179
  low pH, 171–172
  Maillard or browning reaction, 191
  models of, 192–196
  moisture content, 187–189
  neutral detergent fiber (NDF), 184
  normal ensiling dynamics, 179–184
  oxygen effects, 183, 184–186
  products of LAB fermentation, 174–175
  research needs, 196–198
  shifts in dominant LAB, 180–181
  shifts in major microbial groups, 179–180
  sugar content, 189–190, 197
  temperature effects, 190–192
"S organism," 298
Species juxtapositioning, 280–281, 285
Sphaerotilus, 344
SRB (*see* Sulfate-reducing bacteria)
Staphylococcus:
  aureus, 146–147, 150, 155, 156
  carnosus, 145
  typhimurium, 156
  xylosus, 145
Steroids, transformation to drugs, 11–12, 223–227
Streptococcus, 77, 107, 111, 172, 173, 178, 251–252
  bovis, 183
  cremoris, 113
  faecalis, 31, 90, 174
  faecium, 172, 174, 186
  lactis, 113, 212
  thermophilus, 3, 108, 113, 117, 124–125, 125
Streptomyces, 222–223
  lactis, 251
  mycoheptinicum, 222
  roseochromogenes, 11, 224–225
Stuck fermentations, 92
Sulfate-reducing bacteria, 275–277, 409
  in microbially influenced corrosion, 342–346, 352, 354–355, 360

Sulfobacillus thermosulfidooxidans, 388–389, 407
Sulfolobus, 398, 401–402, 407, 417
acicaldarius, 390
ambivalens, 390
brierleyi, 390, 401
solfataricus, 390
"Superbug," 328
Symba fermentation process, 9
Symbiosis (*see* Mutualism)
Syntrophic acetogenic bacteria, 270–272, 274–278, 281–284
Syntrophobacter, 280
Syntrophomonas, 280
wolfei, 284

Thermoactinomyces, 247
Thermoanaerobacter ethanolicus, 216, 240–241, 250
Thermoanaerobium brockii, 213
Thermophilic microbes, 216–217, 296
Thiobacillus:
acidophilus, 392, 406
cuprinus, 391
denitrificans, 400
ferrooxidans, 374, 378, 380, 384–387, 391–397, 399–403, 406–409, 411–412, 415–419
intermedius, 345
neopolitanus, 345
novellus, 345
organoparus, 406
organovorus, 392
prosperus, 396
thiooxidans, 345, 388–389, 396, 400, 403, 407
Torula utilis, 222
Torulopsis, 41, 43, 45, 51, 156, 177
delbrueckii, 45
holmii, 29
Transductions, 108, 113, 328
Trichoderma, 222, 235–237, 240
harzinum, 239
longibrachiatum, 237
reesei, 213, 215, 234–238
viride, 237, 240
Trichosporon fermentans, 247

Undefined or defined cultures, 126–127, 206, 284
Unimulticulture, 8

Vegetable fermentations, 69–103
cabbage, 75–76
cucumbers, 71–75
genera of LAB involved, 77–78
home versus commercial pickling, 153
Vegetable fermentations (*Cont.*):
initiation of fermentation, 71–77
lactic acid bacteria, 71, 73, 75–81, 83–88, 90–93
need for controlled fermentation, 70–71
olives, 76–77
postfermentation, 95–98
primary fermentation, 77–93
safety of, 153–154
secondary fermentation, 93–95
Venturia inaequalis, 42
Vibrio:
anguillarum, 354
natriegens, 354
Vibrus anguillarum, 343
Viruses:
canine hepatitis, 149
diarrheal bovine, 149
and foodborne disease, 137, 141
foot-and-mouth, 159
inactivation of, 149
influenza, 159
Newcastle disease, 149
poliovirus, 159
porcine picornavirus, 149
survival in fermented foods, 159

Wine production and microbial interaction, 37–68
addition of fumaric acid, 58
climate and altitude, 43
cluster dynamics, 43
effect of fungicides, 42–43, 44
effect of vineyard fertilization, 42
effects of antimicrobial agents, 55–57
ethanol tolerance, 47, 49
flavor and wild yeasts, 46
"geranium tone," 58
inhibiting effect of ethanol, 38, 39
inhibition of yeast by bacteria, 53–54
insect vectors, 42
karyotypic identification of yeasts, 46
low pH and high sugar, 38, 51
malolactic fermentation, 37–39, 41, 53–54, 56–57
methode champanoise, 60
microbial spoilage, 57–61
microorganisms in grape juice, 45–47
microorganisms on grapes, 40–41
"mousiness," 53, 58, 59
natural fermentation versus inoculation, 44, 45
natural flora of grapes, 39–44
nutrient limitation, 48–50
population dynamics of bacterial species, 52–55
population dynamics of yeast species, 44–45, 48–52

Wine production and microbial interaction (*Cont.*):
- postfermentation flora, 57–61
- rapid acetification, 52
- *Saccharomyces cerevisae*, 37–39, 45–48, 50–52, 55–57, 59–61
- spoilage microorganisms, 38, 39, 47, 59–60
- temperature effects, 39, 50–51
- use of genetically marked yeast, 46
- wild yeasts in grapes, 38, 39, 41
- yeast killer factors, 47, 51–52, 56

Wolinella succinogenes, 265

Xanthobacter autotrophicus, 309
Xanthomonas, 240, 298
- *campestris*, 13

Xenobiotics (synthetic chemicals), 293–294, 297, 327

Yeast killer factors, 47, 51–52, 56
Yeasts, 1–2, 9, 10, 154, 163
- aerobic versus anaerobic fermentation, 79, 177
- apiculate, 50
- bacteria and, 14, 20, 22–24, 52–55, 124
- baker's yeast, 19–27, 30–33
- biomass production, 248
- in commercial chemical production, 208–209, 214–215
- in dairy fermentation, 124
- in silage fermentation, 177–178
- wine production, 12, 20, 37–52, 55–61
- "yeast spots," 90–91

Yersinia enterocolitica, 155, 156
Yersiniosis, 156

Zapatera spoilage, 94
Zygosaccharomyces, 47, 59
- *bailii*, 59–60, 61
- *bisporus*, 7

Zymomonas mobilis, 47, 214–215, 249–250